CONTROL THEORY AND DESIGN

To our wives, Laura
Bruna
Franca

CONTROL THEORY AND DESIGN

A RH_2 AND RH_∞ VIEWPOINT

PATRIZIO COLANERI
Milan Polytechnic Institute, Italy

JOSÉ C. GEROMEL
University of Campinas, Brazil

ARTURO LOCATELLI
Milan Polytechnic Institute, Italy

ACADEMIC PRESS

San Diego · London · Boston
New York · Sydney · Tokyo · Toronto

Copyright © 1997 by ACADEMIC PRESS

Academic Press, Inc.
525 B Street, Suite 1900, San Diego, California 92101-4495, USA
http://www.apnet.com

Academic Press Limited
24–28 Oval Road, London NW1 7DX, UK
http://www.hbuk.co.uk/ap/

ISBN 0-12-179190-4

Library of Congress Cataloging-in-Publication Data

Colaneri, Patrizio.
    Control theory and design: an RH_2 and RH_∞ viewpoint / by
Patrizio Colaneri, José C. Geromel, Arturo Locatelli.
        p. cm.
    Includes index.
    ISBN 0-12-179190-4 (alk. per)
    1. Automatic control. 2. Control theory. I. Geromel, José C. II. Locatelli, Arturo.
III. Title.
    TJ213.C563   1997
    629.8–dc21   96-49595
                                                                                    CIP

A catalogue record for this book is available from the British Library

Printed in Great Britain by Cambridge University Press, Cambridge.

97 98 99 00 01 02 EB 9 8 7 6 5 4 3 2 1

Contents

Preface & Acknowledgments

Robust control theory has been the object of much of the research activity developed in the last fifteen years within the context of linear systems control. At the stage, the results of these efforts constitute a fairly well established part of the scientific community background, so that the relevant techniques can reasonably be exploited for practical purposes. Indeed, despite their complex derivation, these results are of simple implementation and capable of accounting for a number of interesting real life applications. Therefore the demand of including these topics in control engineering courses is both timely and suitable and motivated the birth of this book which covers the basic facts of robust control theory, as well as more recent achievements, such as robust stability and robust performance in presence of parameter uncertainties. The book has been primarily conceived for graduate students as well as for people first entering this research field. However, the particular care which has been dedicated to didactic instances renders the book suited also to undergraduate students who are already acquainted with basic system and control. Indeed, the required mathematical background is supplied where necessary.

Part of the here collected material has been structured according to the textbook *Controllo in RH_2-RH_∞* (in Italian) by the authors. They are deeply indebted to the publisher Pitagora for having kindly permitted it. The first five chapters introduces the basic results on RH_2 and RH_∞ theory whereas the last two chapters are devoted to present more recent results on robust control theory in a general and self-contained setting. The authors gratefully acknowledge the financial support of *Centro di Teoria dei Sistemi* of the Italian National Research Council - CNR, the Brazilian National Research Council - CNPq (under grant 301373/80) and the Research Council of the State of São Paulo, Brazil - FAPESP (under grant $90/3607 - 0$).

This book is a result of a joint, fruitful and equal scientific cooperation. For this reason, the authors' names appear in the front page in alphabetical order.

Patrizio Colaneri   Milan, Italy
José C. Geromel   Campinas, Brazil
Arturo Locatelli   Milan, Italy

Chapter 1

Introduction

Frequency domain techniques have longly being proved to be particularly fruitful and simple in the design of (linear time invariant) SISO [1] control systems. Less appealing have appeared for many years the attempts of generalizing such nice techniques to the MIMO [2] context. This partially motivated the great deal of interest which has been devoted to time domain design methodologies starting in the early 60's. Indeed, this stream of research originated a huge number of results both of remarkable conceptual relevance and practical impact, the most celebrated of which is probably the LQG [3] design. Widely acknowledged are the merits of such an approach: among them the relatively small computational burden involved in the actual definition of the controller and the possibility of affecting the dynamical behavior of the control system through a guided sequence of experiments aimed at the proper choice of the parameters of both the performance index (weighting matrices) and uncertainty description (noises intensities). Equally well known are the limits of the LQG design methodology, the most significant of which is the possible performance decay caused by operative conditions even slightly differing from the (nominal) ones referred to in the design stage. Specifically, the lack of robustness of the classical LQG design originates from the fact that it does not account for the uncertain knowledge or unexpected perturbations of the plant, actuators and sensors parameters.

The need of simultaneously complying with design requirements naturally specified in the frequency domain and guaranteeing robustness of the control system in the face of uncertainties and/or parameters deviations, focused much of the research activity on the attempt of overcoming the traditional and myopic dichotomy between time and frequency domain approaches. At the stage, after about two decades of intense efforts on these lines, the control system designer can rely on a set of well established results which give proper answers to the significant instances of performance and stability robustness. The value of the results achieved so far partially stems in the construction of a unique formal theoretical picture which naturally includes both the classical LQG design (RH_2 design), revisited at the light of a transfer function-like approach, and the new challenging developments of the so called robust design (RH_∞ design), which encompasses most of the above mentioned robustness instances.

The design methodologies which are presented in the book are based on the minimization of a performance index, simply consisting of the norm of a suitable transfer

---

[1] Single-input single-output

[2] Multi-input multi-output

[3] Linear quadratic gaussian

function. A distinctive feature of these techniques is the fact that they do not come up with a unique solution to the design problem; rather, they provide a whole set of (admissible) solutions which satisfy a constraint on the maximum deterioration of the performance index. The attitude of focusing on the class of admissible controllers instead of determining just one of them can be traced back to a fundamental result which concerns the parametrization of the class of controllers stabilizing a given plant. Chapter 3 is actually dedicated to such a result and deals also with other questions on feedback systems stability. In subsequent Chapters 4 and 5 the main results of RH_2 and RH_∞ design are presented, respectively. In addition, a few distinguishing aspects of the underlying theory are emphasized as well, together with particular, yet significant, cases of the general problem. Chapter 5 contains also a preliminary discussion on the robustness requirements which motivate the formulation of the so called standard RH_∞ control problem. Chapter 6 and 7 go beyond the previous ones in the sense that the design problems to be dealt with are setting in a more general framework. One of the most interesting examples of this situation is the so called mixed RH_2/RH_∞ problem which is expressed in terms of both RH_2 and RH_∞ norms of two transfer functions competing with each other to get the best tradeoff between performance and robustness. Other problems that fall into this framework are those related to regional pole placement, time-domain specification and structural constraints. All of them share basically the same difficulty to be faced numerically. Indeed, they can not be solved by the methodology given in the previous Chapters but by means of mathematical programming methods. More specifically, all can (after a proper change of variables) be converted into convex problems. This feature is important in both practical and theoretical points of view since numerical efficiency allows the treatment of real-word problems of generally large dimension while global optimality is always assured. Chapter 7 is devoted to the controllers design for systems subject to structured convex bounded uncertainties which models in an adequate and precise way many classes of parametric uncertainties with practical appealing. The associated optimal control problems are formulated and solved jointly with respect to the controller transfer function and the feasible uncertainty in order to guarantee minimum loss in the performance index. One of such situation of great importance for its own is the design problem involving actuators failure. Robust stability and performance are addressed for two classes of nonlinear perturbations, leading to what are called Persidiskii and Lur'e design. In general terms, the same technique involving the reduction of the related optimal control design problems to convex programming problems is again used. The main point to be remarked is that the two classes of nonlinear perturbations considered impose additional linear and hence convex constraints, to the matrices variables to be determined.

Treating these arguments requires a fairly deep understanding of some facts from mathematics not so frequently included in the curricula of students in Engineering. Covering the relevant mathematical background is the scope of Chapter 2, where the functional (Hardy) spaces which permeate all over the book are characterized. Some miscellaneous facts on matrix algebra, system and control theory and convex optimization are collected in Appendix A through I.

Chapter 2

Preliminaries

2.1 Introduction

The scope of this chapter is twofold: on one hand it is aimed at presenting the extension of the concepts of *poles* and *zeros*, well known for single-input single-output (SISO) systems, to the multivariable case; on the other, it is devoted to the introduction of the basic notions relative to some *functional spaces* whose elements are matrices of *rational functions* (spaces RL_2, RL_∞, RH_2, RH_∞). The reason of this choice stems from the need of presenting a number of results concerning significant control problems for *linear, continuous-time, finite dimensional and time-invariant* systems.

The derivation of the related results takes substantial advantage on the nature of the analysis and design methodology adopted; such a methodology was actually developed so as to take into account state-space and frequency based techniques at the same time.

For this reason, it should not be surprising the need of carefully extending to multi-input multi-output (MIMO) systems the notions of zeros and poles, which proved so fruitful in the context of SISO systems. In Section 2.5, where this attempt is made, it will be put into sharp relief few fascinating and in some sense unexpected relations between *poles, zeros, eigenvalues, time responses and ranks of polynomial matrices*.

Analogously, it should be taken for granted the opportunity of going in depth on the characterization of *transfer matrices* (transfer functions for MIMO systems) in their natural embedding, namely, in the complex plane. The systems considered hereafter obviously have rational transfer functions. This leads to the need of providing, in Section 2.8 the basic ideas on suitable *functional spaces* and *linear operators* so as to throw some light on the connections between facts which naturally lie in time-domain with others more suited with the frequency-domain setting.

Although the presentation of these two issues is intentionally limited to few basic aspects, nevertheless it requires some knowledge on matrices of *polynomials*, matrices of *rational functions, singular values* and *linear operators*. To the acquisition of such notions are dedicated Sections 2.3-2.7.

2.2   Notation and terminology

The continuous-time linear time-invariant dynamic systems, object of the present text, are described, depending on circumstances, by a *state space representation*

$$\dot{x} = Ax + Bu$$
$$y = Cx + Du$$

or by their *transfer function*

$$G(s) = C(sI - A)^{-1}B + D$$

The signals which refer to a system are indifferently intended to be in time-domain or in frequency-domain all the times the context does not lead to possible misunderstandings. Sometimes, it is necessary to explicitly stress that the derivation is in frequency-domain. In this case, the subscript "L" indicates the *Laplace transform* of the considered signal, whereas the subscript "L0" denotes the Laplace transform when the system state at the *initial time is zero* (typically, this situation occurs when one thinks in terms of transfer functions). For instance, with reference to the above system, one may write

$$y_{L0} = G(s)u_L$$
$$y_L = y_{L0} + C(sI - A)^{-1}x(0)$$

Occasionally, the transfer function $G(s)$ of a system Σ is explicitly related to one of its realizations by writing

$$G(s) = \Sigma(A, B, C, D)$$

or

$$G(s) = \left[\begin{array}{c|c} A & B \\ \hline C & D \end{array} \right]$$

The former notation basically has a compactness value, whereas the latter is mainly useful when one wants to display possible partitions in the input and/or output matrices. For example, the system

$$\dot{x} = Ax + B_1 w + B_2 u$$
$$z = C_1 x + D_{12} u$$
$$y = C_2 x + D_{21} w$$

is related to its transfer function $G(s)$ by writing

$$G(s) = \left[\begin{array}{c|cc} A & B_1 & B_2 \\ \hline C_1 & 0 & D_{12} \\ C_2 & D_{21} & 0 \end{array} \right]$$

When a purely *algebraic* (i.e. nondynamic) system is considered, these notations become

$$G(s) = \Sigma(\emptyset, \emptyset, \emptyset, D) = \left[\begin{array}{c|c} \emptyset & \emptyset \\ \hline \emptyset & D \end{array} \right]$$

Referring to the class of systems considered here, the transfer functions are in fact *rational matrices* of complex variable, namely, matrices whose generic element is a rational function, i.e., a ratio of polynomials with real coefficients. The transfer function is said to be *proper* when each element is a proper rational function, i.e., a ratio of polynomials with the degree of the numerator not greater than the degree of the denominator. When this inequality holds in a strict sense for each element of the matrix, the transfer function is said to be *strictly proper*. Briefly, $G(s)$ is proper if

$$\lim_{s \to \infty} G(s) = K < \infty$$

where the notation $K < \infty$ means that each element of matrix K is finite. Analogously, $G(s)$ is strictly proper if

$$\lim_{s \to \infty} G(s) = 0$$

A rational matrix $G(s)$ is said to be *analytic* in $Re(s) \geq 0$ (resp. ≤ 0) if all the elements of the matrix are bounded functions in the closed right (resp. left) half plane.

In connection with a system characterized by the transfer function

$$G(s) = \left[\begin{array}{c|c} A & B \\ \hline C & D \end{array} \right] \tag{2.1}$$

the so-called *adjoint system* has transfer function

$$G^{\sim}(s) := G'(-s) = \left[\begin{array}{c|c} -A' & -C' \\ \hline B' & D' \end{array} \right]$$

whereas the transfer function of the so-called *transpose system* is

$$G'(s) := \left[\begin{array}{c|c} A' & C' \\ \hline B' & D' \end{array} \right]$$

System (2.1) is said to be *input-output stable* if its transfer function $G(s)$ is analytic in $Re(s) \geq 0$ ($G(s)$ is stable, by short). It is said to be *internally stable* if matrix A is stable, i.e., if all its eigenvalues have negative real parts.

Now observe that a system is input-output stable *if and only if* all elements of $G(s)$, whenever expressed as ratio of polynomials without common roots, have their poles in the open left half plane only. If the realization of system (2.1) is *minimal,* the system is input-output stable *if and only if* it is internally stable.

Finally, the conjugate transpose of the generic (complex) matrix A is denoted by $A^{\sim}$ and, if it is square, $\lambda_i(A)$ is its i-th eigenvalue, while

$$r_s(A) := \max_i |\lambda_i(A)|$$

denotes its *spectral radius*.

2.3   Polynomial matrices

A *polynomial matrix* is a matrix whose elements are polynomials in a unique unknown. Throughout the book, such an unknown is denoted by the letter s. All the polynomial coefficients are *real*. Hence, the element $n_{ij}(s)$ in position (i,j) in the polynomial matrix $N(s)$ takes the form

$$n_{ij}(s) = \alpha_\nu s^\nu + \alpha_{\nu-1} s^{\nu-1} + \cdots + \alpha_1 s + \alpha_0, \; \alpha_k \in R, \; \forall k$$

The *degree* of a polynomial $p(s)$ is denoted by $\deg[p(s)]$. If the leading coefficient α_ν is equal to one, the polynomial is said to be *monic*.

The *rank* of a polynomial matrix $N(s)$, denoted by $\text{rank}[N(s)]$, is defined by analogy from the definition of the rank of a numeric matrix, i.e., it is the dimension of the largest square matrix which can be extracted from $N(s)$ with determinant not identically zero.

A square polynomial matrix is said to be *unimodular* if it has *full rank* (it is invertible) and its determinant is *constant*.

Example 2.1 The polynomial matrices

$$N_1(s) = \begin{bmatrix} 1 & s+1 \\ 0 & 3 \end{bmatrix}, \quad N_2(s) = \begin{bmatrix} s+1 & s-2 \\ s+2 & s-1 \end{bmatrix}$$

are unimodular since $\det[N_1(s)]=\det[N_2(s)]=3$.   □

A very peculiar property of a unimodular matrix is that its inverse is still a polynomial (and obviously unimodular) matrix. Not differently from what is usually done for polynomials, the polynomial matrices can be given the concepts of *divisor* and *greatest common divisor* as well.

Definition 2.1 (Right divisor) *Let $N(s)$ be a polynomial matrix. A square polynomial matrix $R(s)$ is said to be a right divisor of $N(s)$ if it is such that*

$$N(s) = \bar{N}(s)R(s)$$

with $\bar{N}(s)$ a suitable polynomial matrix.   □

An analogous definition can be formulated for the *left divisor*.

Definition 2.2 (Greatest common right divisor) *Let $N(s)$ and $D(s)$ be polynomial matrices with the same number of columns. A square polynomial matrix $R(s)$ is said to be a Greatest Common Right Divisor (GCRD) of $(N(s), D(s))$ if it is such that*

i) $R(s)$ is a right divisor of $D(s)$ and $N(s)$, i.e.

$$N(s) = \bar{N}(s)R(s)$$
$$D(s) = \bar{D}(s)R(s)$$

with $\bar{N}(s)$ and $\bar{D}(s)$ suitable polynomial matrices

ii) For each polynomial matrix $\hat{R}(s)$ such that

$$N(s) = \hat{N}(s)\hat{R}(s)$$
$$D(s) = \hat{D}(s)\hat{R}(s)$$

with $\hat{N}(s)$ and $\hat{D}(s)$ polynomial matrices, it turns out that $R(s) = W(s)\hat{R}(s)$ where $W(s)$ is again a suitable polynomial matrix.   □

A similar definition can be formulated for the *Greatest Common Left Divisor* (GCLD).

It is easy to see, by exploiting the properties of unimodular matrices, that, given two polynomial matrices $N(s)$ and $D(s)$, there exist infinite GCRD's (and obviously GCLD's). A way to compute a GCRD (resp. GCLD) of two assigned polynomial matrices $N(s)$ and $D(s)$ relies on their manipulation through a unimodular matrix which represents a sequence of suitable *elementary operations* on their rows (resp. columns). The elementary operations on the rows (resp. columns) of a polynomial matrix $N(s)$ are

1) Interchange of the i-th row (resp. i-th column) with the j-th row (resp. j-th column)

2) Multiplication of the i-th row (resp. i-th column) by a nonzero scalar

3) Addition of a polynomial multiple of the i-th row (resp. i-th column) to the j-th row (resp. j-th column).

It is readily seen that each elementary operation can be performed premultiplying (resp. postmultiplying) $N(s)$ by a suitable polynomial and unimodular matrix $T(s)$. Moreover, matrix $T(s)N(s)$ (resp. $N(s)T(s)$) turns out to have the same rank as $N(s)$.

Remark 2.1 Notice that, given two polynomials $r_0(s)$ and $r_1(s)$ with $\deg[r_0(s)] \geq \deg[r_1(s)]$, it is always possible to define two sequences of polynomials $\{r_i(s), \ i = 2, 3, \cdots, p+2\}$ and $\{q_i(s), \ i = 1, 2, \cdots, p+1\}$, with $0 \leq p \leq \deg[r_1(s)]$, such that

$$r_i(s) = q_{i+1}(s)r_{i+1}(s) + r_{i+2}(s) , \quad i = 0, 1, \cdots, p$$
$$\deg[r_{i+2}(s)] < \deg[r_{i+1}(s)] , \quad i = 0, 1, \cdots, p$$
$$r_{p+2}(s) = 0$$

Letting

$$T_i(s) := \begin{bmatrix} 1 & -q_i(s) \\ 0 & 1 \end{bmatrix} , \quad n_i(s) := \begin{bmatrix} r_{i-1}(s) \\ r_i(s) \end{bmatrix} , \quad i = 1, 3, 5, \cdots$$

$$T_i(s) := \begin{bmatrix} 1 & 0 \\ -q_i(s) & 1 \end{bmatrix} , \quad n_i(s) := \begin{bmatrix} r_i(s) \\ r_{i-1}(s) \end{bmatrix} , \quad i = 2, 4, 6, \cdots$$

$$T(s) := \prod_{i=0}^{p} T_{p+1-i}(s)$$

and noticing that $T(s)$ is unimodular (product of unimodular matrices), it turns out that

$$T(s)n_1(s) = \begin{bmatrix} r_{p+1}(s) \\ 0 \end{bmatrix} , \quad p = 1, 3, 5, \cdots$$

$$T(s)n_1(s) = \begin{bmatrix} 0 \\ r_{p+1}(s) \end{bmatrix} , \quad p = 2, 4, 6, \cdots$$

For instance, take $r_0(s) = s^4 + 2s^2 - s + 2$, $r_1(s) = s^3 + s - 2$. It follows that $q_1(s) = s$, $q_2(s) = s - 1$, $r_2(s) = s^2 + s + 2$ and $r_3(s) = 0$. □

By repeatedly exploiting the facts shown in Remark 2.1, it is easy to verify that, given a polynomial matrix $N(s)$ with the number of rows not smaller than the number of columns, there exists a suitable polynomial and unimodular matrix $T(s)$ such that

$$T(s)N(s) = \begin{bmatrix} R(s) \\ 0 \end{bmatrix}$$

where $R(s)$ is a square polynomial matrix.

Algorithm 2.1 (GCRD of two polynomial matrices) Let $N(s)$ and $D(s)$ be two polynomial matrices with the same number, say m, of columns and with n_n and n_d rows, respectively.

1) Assume that $m \leq n_d + n_n$, otherwise go to point 4). Let $P(s) := [D'(s)\ N'(s)]'$ and determine a polynomial and unimodular matrix $T(s)$ such that

$$T(s)P(s) = \left[\begin{array}{c} R(s) \\ 0 \end{array} \right] \} \ m \text{ rows}$$

Notice that $T(s)$ can be partitioned as follows

$$T(s) := \left[\begin{array}{cc} T_{d1}(s) & T_{n1}(s) \\ T_{d2}(s) & T_{n2}(s) \end{array} \right] \} \ m \text{ rows}$$

$$\underbrace{\phantom{T_{d1}(s) \quad T_{n1}(s)}}_{n_d \text{ columns}}$$

2) Letting $S(s) := T^{-1}(s)$ and writing

$$S(s) := \left[\begin{array}{cc} S_{d1}(s) & S_{d2}(s) \\ S_{n1}(s) & S_{n2}(s) \end{array} \right] \} \ m \text{ rows}$$

$$\underbrace{\phantom{S_{d1}(s) \quad S_{d2}(s)}}_{n_d \text{ columns}}$$

it turns out

$$D(s) = S_{d1}(s)R(s)$$
$$N(s) = S_{n1}(s)R(s)$$

so that $R(s)$ is a right divisor of both $D(s)$ and $N(s)$.

3) It also holds that
$$R(s) = T_{d1}(s)D(s) + T_{n1}(s)N(s) \tag{2.2}$$
Hence, suppose that $\hat{R}(s)$ is any other right divisor of both $D(s)$ and $N(s)$. Therefore, for some polynomial matrices $\bar{D}(s)$ and $\bar{N}(s)$ it follows that $D(s) = \bar{D}(s)\hat{R}(s)$ and $N(s) = \bar{N}(s)\hat{R}(s)$. The substitution of these two expressions in eq. (2.2) leads to $R(s) = [T_{d1}(s)\bar{D}(s) + T_{n1}(s)\bar{N}(s)]\hat{R}(s)$ so that $R(s)$ is a GCRD of $(N(s), D(s))$.

4) If $m > n_d + n_n$, take two matrices $\hat{D}(s)$ and $\hat{N}(s)$ both with m columns and n_d and n_n rows, respectively

$$\hat{D}(s) := \left[\begin{array}{ccc} I & 0 & 0 \end{array} \right]$$

$$\hat{N}(s) := \left[\begin{array}{ccc} I & 0 & 0 \end{array} \right]$$

and let

$$R(s) := \left[\begin{array}{c} D(s) \\ N(s) \\ 0 \end{array} \right] \} \ m - n_d - n_n \text{ rows} \tag{2.3}$$

Thus, $D(s) = \hat{D}(s)R(s)$ and $N(s) = \hat{N}(s)R(s)$. Hence, $R(s)$ is a right divisor of both $D(s)$ and $N(s)$. Assume now that $\hat{R}(s)$ is any other right divisor, i.e. there exist two polynomial matrices $\bar{D}(s)$ and $\bar{N}(s)$ such that $D(s) = \bar{D}(s)\hat{R}(s)$ and $N(s) = \bar{N}(s)\hat{R}(s)$. By substituting these two last expressions in eq. (2.3) one obtains

$$R(s) := \begin{bmatrix} \bar{D}(s) \\ \bar{N}(s) \\ 0 \end{bmatrix} \hat{R}(s),$$

so leading to the conclusion that $R(s)$ is a GCRD of $(N(s), D(s))$.

□

Example 2.2 Consider the matrices

$$D(s) = \begin{bmatrix} s^2 - s & s^2 \\ 2s^2 + 9s + 5 & 2s^2 + 5s + 5 \end{bmatrix}$$

$$N(s) = \begin{bmatrix} s^2 + 1 & s^2 + 2s + 1 \end{bmatrix}$$

Now take

$$T_1(s) = \begin{bmatrix} 1 & 0 & 0 \\ -2 & 1 & 0 \\ -1 & 0 & 1 \end{bmatrix} \quad , \quad T_2(s) = \begin{bmatrix} 1 & 0 & -s \\ 0 & 1 & -11 \\ 0 & 0 & 1 \end{bmatrix}$$

$$T_3(s) = \begin{bmatrix} 1 & -s/3 & 0 \\ 0 & 1 & 0 \\ 0 & s/6 & 1 \end{bmatrix} \quad , \quad T_4(s) = \begin{bmatrix} 1 & 0 & 0 \\ 0 & 1 & 6 \\ 0 & 0 & 1 \end{bmatrix}$$

$$T_5(s) = \begin{bmatrix} 0 & 0 & 1 \\ 0 & -1 & 0 \\ 1 & 0 & 0 \end{bmatrix} \quad , \quad T_6(s) = \begin{bmatrix} 1 & 0 & 0 \\ 0 & 1 & -51/14 \\ 0 & 0 & 1 \end{bmatrix}$$

$$T_7(s) = \begin{bmatrix} 1 & 0 & 0 \\ 0 & 14/103 & 0 \\ 0 & -196s/309 & 1 \end{bmatrix} \quad , \quad T_8(s) = \begin{bmatrix} 1 & 0 & 0 \\ 0 & 1 & 0 \\ 0 & -1 & 1 \end{bmatrix}$$

Then

$$T(s) = \prod_{i=0}^{7} T_{8-i}(s) =$$

$$= \begin{bmatrix} (3s-2)/2 & s/6 & (6-11s)/6 \\ -(24s+93)/103 & (3s-14)/103 & (18s+70)/103 \\ (112s^2 + 252s + 196)/103 & (-14s^2 + 28s + 14)/103 & -(84s^2 + 70s + 70)/103 \end{bmatrix}$$

so that

$$T(s)P(s) := T(s) \begin{bmatrix} D(s) \\ N(s) \end{bmatrix} = \begin{bmatrix} 1 & (-17s^2 + 6s + 6)/6 \\ 0 & s \\ 0 & 0 \end{bmatrix}$$

and

$$R(s) = \begin{bmatrix} 1 & (-17s^2 + 6s + 6)/6 \\ 0 & s \end{bmatrix}$$

Finally, notice that

$$T^{-1}(s) := S(s) = \begin{bmatrix} S_{d1}(s) & S_{d2}(s) \\ S_{n1}(s) & S_{n2}(s) \end{bmatrix}$$

with

$$S_{d1}(s) = \begin{bmatrix} s(s-1) & (17s^3 - 23s^2 + 6s + 6)/6 \\ 2s^2 + 9s + 5 & (34s^3 + 141s^2 + 31s - 54)/6 \end{bmatrix}$$

$$S_{n1}(s) = \begin{bmatrix} s^2 + 1 & (17s^3 - 6s^2 + 17s + 6)/6 \end{bmatrix}$$

It is then easy to verify that $D(s) = S_{d1}(s)R(s)$ and that $N(s) = S_{n1}(s)R(s)$.   □

The familiar concept of *coprimeness*, easily introduced for polynomials, can be properly extended to polynomial matrices as follows.

Definition 2.3 (Right coprimeness) *Two polynomial matrices $N(s)$ and $D(s)$ having the same number of columns, are said to be right coprime if the two equations*

$$N(s) = \bar{N}(s)T(s)$$
$$D(s) = \bar{D}(s)T(s),$$

where $\bar{N}(s)$ and $\bar{D}(s)$ are suitable polynomial matrices, are verified by a unimodular polynomial matrix $T(s)$ only.   □

Example 2.3 The matrices

$$N(s) = \begin{bmatrix} s^2 - 1 \\ s^3 + 3s^2 - s - 3 \end{bmatrix} , \quad D(s) = \begin{bmatrix} 2s^2 + 6s + 4 \\ s^3 + 3s^2 + 2s \end{bmatrix}$$

are not right coprime. Actually, it turns out that

$$N(s) = \bar{N}(s)R(s) , \quad \bar{N}(s) = \begin{bmatrix} s - 1 \\ s^2 + 2s - 3 \end{bmatrix} , \quad R(s) = s + 1$$

$$D(s) = \bar{D}(s)R(s) , \quad \bar{D}(s) = \begin{bmatrix} 2s + 4 \\ s^2 + 2s \end{bmatrix}$$

and $R(s)$ is not unimodular $(\det[R(s)] = s + 1)$.   □

Of course, an analogous definition can be stated for the left coprimeness. Definitions 2.1-2.3 also yield that two matrices are right (resp. left) coprime if all their common right (resp. left) divisors are actually unimodular. In particular, each GCRD (resp. GCLD) of two right (resp. left) coprime matrices must be unimodular. In view of Algorithm 2.1, this entails that a possible way to verify whether or not two matrices are right (resp. left) coprime, is computing and evaluating the determinant of a greatest common divisor. As a matter of fact, if a GCRD (resp. GCLD) is unimodular, then *all* other greatest common divisors are unimodular as well. More precisely, if $R_1(s)$ and $R_2(s)$ are two GCRD's and $R_1(s)$ is unimodular, it results $R_1(s) = W(s)R_2(s)$, with $W(s)$ polynomial. Since $\det[R_1(s)] \neq 0$ it follows that $\det[R_2(s)] \neq 0$ as well.

Again from Algorithm 2.1 (step 1) it can be concluded that two polynomial matrices $D(s)$ and $N(s)$ with the same number of columns, say m, are right coprime if the rank of $P(s) := [D'(s) \ N'(s)]'$ is m for any s. As a matter of fact, coprimeness is equivalent to $R(s)$ being unimodular so that $\text{rank}[T(s)P(s)]$ must be constant and equal to m. Since $T(s)$ is unimodular, $\text{rank}[P(s)]$ must be constant and equal to m as well.

Lemma 2.1 *Let $N(s)$ and $D(s)$ be two polynomial matrices with the same number of columns and let $R(s)$ be a GCRD of $(N(s), D(s))$. Then,*

i) $T(s)R(s)$ is a GCRD of $(N(s), D(s))$ for any polynomial and unimodular matrix $T(s)$

ii) If $\bar{R}(s)$ is an arbitrary GCRD of $(N(s), D(s))$, then there exists a polynomial and unimodular matrix $T(s)$ such that

$$\bar{R}(s) = T(s)R(s)$$

Proof *Point i)* Being $R(s)$ a GCRD of $(N(s), D(s))$ it follows that

$$N(s) = \widetilde{N}(s)R(s) \ , \quad D(s) = \widetilde{D}(s)R(s)$$
$$N(s) = \bar{N}(s)\bar{R}(s) \ , \quad D(s) = \bar{D}(s)\bar{R}(s)$$
$$R(s) = W(s)\bar{R}(s)$$

where $\widetilde{N}(s)$, $\widetilde{D}(s)$, $\bar{N}(s)$, $\bar{D}(s)$ and $W(s)$ are suitable polynomial matrices. Taken an arbitrary polynomial and unimodular matrix $T(s)$, let $\hat{R}(s) := T(s)R(s)$. It follows that $N(s) = \hat{N}(s)\hat{R}(s)$, $\hat{N}(s) = \widetilde{N}(s)T^{-1}(s)$, $D(s) = \hat{D}(s)\hat{R}(s)$, $\hat{D}(s) = \widetilde{D}(s)T^{-1}(s)$. Furthermore, it is $\hat{R}(s) = \hat{W}(s)\bar{R}(s)$, $\hat{W}(s) = T(s)W(s)$. Hence, $\hat{R}(s)$ is a GCRD of $(N(s), D(s))$ as well.

Point ii) If $R(s)$ and $\bar{R}(s)$ are two GCRD's of $(N(s), D(s))$, then, for two suitable polynomial matrices $W(s)$ and $\bar{W}(s)$, it results $R(s) = W(s)\bar{R}(s)$ and $\bar{R}(s) = \bar{W}(s)R(s)$. From these relations it follows that $\mathrm{rank}[\bar{R}(s)] \leq \mathrm{rank}[R(s)] \leq \mathrm{rank}[\bar{R}(s)]$. Therefore, $\mathrm{rank}[R(s)] = \mathrm{rank}[\bar{R}(s)]$. Let now $U(s)$ and $\bar{U}(s)$ be two polynomial and unimodular matrices such that

$$U(s)R(s) = \left[\begin{array}{c} H(s) \\ 0 \end{array} \right] \ , \quad \bar{U}(s)\bar{R}(s) = \left[\begin{array}{c} \bar{H}(s) \\ 0 \end{array} \right]$$

where the two submatrices $H(s)$ and $\bar{H}(s)$ have the same number of rows equal to $\mathrm{rank}[R(s)]$. Consequently,

$$\left[\begin{array}{c} H(s) \\ 0 \end{array} \right] = U(s)R(s)$$
$$= U(s)W(s)\bar{R}(s)$$
$$= U(s)W(s)\bar{U}^{-1}(s) \left[\begin{array}{c} \bar{H}(s) \\ 0 \end{array} \right]$$
$$= \Gamma(s) \left[\begin{array}{c} \bar{H}(s) \\ 0 \end{array} \right]$$
$$= \left[\begin{array}{cc} \Gamma_{11}(s) & \Gamma_{12}(s) \\ \Gamma_{21}(s) & \Gamma_{22}(s) \end{array} \right] \left[\begin{array}{c} \bar{H}(s) \\ 0 \end{array} \right] \tag{2.4}$$

and

$$\left[\begin{array}{c} \bar{H}(s) \\ 0 \end{array} \right] = \bar{U}(s)\bar{R}(s)$$
$$= \bar{U}(s)\bar{W}(s)R(s)$$
$$= \bar{U}(s)\bar{W}(s)U^{-1}(s) \left[\begin{array}{c} H(s) \\ 0 \end{array} \right]$$

$$= \bar{\Gamma}(s) \begin{bmatrix} H(s) \\ 0 \end{bmatrix}$$

$$= \begin{bmatrix} \bar{\Gamma}_{11}(s) & \bar{\Gamma}_{12}(s) \\ \bar{\Gamma}_{21}(s) & \bar{\Gamma}_{22}(s) \end{bmatrix} \begin{bmatrix} H(s) \\ 0 \end{bmatrix} \tag{2.5}$$

From eq.(2.4), (2.5) it follows that

$$0 = \Gamma_{21}(s)\bar{H}(s) \tag{2.6}$$

$$0 = \bar{\Gamma}_{21}(s)H(s) \tag{2.7}$$

Being $R(s)$ and $\bar{R}(s)$ square, matrices $H(s)$ and $\bar{H}(s)$ have ranks equal to the number of their rows, which is obviously not greater than the number of their columns. Therefore, from eqs. (2.6) and (2.7) it follows that $\Gamma_{21}(s) = \bar{\Gamma}_{21}(s)$. Equations (2.4),(2.5) outline that matrices $\Gamma_{12}(s)$, $\Gamma_{22}(s)$, $\bar{\Gamma}_{12}(s)$ and $\bar{\Gamma}_{21}(s)$ are in fact arbitrary. Hence, one can set $\Gamma_{12}(s)=\bar{\Gamma}_{12}(s) = 0$ and $\Gamma_{22}(s) = \bar{\Gamma}_{22}(s) = I$. Based on these considerations, one can henceforth assume that $\Gamma(s)$ and $\bar{\Gamma}(s)$ have the form

$$\Gamma(s) = \begin{bmatrix} \Gamma_{11} & 0 \\ 0 & I \end{bmatrix} , \quad \bar{\Gamma}(s) = \begin{bmatrix} \bar{\Gamma}_{11} & 0 \\ 0 & I \end{bmatrix}$$

so that, from eqs. (2.4) and (2.5) it follows

$$\begin{bmatrix} H(s) \\ 0 \end{bmatrix} = \begin{bmatrix} \Gamma_{11} & 0 \\ 0 & I \end{bmatrix} \begin{bmatrix} \bar{H}(s) \\ 0 \end{bmatrix}$$

$$= \begin{bmatrix} \Gamma_{11} & 0 \\ 0 & I \end{bmatrix} \begin{bmatrix} \bar{\Gamma}_{11} & 0 \\ 0 & I \end{bmatrix} \begin{bmatrix} H(s) \\ 0 \end{bmatrix}$$

In particular, it is $H(s) = \Gamma_{11}(s)\bar{\Gamma}_{11}(s)H(s)$, so that, recalling the properties of $H(s)$, it results $I = \Gamma_{11}(s)\bar{\Gamma}_{11}(s)$. Hence, both $\Gamma_{11}(s)$ and $\bar{\Gamma}_{11}(s)$ are unimodular, since their inverses are still polynomial matrices. The same holds for $\Gamma(s)$ and $\bar{\Gamma}(s)$ as well. Finally,

$$\bar{R}(s) = \bar{U}^{-1}(s) \begin{bmatrix} \bar{H}(s) \\ 0 \end{bmatrix}$$

$$= \bar{U}^{-1}(s)\bar{\Gamma}(s) \begin{bmatrix} H(s) \\ 0 \end{bmatrix}$$

$$= \bar{U}^{-1}(s)\bar{\Gamma}(s)U(s)R(s)$$

$$:= T(s)R(s)$$

and $T(s)$ is actually unimodular since it is the product of unimodular matrices. □

Remark 2.2 In view of the results now proved and the given definitions, it is apparent that when the matrices are in fact scalars it results: (i) a right divisor is also a left divisor and vice-versa; (ii) two GCRD's differ only for a multiplicative scalar, since all unimodular polynomials $p(s)$ take the form $p(s) := \alpha \in R$, $\alpha \neq 0$; (iii) two polynomials are coprime if and only if they do not have common roots. □

Right coprime polynomial matrices enjoy the property stated in the following lemma, which provides the generalization of a well known result relative to integers and polynomials (see also Theorem A.1).

Lemma 2.2 Let $N(s)$ and $D(s)$ be two polynomial matrices with the same number of columns. Then, they are right coprime if and only if there exist two polynomial matrices $X(s)$ and $Y(s)$ such that

$$X(s)N(s) + Y(s)D(s) = I \qquad (2.8)$$

Proof Based on the results illustrated in Algorithm 2.1, it is always possible to write a generic GCDR $R(s)$ of $(N(s), D(s))$ as $R(s) = \hat{X}(s)N(s) + \hat{Y}(s)D(s)$. Moreover, if $N(s)$ and $D(s)$ are coprime, $R(s)$ must be unimodular so that

$$
\begin{aligned}
I &= R^{-1}(s)R(s) \\
&= R^{-1}(s)[\hat{X}(s)N(s) + \hat{Y}(s)D(s)] \\
&= X(s)N(s) + Y(s)D(s),
\end{aligned}
$$

where $X(s) := R^{-1}(s)\hat{X}(s)$, $Y(s) := R^{-1}(s)\hat{Y}(s)$.

Conversely, suppose that there exist two matrices $X(s)$ and $Y(s)$ satisfying eq. (2.8) and let $R(s)$ be a GCRD of $(N(s), D(s))$, i.e.

$$N(s) = \hat{N}(s)R(s) , \quad D(s) = \hat{D}(s)R(s)$$

It is then possible to write $I = [X(s)\hat{N}(s) + Y(s)\hat{D}(s)]R(s)$ yielding

$$R^{-1}(s) = X(s)\hat{N}(s) + Y(s)\hat{D}(s) \qquad (2.9)$$

The right side of equation (2.9) is a polynomial matrix. This entails that $R(s)$ is a unimodular matrix so that $N(s)$ and $D(s)$ are right coprime. □

Example 2.4 Consider the two polynomial matrices

$$N(s) = \begin{bmatrix} 2s^2 + 1 & 2s \\ s & 1 \end{bmatrix} , \quad D(s) = \begin{bmatrix} 2s^3 + s & 2s^2 \end{bmatrix}$$

They are right coprime. As a matter of fact

$$X(s) := \begin{bmatrix} -2s^3 - 2s^2 + 1 & -2s \\ 2s^4 + 2s^3 + s^2 & 2s^2 + 1 \end{bmatrix}$$

$$Y(s) := \begin{bmatrix} 2s^2 + 2s \\ -2s^3 - 2s^2 - s - 1 \end{bmatrix}$$

are such that $X(s)N(s) + Y(s)D(s) = I$. Moreover, taken

$$T(s) := \begin{bmatrix} -s^3 - s^2 + 1 & -s & s^2 + s \\ s^2 + s & 1 & -s - 1 \\ -s & 0 & 1 \end{bmatrix}$$

one can easily realize that it is unimodular and that

$$T(s) \begin{bmatrix} N(s) \\ D(s) \end{bmatrix} = \begin{bmatrix} R(s) \\ 0 \end{bmatrix}$$

with

$$R(s) = \begin{bmatrix} s^2 + 1 & s \\ s & 1 \end{bmatrix}$$

which is unimodular as well. □

Of course, a version of the result provided in Lemma 2.2 can be stated for left co-primeness as well. An important and significant *canonical form* can be associated with a polynomial matrix, namely the so called *Smith form*. This canonical form is formally defined in the following theorem whose proof also provides a *systematic* procedure for its computation.

Theorem 2.1 (Smith form) *Let $N(s)$ be a $n \times m$ polynomial matrix and consider that* $\text{rank}[N(s)] := r \leq \min[n, m]$. *Then two polynomial and unimodular matrices $L(s)$ and $R(s)$ exist such that $N(s) = L(s)S(s)R(s)$ with*

$$
S(s) = \begin{bmatrix} \alpha_1(s) & 0 & \cdots & 0 & 0 \\ 0 & \alpha_2(s) & \cdots & 0 & 0 \\ \vdots & \vdots & \ddots & \vdots & \vdots \\ 0 & 0 & \cdots & \alpha_r(s) & 0 \\ 0 & 0 & \cdots & 0 & 0 \end{bmatrix} \left.\vphantom{\begin{bmatrix}0\\0\end{bmatrix}}\right\} n - r \text{ rows}
$$

$$\underbrace{}_{m - r \text{ columns}}$$

where each polynomial $\alpha_i(s)$ is monic and divides the next one, i.e. $\alpha_i(s) | \alpha_{i+1}(s)$, $i = 1, 2, \cdots, r - 1$. Matrix $S(s)$ is said to be the Smith form of $N(s)$.

Proof The proof of the theorem is constructive since the procedure to be described leads to the determination of the matrices $S(s)$, $L(s)$ and $R(s)$. In the various steps which characterize such a procedure, the matrix $N(s)$ is subject to a number of manipulations resulting from suitable elementary operations on its rows and columns, i.e. pre-multiplications or post-multiplications by unimodular matrices. These operations determine the matrices $L(s)$ and $R(s)$. For simplicity, let $n_{ij}(s)$ be the (i, j) element of the matrix which is presently considered.

1) Through two elementary operations on the rows and the columns of $N(s)$, bring a nonzero and minimum degree polynomial of $N(s)$ in position $(1, 1)$.

2) Write the element $(2, 1)$ of $N(s)$ as $n_{21}(s) = n_{11}(s)\gamma(s) + \beta(s)$, with $\beta(s)$ such that $\deg[\beta(s)] < \deg[n_{11}(s)]$. Now multiply the first row by $\gamma(s)$ and subtract the result from the second row. In this way the $(2, 1)$ element becomes $\beta(s)$. Now, if $\beta(s) = 0$ go to step 3), otherwise interchange the first row with the second one and repeat again this step. This causes a continuous reduction of the degree of the element $(2, 1)$ so that, in a finite number of iterations, it results $n_{21}(s) = 0$.

3) As exactly done in step 2), bring all the elements of the first column but element $(1, 1)$ to zero.

4) Through elementary operations on the columns bring all the elements of the first row but $n_{11}(s)$ to zero.

5) If step 4) brought the elements of the first column under $n_{11}(s)$ to be nonzero, then go back to step 2). Notice that a finite number of operations through steps 2)-4) leads to a situation in which $n_{11}(s) \neq 0$, $n_{i1}(s) = 0$, $i = 2, 3, \cdots, n$, $n_{1j}(s) = 0$, $j = 2, 3, \cdots, m$. If, in one of the columns aside the first one an element is not divisible by $n_{11}(s)$, add this column to the first one and go back to step 2). At each iteration of the cycle 2)-5) the degree of $n_{11}(s)$ decreases. Hence, in a finite number of cycles one arrives to the situation reported above where $n_{11}(s)$ divides each element of the submatrix $N_1(s)$ constituted by the last $m - 1$ columns and $n - 1$ rows. Assume that $n_{11}(s)$ is monic (otherwise perform an obvious elementary

operation) and let $\alpha_1(s) = n_{11}(s)$. Now apply to the submatrix $N_1(s)$ (obviously assumed to be nonzero) the entire procedure performed for matrix $N(s)$. The $(1,1)$ element will be now $\alpha_2(s)$ and, in view of the adopted procedure $\alpha_1(s)$ will be a divisor of $\alpha_2(s)$. Finally $\alpha_i(s) \neq 0$, $i = 1, \cdots, r$ since an elementary operation does not affect the matrix rank. □

Example 2.5 Consider the polynomial matrix

$$N(s) = \begin{bmatrix} s^2 + 1 & s^3 + s \\ s + 1 & s \end{bmatrix}$$

and the two polynomial and unimodular matrices

$$L(s) = \begin{bmatrix} (s^2 + 1)/2 & s - 1 \\ (s + 1)/2 & 1 \end{bmatrix}, \quad R(s) = \begin{bmatrix} 2 & s^3 - s^2 + 2s \\ 0 & -1/2 \end{bmatrix}$$

It follows that $N(s) = L(s)S(s)R(s)$ where

$$S(s) := \begin{bmatrix} 1 & 0 \\ 0 & s^4 + s^2 \end{bmatrix}$$

□

2.4 Proper rational matrices

This section deals with matrices $F(s)$ whose elements are ratios of polynomials in the same unknown. Therefore, the generic element $f_{ij}(s)$ of $F(s)$ has the form

$$f_{ij}(s) = \frac{a(s)}{b(s)} := \frac{\alpha_\nu s^\nu + \alpha_{\nu-1} s^{\nu-1} + \cdots + \alpha_1 s + \alpha_0}{\beta_\mu s^\mu + \beta_{\mu-1} s^{\mu-1} + \cdots + \beta_1 s + \beta_0}$$

$$\alpha_i \in R, \ i = 0, 1, \cdots, \nu, \ \beta_i \in R, \ i = 0, 1, \cdots \mu$$

The *relative degree* reldeg$[f_{ij}(s)]$ of $f_{ij}(s)$ is defined as the difference between the degree of the two polynomials which constitute the denominator and numerator of $f_{ij}(s)$, respectively. Specifically, with reference to the above function, and assuming $\alpha_\nu \neq 0$ and $\beta_\mu \neq 0$, it is

$$\text{reldeg}[f_{ij}(s)] := \deg[b(s)] - \deg[a(s)] = \mu - \nu$$

A rational matrix $F(s)$ is said to be *proper* (resp. *strictly proper*) if reldeg$[f_{ij}(s)] \geq 0$ (resp. reldeg$[f_{ij}(s)] > 0$) for all i, j. Throughout the section it is implicitly assumed that the rational matrices considered herein are always either proper or strictly proper.

The *rank* of a rational matrix $F(s)$ is, in analogy with the definition given for a polynomial matrix, the dimension of the largest square submatrix in $F(s)$ with determinant not identically equal to zero.

A rational square matrix is said to be *unimodular* if it has *maximum* rank and its determinant is a *rational* function with *zero* relative degree. Hence, a unimodular rational matrix admits a unimodular rational inverse and vice-versa.

Example 2.6 The matrix

$$F(s) = \begin{bmatrix} (s^2 + 2s + 3)/(s^2 - 1) & 2 \\ s/(s + 1) & 1 \end{bmatrix}$$

is unimodular. Actually, $\det[F(s)] = (-s^2 + 4s + 3)/(s^2 - 1)$. Moreover,

$$F^{-1}(s) = \begin{bmatrix} (s^2 - 1)/(-s^2 + 4s + 3) & (-2s^2 + 2)/(-s^2 + 4s + 3) \\ (-s^2 + s)/(-s^2 + 4s + 3) & (s^2 + 2s + 3)/(-s^2 + 4s + 3) \end{bmatrix}$$

□

The concepts of *divisor* and *greatest common divisor*, already given for polynomial matrices, are now extended to rational matrices in the definitions below.

Definition 2.4 (Right divisor) *Let $F(s)$ be a rational matrix. A rational square matrix $R(s)$ is said to be a right divisor of $F(s)$ if*

$$F(s) = \bar{F}(s)R(s)$$

with $\bar{F}(s)$ rational.   □

A similar definition could be given for a left divisor as well.

Definition 2.5 (Greatest common right divisor) *Consider two rational matrices $F(s)$ and $G(s)$ with the same number of columns. A Greatest Common Right Divisor (GCRD) of $(F(s), G(s))$ is a square rational matrix $R(s)$ such that*

i) $R(s)$ is a right divisor of $(F(s), G(s))$, i.e.

$$F(s) = \bar{F}(s)R(s)$$
$$G(s) = \bar{G}(s)R(s)$$

with $\bar{F}(s)$ and $\bar{G}(s)$ rational.

ii) If $\hat{R}(s)$ is any other right divisor of $(F(s), G(s))$, then $R(s) = W(s)\hat{R}(s)$ with $W(s)$ rational.

□

A similar definition holds for a Greatest Common Left Divisor (GCLD). By exploiting the properties of the rational unimodular matrices, it is easy to see that, given two rational matrices $F(s)$ and $G(s)$, there exist more than one GCRD (and GCLD). A way to compute a GCRD (resp. GCLD) of an assigned pair of rational matrices $F(s)$ and $G(s)$ calls for their manipulation via a rational unimodular matrix resulting from a sequence of *elementary operations* on their rows (resp. columns). The elementary operations on the rows (resp. columns) of a rational matrix $F(s)$ are:

1) Interchange of the i-th row (resp. i-th column) with the j-th row (resp. j-th column) and vice-versa

2) Multiplication of the i-th row (resp. i-th column) by a non zero rational function with zero relative degree

3) Addition of the i-th row (resp. i-th column) to the j-th row (resp. j-th column) multiplied by a rational function with zero relative degree

Obviously, each of these elementary operations reduces to premultiplying (resp. post-multiplying) matrix $F(s)$ by a suitable rational unimodular matrix $T(s)$. Moreover, matrix $T(s)F(s)$ (resp. $F(s)T(s)$) has the same rank as $F(s)$.

Remark 2.3 Given two scalar rational functions $f(s)$ and $g(s)$ with relative degree such that $\mathrm{reldeg}[f(s)] \leq \mathrm{reldeg}[g(s)]$, then $\mathrm{reldeg}[g(s)/f(s)] \geq 0$. Hence, considering the rational unimodular matrix

$$T(s) := \begin{bmatrix} 1 & 0 \\ -\dfrac{g(s)}{f(s)} & 1 \end{bmatrix}$$

it follows that

$$T(s) \begin{bmatrix} f(s) \\ g(s) \end{bmatrix} = \begin{bmatrix} f(s) \\ 0 \end{bmatrix}$$

By recursively exploiting this fact, it is easy to convince oneself that, if a rational matrix $F(s)$ does not have more columns than rows, it is always possible to build up a rational unimodular matrix $T(s)$ such that

$$T(s)F(s) = \begin{bmatrix} R(s) \\ 0 \end{bmatrix}$$

with $R(s)$ square and rational. Moreover, the null matrix vanishes when $F(s)$ is square. □

A GCRD of two rational matrices can be computed in the way described in the following algorithm, which relies on the same arguments as in Algorithm 2.1.

Algorithm 2.2 Let $F(s)$ and $G(s)$ be two rational matrices with the same number, say m, of columns and possibly different numbers of rows, say n_f and n_g, respectively.

1) Assume first that $m \leq n_f + n_g$, otherwise go to point 2). Let $H(s) := [F'(s)\ G'(s)]'$ and determine a rational unimodular matrix $T(s)$ such that

$$T(s)H(s) = \begin{bmatrix} R(s) \\ 0 \end{bmatrix}$$

   Then $R(s)$ is a GCRD

2) If $m > n_f + n_g$, then

$$R(s) = \begin{bmatrix} F(s) \\ G(s) \\ 0 \end{bmatrix}$$

   is a GCRD   □

Example 2.7 Consider the two rational matrices

$$F(s) = \begin{bmatrix} (s+1)/s & 1/(s+2) \end{bmatrix}$$

$$G(s) = \begin{bmatrix} (s+1)/(s-1) & s/(s^2+1) \\ 1 & s/(s+1) \end{bmatrix}$$

Take now

$$T_1(s) = \begin{bmatrix} 1 & 0 & -(s+1)/s \\ 0 & 1 & -(s+1)/(s-1) \\ 0 & 0 & 1 \end{bmatrix}, \quad T_2(s) = \begin{bmatrix} 1 & 0 & 1 \\ 1 & 0 & 0 \\ 0 & 1 & 0 \end{bmatrix}$$

$$T_3(s) = \begin{bmatrix} 1 & 0 & 0 \\ 0 & 1 & 0 \\ 0 & (s^4+s^3+4s)/(1-s^4) & 1 \end{bmatrix}$$

It follows that

$$T_3(s)T_2(s)T_1(s) \begin{bmatrix} F(s) \\ G(s) \end{bmatrix} = \begin{bmatrix} 1 & s/(s+1) \\ 0 & -(s+1)/(s+2) \\ 0 & 0 \end{bmatrix}$$

so that

$$R(s) = \begin{bmatrix} 1 & s/(s+1) \\ 0 & -(s+1)/(s+2) \end{bmatrix}, \quad \bar{F}(s) = \begin{bmatrix} (s+1)/s & 1 \end{bmatrix}$$

$$\bar{G}(s) = \begin{bmatrix} (s+1)/(s-1) & (s^4+s^3+4s)/(s^4-1) \\ 1 & 0 \end{bmatrix}$$

and $F(s) = \bar{F}(s)R(s)$, $G(s) = \bar{G}(s)R(s)$. □

Also for rational matrices it is possible to consistently introduce the concept of *co-primeness*.

Definition 2.6 (Right coprimeness) *Two rational matrices $F(s)$ and $G(s)$ with the same number of columns are said to be right coprime if the relations*

$$F(s) = \bar{F}(s)T(s)$$
$$G(s) = \bar{G}(s)T(s)$$

with $\bar{F}(s)$ and $\bar{G}(s)$ rational matrices, are verified only if $T(s)$ is a rational unimodular matrix. □

Example 2.8 The two matrices

$$F(s) = \begin{bmatrix} (s-1)/(s+1)^2 & (s^2+1)/(s+1)(s^2+3s) \\ 1/(s+1) & s/(s^2-1) \end{bmatrix}$$

$$G(s) = \begin{bmatrix} 1/(s+1) & (2s+1)/(s+1)(s+3) \end{bmatrix}$$

are not right coprime since it results

$$F(s) = \begin{bmatrix} (s-1)/(s+1) & (s^2+1)/(s^2+3s) \\ 1 & s/(s-1) \end{bmatrix} R(s)$$

$$G(s) = \begin{bmatrix} 1 & (2s+1)/(s+3) \end{bmatrix} R(s)$$

with $R(s) = 1/(s+1)$ which is not unimodular. □

An analogous definition holds for left coprimeness. From Definition 2.6 it follows that two rational matrices are right (resp. left) coprime if all their common right (resp. left) divisors are unimodular. In particular, each GCRD (resp. GCLD) of two rational right (resp. left) coprime matrices must be unimodular. Therefore, a necessary condition for matrices $F(s)$ and $G(s)$ to be right (resp. left) coprime is that the number of their columns be not greater than the sum of the number of their rows (resp. the number of their rows be not greater than the sum of the number of their columns), since, from Algorithm 2.2 point 2) in the opposite case one of their GCRD would not be unimodular. Moreover, a way to verify whether or not two rational matrices are right (resp. left) coprime consists in the computation through Algorithm 2.2 of a greatest common divisor and evaluation of its determinant. As a matter of fact, as stated in the next lemma, if a greatest common divisor is unimodular then all the greatest common divisors are unimodular as well.

Lemma 2.3 *Let $F(s)$ and $G(s)$ be two rational matrices with the same number of columns. Let $R(s)$ be a GCRD of $(F(s), G(s))$. Then,*

  i) $T(s)R(s)$ is a GCRD for any rational unimodular $T(s)$

  ii) If $\bar{R}(s)$ is a GCRD of $(F(s), G(s))$, then there exists a rational unimodular matrix $T(s)$ such that $\bar{R}(s) = T(s)R(s)$

Proof The proof follows from that of Lemma 2.1 by substituting there the term "rational" in place of the term "polynomial" and symbols $F(s)$ and $G(s)$ in place of $N(s)$ and $D(s)$, respectively. □

A further significant property of a GCRD of a pair of rational matrices $F(s)$ and $G(s)$ is stated in the following lemma, whose proof hinges on Algorithm 2.2.

Lemma 2.4 *Consider two rational matrices $F(s)$ and $G(s)$ with the same number of columns and let $R(s)$ be a GCRD of $(F(s), G(s))$. Then, there exist two rational matrices $X(s)$ and $Y(s)$ such that*

$$X(s)F(s) + Y(s)G(s) = R(s)$$

Proof Let n_f and n_g be the number of rows of $F(s)$ and $G(s)$, respectively, and m the number of their columns. Preliminarily, assume that $n_f + n_g \geq m$ and let $T(s)$ be a unimodular matrix such that

$$T(s) \begin{bmatrix} F(s) \\ G(s) \end{bmatrix} = \begin{bmatrix} T_{11}(s) & T_{12}(s) \\ T_{21}(s) & T_{22}(s) \end{bmatrix} \begin{bmatrix} F(s) \\ G(s) \end{bmatrix} = \begin{bmatrix} \bar{R}(s) \\ 0 \end{bmatrix} \tag{2.10}$$

Based on Algorithm 2.2, matrix $\bar{R}(s)$ turns out to be a GCRD of $(F(s), G(s))$. Hence, thanks to Lemma 2.3, there exist a unimodular matrix $U(s)$ such that $R(s) = U(s)\bar{R}(s)$, that is, in view of eq. (2.10),

$$R(s) = U(s)T_{11}(s)F(s) + U(s)T_{12}(s)G(s) = X(s)F(s) + Y(s)G(s)$$

On the contrary, if $m > n_f + n_g$, Algorithm 2.2 entails that

$$\bar{R}(s) := \begin{bmatrix} F(s) \\ G(s) \\ 0 \end{bmatrix}$$

is a GCRD of $(F(s), G(s))$. In view of Lemma 2.3, it is possible to write

$$R(s) = U(s)\bar{R}(s)$$

$$= U(s) \begin{bmatrix} I \\ 0 \\ 0 \end{bmatrix} F(s) + U(s) \begin{bmatrix} 0 \\ I \\ 0 \end{bmatrix} G(s)$$

$$:= X(s)F(s) + Y(s)G(s)$$

where $U(s)$ is a suitable rational and unimodular matrix. □

The following result, that parallels the analogous one presented in Lemma 2.2, can now be stated.

Lemma 2.5 *Let $F(s)$ and $G(s)$ be two rational matrices with the same number of columns. Then, $F(s)$ and $G(s)$ are right coprime if and only if there exist two rational matrices $X(s)$ and $Y(s)$ such that*

$$X(s)F(s) + Y(s)G(s) = I$$

Proof Recall that two matrices are right coprime if each one of their GCRD's is unimodular. Hence, if $R(s)$ is a GCRD of $(F(s), G(s))$, thanks to Lemma 2.4, it results $R(s) = \bar{X}(s)F(s) + \bar{Y}(s)G(s)$, with $\bar{X}(s)$ and $\bar{Y}(s)$ suitable rational matrices. From this last equation it follows

$$I = R^{-1}(s)\bar{X}(s)F(s) + R^{-1}(s)\bar{Y}(s)G(s) := X(s)F(s) + Y(s)G(s)$$

Conversely, let $R(s)$ be a GCRD of $(F(s), G(s))$ derived according to Algorithm 2.2 point 1), as the number of their columns must be not greater than the sum of the numbers of their rows, so that

$$T(s)\left[\begin{array}{c} F(s) \\ G(s) \end{array}\right] = \left[\begin{array}{c} R(s) \\ 0 \end{array}\right]$$

where $T(s)$ is a suitable rational and unimodular matrix. Hence,

$$\left[\begin{array}{c} F(s) \\ G(s) \end{array}\right] = T^{-1}(s)\left[\begin{array}{c} R(s) \\ 0 \end{array}\right] := \left[\begin{array}{cc} S_{11}(s) & S_{12}(s) \\ S_{21}(s) & S_{22}(s) \end{array}\right]\left[\begin{array}{c} R(s) \\ 0 \end{array}\right]$$

so that

$$I = X(s)F(s) + Y(s)G(s) = [X(s)S_{11}(s) + Y(s)S_{21}(s)]R(s)$$

shows that $R(s)$ is unimodular (its inverse is rational). Therefore, $(F(s), G(s))$ are right coprime.   □

Example 2.9 Consider two rational matrices

$$F(s) = \left[\begin{array}{cc} 1/(s^2 + s) & (2s^2 + 2s - 2)/(s^2 + 2s) \end{array}\right]$$

$$G(s) = \left[\begin{array}{cc} -s/(s+1) & s/(s+1) \\ (s^2 - s - 1)/(s^2 + s) & (s^2 + 2s + 2)/(s^2 + 2s) \end{array}\right]$$

These matrices are right coprime. Actually,

$$T(s)\left[\begin{array}{c} F(s) \\ G(s) \end{array}\right] = \left[\begin{array}{c} R(s) \\ 0 \end{array}\right]$$

where

$$T(s) = \frac{1}{p(s)}\left[\begin{array}{ccc} s^3 - s^2 - 2s - 1 & -s^3 + s & s^3 + s^2 - s - 1 \\ s^3 + 2s^2 + s & s^3 + s^2 & -s^3 - 2s^2 - s \\ -s^3 & s^3 - s^2 - 2s & s^3 \end{array}\right]$$

with $p(s) := 2s^3 - 3s - 2$ and

$$R(s) = \left[\begin{array}{cc} s/(s+1) & (s+1)/(s+2) \\ -1 & 1 \end{array}\right]$$

that is a rational and unimodular matrix. Moreover, taking

$$X(s) = \frac{1}{p(s)q(s)}\left[\begin{array}{c} -2s^4 - 9s^3 - 13s^2 - 8s - 2 \\ 2s^5 + 6s^4 + 2s^3 - 7s^2 - 7s - 2 \end{array}\right]$$

and

$$Y(s) = \frac{1}{p(s)q(s)} \begin{bmatrix} -2s^5 - 6s^4 - 4s^3 + 2s^2 + 2s & 2s^5 + 8s^4 + 10s^3 + 2s^2 - 4s - 2 \\ s^3 + 3s^2 + 2s & -s^3 - 4s^2 - 5s - 2 \end{bmatrix}$$

with $q(s) := 2s^2 + 4s + 1$, it follows $X(s)F(s) + Y(s)G(s) = I$. □

Also for rational matrices there exist a particularly useful *canonical form*, which is called *Smith-McMillan form*. This form is precisely defined in the following theorem, whose proof also provides a procedure for its computation.

Theorem 2.2 (Smith-McMillan form) *Let $G(s)$ be a proper rational matrix with n rows, m columns and $\text{rank}[G(s)] = r \leq \min[n, m]$. Then there exist two polynomial and unimodular matrices $L(s)$ and $R(s)$ such that $G(s) = L(s)M(s)R(s)$, where*

$$M(s) = \begin{bmatrix} f_1(s) & 0 & \cdots & 0 & 0 \\ 0 & f_2(s) & \cdots & 0 & 0 \\ \vdots & \vdots & \ddots & \vdots & \vdots \\ 0 & 0 & \cdots & f_r(s) & 0 \\ 0 & 0 & \cdots & 0 & 0 \end{bmatrix} \Big\} \, n - r \text{ rows}$$

$$\underbrace{}_{m - r \text{ columns}}$$

with

$$f_i(s) = \frac{\varepsilon_i(s)}{\psi_i(s)}, \quad i = 1, 2, \cdots, r$$

and

- $\varepsilon_i(s)$ *and* $\psi_i(s)$ *are monic,* $i = 1, 2, \cdots r$
- $\varepsilon_i(s)$ *and* $\psi_i(s)$ *are coprime* $i = 1, 2, \cdots r$
- $\varepsilon_i(s)$ *divides* $\varepsilon_{i+1}(s)$, $i = 1, 2, \cdots r$
- ψ_{i+1} *divides* ψ_i, $i = 1, 2, \cdots r$

Matrix $M(s)$ is the Smith-McMillan form of $F(s)$.

Proof Let $\psi(s)$ be the least common multiple of all polynomials at the denominators of the elements of $F(s)$. Therefore, matrix $N(s) := \psi(s)F(s)$ is polynomial. If $S(s)$ is the Smith form of $N(s)$ (recall Theorem 2.1) it follows that

$$F(s) = \frac{1}{\psi(s)} N(s) = \frac{1}{\psi(s)} L(s)S(s)R(s)$$

Hence,

$$M(s) = \frac{S(s)}{\psi(s)}$$

once all the possible simplifications between the elements of $S(s)$ and the polynomial $\psi(s)$ have been performed. This matrix obviously has the properties claimed in the statement. □

Remark 2.4 The result stated in Theorem 2.2 allows one to represent a generic rational $p \times m$ matrix $G(s)$ with $\text{rank}[G(s)] := r \leq \min[m, p]$ in the two forms

$$G(s) = N(s)D^{-1}(s) = \hat{D}^{-1}(s)\hat{N}(s)$$

where the *polynomial* matrices $N(s)$ and $D(s)$ are right coprime, while the *polynomial* matrices $\hat{D}(s)$ and $\hat{N}(s)$ are left coprime. Actually, observe that letting

$$\Psi(s) := \begin{bmatrix} \psi_1(s) & 0 & \cdots & 0 \\ 0 & \psi_2(s) & \cdots & 0 \\ \vdots & \vdots & \ddots & \vdots \\ 0 & 0 & \cdots & \psi_r(s) \end{bmatrix}$$

$$E(s) := \begin{bmatrix} \varepsilon_1(s) & 0 & \cdots & 0 \\ 0 & \varepsilon_2(s) & \cdots & 0 \\ \vdots & \vdots & \ddots & \vdots \\ 0 & 0 & \cdots & \varepsilon_r(s) \end{bmatrix}$$

it follows

$$M(s) = \begin{bmatrix} \Psi^{-1}(s) & 0 \\ 0 & I \end{bmatrix} \begin{bmatrix} E(s) & 0 \\ 0 & 0 \end{bmatrix}$$

$$= \begin{bmatrix} E(s) & 0 \\ 0 & 0 \end{bmatrix} \begin{bmatrix} \Psi^{-1}(s) & 0 \\ 0 & I \end{bmatrix}$$

Now, defining

$$N(s) := L(s) \begin{bmatrix} E(s) & 0 \\ 0 & 0 \end{bmatrix}, \quad D(s) := R^{-1}(s) \begin{bmatrix} \Psi(s) & 0 \\ 0 & I \end{bmatrix}$$

$$\hat{N}(s) := \begin{bmatrix} E(s) & 0 \\ 0 & 0 \end{bmatrix} R(s), \quad \hat{D}(s) := \begin{bmatrix} \Psi(s) & 0 \\ 0 & I \end{bmatrix} L^{-1}(s)$$

one can easily check that $G(s) = N(s)D^{-1}(s) = \hat{D}^{-1}(s)\hat{N}(s)$. In order to verify that $N(s)$ and $D(s)$ are right coprime, one can resort to Lemma 2.2. Actually, considering the two matrices $X(s)$ and $Y(s)$ defined by

$$X(s) := \begin{bmatrix} x_1(s) & 0 & \cdots & 0 & 0 \\ 0 & x_2(s) & \cdots & 0 & 0 \\ \vdots & \vdots & \ddots & x_r(s) & 0 \\ 0 & 0 & \cdots & 0 & 0 \end{bmatrix} L^{-1}(s)$$

$$Y(s) := \begin{bmatrix} y_1(s) & 0 & \cdots & 0 & 0 \\ 0 & y_2(s) & \cdots & 0 & 0 \\ \vdots & \vdots & \ddots & y_r(s) & 0 \\ 0 & 0 & \cdots & 0 & I \end{bmatrix} R(s)$$

it turns out that $X(s)N(s) + Y(s)D(s) = I$ for a suitable choice of the polynomials $x_i(s)$ and $y_i(s)$ (recall that the polynomials $\psi_i(s)$ and $\varepsilon_i(s)$ are coprime and take in mind Theorem A.1). From Lemma 2.2 it follows that the two matrices $N(s)$ and $D(s)$ are right coprime. Analogously, one can verify that $\hat{N}(s)$ and $\hat{D}(s)$ are left coprime. $\qquad\square$

Example 2.10 Consider the rational matrix

$$F(s) = \frac{1}{s^3 - s^2} \begin{bmatrix} 3s^3 + 7s^2 + 3s - 1 & 4s^3 + 8s^2 + 2s - 2 \\ 2s^3 + 4s^2 + s - 1 & 3s^3 + 5s^2 - 2 \end{bmatrix}$$

Now applying what has been indicated in the proof of Theorems 2.1, 2.2 it follows that $F(s) = L(s)M(s)R(s)$, with

$$L(s) = \begin{bmatrix} (3s^2 + 4s - 1)/3 & (6s + 5)/12 \\ (2s^2 + 2s - 1)/3 & (2s + 1)/6 \end{bmatrix}$$

$$R(s) = \begin{bmatrix} 3 & -2s^3 - 3s^2 + 2s + 6 \\ 0 & 4 \end{bmatrix}$$

$$M(s) = \begin{bmatrix} (s + 1)/(s^3 - s^2) & 0 \\ 0 & (s + 1)^2(s + 2)/s \end{bmatrix}$$

Moreover, by following the arguments in Remark 2.4,

$$E(s) = \begin{bmatrix} s + 1 & 0 \\ 0 & (s + 1)^2(s + 2) \end{bmatrix}$$

$$\Psi(s) = \begin{bmatrix} s^2(s - 1) & 0 \\ 0 & s \end{bmatrix}$$

it turns out that

$$N(s) = L(s)E(s) = \frac{1}{12} \begin{bmatrix} 4(3s^3 + 7s^2 + 3s - 1) & 6s^4 + 29s^3 + 50s^2 + 37s + 10 \\ 4(2s^3 + 4s^2 + s - 1) & 2(2s^4 + 9s^3 + 14s^2 + 9s + 2) \end{bmatrix}$$

$$D(s) = R^{-1}(s)\Psi(s) = \frac{1}{12} \begin{bmatrix} 4(s^3 - s) & s(2s^3 + 3s^2 - 2s - 6) \\ 0 & 3s \end{bmatrix}$$

$$\hat{N}(s) = E(s)R(s) = \begin{bmatrix} 3(s + 1) & -2s^4 - 5s^3 - s^2 + 8s + 6 \\ 0 & 4(s^3 + 4s^2 + 5s + 2) \end{bmatrix}$$

$$\hat{D}(s) = \Psi(s)L^{-1}(s) = \begin{bmatrix} 2s^2(2s^2 - s - 1) & -s^2(6s^2 - s - 5) \\ -4s(2s^2 + 2s - 1) & 4s(3s^2 + 4s - 1) \end{bmatrix}$$

□

Many of the results provided till now can be straightforwardly extended to the subset of proper, rational and stable functions, namely the subset constituted by the matrices whose generic element $f_{ij}(s)$ is a proper, rational function with poles in the open left half plane only. This extension calls for the introduction of a suitable scalar associated with a generic proper rational scalar function $f(s)$. Precisely, rhpdeg$[f(s)]$ will indicate the number of finite nonnegative real part zeros of $f(s)$ plus reldeg$[f(s)]$. For example

$$f(s) = \frac{(s - 1)(s + 2)}{(s + 4)^2}$$

is such that rhpdeg$[f(s)] = 3$ since reldeg$[f(s)] = 2$ and $f(s)$ has a zero in $s = 1$. Moreover, the function

$$f(s) = \frac{s}{s + 1}$$

is such that rhpdeg$[f(s)] = 1$, since $f(s)$ has a zero in $s = 0$ and reldeg$[f(s)] = 0$. It will conventionally be set rhpdeg$[0] = -\infty$. Preliminarily, observe that the definitions of divisor and greatest common divisors of rational matrices (Definitions 2.4 and 2.5) can be trivially generalized to the subset of stable matrices by actually requiring the stability property. As for the generalization of the concept of unimodular matrix $T(s)$, it suffices requiring, besides the stability of $T(s)$, also that rhpdeg$[\det[T(s)]] = 0$. In this way, the stable matrix $T(s)$ has a stable inverse as well.

The three elementary operations on the rows (columns) of a rational matrix are extended to stable rational matrices by simply requiring that in the second operation the multiplying function $f(s)$ be stable with rhpdeg$[f(s)] = 0$ and simply that in the third operation $f(s)$ be stable.

Lemma 2.6 *Let $f(s)$ and $g(s)$ be two stable rational scalar functions with $g(s) \neq 0$ and rhpdeg$[f(s)] \geq$ rhpdeg$[g(s)]$. Then, there exists a stable rational function $q(s)$ such that*

$$\text{rhpdeg}[f(s) - g(s)q(s)] < \text{rhpdeg}[g(s)] \tag{2.11}$$

Proof If rhpdeg$[g(s)] = 0$, then $g^{-1}(s)$ is rational, proper and stable so that equation (2.11) is obviously satisfied with $q(s) = g^{-1}(s)f(s)$. Therefore, suppose that rhpdeg$[g(s)] := \nu \neq 0$ and write

$$g(s) := \frac{ng(s)}{dg(s)} := \frac{ng^+(s)ng^-(s)}{dg(s)}$$

where the polynomials $ng(s)$ and $dg(s)$ are coprime whereas $ng^+(s)$ has roots in the closed right half plane only and $ng^-(s)$ in the open left half plane only. Moreover, let

$$h(s) := \frac{(s+1)^\nu ng^-(s)}{dg(s)}$$

so that both $h(s)$ and $h^{-1}(s)$ are proper stable rational functions. Of course, it results

$$g(s) = \frac{h(s)ng^+(s)}{(s+1)^\nu}$$

Also, write

$$f(s) := \frac{nf(s)}{df(s)}$$

where the two polynomials $nf(s)$ and $df(s)$ are coprime. Notice that, being $f(s)$ stable, the zeros of $df(s)$ are in the open left half plane. This entails that $df(s)$ and $ng^+(s)$ are coprime. By exploiting Lemma A.2, one can claim that there exist two polynomials $\varphi(s)$ and $\psi(s)$ with $\deg[\varphi(s)] < \deg[df(s)]$ such that

$$ng^+(s)\varphi(s) + df(s)\psi(s) = (s+1)^{\nu-1}nf(s)$$

From this relation it follows

$$\frac{ng^+(s)\varphi(s)}{df(s)(s+1)^{\nu-1}} + \frac{\psi(s)}{(s+1)^{\nu-1}} = \frac{nf(s)}{df(s)}$$

Let now

$$q(s) := \frac{(s+1)\phi(s)}{df(s)h(s)}$$

and observe that such function is rational, proper and stable. This conclusion derives from properness and stability of $h^{-1}(s)$ and the fact that $\varphi(s)/df(s)$ is strictly proper and stable. Moreover, let

$$r(s) := \frac{\psi(s)}{(s+1)^{\nu-1}}$$

then,

$$
q(s)g(s) + r(s) = \frac{(s+1)\varphi(s)dg(s)ng^+(s)ng^-(s)}{(s+1)^\nu df(s)ng^-(s)dg(s)} + \frac{\psi(s)}{(s+1)^{\nu-1}}
$$

$$
= \frac{(s+1)\varphi(s)ng^+(s)}{(s+1)^\nu df(s)} + \frac{\psi(s)}{(s+1)^{\nu-1}}
$$

$$
= \frac{nf(s)}{df(s)} = f(s)
$$

Thus, being $f(s)$, $g(s)$ and $q(s)$ proper, rational and stable, the function $r(s) = f(s) - q(s)g(s)$ is rational, proper and stable as well. Finally, recalling the definition of $r(s)$, one can conclude that

$$\text{rhpdeg}[r(s)] \leq \nu - 1 < \nu = \text{rhpdeg}[g(s)]$$

□

Remark 2.5 Lemma 2.6 allows one to discuss further what has been shown in Remark 2.3, in the context of stable matrices. As a matter of fact, let $f(s)$ and $g(s)$ be two rational stable scalar functions with

$$\text{rhpdeg}[f(s)] \geq \text{rhpdeg}[g(s)]$$

In view of Lemma 2.6 there exists a stable rational matrix $q(s)$ such that

$$\text{rhpdeg}[f(s) - q(s)g(s)] < \text{rhpdeg}[g(s)]$$

Then, the unimodular stable matrix

$$T_1(s) = \begin{bmatrix} 1 & -q(s) \\ 0 & 1 \end{bmatrix}$$

is such that

$$T_1(s) \begin{bmatrix} f(s) \\ g(s) \end{bmatrix} = \begin{bmatrix} f(s) - q(s)g(s) \\ g(s) \end{bmatrix} := \begin{bmatrix} f^1(s) \\ g^1(s) \end{bmatrix}$$

By iterating this operation with stable unimodular matrices of the given form (or, alternatively, of the form corresponding to its transpose), one get

$$\prod_{i=0}^{n-1} T_{n-i}(s) \begin{bmatrix} f(s) \\ g(s) \end{bmatrix} = \begin{bmatrix} f^n(s) \\ g^n(s) \end{bmatrix}$$

with either $\text{rhpdeg}[f^n(s)] = 0$ or $\text{rhpdeg}[g^n(s)] = 0$. Assuming, for instance, that the first situation has occurred, it follows that

$$\prod_{i=-1}^{n-1} T_{n-i}(s) \begin{bmatrix} f(s) \\ g(s) \end{bmatrix} = \begin{bmatrix} f^n(s) \\ 0 \end{bmatrix}$$

where

$$T_{n+1} = \begin{bmatrix} 1 & 0 \\ -(f^n(s))^{-1}g^n(s) & 1 \end{bmatrix}$$

Then, one can conclude that, given two stable rational scalar functions $f(s)$ and $g(s)$, there exists a unimodular stable matrix $T(s)$ such that

$$T(s)\begin{bmatrix} f(s) \\ g(s) \end{bmatrix} = \begin{bmatrix} r(s) \\ 0 \end{bmatrix}$$

with $r(s)$ stable and rhpdeg$[r(s)] = 0$.   □

The arguments used in Remark 2.5 are well suited to be extended in the context of proper stable rational matrices. The same occurs for Lemmas 2.3, 2.4 provided that Definition 2.6 on coprimeness is adapted to this new setting. With these arguments in mind, it is possible to state the following lemma, which specializes Lemma 2.5 to the case of stable rational matrices.

Lemma 2.7 *Let $F(s)$ and $G(s)$ be two stable rational matrices with the same number of columns. Then $F(s)$ and $G(s)$ are right coprime (in the setting of proper stable matrices) if and only if there exist two stable rational matrices $X(s)$ and $Y(s)$ such that*

$$X(s)F(s) + Y(s)G(s) = I$$

Example 2.11 Consider the two stable rational functions

$$f(s) = \frac{s+2}{s+1} , \quad g(s) = \frac{(s+2)^2}{(s+1)^2}$$

They are (right) coprime. Actually, taking

$$x(s) := \frac{s+1}{2(s+2)} , \quad y(s) := \frac{(s+1)^2}{2(s+2)^2}$$

it follows

$$x(s)f(s) + y(s)g(s) = 1$$

On the other side, consider the stable unimodular matrix

$$T(s) := \begin{bmatrix} 1 & 0 \\ -(f(s))^{-1}g(s) & 1 \end{bmatrix}$$

Then

$$T(s)\begin{bmatrix} f(s) \\ g(s) \end{bmatrix} = \begin{bmatrix} f(s) \\ 0 \end{bmatrix}$$

with $f(s)$ stable and unimodular.   □

2.5   Poles and zeros

This section is devoted to a schematic presentation of the main properties of the poles and zeros of a linear and time-invariant dynamic system Σ,

$$\dot{x} = Ax + Bu$$
$$y = Cx + Du$$

with n states, m inputs and p outputs. It will be of main importance in the sequel to distinguish between two cases. In the first one, reference is only made to an input-output description of the system, i.e to its transfer function

$$G(s) = C(sI - A)^{-1}B + D$$

whereas in the second case a state-space description of the system is considered. In order to rule out trivialities and make simpler the exposition, it is assumed, throughout all the section, that $G(s)$ has full rank, i.e. $\text{rank}[G(s)]=\min[p, m]:=\text{r}$.

Definition 2.7 (Zeros and poles of a rational matrix) *Consider the rational matrix $G(s)$ and its associated Smith-McMillan form*

$$M(s) = \begin{bmatrix} f_1(s) & 0 & \cdots & 0 & 0 \\ 0 & f_2(s) & \cdots & 0 & 0 \\ \vdots & \vdots & \ddots & \vdots & \vdots \\ 0 & 0 & \cdots & f_r(s) & 0 \\ 0 & 0 & \cdots & 0 & 0 \end{bmatrix}$$

with

$$f_i(s) := \frac{\varepsilon_i(s)}{\psi_i(s)}, \quad i = 1, 2, \cdots, r$$

and define the polynomials $\pi_p(s)$ and $\pi_{zt}(s)$ as

$$\pi_p(s) := \psi_1(s)\psi_2(s)\cdots\psi_r(s)$$
$$\pi_{zt}(s) := \varepsilon_1(s)\varepsilon_2(s)\cdots\varepsilon_r(s)$$

The poles of $G(s)$ are defined as the roots of $\pi_p(s)$ and the zeros of $G(s)$ as those of $\pi_{zt}(s)$. □

The definition of zeros and poles of $G(s)$ coincides with that of *transmission* zeros and *transmission* poles of a system having $G(s)$ as transfer function. As customary, the transmission poles will be simply referred to as poles of the system.

Definition 2.8 *Consider a linear time invariant system Σ with transfer function $G(s)$. The transmission zeros (resp. poles) of Σ are the zeros (resp. poles) of $G(s)$.*□

Example 2.12 Consider the linear system $\Sigma(A, B, C, D)$ defined by matrices

$$A = \begin{bmatrix} 0 & 1 & 0 & 0 \\ -10 & 7 & 0 & 0 \\ 0 & 0 & 5 & 0 \\ 1 & -1 & 1 & 0 \end{bmatrix}, \quad B = \begin{bmatrix} 0 \\ 1 \\ 2 \\ 0 \end{bmatrix}$$

$$C = \begin{bmatrix} -13 & 5 & 0 & 0 \\ 0 & 0 & 1 & 0 \end{bmatrix}, \quad D = \begin{bmatrix} 1 \\ 1 \end{bmatrix}$$

The transfer function and its Smith-McMillan form are given by

$$G(s) = \begin{bmatrix} (s-3)(s+1)/(s-5)(s-2) \\ (s-3)/(s-5) \end{bmatrix} = \begin{bmatrix} s+1 & 1 \\ s-2 & 1 \end{bmatrix} \begin{bmatrix} (s-3)/(s-5)(s-2) \\ 0 \end{bmatrix}$$

Therefore there is only one transmission zero in $s = 3$. □

Remark 2.6 In general, two polynomials $\varepsilon_j(s)$ and $\psi_i(s)$, $i \neq j$, can have common roots. This entails that a *multivariable* system may admit coincident poles and zeros even in case of minimallity of its state-space description. □

The Smith-McMillan form of $G(s)$ allows one for an alternative characterization of the transmission zeros in terms of vectors belonging to the kernel of $G(s)$ or $G'(s)$, as proved in the following lemma.

Lemma 2.8 (Rank property of transmission zeros) *Consider a transfer function $G(s)$ of rank $r := \min[p, m]$. The complex number λ is a transmission zero of $G(s)$ if and only if there exists a non zero vector z such that*

$$
\begin{cases}
\lim_{s \to \infty} \; G(s)z = 0 & \text{if } p \geq m \\[2mm]
\lim_{s \to \infty} \; G'(s)z = 0 & \text{if } p \leq m
\end{cases}
$$

Proof Consider first the case $p \geq m$ and let $M(s)$ be the Smith-McMillan form of $G(s)$, so that $G(s) = L(s)M(s)R(s)$, where $L(s)$ and $R(s)$ are suitable polynomial and unimodular matrices of dimensions p and m, respectively. It is possible to write (recall Remark 2.4)

$$
M(s) = \left[\begin{array}{c} E(s)\Psi^{-1}(s) \\ 0 \end{array} \right]
$$

where $E(s) := \text{diag}\{\varepsilon_1(s), \cdots, \varepsilon_m(s)\}$, $\Psi(s) := \text{diag}\{\psi_1(s), \cdots, \psi_m(s)\}$. Further, denote by $e_k(h)$ the k-th column of the h-dimensional identity matrix and let λ be a zero of $G(s)$, root of the polynomial $\varepsilon_k(s)$ of $E(s)$. Since $\psi_k(\lambda) \neq 0$ and $R(\lambda)$ is nonsingular, it then follows that $z = R^{-1}(\lambda)e_k(m)$ satisfies the condition of the theorem.

Conversely, if there exists $z \neq 0$ such that $G(s)z \to 0$ as $z \to \lambda$, then, necessarily,

$$
E(\lambda) \lim_{s \to \lambda} \Psi^{-1}(s)R(\lambda)z = 0
$$

so that λ is a root of at least one of the polynomials $\varepsilon_i(s)$, $i = 1, \cdots r$. The proof of the lemma in the converse case $(p \leq m)$ formally proceeds along the same route by replacing $G(s)$ with $G'(s)$. □

A quite different definition of transmission zeros and poles makes reference to the *minors* of $G(s)$. A k-degree minor of a matrix A is the determinant of any square k-dimensional submatrix of A. It is possible to prove that the polynomial $\pi_p(s)$ of the poles of $G(s)$ is given by the *least common denominator of all the non zero minors of any order of $G(s)$*. Analogously, the polynomial $\pi_{zt}(s)$ of the transmission zeros of $G(s)$ is the *greatest common divisor of all numerators of the minors of order r of $G(s)$ provided that they have been adjusted so as to present the polynomial $\pi_p(s)$ as their denominator*.

Remark 2.7 (Transmission zeros of a square system) In the particular case where $G(s)$ is square, it follows (recall Theorem 2.2)

$$
\det[G(s)] = \frac{\pi_{zt}(s)}{\pi_p(s)}
$$

Notice, however, that the presence of possible cancellations avoids in general to catch all transmission zeros and poles of $G(s)$ from its determinant. □

The transmission poles and zeros of a system $\Sigma(A, B, C, D)$ enjoy an important input output characterization. As for the transmission zeros, reference is made to the so called *blocking property*, which deals with the possibility of getting identically zero forced output when the input is suitably chosen in the class of exponential and impulsive signals. Before formally stating the relevant result, it is advisable to stress that the transmission zeros of $G(s)$ and $G'(s)$ actually coincides (recall the appropriate definition and Lemma 2.8). The same occurs for (transmission) poles. Let now indicate with $\delta(t) := \delta^{(0)}(t)$ the impulsive "function" and with $\delta^{(k)}(t)$ its k-th order derivative (recall that $\delta_L^{(k)} = s^k$). Moreover define

$$\hat\Sigma := \begin{cases} \Sigma & \text{if } p \geq m \\ \Sigma' & \text{if } p \leq m \end{cases} \quad , \quad \hat{G}(s) := \begin{cases} G(s) & \text{if } p \geq m \\ G'(s) & \text{if } p \leq m \end{cases} \tag{2.12}$$

Theorem 2.3 (Time domain characterization of transmission zeros and poles)
Let $G(s)$ be the transfer function of a system Σ. Then

i) *The complex number λ is a pole of Σ if and only if there exists an impulsive input*

$$u(t) = \sum_{i=0}^{\nu} \alpha_i \delta^{(i)}(t)$$

of $\hat\Sigma$, with α_i suitable constants, $\nu \geq 0$, such that the forced output $y_f(\cdot)$ of $\hat\Sigma$ is

$$y_f(t) = y_0 e^{\lambda t}, \quad \forall t > 0$$

ii) *The complex number λ is a transmission zero of Σ if and only if there exists an exponential/impulsive input*

$$u(t) = u_0 e^{\lambda t} + \sum_{i=0}^{\nu} \alpha_i \delta^{(i)}(t)$$

of $\hat\Sigma$, with α_i suitable constants, $\nu \geq 0$, $u_0 \neq 0$, such that the forced output $y_f(\cdot)$ of $\hat\Sigma$ is

$$y_f(t) = 0, \quad \forall t > 0$$

Proof Consider the Smith-McMillan form $\hat{M}(s)$ of $\hat{G}(s)$, introduced in Theorem 2.2, so that $\hat{G}(s) = L(s)\hat{M}(s)R(s)$, where $\hat{M}(s)$ is the $\max[m, p] \times r$ matrix $\hat{M}(s) = [\text{diag}\{\varepsilon_i(s)/\psi_i(s)\} \ 0]'$. The polynomial matrices $L(s)$ and $R(s)$ are unimodular. Denote by $l_i(s)$ and $r_i'(s)$ the i-th column and i-th row of $L(s)$ and $R(s)$, respectively. It turns out that

$$y_{fL0} = \sum_{i=1}^{r} l_i(s) \frac{\varepsilon_i(s)}{\psi_i(s)} r_i'(s) u_L$$

Point i) Assume that λ is a pole of $\hat{G}(s)$, i.e. a root of the polynomial $\psi_k(s)$. Hence $\gamma(s) := (s - \lambda)^{-1} \psi_k(s)$ is a polynomial. Then the input u_L defined as

$$u_L := R(s)^{-1} \begin{bmatrix} 0 \\ \vdots \\ 0 \\ \gamma(s) \\ 0 \\ \vdots \\ 0 \end{bmatrix} \quad \longleftarrow \quad k\text{-th row} \tag{2.13}$$

is a polynomial vector since $R^{-1}(s)$ is a polynomial matrix. Therefore $r'_i(s)u_L = 0, i \neq k$ and $r'_k(s)u_L = \gamma(s)$. It then turns out that $y_{fL0} = (s-\lambda)^{-1}l_k(s)\varepsilon_k(s)$. Since λ is not a root of $\varepsilon_k(s)$,

$$y_{fL0} = y_0(s-\lambda)^{-1} + \beta(s)$$

where y_0 is a suitable constant vector and $\beta(s)$ a suitable polynomial vector. Transforming back this expression in the time domain for $t > 0$, the conclusion follows.

Conversely, assume that there exists an input, with polynomial Laplace transform u_L, such that the Laplace transform of the forced output is $y_{fL0} = y_0(s-\lambda)^{-1} + \beta(s)$, being $\beta(s)$ a polynomial vector. Then

$$y_{fL0} = \sum_{i=1}^{r} l_i(s)\frac{\varepsilon_i(s)}{\psi_i(s)}r'_i(s)u_L = y_0(s-\lambda)^{-1} + \beta(s)$$

This means that at least one polynomial $\psi_i(s)$ must possess λ as a root.

Point ii) Assume now that λ is a zero of $\hat{G}(s)$, root of the polynomial $\varepsilon_k(s)$. Choose u_L as in eq. (2.13) with $\gamma(s) := (s-\lambda)^{-1}\psi_k(s)$. Since λ is not a root of $\psi_k(s)$ such an input matches the form given in the statement. Moreover $r'_i(s)u_L = 0, i \neq k$ and $r'_k(s)u_L = (s-\lambda)^{-1}\psi_k(s)$, so that $y_{fL0} = (s-\lambda)^{-1}l_k(s)\varepsilon_k(s)$ is a polynomial vector whose inverse Laplace transform is zero for strictly positive time instants.

Conversely assume that there exists an input of the form $u_L = u_0(s-\lambda)^{-1} + \beta(s)$, with $u_0 \neq 0$ constant and $\beta(s)$ polynomial such that

$$y_{fL0} = \sum_{i=1}^{r} l_i(s)\frac{\varepsilon_i(s)}{\psi_i(s)}r'_i(s)(u_0(s-\lambda)^{-1} + \beta(s))$$

is polynomial. A little thought shows that, besides other things, y_{fL0} may well be polynomial only if λ is a root of at least one polynomial $\varepsilon_i(s)$.   □

The terminology adopted for the transmission zeros derives from the fact that they basically make reference to the transfer function (transmittance) of the system at hand. In the simple case of single input single output systems, the transfer function can be given the form

$$G(s) = \frac{C\mathrm{adj}[sI - A]B}{\det[sI - A]} + D$$

where $\mathrm{adj}[sI-A]$ is the matrix whose generic element (i,j) is given by the determinant, multiplied by $(-1)^{i+j}$, of the matrix obtained by $(sI - A)$ ruling out its j-th row and i-th column. The transmission zeros coincide with the roots of the numerator once all the possible cancellations between the polynomial $C\mathrm{adj}[sI-A]B$ and the characteristic polynomial have been actually performed. As shown in the sequel, all the roots of $C\mathrm{adj}[sI - A]B + D\det[sI - A]$ are still properly called zeros of the system. These roots actually constitutes the so called *invariant zeros*. In the general multivariable framework, the definition of such zeros calls for the introduction of the polynomial matrix

$$P(s) := \begin{bmatrix} sI - A & -B \\ C & D \end{bmatrix}$$

which is referred to as *system matrix*.

Definition 2.9 (Invariant zeros) *Consider system* Σ *and let* $P(s)$ *be the associated system matrix with* $v :=\mathrm{rank}[P(s)]$. *Moreover, let* $S(s)$ *be the Smith form of* $P(s)$, *i.e.*

$$S(s) = \begin{bmatrix} \alpha_1(s) & 0 & \cdots & 0 & 0 \\ 0 & \alpha_2(s) & \cdots & 0 & 0 \\ \vdots & \vdots & \ddots & \vdots & \vdots \\ 0 & 0 & \cdots & \alpha_v(s) & 0 \\ 0 & 0 & \cdots & 0 & 0 \end{bmatrix}$$

A complex number λ *is said to be an invariant zero of* Σ *if it is a root of the polynomial*

$$\pi_{zi}(s) := \alpha_1(s)\alpha_2(s)\cdots\alpha_v(s)$$

$\square$

Remark 2.8 Notice that $v =\mathrm{rank}[P(s)]= n+\mathrm{rank}[G(s)]$. As a matter of fact, simple computations show that

$$P(s) := \begin{bmatrix} sI - A & -B \\ C & D \end{bmatrix}$$
$$= \begin{bmatrix} sI - A & 0 \\ C & I \end{bmatrix} \begin{bmatrix} (sI - A)^{-1} & 0 \\ 0 & G(s) \end{bmatrix} \begin{bmatrix} sI - A & -B \\ 0 & I \end{bmatrix}$$

Therefore, the claim on the rank of $P(s)$ is proved by noticing that in the right hand side of the above equation the first and last matrices above are nonsingular, while the remaining one has rank equal to $n + \mathrm{rank}[G(s)]$. $\square$

Like the transmission zeros, also the invariant zeros admit a rank characterization, which in this case concerns the kernel of either $P(s)$ or $P'(s)$.

Lemma 2.9 (Rank property of invariant zeros) *Let* $P(s)$ *be the system matrix of a system* Σ *with transfer function* $G(s) = C(sI - A)^{-1}B + D$ *with* $\mathrm{rank}[G(s)] = \min[p, m]$. *The complex number* λ *is an invariant zero of the system if and only if* $P(s)$ *looses rank in* $s = \lambda$, *i.e. if and only if there exists a nonzero vector* z *such that*

$$\begin{cases} P(\lambda)z = 0 & \text{if } p \geq m \\ P'(\lambda)z = 0 & \text{if } p \leq m \end{cases}$$

Proof Consider first the case $p \geq m$ and let $S(s)$ be the Smith form of $P(s)$, so that $P(s) = L(s)S(s)R(s)$, where $L(s)$ and $R(s)$ are suitable polynomial and unimodular matrices of dimensions $p+n$ and $n+m$, respectively, whereas $S(s)$ is as in Definition 2.9. Let $e_k(h)$ be the k-th column of the h-dimensional identity matrix. Recall also that matrix $R^{-1}(s)$ is polynomial and unimodular as well. Hence, if λ is an invariant zero of the system, root of the polynomial $\alpha_k(s)$ in $S(s)$, then it is $P(\lambda)z = 0$ with $z = R^{-1}(\lambda)e_k(n + m)$. Conversely, if there exists $z \neq 0$ such that $P(\lambda)z = 0$, then, necessarily, $S(\lambda)R(\lambda)z = 0$, which in turn implies that λ is a root of at least one polynomial $\alpha_i(s)$, since $R(\lambda)z \neq 0$.

The proof in the case $p \leq m$ develops along the same lines, once $P(s)$ has been replaced by $P'(s)$. $\square$

Example 2.13 Consider again the system defined in Example 2.12. It results

$$
P(s) = \begin{bmatrix}
s & -1 & 0 & 0 & 0 \\
10 & s-7 & 0 & 0 & -1 \\
0 & 0 & s-5 & 0 & -2 \\
-1 & 1 & -1 & s-6 & 0 \\
-13 & 5 & 0 & 0 & 1 \\
0 & 0 & 1 & 0 & 1
\end{bmatrix}
$$

As for the effective computation of the invariant zeros, one can actually utilize the result stated in Lemma 2.9, by looking for the vectors $z = [z_1 \; z_2 \; \cdots \; z_5]'$ in the kernel of $P(\lambda)$, i.e. those vectors z such that $P(\lambda)z = 0$. A nonzero solution of the relevant equations can be found only if $\lambda = 3$ or $\lambda = 6$, which are therefore invariant zeros. Recall (Example 2.12) that only $\lambda = 3$ is a transmission zero.   □

As apparent from their definition, the invariant zeros are not affected by a change of basis in the state space, as stated in the following lemma.

Lemma 2.10 (Invariant zeros vs. changes of basis) *The set of the invariant zeros of a system Σ is invariant with respect to a change of basis.*

Proof If the triple $(\bar{A}, \bar{B}, \bar{C})$, with $\bar{A} = TAT^{-1}$, $\bar{B} = TB$, $\bar{C} = CT^{-1}$, describes, together with matrix D, system Σ in a new basis, it follows

$$
\bar{P}(s) = \begin{bmatrix} sI - \bar{A} & -\bar{B} \\ \bar{C} & D \end{bmatrix} = \begin{bmatrix} T & 0 \\ 0 & I \end{bmatrix} \begin{bmatrix} sI - A & -B \\ C & D \end{bmatrix} \begin{bmatrix} T^{-1} & 0 \\ 0 & I \end{bmatrix}
$$

so that $\bar{P}(s)$ and $P(s)$ have the same Smith form. Hence both systems $\Sigma(A, B, C, D)$ and $\Sigma(\bar{A}, \bar{B}, \bar{C}, D)$ have the same invariant zeros.   □

Also the invariant zeros enjoy a *blocking property*, which stems on the existence of an exponential input yielding identically zero forced output.

Theorem 2.4 (Time domain characterization of invariant zeros) *The complex number λ is an invariant zero of Σ if and only if at least one of the two following conditions holds*

i) λ is an eigenvalue of the unobservable part of $\hat{\Sigma}$;

ii) there exist two vectors x_0 and $u_0 \neq 0$ such that the forced output of $\hat{\Sigma}$ corresponding to the input $u = u_0 e^{\lambda t}, t \geq 0$ and initial state $x(0) = x_0$ is identically zero for $t \geq 0$.

Proof It is sufficient to prove the theorem in the case $p \geq m$, since the proof in the converse case easily follows by replacing Σ with Σ'. Hence assume $\hat{\Sigma} = \Sigma$.

Let now λ be an invariant zero of Σ and let $P(s)$ be the system matrix. Thanks to Lemma 2.9 there exists a non zero vector $z = [v' \; w']'$ such that $P(\lambda)z = 0$, i.e.

$$(\lambda I - A)v = Bw \tag{2.14}$$

$$Cv + Dw = 0 \tag{2.15}$$

Letting $x(0) := v$ and $u_0 := w$, it is now verified that the input $u(t) = we^{\lambda t}$ produces, together with the initial state $x(0) = v$, an identically zero output for $t \geq 0$. The Laplace transform of the input is $u_L = (s - \lambda)^{-1}w$ and that of the state is $x_L = (sI - A)^{-1}[v + Bw(s - \lambda)^{-1}]$. Eq. (2.14) entails $(s - \lambda)^{-1}Bw = (s - \lambda)^{-1}(\lambda I - A)v$, so that $x_L = (s-\lambda)^{-1}v$ and, from eq. (2.15), $[Cx_L + Dw(s-\lambda)^{-1}](s-\lambda) = y_L(s-\lambda) = 0$. Since $y(0) = 0$ (eq. (2.15)), the conclusion is drawn that $y(t) = 0, t \geq 0$. In particular, if $w \neq 0$ then condition $ii)$ is verified, whereas, if $w = 0$, then eqs. (2.14),(2.15) entail, in view of the PBH test, that condition $i)$ holds.

Conversely, assume without any loss of generality (recall Lemma 2.10), that the system at hand is from the very beginning in the standard Kalman canonical form for observability, i.e.

$$A = \begin{bmatrix} A_1 & 0 \\ A_2 & A_3 \end{bmatrix}, \quad B = \begin{bmatrix} B_1 \\ B_2 \end{bmatrix}, \quad C = \begin{bmatrix} C_1 & 0 \end{bmatrix}$$

where the pair (A_1, C_1) is observable. If condition $i)$ holds (namely λ is an eigenvalue of A_3) choose $z = [0 \ \xi' \ 0]'$, where $\xi \neq 0$ is such that $(\lambda I - A_3)\xi = 0$. Then, obviously, $P(\lambda)z = 0$ so that λ is an invariant zero of Σ. If condition $ii)$ holds, let, according to the structure of A, $x_0 := [x'_{01} \ x'_{02}]'$. Being $y_L = 0$, it follows

$$C_1(sI - A_1)^{-1}\left[x_{01} + \frac{B_1 u_0}{(s-\lambda)}\right] + \frac{Du_0}{(s-\lambda)} = 0$$

By noticing that

$$(sI - A_1)^{-1}x_{01} = \frac{I - (sI - A_1)^{-1}(\lambda I - A_1)}{(s-\lambda)}x_{01}$$

$$y(0) = C_1 x_{01} + Du_0$$

it then follows

$$C_1(sI - A_1)^{-1}[(\lambda I - A_1)x_{01} - B_1 u_0] = 0$$

The first term of such an equation is the Laplace transform of the (free) output of the system $\Sigma(A_1, 0, C_1, 0)$ when the initial state is $(\lambda I - A_1)x_{01} - B_1 u_0$. Since this system is observable it follows that $(\lambda I - A_1)x_{01} - B_1 u_0 = 0$. Choose

$$z := \begin{bmatrix} x'_{01} & \xi' & u'_0 \end{bmatrix}$$

where $\xi := (\lambda I - A_3)^{-1}(A_2 x_{01} + B_2 u_0)$. Obviously, $P(\lambda)z = 0$ so that λ is an invariant zero of Σ. □

The theorem above points out the circumstances under which the output of $\hat{\Sigma}$ is zero for all $t \geq 0$. Actually, a part from the trivial case of zero initial state and input, the output of $\hat{\Sigma}$ can be such if and only if Σ possesses invariant zeros. As already said, for SISO systems the invariant zeros are the roots of $C\text{adj}[(sI - A)^{-1}]B + D\det[(sI - A)]$ whereas (in general) only a part of these roots are transmission zeros. This relationship holds for MIMO systems as well.

Theorem 2.5 (Invariant vs. transmission zeros) *A transmission zero of a system Σ is also an invariant zero of Σ.*

Proof Consider first the case $p \geq m$ and let λ be a transmission zero of Σ. Thanks to Theorem 2.3 there exists an exponential/impulsive input

$$u(t) = u_0 e^{\lambda t} + \sum_{i=0}^{\nu} \alpha_i \delta^{(i)}(t)$$

with α_i and $u_0 \neq 0$ suitable constants, such that the forced output of Σ is, $\forall t > 0$,

$$y_f(t) = C \int_0^t e^{A(t-\tau)} B[u_0 e^{\lambda \tau} + \sum_{i=0}^{\nu} \alpha_i \delta^{(i)}(\tau)] d\tau + D u_0 e^{\lambda t} = 0$$

Letting

$$x_0 := \begin{bmatrix} B & AB & \cdots & A^\nu B \end{bmatrix} \begin{bmatrix} \alpha_0 \\ \alpha_1 \\ \vdots \\ \alpha_\nu \end{bmatrix}$$

it follows that

$$C e^{At} x_0 = C \int_0^t e^{A(t-\tau)} B \sum_{i=0}^{\nu} \alpha_i \delta^{(i)}(\tau) d\tau$$

so that

$$y_f(t) = C e^{At} x_0 + \int_0^t e^{A(t-\tau)} B u_0 e^{\lambda \tau} d\tau + D u_0 e^{\lambda t} = 0 , \quad \forall t > 0$$

This last expression coincides with the output response of system Σ when the initial state is $x(0) = x_0$ and the input is $u(t) = u_0 e^{\lambda t}$. Such response is obviously continuous from the right, so that the output is zero at $t = 0$ as well. Theorem 2.4 ensures that λ is an invariant zero of Σ.

The proof of the Theorem in the case $p \leq m$ can be derived in complete analogy by considering system Σ' instead of Σ.   □

The following result clarifies and completes the relationships between transmission and invariant zeros.

Theorem 2.6 (Transmission vs. invariant zeros of a system in minimal form)
The transmission and invariant zeros of a reachable and observable system do coincide.

Proof As already seen in Theorem 2.5, a transmission zero is also an invariant zero. It is then left to show the converse statement when (A, B) is reachable and (A, C) is observable. Consider the case $p \geq m$ since the other case is easily proved by transposition. Let λ be an invariant zero. Thanks to Lemma 2.9 there exists a nonzero vector $z = [v'\ w']'$ such that $P(\lambda)z = 0$. Notice that if $w = 0$ and $v \neq 0$, then this condition implies that $Av = \lambda v$ and $Cv = 0$, contrary to the observability assumption of (A, C) (recall Lemma D.1). Hence $w \neq 0$. Moreover, thanks to Theorem 2.4, the system response when the initial state is $x(0) = v$ and the input is $u(t) = w e^{\lambda t}$ is identically zero, i.e.

$$C \left[e^{At} v + \int_0^t e^{A(t-\tau)} B w e^{\lambda \tau} d\tau \right] + D w e^{\lambda t} = 0 , \quad \forall t \geq 0 \qquad (2.16)$$

Recalling Theorem 2.3, λ is a transmission zero if there exists an input of the form

$$u(t) = u_0 e^{\lambda t} + \sum_{i=0}^{\nu} \alpha_i \delta^{(i)}(t)$$

which yields an identically zero forced output (for $t > 0$), i.e. if

$$C \int_0^t e^{A(t-\tau)} B[u_0 e^{\lambda\tau} + \sum_{i=0}^{\nu} \alpha_i \delta^{(i)}(\tau)]d\tau + Du_0 e^{\lambda t} = 0 , \quad \forall t > 0$$

Hence the theorem is proved if one shows that there exist real coefficients α_i such that

$$C e^{At} v = \int_0^t C e^{A(t-\tau)} B \sum_{i=0}^{\nu} \alpha_i \delta^{(i)}(\tau) d\tau$$

$$= C e^{At} \sum_{i=0}^{\nu} A^i B \alpha_i , \quad \forall t > 0 \tag{2.17}$$

Any vector $\alpha := [\alpha_0' \; \alpha_1' \; \cdots \; \alpha_\nu']'$ such that

$$\begin{bmatrix} B & AB & \cdots & AB^\nu \end{bmatrix} \alpha = v$$

satisfies eq. (2.17). Notice that a vector α exists corresponding to $\nu = n - 1$ since (A, B) is reachable so that the Grammian matrix $[B \; AB \; \cdots \; AB^{n-1}]$ has full row rank. $\qquad\square$

The invariant and transmission zeros do not exhaust the totality of zeros which can be defined for a system. Actually, consider an unobservable system with $p < m$. It may well happen that an eigenvalue of the unobservable part, say λ, is such that $P(\lambda)$ does not loose rank. Associated with such an eigenvalue there exists an eigenvector (initial state) $x(0)$ such that the free motion of the output $y(\cdot)$ is identically zero. Therefore the complex number λ can be still considered as a zero of the system, whose nature is different from that of the zeros previously introduced . In complete analogy, an unreachable system Σ, with $p > m$, can admit an eigenvalue of the unreachable part, say λ, which is such that the associated system matrix $P(\lambda)$ does not loose rank. Hence λ is not an invariant zero. However, it is well known that there exists an initial state for Σ' (eigenvector associated with λ) capable of zeroing the free output of Σ'. Again, λ can be fairly considered as a zero of system Σ. Such zeros will be referred to as *decoupling zeros*.

Definition 2.10 (Output decoupling zeros) *Consider* $\Sigma(A, B, C, D)$ *a n-dimensional system and the polynomial matrix*

$$P_C(s) := \begin{bmatrix} sI - A \\ C \end{bmatrix}$$

Let

$$S_C(s) = \begin{bmatrix} \text{diag}\{\alpha_i^C(s)\} \\ 0 \end{bmatrix}$$

be the Smith form of $P_C(s)$*. A complex number* λ *is said to be an output decoupling zero if it is a root of the polynomial*

$$\alpha^C(s) := \alpha_1^C(s) \alpha_2^C(s) \cdots \alpha_n^C(s)$$

$\qquad\square$

Definition 2.11 (Input decoupling zeros) *Consider* $\Sigma(A, B, C, D)$ *a n-dimension-al system and the polynomial matrix*

$$P_B(s) := \left[\; sI - A \quad -B \;\right]$$

Let

$$S_B(s) = \left[\; \mathrm{diag}\{\alpha_i^B(s)\} \quad 0 \;\right]$$

be the Smith form of $P_B(s)$. *A complex number* λ *is said to be an input decoupling zero if it is a root of the polynomial*

$$\alpha^B(s) := \alpha_1^B(s)\alpha_2^B(s)\cdots\alpha_n^B(s)$$

□

Definition 2.12 (Input-output decoupling zeros) *Consider* $\Sigma(A, B, C, D)$ *a system, its associated polynomial matrices* $P_C(s)$, $P_B(s)$ *with their Smith forms* $S_C(s)$ *and* $S_B(s)$, *and the polynomials* $\alpha^C(s)$ *and* $\alpha^B(s)$, *respectively. A complex number* λ *is said to be an input-output decoupling zero if it is a root of both polynomials* $\alpha^C(s)$ *and* $\alpha^B(s)$.

□

The decoupling zeros are not affected by a change of basis in the state-space of the system as it can be checked by resorting to the same arguments exploited in the proof of Lemma 2.10. Further, they can be characterized in terms of the kernels of $P_C(\lambda)$ and $P'_B(\lambda)$. The relevant results are presented in the following lemmas given without proofs since completely similar to that of Lemma 2.9.

Lemma 2.11 (Rank property of the output decoupling zeros) *A complex number* λ *is an output decoupling zero if and only if there exists* $z \neq 0$ *such that*

$$P_C(\lambda)z = 0$$

Lemma 2.12 (Rank property of the input decoupling zeros) *A complex number* λ *is an input decoupling zero if and only if there exists* $w \neq 0$ *such that*

$$P'_B(\lambda)w = 0$$

Lemma 2.13 (Rank property of the input-output decoupling zeros) *A complex number* λ *is an input-output decoupling zero if and only if there exist* $z \neq 0$ *and* $w \neq 0$ *such that*

$$P_C(\lambda)z = 0$$
$$P'_B(\lambda)w = 0$$

In Tables 2.1 and 2.2 the definitions and basic properties of the zeros introduced so far are schematically illustrated.

Remark 2.9 Based on Lemmas 2.11-2.13, and on the PBH tests relative to observability and reachability (Lemmas D.1- D.2), it is straightforward to realize that a system in minimal form does not possess decoupling zeros.   □

It is worth pointing out that the given definitions put in relief possible relations between invariant and decoupling zeros. In fact, if the number of inputs does not exceed the number of outputs, it is immediately seen that the output decoupling zeros are invariant zeros as well. Analogously, in the converse case, i.e. when the number of outputs is not greater than the number of inputs, the input decoupling zeros are particular invariant zeros. However, as shown in the example below, there may well happen that a system has decoupling zeros which are not invariant.

Type	Definition	Rank property
Transmission	$\pi_{zt}(\lambda) = 0$	$\lim\limits_{s \to \lambda} \quad G(s)z = 0 \quad$ if $p \geq m$ $\lim\limits_{s \to \lambda} \quad G'(s)z = 0 \quad$ if $p \leq m$
Invariant	$\pi_{zi}(\lambda) = 0$	$P(\lambda)z = 0 \quad$ if $p \geq m$ $P'(\lambda)z = 0 \quad$ if $p \leq m$
Output decoupling	$\alpha^C(\lambda) = 0$	$P_C(\lambda)z = 0$
Input decoupling	$\alpha^B(\lambda) = 0$	$P'_B(\lambda)z = 0$
Input – output decoupling	$\alpha^C(\lambda) = 0$ $\alpha^B(\lambda) = 0$	$P_C(\lambda)z = 0$ $P'_B(\lambda)z = 0$

Table 2.1: Rank properties of the zeros

Type	Definition	Output property
Transmission	$\pi_{zt}(\lambda) = 0$	$x(0) = 0 \ , \ \exists u(\cdot)$ $y_f(t) = 0 \ , \ \forall t > 0$ on $\hat{\Sigma}$
Invariant	$\pi_{zi}(\lambda) = 0$	$x(0) = 0 \ , \ \exists u(\cdot)$ $y(t) = 0 \ , \ \forall t \geq 0$ on $\hat{\Sigma}$
Output decoupling	$\alpha^C(\lambda) = 0$	$\exists x(0) \neq 0 \ , \ u(\cdot) = 0$ $y(t) = 0 \ , \ \forall t \geq 0$ on Σ
Input decoupling	$\alpha^B(\lambda) = 0$	$\exists x(0) \neq 0 \ , \ u(\cdot) = 0$ $y(t) = 0 \ , \ \forall t \geq 0$ on Σ'
Input – output decoupling	$\alpha^C(\lambda) = 0$ $\alpha^B(\lambda) = 0$	$\exists x(0) \neq 0 \ , \ u(\cdot) = 0$ $y(t) = 0 \ , \ \forall t \geq 0$ on $\begin{matrix}\Sigma\\\Sigma'\end{matrix}$

Table 2.2: Output properties of the zeros

Example 2.14 Consider again the system defined in Example 2.13. It is obvious that the invariant zero $\lambda = 6$ is also an output decoupling zero. However, there exists an input decoupling zero, $\lambda = 5$, which is not invariant. Actually, matrix $P(5)$ has full rank (equal to five), even though the first four rows are linearly dependent, so that $P_B(s) = [sI - A \ - B]$ looses rank for $s = 5$. $\qquad\square$

It should be now evident the relation existing between invariant and decoupling zeros when the system at hand is square.

Lemma 2.14 (Decoupling vs. invariant zeros for square systems) *Consider a system with the same number of inputs and outputs (square system). Then the set of decoupling zeros is a subset of the set of invariant zeros.*

Proof If λ is a decoupling zero, one or both of the two matrices $P_C(s)$ and $P_B(s)$ must loose rank for $s = \lambda$. Hence matrix $P(s)$ looses rank in $s = \lambda$ as well. $\qquad\square$

In case of nonminimal systems, the set of invariant zeros does not coincide with that of transmission zeros. Moreover, there may be decoupling zeros which are not invariant. These fact motivates the definition below.

Definition 2.13 (System zeros) *The set $\mathcal{Z}_s$ of system zeros is defined as*

$$\mathcal{Z}_s = \mathcal{Z}_t \cup \mathcal{Z}_{i-o} \cup \mathcal{Z}_{o-i} \cup \mathcal{Z}_{io}$$

where $\mathcal{Z}_t$ is the set of transmission zeros, $\mathcal{Z}_{i-o}$ is the set of input decoupling zeros which are not also output decoupling, $\mathcal{Z}_{o-i}$ is the set of output decoupling zeros which are not also input decoupling, and $\mathcal{Z}_{io}$ is the set of input-output decoupling zeros. □

Lemma 2.15 *The set of invariant zeros of a system $\Sigma(A, B, C, D)$ is a subset of the set of the system zeros.*

Proof The lemma will be proved in the case $p \geq m$, as the converse case being easily handled by transposition. One has to show that if λ is an invariant zero, it is also either a transmission or a decoupling zero. Without any loss of generality, assume that the system is decomposed accordingly to the Kalman canonical decomposition, i.e.

$$A = \begin{bmatrix} A_1 & A_2 & A_3 & A_4 \\ 0 & A_5 & 0 & A_6 \\ 0 & 0 & A_7 & A_8 \\ 0 & 0 & 0 & A_9 \end{bmatrix} , \quad B = \begin{bmatrix} B_1 \\ B_2 \\ 0 \\ 0 \end{bmatrix} , \quad C = \begin{bmatrix} 0 & C_1 & 0 & C_2 \end{bmatrix}$$

where $\Sigma(A_5, B_2, C_1, D)$ constitutes a subsystem which is completely observable and reachable. Of course, if λ belongs to the unreachable and/or unobservable part of the system, then it is a decoupling zero, so that the proof would be over.

Let λ be an invariant zero and assume that it is not an eigenvalue of any of the matrices A_1, A_7, A_9. Then, there exists $z \neq 0$ such that $P(\lambda)z = 0$, with $z := [z_1' \; z_2' \; z_3' \; z_4' \; u_0']'$. Since λ is not an eigenvalue of A_9, it turns out that $z_4 = 0$ so that

$$(\lambda I - A_5)^{-1} z_2 - B_2 u_0 = 0 , \quad C_1 z_2 + D u_0 = 0$$

Hence λ is a transmission zero of the reachable and observable part $\Sigma(A_5, B_2, C_1, D)$ of the system, and hence of the system itself, provided that $[z_2' \; u_0']' \neq 0$. If it were not so, it would happen that $(\lambda I - A_1)z_1 - A_3 z_3 = 0$ and $(\lambda I - A_7)z_3 = 0$. Since λ is not an eigenvalue neither of A_7 nor of A_1, it would follow $z_1 = 0$ and $z_3 = 0$ so that $z = 0$, a contradiction. □

Remark 2.10 In view of Lemma 2.15 and Theorem 2.4 it is easy to conclude that the set of systems zeros coincides with that of invariant zeros, relative to square systems. If in addition the system is in minimal form, the three sets (of transmission, invariant and system zeros) do actually coincide. □

At the light of what has previously been said, the set of system zeros constitutes the totality of the zeros defined till now. It can be expressively partitioned in its subsets, as shown in fig. 2.5 with reference to the case $m > p$, $m = p$, and $m < p$. In the figure, the symbols t, i, d_i, d_o, d_{io} denote, respectively, the transmission, invariant, input decoupling, output decoupling and input-output decoupling zeros. The presence of one of this symbols in a part of the figure indicates that this part is contained in the set of the zeros under consideration.

This section ends with a brief discussion on the concept of *inverse system* in the simple case where $p = m$ and $\det[D] \neq 0$. For the more general case, the reader is referred to specialized texts.

$m > p$	t , i	i , d_i	
		i , d_{io}	
		i , d_o	d_o

$m = p$	t , i	i , d_i
		i , d_{io}
		i , d_o

$m < p$	t , i	i , d_i	d_i
		i , d_{io}	
		i , d_o	

Figure 2.1: The zeros of a system

Definition 2.14 (Inverse system) *Consider a square system* $\Sigma(A, B, C, D)$ *such that* $\det[D] \neq 0$. *Then, the inverse system is*

$$\Sigma_{inv} := \left[\begin{array}{c|c} F & G \\ \hline H & E \end{array} \right]$$

where $F := A - BD^{-1}C$, $G := BD^{-1}$, $H := -D^{-1}C$, $E := D^{-1}$.   □

The reason why system Σ_{inv} is called the inverse system can be simply explained as follows. Let Σ and Σ_{inv} be described by

$$\dot{x} = Ax + Bu$$
$$y = Cx + Du$$

and

$$\dot{\xi} = (A - BD^{-1}C)\xi + BD^{-1}\nu$$
$$\theta = -D^{-1}C\xi + D^{-1}\nu$$

respectively. Now, build up the series connection of the systems, according to the following two cases:

i) $\nu = y$, so that Σ_{inv} follows Σ (system $\Sigma_{inv}\Sigma$)

ii) $u = \theta$, so that Σ_{inv} precedes Σ (system $\Sigma\Sigma_{inv}$)

In the first case, letting $z := \xi - x$ it follows

$$\dot{z} = (A - BD^{-1}C)z$$
$$\theta = -D^{-1}Cz + u$$

whereas in the second

$$\dot{z} = Az$$
$$y = -Cz + \nu$$

In both cases, the transfer function of the resulting system is the identity. Hence, if $G(s)$ is the transfer function of Σ, the transfer function of Σ_{inv} is exactly $G^{-1}(s)$.

Theorem 2.7 *Consider a square system $\Sigma(A, B, C, D)$ where D is nonsingular and let Σ_{inv} be its associated inverse system. Then:*

   i) *The set of transmission zeros of Σ coincides with the set of poles of Σ_{inv} and, conversely, the set of poles of Σ coincides with the set of transmission zeros of Σ_{inv}.*

   ii) *The set of eigenvalues of Σ coincides with the set of invariant zeros of Σ_{inv} and, conversely, the set of invariant zeros of Σ coincides with the set of eigenvalues of Σ_{inv}.*

Proof *Point i)* It suffices to verify that, if $M(s)$ is the Smith-McMillan form of the transfer function of Σ, then $TM^{-1}(s)T$ is the Smith-McMillan form of the transfer function of Σ_{inv}, where

$$T := \begin{bmatrix} 0 & 0 & \cdots & 0 & 1 \\ 0 & 0 & \cdots & 1 & 0 \\ \vdots & \vdots & \ddots & \vdots & \vdots \\ 0 & 1 & \cdots & 0 & 0 \\ 1 & 0 & \cdots & 0 & 0 \end{bmatrix}$$

The conclusion then follows from the definition of poles and transmission zeros (Definitions 2.7 and 2.8).
   Point ii) Assume that λ is an invariant zero of Σ_{inv}, i.e. $(\lambda I - (A - BD^{-1}C))w_1 - BD^{-1}w_2 = 0$ and $-D^{-1}Cw_1 + D^{-1}w_2 = 0$, with $w := [w_1'\ w_2']' \neq 0$. These equations imply that $w_1 \neq 0$ and $Aw_1 = \lambda w_1$, so that λ is an eigenvalue of Σ. Also these considerations can be easily reversed. Finally, assume that λ is an invariant zero of Σ, i.e. $(\lambda I - A)w_1 - Bw_2 = 0$ and $Cw_1 + Dw_2 = 0$ with $w := [w_1'\ w_2']' \neq 0$. Hence $w_2 = -D^{-1}Cw_1$, $w_1 \neq 0$, so that $(A - BD^{-1}C)w_1 = \lambda w_1$ implies that λ is an eigenvalue of Σ_{inv}. Reversing the procedure proves the validity of the converse statement as well.   □

2.6   Singular values

In this section, some of the most significant properties of the *singular values* of a matrix are reported along with the so-called *singular value decomposition*. Reference is made to constant vectors and matrices with complex elements. Recall that the symbol "~" denotes the operation of conjugate transposition. The norm adopted for vectors is the one induced by the usual inner product in C^n.

Definition 2.15 (Singular values of a matrix) *Let A be a $n \times m$ complex matrix. The square roots $\sigma_i(A), i = 1, \cdots m$ of the eigenvalues of $A^\sim A$ are called singular values of A.*   □

Notice that the singular values of A are real and nonnegative, since $A^\sim A$ is a hermitian and positive semidefinite matrix. Of course, $m-\text{rank}[A]$ singular values of A are in fact zero. The singular values of A can be put into evidence by the so-called *singular value decomposition* of A, whose existence is established in the theorem below. Its proof, which is reported in Appendix B, is based on a few preliminarily steps that provide a way, although not optimal for the computational burden, to determine such a decomposition.

Theorem 2.8 (Singular value decomposition) *Let A be a $n \times m$ matrix such that* rank$[A]= k$. *Then, there exist two unitary matrices U and V such that*

$$U^\sim AV = S$$

where the only nonzero elements of S are those in positions $(i,i), i = 1, \cdots k$. Such elements are positive and nonincreasing.

The elements of S in positions $i = 1, \cdots$, min$[n, m]$, are the singular values of A. Further,

$$A^\sim A = VS^\sim U^\sim USV^\sim = VS^\sim SV^\sim = VDV^\sim$$

where $D = \text{diag}\{\sigma_i^2(A), i = 1, \cdots, m\}$. Of course, if $m > n$, matrix A possesses at least $m - n$ singular values at the origin. The *greatest* and *least* singular values of A are indicated with $\bar{\sigma}(A)$ and $\underline{\sigma}(A)$, respectively. The unitary matrices U and V specify the singular value decomposition of A.

Remark 2.11 Based on Theorem 2.8, matrix S has the same dimensions as A and exhibits the following structure

$$S = \begin{cases} \begin{bmatrix} \Delta & 0 \end{bmatrix} & \text{if} \quad n < m \\ \\ \Delta & \text{if} \quad n = m \\ \\ \begin{bmatrix} \Delta \\ 0 \end{bmatrix} & \text{if} \quad n > m \end{cases}$$

The "meaningful" part of S is therefore constituted by the diagonal matrix Δ with dimension min$[n, m]$. For such a reason, the singular value decomposition is sometimes presented in a different way, distinguishing two different situations:

1) **Case $n > m$:** There exist two matrices U_1 and V of dimensions $n \times m$ and $m \times m$, respectively, such that $A = U_1 \Delta V^\sim$, with $V^\sim V = VV^\sim = I$ and $U_1^\sim U_1 = I$. The diagonal matrix Δ is m-dimensional and contains the singular values of A.

2) **Case $n < m$:** There exist two matrices U and V_1 of dimensions $n \times n$ and $m \times n$, respectively, such that $A = U\Delta V_1^\sim$, with $U^\sim U = UU^\sim = I$ and $V_1 V_1^\sim = I$. The diagonal matrix Δ is n-dimensional and contains the singular values of A.

With reference to what said in Theorem 2.8, it is immediate to check that $U = [U_1 \ U_2]$, $S = [\Delta \ 0]'$, if $n > m$ whereas $V = [V_1 \ V_2]$, $S = [\Delta \ 0]$, if $n < m$. $\qquad \square$

Remark 2.12 The *pseudoinverse* of A can be found in a very simple way once the singular value decomposition of A is known. Actually, let

$$U^\sim AV = S = \begin{bmatrix} \Sigma & 0 \\ 0 & 0 \end{bmatrix},$$

with Σ square and nonsingular. Letting

$$B := V \begin{bmatrix} \Sigma^{-1} & 0 \\ 0 & 0 \end{bmatrix} U^\sim := VTU^\sim$$

it follows that

$$BAB = VTU^\sim USV^\sim VTU^\sim = VTSTU^\sim = VTU^\sim = B$$
$$ABA = USV^\sim VTU^\sim USV^\sim = USTSV^\sim = USV^\sim = A$$

These relations show that B is the pseudoinverse of A. $\qquad \square$

The most significant properties of the singular values of a matrix are reported in the following lemmas. The symbol $\|x\|$ indicates the usual norm induced by the inner product in C^n, i.e. $\|x\|^2 = x^\sim x$ and recall that $\lambda_i(A)$ is the i-th eigenvalue of A and $r_s(A)$ is the spectral radius of A. The proof of the following lemmas are reported in Appendix B.

Lemma 2.16 *Given a matrix A, the following relations hold:*

1)
$$\bar{\sigma}(A) = \max_{x \neq 0} \frac{\|Ax\|}{\|x\|}$$

2)
$$\underline{\sigma}(A) = \min_{x \neq 0} \frac{\|Ax\|}{\|x\|}$$

Lemma 2.17 *Let A be a matrix. Then $\bar{\sigma}(A) = \bar{\sigma}(A^\sim)$*

Lemma 2.18 *Let A be a square matrix. Then*

1)
$$\underline{\sigma}(A) \leq |\lambda_i(A)| \leq \bar{\sigma}(A)$$

2)
$$r_s(A) \leq \bar{\sigma}(A)$$

Moreover, if A is nonsingular, then

3)
$$\underline{\sigma}(A) = \frac{1}{\bar{\sigma}(A^{-1})}$$

4)
$$\bar{\sigma}(A) = \frac{1}{\underline{\sigma}(A^{-1})}$$

Finally, if A is hermitian, then

5)
$$r_s(A) = \bar{\sigma}(A)$$

Lemma 2.19 *Let $\alpha \in C$ be an arbitrary scalar and A a matrix. Then,*
$$\sigma_i(\alpha A) = |\alpha|\sigma_i(A)$$

Lemma 2.20 *Let A and B be two matrices with the same dimensions. Then,*
$$\bar{\sigma}(A + B) \leq \bar{\sigma}(A) + \bar{\sigma}(B)$$

Lemma 2.21 *Let A and B be two matrices such that AB makes sense. Then,*
$$\bar{\sigma}(AB) \leq \bar{\sigma}(A)\bar{\sigma}(B)$$

Lemma 2.22 *Let A and B be two matrices with the same dimensions. Then,*
$$\underline{\sigma}(A) - \bar{\sigma}(B) \leq \underline{\sigma}(A + B) \leq \underline{\sigma}(A) + \bar{\sigma}(B)$$

Lemma 2.23 *Let A and B be two matrices with the same number of rows. Then,*

$$\max[\bar{\sigma}(A), \bar{\sigma}(B)] \leq \bar{\sigma}([\ A \quad B\ ]) \leq \sqrt{2}\max[\bar{\sigma}(A), \bar{\sigma}(B)]$$

Lemma 2.24 *Let A be a square matrix. Then,*

$$\sum_i \sigma_i^2(A) = \text{trace}[A^\sim A]$$

The quantity $\|A\|_F := \sqrt{\text{trace}(A^\sim A)}$ is the so called Frobenius norm of A.

Lemma 2.25 *Let m be the number of columns of a matrix A and denote by A_{ij} its element in position (i, j). Then,*

$$\max_{i,j} |A_{ij}| \leq \bar{\sigma}(A) \leq m \max_{i,j} |A_{ij}|$$

2.7 Basic facts on linear operators

In this section some facts on the theory of linear operators are recalled. Since no confusion can arise in the present context, the term linear will be often disregarded.

Definition 2.16 (Operator) *A (linear) operator T is a linear map acting between two linear spaces.* □

For an operator $T : X \to Y$ acting between two normed linear spaces X and Y, it is possible to introduce the notion of *boundedness*, *norm* and *continuity*.

Definition 2.17 (Boundedness of an operator) *Let X and Y be two normed linear spaces and $T : X \to Y$ an operator. This operator is said to be bounded if there exists a constant M such that*

$$\|Tx\| \leq M\|x\|, \ \forall x \in X \tag{2.18}$$

□

Definition 2.18 *Let X and Y be two normed linear spaces and $T : X \to Y$ a bounded operator. The smallest constant M satisfying eq. (2.18) is called norm of T and denoted by $\|T\|$.* □

Remark 2.13 Thanks to linearity and Definitions 2.17, 2.18, the norm of an operator can be significantly characterized as follows

$$\|T\| = \sup_{x \neq 0} \frac{\|Tx\|}{\|x\|} = \sup_{\|x\|=1} \|Tx\|$$

In the particular case where T is actually a complex matrix, it follows, in view of Lemma 2.16, that $\|T\| = \bar{\sigma}(T)$. □

Definition 2.19 (Continuous operator) *Let X and Y be two normed linear spaces and $T : X \to Y$ an operator. If, chosen a vector $\xi \in X$, for any $\varepsilon > 0$ there exists a $\delta > 0$ such that*

$$\|x - \xi\| < \delta \Longrightarrow \|Tx - T\xi\| < \varepsilon$$

then T is said to be continuous. □

The concept of *rank*, which is well understood for operators defined in finite dimensional spaces, can be extended to the general case as follows.

Definition 2.20 (Rank of an operator) *Let X and Y be two normed linear spaces and $T : X \to Y$ an operator. The rank of T is the dimension of the closure of the range of T.*   □

Also the concept of *eigenvalue* and *eigenvector*, which are familiar in the context of linear operators in finite dimensional spaces (matrices), can be extended in an analogous way.

Definition 2.21 (Eigenvalue and eigenvector of an operator) *Let $T : X \to X$ be an operator and λ a complex number. If*

$$Tx = \lambda x, \ x \neq 0$$

then λ and x are called eigenvalue and eigenvector of T, respectively.   □

Remark 2.14 It is well known that, in the case of finite dimensional spaces over the field of complex numbers, an operator admits at least one eigenvalue. This is not true, in general, if X is not finite dimensional. For example, the operator J (integral) acting on the space of polynomials $p(s)$ with real coefficients defined as

$$J : \sum_{i=0}^{n} \alpha_i s^i \mapsto \sum_{i=0}^{n} \alpha_i \frac{s^{i+1}}{i+1}$$

does not admit eigenvalues. Actually $Jp(s) = \lambda p(s)$ holds only for $p(s) = 0$. On the contrary, the (derivative) operator D acting again in the space of polynomial with real coefficients, defined as

$$D : \sum_{i=0}^{n} \alpha_i s^n \mapsto \sum_{i=0}^{n} i\alpha_i s^{i-1}$$

admits $\lambda = 0$ as an eigenvalue, since $Dp(s) = 0$ for $p(s) = \alpha_0 \neq 0$. Furthermore, if the rank of the operator is *finite*, then it admits at least one eigenvalue even if X is not finite dimensional.   □

If $T : X \to Y$ is a bounded operator and X and Y are Hilbert spaces (so that suitable inner products are there defined along with the induced norms), it is possible to define the *adjoint* operator of T, hereafter indicated with T^*. To this aim, consider an operator $T : X \to Y$ and the map $S := Y \to X$ defined by the equation

$$< Tx, y > = < x, Sy >, \ \forall x \in X, \ \forall y \in Y$$

The map S is easily shown to be a *linear* and *bounded* operator. Hence the following definition is in order.

Definition 2.22 (Adjoint operator) *Let X and Y be two normed linear Hilbert spaces with the norm induced by the (relevant) inner products. Let $T : X \to Y$ be a bounded operator. The adjoint operator T^* is defined by*

$$< Tx, y > = < x, T^*y >, \ \forall x \in X, \ \forall y \in Y$$

                                                                                    □

The linear operators enjoy a number of useful properties, some of them gathered in the following theorem, whose proof can be easily found in any specialized text.

Theorem 2.9 *Let X and Y be two linear Hilbert spaces and let $T : X \to Y$ and $S : Y \to X$ two bounded operators. Then,*

1) T^* is bounded and $\|T\| = \|T^*\|$.

2) $\|T^*T\| = \|TT^*\| = \|T\|^2$.

3) $(T + S)^* = T^* + S^*$.

4) If $\alpha \in C$, $(\alpha T)^* = \alpha^\sim T^*$.

5) $(TS)^* = S^*T^*$.

6) If T^{-1} exists and is bounded, then $(T^{-1})^* = (T^*)^{-1}$.

7) $(T^*)^* = T$.

A particular case occurs when an operator $T : X \to X$ coincides with its adjoint. In this case T is said to be *self-adjoint*. For example, the operator T^*T is self-adjoint: its eigenvalues, if any, are real and nonnegative.

2.8 Functional spaces of rational matrices

The present book often refers to some *functional* spaces whose elements are proper rational matrices endowed with peculiar properties. To say the true, these elements could be viewed as belonging to suitable subspaces of more general linear spaces whose elements are not necessarily rational functions. However, undertaking this broader perspective is not strictly necessary in the present context, so that the exposition is restricted to the smaller world of rational functions.

Being the elements of the spaces under consideration rational matrices, it should be necessary, for a more rigorous notation, to indicate time by time the dimensions of the relevant matrices. However, this choice is completely useless, since the context widely clarifies the dimensions of the matrices under consideration. Therefore, it will be said that the $n \times m$ matrix A belongs to X instead of $X^{n \times m}$, and so on and so forth. Consistently, the identity (resp. null) matrix will be simply indicated by I (resp. 0).

Definition 2.23 (The space RL_∞) *The set of the rational matrices $F(s)$ such that*

$$\sup_\omega \|F(j\omega)\| = \sup_\omega \bar{\sigma}[F(j\omega)] < \infty$$

constitutes the space RL_∞. □

Definition 2.24 (Norm in RL_∞) *Let $F(s)$ be an element of RL_∞. The norm of $F(s)$ is the scalar*

$$\|F(s)\|_\infty := \sup_\omega \bar{\sigma}[F(j\omega)]$$

□

Remark 2.15 Based on the given definitions it should be obvious that a matrix belongs to RL_∞ if any element is a proper rational function without poles on the imaginary axis. In the scalar case, the norm coincides with the peak value of the frequency gain. □

An important subspace of RL_∞ is that of the rational matrices which are bounded in the right half plane.

Definition 2.25 (The space RH_∞) *The set of rational functions $F(s)$ such that*

$$\sup_{Re(s)\geq 0} \|F(s)\| < \infty$$

constitutes the space RH_∞.   □

Remark 2.16 In view of the given definitions, it results that the elements of RH_∞ are rational, proper and stable matrices. Moreover, being RH_∞ a subspace of RL_∞, the norm adopted for the former can be the same as that utilized for the elements of the latter. Further, thanks to a well known property of analytic functions, if $F(s) \in RH_\infty$, then

$$\|F(s)\|_\infty = \sup_{Re(s)\geq 0} \|F(s)\|$$

□

Example 2.15 Let consider the functions

$$F_1(s) = \frac{s+1}{(s+2)(s-3)} \ , \quad F_2(s) = \frac{s-1}{s} \ , \quad F_3(s) = \frac{1}{s-1}$$

$$F_4(s) = \frac{s}{(s+2)(s+3)} \ , \quad F_5(s) = \frac{s-1}{s+1} \ , \quad F_6(s) = \frac{1}{s+1}$$

$$F_7(s) = \left[\begin{array}{cc} (s+1)/(s^2-4) & (s+1)/(s^4+4) \\ 1 & (s-1)/(s+2) \end{array} \right] \ , \quad F_8(s) = \left[\begin{array}{c} s/(s+3) \\ 1/(s-1) \end{array} \right]$$

$$F_9(s) = \left[\begin{array}{ccc} 1 & (s-1)/(s+1) & 1/(s+2) \end{array} \right]$$

Among these functions, $F_1(s)$, $F_3(s)$, $F_4(s)$, $F_5(s)$, $F_6(s)$, $F_8(s)$ and $F_9(s)$ belong to RL_∞, whereas $F_4(s)$, $F_5(s)$, $F_6(s)$ and $F_9(s)$ belong also to RH_∞.   □

In the space RH_∞ there are functions which are particularly meaningful. Among them, the so called *inner* and *outer* functions are the object of the two definitions below.

Definition 2.26 (Inner function) *A function $F(s) \in RH_\infty$ is said to be inner if*

$$F^\sim(s)F(s) = I, \ \forall s$$

□

Definition 2.27 (Outer function) *A function $F(s) \in RH_\infty$ is said to be outer if there exists a function $X(s)$, analytic in $Re(s) > 0$, such that*

$$F(s)X(s) = I, \ \forall s$$

□

Remark 2.17 The definition of outer function can be equivalently formulated by requiring that $F(s)$ has full row rank for any s with $Re(s) > 0$. Obviously, if $F(s)$ is a square matrix belonging to RH_∞ together with its inverse, then it is outer. Moreover, if $F(s)$ and $G(s)$ belong to RH_∞ and are left coprime in the setting of proper stable matrices, then matrix $[F(s) \ G(s)]$ is outer.

In the restricted case of scalar functions, it is readily seen that a function $F(s) \in RH_\infty$ is inner if $\lim_{s\to\infty} F(s) = 1$ and the zeros and poles are symmetrically positioned, with respect to the origin, in the complex plane. On the other hand, a function $F(s) \in RH_\infty$ is outer (minimum phase) if it has no zeros with positive real part.   □

It is possible to prove that any matrix $F(s) \in RH_\infty$ can be decomposed in the product of an inner matrix and an outer matrix (*inner outer factorization*). Precisely, given $F(s)$ it is always possible to find two matrices $F_i(s)$ and $F_o(s)$ such that $F(s) = F_i(s)F_o(s)$. In the scalar case, such a decomposition is simply performed in the way indicated in the proof of the following theorem.

Theorem 2.10 (Inner outer factorization of scalar functions) *Let $F(s) \in RH_\infty$ be a scalar function. Then, there exist a scalar inner function $F_i(s)$ and a scalar outer function $F_o(s)$ such that*

$$F(s) = F_i(s)F_o(s)$$

Moreover, if $F(j\omega) \neq 0$, $0 \leq \omega \leq \infty$, then $F_o^{-1}(s) \in RH_\infty$.

Proof Let z_i, $i = 1, 2, \cdots, \nu$ be the zeros of $F(s)$ with positive real parts (taken with their multiplicity). Moreover, let

$$F_i(s) := \prod_{i=1}^{\nu} \frac{s - z_i}{s + z_i^{\sim}}$$

$$F_o(s) := \frac{F(s)}{F_i(s)}$$

Observe that the zeros of $F_o(s)$ are the zeros of $F(s)$ with nonpositive real part and the opposite of the zeros of $F(s)$ positive real part. Hence, if $F(s)$ does not have zeros on the extended imaginary axis, then rhpdeg$[F_o(s)] = 0$, so that $F_o^{-1}(s) \in RH_\infty$. □

Example 2.16 Let consider the functions

$$F_1(s) = 1, \quad F_2(s) = \frac{s}{s+1}, \quad F_3(s) = \frac{s-1}{s+1}$$

$$F_4(s) = \frac{s-1}{s+2}, \quad F_5(s) = \frac{s+1}{s+3}$$

$F_1(s)$ and $F_3(s)$ are inner, whereas $F_1(s)$, $F_2(s)$ and $F_5(s)$ are outer. The function $F_4(s)$ is neither inner nor outer. For this function, take $F_{4i}(s) = (s-1)/(s+1)$ and $F_{4o} = (s+1)/(s+2)$, so that $F_4(s) = F_{4i}(s)F_{4o}(s)$. Since $F(s)$ has no zeros on the extended imaginary axis, the function $F_{4o}(s)$ admits as inverse an element of RH_∞. □

Another meaningful subspace of RL_∞ is now defined.

Definition 2.28 (The space RL_2) *The set of rational functions $F(s)$ such that*

$$\frac{1}{2\pi} \int_{-\infty}^{\infty} \text{trace}[F^{\sim}(j\omega)F(j\omega)]d\omega < \infty$$

constitutes the space RL_2. □

It is easy to figure out that, taken two rational functions $F(s)$ and $G(s)$, both belonging to RL_2, the scalar

$$\alpha := \frac{1}{2\pi} \int_{-\infty}^{\infty} \text{trace}[G^{\sim}(j\omega)F(j\omega)]d\omega < \infty$$

satisfies all axioms which characterize the inner product, so that the following definition is in order.

Definition 2.29 (Inner product in RL_2) *The inner product of two functions of RL_2 is defined as*

$$< G(s), F(s) > = \frac{1}{2\pi} \int_{-\infty}^{\infty} \text{trace}[G^\sim(j\omega)F(j\omega)]d\omega$$

□

The space RL_2 is therefore a pre-Hilbert space. To say the true, it is possible to show that RL_2 is complete, and this implies that it is actually a Hilbert space. Accordingly to what said before, the norm in RL_2 can be induced by its inner product.

Definition 2.30 (Norm in RL_2) *Let $F(s)$ be an element of RL_2. The norm of $F(s)$ is the scalar*

$$\|F(s)\|_2 := \left[\frac{1}{2\pi} \int_{-\infty}^{\infty} \text{trace}[F^\sim(j\omega)F(j\omega)]d\omega \right]^{1/2}$$

□

The space RL_2 can be decomposed into the *direct sum* of two subspaces, RH_2 and $RH_2^\perp$, i.e

$$RL_2 = RH_2 \oplus RH_2^\perp$$

where the subspaces RH_2 and $RH_2^\perp$ are defined as follows.

Definition 2.31 (The subspaces RH_2 and $RH_2^\perp$) *The subspace RH_2 is constituted by the functions of RL_2 which are analytic in the right half plane. Conversely, the subspace $RH_2^\perp$ is constituted by the functions RL_2 which are analytic in the left half plane.*

□

Remark 2.18 The given definitions imply that a matrix belongs to RL_2 if its elements are strictly proper rational functions without poles on the imaginary axis. It belongs to RH_2 (resp. $RH_2^\perp$) if its elements are strictly proper rational functions without poles in the closed right (resp. left) half plane. Obviously, it turns out that $RH_2 \subset RH_\infty$.

□

Remark 2.19 The elements in the spaces RL_2, RH_2 and $RH_2^\perp$ (which are rational functions of complex variable) can be related to the elements of the spaces $RL_2(-\infty \ \infty)$, $RL_2[0 \ \infty)$, and $RL_2(-\infty \ 0]$, which are functions of the real variable t. Such functions are characterized by having *rational Fourier transform* and being *square integrable* in the intervals $(-\infty \ \infty)$, $[0 \ \infty)$, and $(-\infty \ 0]$, respectively. Moreover, the elements of $RL_2(-\infty \ 0]$ and $RL_2[0 \ \infty)$ are zero for $t > 0$ and $t < 0$, respectively, so that both of them belong to $RL_2(-\infty \ \infty)$. Notice that any element of $RL_2(-\infty \ \infty)$ can be uniquely written as the sum of an element of $RL_2[0 \ \infty)$ and one of $RL_2(-\infty \ 0]$. Hence $RL_2(-\infty \ \infty) = RL_2[0 \ \infty) \oplus RL_2(-\infty \ 0]$.

For example, the matrix of functions for which there exists the integral

$$\int_0^\infty e^{-j\omega} f(t)dt := F(j\omega)$$

and such that

$$\int_0^\infty \text{trace}[f'(t)f(t)]dt < \infty$$

is an element of $RL_2[0 \ \infty)$, which can be put into correspondence with the element $F(s) \in RH_2$, provided that this matrix is rational.

□

Example 2.17 With reference to the functions $F_i(s)$ defined in Example 2.15, it follows that $F_1(s)$, $F_3(s)$, $F_4(s)$ and $F_6(s)$ are elements of RL_2. Moreover, $F_4(s)$ and $F_6(s)$ are also elements of RH_2 and $F_3(s)$ is an element of $RH_2^\perp$. Finally, notice that

$$F_1(s) = \frac{s+1}{(s+2)(s-3)} = \frac{1}{5(s+2)} + \frac{4}{5(s-3)}$$
$$:= F_{1s}(s) + F_{1a}(s)$$

where $F_{1s}(s) \in RH_2$ and $F_{1a}(s) \in RH_2^\perp$. Hence, an element of RL_2 has been shown to equal the sum of a stable function and an antistable one. □

Remark 2.20 In linear system theory, the norm of a function in RH_2 lends itself to particularly significant characterizations. To put them into light, consider the time-invariant system

$$\dot{x} = Ax + Bw, \quad x(0) = 0$$
$$z = Cx$$

and assume that it is completely reachable and observable. If the matrix A is stable, its transfer function $F(s)$ is an element of RH_2. Now, consider the m-dimensional input vector $w^{(i)}(t) = \delta(t)e_i(m)$, where $i = 1, \cdots m$, $e_i(m)$ is the i-th column of the m-dimensional identity matrix and $\delta(t)$ is the "impulsive" function. Let $z^{(i)}$ be the corresponding forced response of the output z and $Z^{(i)}(j\omega)$ its Fourier transform. One wants to evaluate the quantity

$$J_1 := \sum_{i=1}^{m} \int_0^\infty z^{(i)'} z^{(i)} dt$$

By exploiting the well known Parceval theorem and the properties of the trace operator, it follows

$$\|F(s)\|_2^2 = \frac{1}{2\pi} \int_{-\infty}^{\infty} \text{trace}\left[F'(-j\omega)F(j\omega)\right] d\omega$$
$$= \frac{1}{2\pi} \int_{-\infty}^{\infty} \sum_{i=1}^{m} e_i(m)' F'(-j\omega)F(j\omega)e_i(m)d\omega$$
$$= \frac{1}{2\pi} \int_{-\infty}^{\infty} \sum_{i=1}^{m} Z^{(i)'}(-j\omega)Z^{(i)}(j\omega)d\omega$$
$$= \sum_{i=1}^{m} \int_0^\infty z^{(i)'} z^{(i)} dt = J_1$$

As for the computation of the scalar J_1, observe that

$$\int_0^\infty z^{(i)'} z^{(i)} dt = \int_0^\infty \text{trace}\left[e_i(m)' B' e^{A't} C' C e^{At} B e_i(m)\right] dt$$

so that

$$J_1 = \sum_{i=1}^{m} \int_0^\infty \text{trace}\left[e_i(m)' B' e^{A't} C' C e^{At} B e_i(m)\right] dt$$
$$= \sum_{i=1}^{m} \int_0^\infty \text{trace}\left[e^{A't} C' C e^{At} B e_i(m)e_i(m)' B'\right] dt$$
$$= \int_0^\infty \text{trace}\left[e^{A't} C' C e^{At} \sum_{i=1}^{m} B e_i(m)e_i(m)' B'\right] dt$$

$$= \int_0^\infty \text{trace}\left[e^{A't}C'Ce^{At}BB'\right]dt$$

$$= \text{trace}\left[B'\int_0^\infty e^{A't}C'Ce^{At}dt B\right] \tag{2.19}$$

$$= \text{trace}\left[C\int_0^\infty e^{At}BB'e^{A't}dt C'\right] \tag{2.20}$$

As well known, the value of the last two integrals which appear in eqs. (2.19),(2.20) can be computed by resorting to two suitable Lyapunov equations. More precisely, it follows

$$\text{trace}\left[B'\int_0^\infty e^{A't}C'Ce^{At}dt B\right] = \text{trace}\left[B'P_oB\right]$$

$$\text{trace}\left[C\int_0^\infty e^{At}BB'e^{A't}dt C'\right] = \text{trace}\left[CP_rC'\right] \tag{2.21}$$

where P_o and P_r are the unique solutions of the Lyapunov equations (in the unknown P)

$$0 = A'P + PA + C'C$$

$$0 = AP + PA' + BB'$$

respectively. As a matter of fact, stability of A implies that these equations admit a unique solution, which is also positive semidefinite (recall Lemma C.1).

Consider again the system defined at the beginning of this remark and let now the input w be a white noise with identity intensity. Associated with such a system, consider the quantity

$$J_2 := \lim_{t\to\infty} \text{E}\left[z'(t)z(t)\right]$$

It follows that

$$J_2 = \lim_{t\to\infty} \text{E}\left[\text{trace}\left[z'(t)z(t)\right]\right]$$

$$= \lim_{t\to\infty} \text{E}\left[\text{trace}\left[z(t)z'(t)\right]\right]$$

$$= \lim_{t\to\infty} \text{trace}\left[\text{E}\left[\int_0^t \Phi(t,\tau)w(\tau)d\tau \int_0^t w'(\sigma)\Phi'(t,\sigma)d\sigma\right]\right]$$

$$= \lim_{t\to\infty} \text{trace}\left[\int_0^t \Phi(t,\tau)\int_0^t \text{E}\left[w(\tau)w'(\sigma)\right]\Phi'(t,\sigma)d\sigma d\tau\right]$$

$$= \lim_{t\to\infty} \text{trace}\left[\int_0^t \Phi(t,\tau)\Phi'(t,\tau)d\tau\right]$$

where $\Phi(\xi,\vartheta) := Ce^{A(\xi-\vartheta)}B$. Letting $\eta = t - \tau$ one obtains

$$J_2 = \lim_{t\to\infty} \text{trace}\left[C\int_0^t e^{A\eta}BB'e^{A'\eta}d\eta C'\right]$$

$$= \text{trace}\left[C\int_0^\infty e^{A\eta}BB'e^{A'\eta}d\eta C'\right] = J_1$$

where the last equality follows from what previously shown. Finally, consider again the system fed by a white noise w with identity intensity, and let

$$J_3 := \lim_{T\to\infty} \frac{1}{T}\text{E}\left[\int_0^T z'z\,dt\right]$$

From the analysis performed for the scalar J_2, one can easily check that

$$E\left[z'(t)z(t)\right] = \text{trace}\left[C\int_0^t e^{A'\eta}BB'e^{A\eta}d\eta C'\right] = \text{trace}\left[CP(t)C'\right]$$

with

$$P(t) := \int_0^t e^{A\eta}BB'e^{A'\eta}d\eta \tag{2.22}$$

Therefore,

$$P(0) = 0$$
$$\dot{P}(t) = e^{At}BB'e^{A't}$$

so that, taking into account eq. (2.22), it follows

$$\dot{P}(t) = AP(t) + P(t)A' + BB'$$

By integrating both sides of this equation from 0 to T and recalling that $P(0) = 0$, one gets

$$P(T) = \int_0^T \dot{P}(t)dt = A\int_0^T P(t)dt + \int_0^T P(t)dtA' + BB'T$$

Letting

$$X(T) := \int_0^T P(t)dt$$

it then follows that

$$\frac{P(T)}{T} = A\frac{X(T)}{T} + \frac{X(T)}{T}A' + BB' \tag{2.23}$$

Observe that, thanks to eq. (2.22), it is

$$\lim_{T\to\infty} P(T) = \lim_{T\to\infty}\int_0^T e^{A\eta}BB'e^{A'\eta}d\eta = P_r$$

so that, taking the limit as $T \to \infty$ of both members of eq. (2.23), it turns out that

$$0 = AY + YA' + BB' \tag{2.24}$$

where

$$Y := \lim_{T\to\infty}\frac{X(T)}{T}$$

Since the solution of eq. (2.24) is unique, one can conclude that $Y = P_r$ and consequently $J_3 = J_1$. □

Remark 2.20 also indicates how the computation of the norm of a function in the space RH_2 can be actually performed. As a matter of fact, it is sufficient to solve a Lyapunov equation. Of course, the problem of computing the norm of a function $F(s)$ in $RH_2^\perp$, is easily solved by observing that $G(s) := F^\sim(s) \in RH_2$ and that

$$\|F(s)\|_2^2 = \frac{1}{2\pi}\int_{-\infty}^{\infty} \text{trace}[F^\sim(j\omega)F(j\omega)]d\omega$$
$$= \frac{1}{2\pi}\int_{-\infty}^{\infty} \text{trace}[F(j\omega)F^\sim(j\omega)]d\omega$$
$$= \frac{1}{2\pi}\int_{-\infty}^{\infty} \text{trace}[G^\sim(j\omega)G(j\omega)]d\omega$$
$$= \|G(s)\|_2^2$$

Hence, the norm of $F(s) \in RH_2$ can be computed by solving a Lyapunov equation associated with $G(s) = F^\sim(s) \in RH_2$.

Finally, the problem of computing the norm of a function $F(s)$ in the space RL_2 is easily solvable by writing a decomposition of such a function into the sum of an element of RH_2 and another one in $RH_2^\perp$. The following result is then provided, whose proof is obvious and then omitted.

Theorem 2.11 Let $F(s) = F_a(s) + F_s(s)$ with $F_a(s) \in RH_2^\perp$ and $F_s(s) \in RH_2$. Then,

  i) $< F_a(s), F_s(s) >= 0$

  ii) $\|F(s)\|_2^2 = \|F_a(s)\|_2^2 + \|F_s(s)\|_2^2$

Example 2.18 Consider the function

$$F(s) = \frac{s}{(s+1)(s-1)} \in RL_2$$

Letting

$$F_a(s) := \frac{1}{2(s-1)} , \quad F_s(s) := \frac{1}{2(s+1)}$$

it follows that $F(s) = F_s(s) + F_a(s)$ and $G_a(s) := F_a^\sim(s) = -1/2(s+1)$. With the functions $F_s(s)$ and $G_a(s)$ let associate the two (minimal) realizations $\Sigma(A_s, B_s, C_s, D_s)$ and $\Sigma(A_a, B_a, C_a, D_a)$, respectively, with $A_a = A_s = -1$, $B_a = B_s = 0.5$, $C_s = 1$, $C_a = -1$ and $D_s = D_a = 0$. It turns out that

$$\|F_s(s)\|_2^2 = \|G_a(s)\|_2^2 = \frac{1}{8}$$

so that

$$\|F(s)\|_2 = \frac{1}{2}$$

$\square$

Remark 2.21 It is worth noticing that the norm of a generic function $F(s) \in RL_\infty$ coincides with that of a suitable function $\hat{F}(s) \in RH_\infty$, which is easily derived from $F(s)$. Actually, notice that the least common multiple of all denominators of the elements of $F(s)$, denoted by $\psi(s)$, can be always factorized as $\psi(s) := \psi_a(s)\psi_s(s)$, where, since $F(s) \in RL_\infty$, the polynomial $\psi_a(s)$ has all its roots in the open right half plane and $\psi_s(s)$ has all its roots in the open left half plane. Hence, if

$$F(s) := \frac{P(s)}{\psi_a(s)\psi_s(s)}$$

being $P(s)$ a polynomial matrix, then

$$\hat{F}(s) := \frac{P(s)}{\psi_a(-s)\psi_s(s)}$$

is such that, $\hat{F}(s) \in RH_\infty$ and $F^\sim(s)F(s) = \hat{F}^\sim(s)\hat{F}(s)$. Consequently, $\|F(s)\|_\infty = \|\hat{F}(s)\|_\infty$. $\square$

The norm of a function $F(s) \in RL_\infty$, thought of as the transfer function of a linear system, can be meaningfully related to the norm of the input and output signals, both assumed to be square integrable. Precisely, the following result, besides further characterizing the concept of norm in RL_∞ for dynamical systems, also provides an alternative definition of such a norm.

Theorem 2.12 *Let $F(s) \in RL_\infty$. Then*

$$\|F(s)\|_\infty = \sup_{\substack{u \neq 0 \\ u \in RL_2}} \frac{\|F(s)u\|_2}{\|u\|_2}$$

Proof Observe first that, thanks to Remark 2.21, one can assume, without any loss of generality, that $F(s) \in RH_\infty$. As a consequence, it results that $F(s)u \in RL_2$ if $u \in RL_2$. Now, notice that

$$\sup_{\substack{u \neq 0 \\ u \in RL_2}} \frac{\|F(s)u\|_2^2}{\|u\|_2^2} \geq \sup_{\substack{Re(\lambda) > 0 \\ \mu \neq 0}} \frac{\|F(s)\frac{\mu}{s-\lambda}\|_2^2}{\|\frac{\mu}{s-\lambda}\|_2^2} \qquad (2.25)$$

where μ is a generic constant vector with suitable dimension. Recall that the time response of a system with transfer function $F(s)$ fed by the input $u = \mu e^{\lambda t}$ is, if λ does not coincide with any poles of $F(s)$, $y = F(\lambda)\mu e^{\lambda t}$, provided that a suitable initial state is chosen. Then,

$$F(s)\frac{\mu}{s-\lambda} = F(\lambda)\frac{\mu}{s-\lambda} - f(s)$$

where $f \in RH_2$ is the Laplace transform of the output free response, whereas the left hand side of the equation corresponds to the output forced response. Since $F(\lambda)\frac{\mu}{s-\lambda} \in RH_2^\perp$, it follows that (recall Theorem 2.11)

$$\|F(s)\frac{\mu}{s-\lambda}\|_2^2 = \|F(\lambda)\frac{\mu}{s-\lambda}\|_2^2 + \|f(s)\|_2^2 \geq \|F(\lambda)\frac{\mu}{s-\lambda}\|_2^2 \qquad (2.26)$$

From eqs. (2.25), (2.26) it follows that

$$\begin{aligned}
\sup_{\substack{u \neq 0 \\ u \in RL_2}} \frac{\|F(s)u\|_2^2}{\|u\|_2^2} &\geq \sup_{\substack{Re(\lambda) > 0 \\ \mu \neq 0}} \frac{\|F(s)\frac{\mu}{s-\lambda}\|_2^2}{\|\frac{\mu}{s-\lambda}\|_2^2} \\
&\geq \sup_{\substack{Re(\lambda) > 0 \\ \mu \neq 0}} \frac{\|F(\lambda)\mu\|^2}{\|\mu\|^2} \\
&\geq \sup_{Re(\lambda) > 0} \|F(\lambda)\|^2 \\
&\geq \|F(s)\|_\infty \qquad (2.27)
\end{aligned}$$

In getting the above expression, the following two facts have been exploited:

1) If K is a constant, then $\|\frac{K}{s-\lambda}\|_2^2 = K^\sim K \|\frac{1}{s-\lambda}\|_2^2 = \|K\|^2\|\frac{1}{s-\lambda}\|_2^2$

2) $\sup_{\mu \neq 0} \frac{\|F(\lambda)\mu\|^2}{\|\mu\|^2} = \|F(\lambda)\|^2$

On the other hand, letting e_i denote the i-th column of the identity matrix (the context will clarify the relevant dimension), and defining

$$u := \begin{bmatrix} ue_1 & \cdots & ue_\nu \end{bmatrix}$$

it follows that

$$\sup_{\substack{u \neq 0 \\ u \in RL_2}} \frac{\|F(s)u\|_2^2}{\|u\|_2^2} = \sup_{\substack{u \neq 0 \\ u \in RL_2}} \frac{\int_{-\infty}^\infty \text{trace}\left[u^\sim(j\omega)F^\sim(j\omega)F(j\omega)u(j\omega)\right] d\omega}{\int_{-\infty}^\infty \text{trace}\left[u^\sim(j\omega)u(j\omega)\right] d\omega}$$

$$= \sup_{\substack{u \neq 0 \\ u \in RL_2}} \frac{\int_{-\infty}^{\infty} \sum_{i=1}^{\nu} e_i' u^\sim(j\omega) F^\sim(j\omega) F(j\omega) u(j\omega) e_i d\omega}{\int_{-\infty}^{\infty} \sum_{i=1}^{\nu} e_i' u^\sim(j\omega) u(j\omega) e_i d\omega}$$

$$= \sup_{\substack{u \neq 0 \\ u \in RL_2}} \frac{\sum_{i=1}^{\nu} \int_{-\infty}^{\infty} \|F(j\omega) u(j\omega) e_i\|^2 d\omega}{\sum_{i=1}^{\nu} \int_{-\infty}^{\infty} \|u(j\omega) e_i\|^2 d\omega}$$

$$\leq \sup_{\substack{u \neq 0 \\ u \in RL_2}} \frac{\sum_{i=1}^{\nu} \int_{-\infty}^{\infty} \|F(j\omega)\|^2 \|u(j\omega) e_i\|^2 d\omega}{\sum_{i=1}^{\nu} \int_{-\infty}^{\infty} \|u(j\omega) e_i\|^2 d\omega}$$

$$\leq \sup_{\substack{u \neq 0 \\ u \in RL_2}} \frac{\|F(s)\|_\infty^2 \sum_{i=1}^{\nu} \int_{-\infty}^{\infty} \|u(j\omega) e_i\|^2 d\omega}{\sum_{i=1}^{\nu} \int_{-\infty}^{\infty} \|u(j\omega) e_i\|^2 d\omega}$$

$$\leq \|F(s)\|_\infty^2 \tag{2.28}$$

From equations (2.27),(2.28), the result follows.   □

The computation of the norm of $F(s) \in RL_\infty$ can be easily performed by exploiting the result provided in the following theorem, which refers to functions in RH_∞, only. As explained in Remark 2.21, this does not entail any loss of generality.

Theorem 2.13 *Let $F(s) = C(sI - A)^{-1}B + D$, with A stable and $\bar{\sigma}(D) < \gamma$. Moreover, let $\hat{A}(\gamma) := A + B(\gamma^2 I - D'D)^{-1}D'C$ and*

$$Z(\gamma) := \begin{bmatrix} \hat{A}(\gamma) & B(\gamma^2 I - D'D)^{-1}B' \\ -C'(I - \gamma^{-2}DD')^{-1}C & -\hat{A}'(\gamma) \end{bmatrix}$$

Then, the following conditions are equivalent:

a) *$\|F(s)\|_\infty < \gamma$*

b) *All the eigenvalues of $Z(\gamma)$ do not lie on the imaginary axis*

c) *The subspace generated by the (generalized) eigenvectors of $Z(\gamma)$, associated with the eigenvalues with negative real parts, is complementary to $\mathrm{Im}[[0\ I]']$*

d) *There exists a symmetric, positive semidefinite and stabilizing solution $S(\gamma)$ of the algebraic Riccati equation (in the unknown S) associated with $Z(\gamma)$*

$$0 = S\hat{A}(\gamma) + \hat{A}'(\gamma)S + C'(I - \gamma^{-2}DD')^{-1}C + SB(\gamma^2 I - D'D)^{-1}B'S$$

namely, such that matrix $\hat{A}(\gamma) + B(\gamma^2 I - D'D)^{-1}B'S(\gamma)$ is stable.

Proof Preliminarily observe that the assumption on the norm of D ensures, thanks to Lemma B.11 that both matrices $(I - \gamma^{-2}DD')$ and $(\gamma^2 I - D'D)$ are positive definite so that the statement of the theorem is well defined.

a) $\implies$ b) Assume, by contradiction, that $Z(\gamma)$ has an eigenvalue in $j\omega$ and let $\xi := [x'\ y']' \neq 0$ be an associated eigenvector. From $Z(\gamma)\xi = j\omega\xi$ it follows

$$(j\omega - \hat{A}(\gamma))x = B(\gamma^2 I - D'D)^{-1}B'y$$
$$-(j\omega + \hat{A}'(\gamma))y = C'(I - \gamma^{-2}DD')^{-1}Cx$$

By recalling the definition of $\hat{A}(\gamma)$ and the identity $(I - \gamma^{-2}DD')^{-1} = I + D(\gamma^2 I - D'D)^{-1}D'$ (Lemma B.9), these expressions can be rewritten as follows

$$(j\omega - A)x = B(\gamma^2 I - D'D)^{-1}(B'y + D'Cx) \tag{2.29}$$
$$-(j\omega + A')y = C'Cx + C'D(\gamma^2 I - D'D)^{-1}(D'Cx + B'y) \tag{2.30}$$

Let now

$$v := (\gamma^2 I - D'D)^{-1}(B'y + D'Cx) \tag{2.31}$$

$$z := Cx + Dv \tag{2.32}$$

In view of eqs. (2.29),(2.30) and thanks to the stability of A, these last expressions become

$$x = (j\omega I - A)^{-1}Bv \tag{2.33}$$

$$y = -(j\omega I + A')^{-1}C'z \tag{2.34}$$

which, substituted into eqs. (2.31),(2.32) yield

$$z = F(j\omega)v, \quad \gamma^2 v = F'(-j\omega)z$$

that is

$$F'(-j\omega)F(j\omega)v = \gamma^2 v$$

Being $v \neq 0$ (otherwise, both x and z would be zero and thus also y), it results

$$
\begin{aligned}
\gamma^2 &= \frac{v^\sim F^\sim(j\omega)F(j\omega)v}{v^\sim v} \\
&= \frac{\|F(j\omega)v\|^2}{\|v\|^2} \\
&\leq \sup_{v \neq 0} \frac{\|F(j\omega)v\|^2}{\|v\|^2} \\
&\leq \|F(j\omega)\|^2 \\
&\leq \sup_\omega \|F(j\omega)\|^2 = \|F(s)\|_\infty^2
\end{aligned}
$$

which is a contradiction.

$b) \implies a)$ Conversely, suppose that $\|F(s)\|_\infty \geq \gamma$. Since, by assumption, $\|D\| = \lim_{\omega \to \infty} \|F(\omega)\| < \gamma$, by a continuity argument there exists a real number ω such that $\|F(j\omega)\| = \gamma$. Therefore, γ^2 is an eigenvalue (actually the maximum one) of $F^\sim(j\omega)F(j\omega)$ so that there exists a nonzero vector v such that $F^\sim(j\omega)F(j\omega)v = \gamma^2 v$. Being $v \neq 0$, it also follows that $z := F(j\omega)v \neq 0$. Now define two vectors x and y as in eqs. (2.33),(2.34). With such definitions, simple computations show that also eqs. (2.29)-(2.32) hold. From eq. (2.31) one can conclude that $\xi := [x'\ y']' \neq 0$ since, on the contrary, v would be zero. Finally, eqs. (2.29),(2.30) imply that $Z(\gamma)\xi = j\omega\xi$, which is a contradiction.

$d) \implies b)$ The existence of a stabilizing solution of the Riccati equation introduced in the statement implies, by a well known result, that the eigenvalues of $Z(\gamma)$ do not lie on the imaginary axis.

$b) \implies d)$ Notice first that, as obvious, stability of A guarantees stabilizability of the pair (A, B) and, recalling the definition of $\hat{A}(\gamma)$, also that of the pair $(\hat{A}(\gamma), B)$. This condition implies also the stabilizability of the pair $(\hat{A}(\gamma), B(\gamma^2 I - D'D)^{-1}B')$. To see this, suppose by contradiction that, for a certain $x \neq 0$ and $\lambda, Re(\lambda) > 0$, it is $\hat{A}'(\gamma)x = \lambda x$ and $B(\gamma^2 I - D'D)^{-1}B'x = 0$ (recall the *PBH* test). It follows that $x^\sim B(\gamma^2 I - D'D)^{-1}B'x = 0$ so that $B'x = 0$ thanks to the fact that $(\gamma^2 I - D'D)^{-1} > 0$. The conditions $x \neq 0$, $Re(\lambda) > 0$, $\hat{A}'(\gamma)x = \lambda x$ and $B'x = 0$, finally violate the stabilizability of $(\hat{A}(\gamma), B)$. Lemma C.3 can now be applied to the Riccati equation, ensuring the existence of its symmetric and stabilizing solution $S(\gamma)$. It is only left to show that such a solution is actually positive semidefinite. To this aim, rewrite the Riccati equation as follows

$$
\begin{aligned}
0 &= S(\gamma)A + A'S(\gamma) + C'C + [S(\gamma)B + C'D]\,(\gamma^2 I - D'D)^{-1}\,[S(\gamma)B + C'D]' \\
&:= S(\gamma)A + A'S(\gamma) + W(\gamma)
\end{aligned}
$$

where $W(\gamma) \geq 0$. Thanks to Lemma C.1, it can be concluded that $S(\gamma) \geq 0$.

   c) $\Longrightarrow$ b) Condition c) implies that the subspace generated by the (generalized) eigenvectors of $Z(\gamma)$ associated with the eigenvalues with negative real part has the same dimension n as the system. Hence $Z(\gamma)$ has n eigenvalues with negative real part and, obviously, has no eigenvalues on the imaginary axis.

   b) $\Longrightarrow$ c) Since condition b) has already been proved to be equivalent to condition d), it follows that the stabilizing solution $S(\gamma)$ is such that $\mathrm{Im}[[I\ \ S(\gamma)]']$, which is obviously complementary to $\mathrm{Im}[[0\ \ I]']$, is actually the subspace generated by n (generalized) linear independent eigenvectors of $Z(\gamma)$ associated with the n eigenvalues with negative real part. $\square$

Theorem 2.14 *Let $F(s) := C(sI - A)^{-1}B + D$ and γ a positive scalar. Then, the following two conditions are equivalent:*

   a) *The matrix A is stable and $\|F(s)\|_\infty < \gamma$*

   b) *$\bar{\sigma}(D) < \gamma$ and there exists the positive semidefinite stabilizing solution $S(\gamma)$ of the algebraic Riccati equation (in the unknown S)*

$$
0 = S\hat{A}(\gamma) + \hat{A}'(\gamma)S + C'(I - \gamma^{-2}DD')^{-1}C + SB(\gamma^2 I - D'D)^{-1}B'S
$$

   i.e. such that

$$
\hat{A}(\gamma) + B(\gamma^2 I - D'D)^{-1}B'S(\gamma) = A + B(\gamma^2 I - D'D)^{-1}(B'S(\gamma) + D'C)
$$

   is stable.

Proof a) $\Longrightarrow$ b) If $\|F(s)\|_\infty < \gamma$, then, obviously $\bar{\sigma}(D) < \gamma$ since $D = \lim_{\omega \to \infty} F(j\omega)$. Therefore, Theorem 2.13 guarantees the existence of $S(\gamma)$.

   b) $\Longrightarrow$ a) Preliminarily observe that $\bar{\sigma}(D) < \gamma$ implies, thanks to Lemma B.11, that matrices $(I - \gamma^{-2}DD')$ and $(\gamma^2 I - D'D)$ are positive definite so that the Riccati equation is well defined. Such an equation can be equivalently rewritten as (recall the definition for $\hat{A}(\gamma)$ and Lemma B.9)

$$
0 = SA + A'S + (SB + C'D)(\gamma^2 I - D'D)^{-1}(B'S + D'C) + C'C
$$

The stabilizing solution $S(\gamma)$ is also a solution of the Lyapunov equation

$$
0 = SA + A'S + \Lambda'(\gamma)\Lambda(\gamma)
$$

where

$$
\Lambda'(\gamma)\Lambda(\gamma) := [S(\gamma)B + C'D](\gamma^2 I - D'D)^{-1}\,[B'S(\gamma) + D'C] + C'C
$$

The proof is concluded by showing that the pair $(A, \Lambda(\gamma))$ is detectable. Actually, in this case, Lemma C.1 implies that A is stable, so that the proof follows directly from

Theorem 2.13. Therefore, assume by contradiction that this pair is not detectable. Based on the *PBH* test (recall Lemma D.2) it follows that

$$Ax = \lambda x, \ Re(\lambda) \geq 0, \ x \neq 0$$
$$\Lambda(\gamma)x = 0$$

Recalling the definition of $\Lambda'(\gamma)\Lambda(\gamma)$, the second equation implies that

$$x^{\sim}\Lambda'(\gamma)\Lambda(\gamma)x = x^{\sim}[S(\gamma)B + C'D](\gamma^2 I - D'D)^{-1}[B'S(\gamma) + D'C]x + x^{\sim}C'Cx = 0$$

so that, in particular,

$$[B'S(\gamma) + D'C]x = 0$$

thanks to the fact that $C'C \geq 0$ and $(\gamma^2 I - D'D) > 0$. Hence,

$$\{A + B(\gamma^2 I - D'D)^{-1}[B'S(\gamma) + D'C]\}x = Ax = \lambda x$$

This is obviously a contradiction, since $x \neq 0$, $Re(\lambda) \geq 0$ and $S(\gamma)$ is stabilizing. □

Example 2.19 Consider the rational function

$$F(s) = \frac{10}{s^2 + 2s + 10} = \left[\begin{array}{cc|c} 0 & 1 & 0 \\ -10 & -2 & 1 \\ \hline 10 & 0 & 0 \end{array}\right]$$

It turns out that $|F(j\omega)|^2 = 100/(\omega^4 - 16\omega^2 + 100)$ so that $\|F(s)\|_\infty = 5/3 \simeq 1.66$. Taken $\gamma = 1.67$, the eigenvalues of $Z(\gamma)$ are $\pm 0.067 \pm 2.83j$ whereas for $\gamma = 1.66$ the eigenvalues of $Z(\gamma)$ are $\pm 2.92j$ and $\pm 2.73j$. □

Remark 2.22 Theorems 2.13 and 2.14 call for a Riccati equation which resembles the one utilized in the context of optimal LQ control. In view of Lemma 2.16 and Remark 2.13, it results that

$$\|F(j\omega)\| = \bar{\sigma}[F(j\omega)] = \bar{\sigma}[F'(-j\omega)] = \|F'(j\omega)\|$$

so that

$$\|F(s)\|_\infty = \|F'(s)\|_\infty$$

Hence, the two relevant theorems can be equivalently reformulated with reference to $F'(s) = B'(sI - A')^{-1}C' + D'$ instead of $F(s)$. As a consequence, the relevant Riccati equation becomes the following

$$0 = S\hat{A}'(\gamma) + \hat{A}(\gamma)S + B(I - \gamma^{-2}D'D)^{-1}B' + SC'(\gamma^2 I - DD')^{-1}CS$$

which resembles the equation involved in the optimal filtering problem. □

The previous results on the characterization of an upper bound of $\|F(s)\|_\infty$ for $F(s)$ being a function in RH_∞ are all expressed in terms of the existence of a symmetric, positive semidefinite and stabilizing solution of an algebraic Riccati equation (recall Theorems 2.13 and 2.14) which are of particular importance to get the results of Chapter 5. However, in Chapter 6 reference is made to the following similar results which are alternatively expressed in terms of Riccati inequalities with no additional requirement concerning the stability of their solutions.

Theorem 2.15 *Let $F(s) := C(sI - A)^{-1}B$ and γ a positive scalar. Then, the following conditions are equivalent:*

a) Matrix A is stable and $\|F(s)\|_\infty < \gamma$

b) *There exists a symmetric and positive definite matrix S satisfying the Riccati inequality*

$$SA + A'S + \gamma^{-2}SBB'S + C'C < 0 \tag{2.35}$$

c) *There exists a symmetric and positive definite matrix P satisfying the Riccati inequality*

$$PA' + AP + \gamma^{-2}PC'CP + BB' < 0 \tag{2.36}$$

Proof That points b) and c) are equivalent is straightforward. Indeed, the positive definite solutions (if any) of both inequalities are related one to the other by $P = \gamma^2 S^{-1}$.

b) $\Longrightarrow$ a) Since there exists $S > 0$ satisfying (2.35), it also satisfies the Riccati equation

$$0 = SA + A'S + \gamma^{-2}SBB'S + \bar{C}'\bar{C} \tag{2.37}$$

where $\bar{C}' = [C'\ E']$ for some matrix E such that $E'E > 0$. From the Extended Lyapunov lemma the stability of matrix A follows. With $S > 0$ being a solution of (2.37), define the auxiliary Riccati equation

$$0 = X(A + \gamma^{-2}BB'S)' + (A + \gamma^{-2}BB'S)'X - \gamma^{-2}XBB'X + E'E \tag{2.38}$$

Using the fact that A is stable and $E'E > 0$, from Lemma C.3 it is readily verified that it admits a symmetric and positive semidefinite solution $X \geq 0$ such that matrix

$$\begin{aligned}
A_X &:= A + \gamma^{-2}BB'S - \gamma^{-2}BB'X \\
&= A + \gamma^{-2}BB'(S - X)
\end{aligned} \tag{2.39}$$

is stable. Defining $W := S - X$ and using (2.37) together with (2.38) we get

$$0 = WA + A'W + \gamma^{-2}WBB'W + C'C$$

Moreover, since A is stable, this means that $W \geq 0$ and from (2.39) matrix $A + \gamma^{-2}BB'W$ is stable. From Theorem 2.14 the conclusion is that part a) holds indeed.

a) $\Longrightarrow$ b) Let us define the transfer function

$$\bar{F}(s) := \left[\begin{array}{c} C \\ \sqrt{\epsilon}I \end{array} \right] (sI - A)^{-1}B$$

where $\epsilon > 0$ is a scalar to be determined. It is a simple matter to verify that for all $\omega \in R$

$$\bar{F}'(-j\omega)\bar{F}(j\omega) = F'(-j\omega)F(j\omega) + \epsilon G'(-j\omega)G(j\omega)$$

where $G(s) := (sI - A)^{-1}B$. Hence

$$\|\bar{F}(s)\|_\infty^2 \leq \|F(s)\|_\infty^2 + \epsilon\|G(s)\|_\infty^2$$

and choosing the scalar ϵ such that

$$0 < \epsilon < \frac{\gamma^2 - \|F(s)\|_\infty^2}{\|G(s)\|_\infty^2}$$

which is always possible since by assumption, matrix A is stable and $\|F(s)\|_\infty < \gamma$ then we get $\|\bar{F}(s)\|_\infty < \gamma$. Using again Theorem 2.14 the conclusion is that there exists a symmetric and positive semidefinite solution $S \geq 0$ to the Riccati equation

$$0 = SA + A'S + \gamma^{-2}SBB'S + C'C + \epsilon I$$

which obviously satisfies the Riccati inequality (2.35) and in view of Lemma C.1 is actually positive definite. The proof of the theorem proposed is complete. $\qquad\square$

Theorem 2.16 *Let $F(s) := C(sI - A)^{-1}B$ and γ a positive scalar. Under the assumption that the pair (A, B) is reachable, the following conditions are equivalent:*

a) *Matrix A is stable and $\|F(s)\|_\infty \leq \gamma$*

b) *There exists a symmetric and positive definite matrix S satisfying the Riccati inequality*

$$SA + A'S + \gamma^{-2}SBB'S + C'C \leq 0 \qquad (2.40)$$

c) *There exists a symmetric and positive definite matrix P satisfying the Riccati inequality*

$$PA' + AP + \gamma^{-2}PC'CP + BB' \leq 0 \qquad (2.41)$$

Proof The equivalence of points b) and c) is immediate. The positive definite matrices satisfying inequalities (2.40) and (2.41), if any, are related one to the other by $S = \gamma^2 P^{-1}$.

b) $\Longrightarrow$ a) Assuming (2.40) admits a positive definite feasible solution then $P = \gamma^2 S^{-1}$ is feasible for inequality (2.41) which together with the reachability of the pair (A, B) implies that A is stable. On the other hand, for all $\omega \in R$, inequality (2.40) can be rewritten as

$$(-j\omega - A')S + S(j\omega - A) - \gamma^{-2}SBB'S - C'C \geq 0$$

from which and $G(s) := \gamma I - \gamma^{-1}B'S(sI - A)^{-1}B$ we have

$$F'(-j\omega)F(j\omega) \leq \gamma^2 I - G'(-j\omega)G(j\omega)$$
$$\leq \gamma^2 I, \quad \forall \omega \in R$$

consequently $\|F(s)\|_\infty \leq \gamma$ which is the desired result.

a) $\Longrightarrow$ b) From Theorem 2.15 we only need to prove that if $\|F(s)\|_\infty = \gamma$ then the Riccati inequality (2.40) is still feasible for some positive definite matrix. To this end, consider the sequence of matrices $C_n := \sqrt{\epsilon_n}C$ with ϵ_n being an arbitrary element of an increasing sequence of scalars such that $0 < \epsilon_n < 1$ and ϵ_n goes to 1 as n goes to infinite. By virtue of

$$\|C_n(sI - A)^{-1}B\|_\infty = \sqrt{\epsilon_n}\,\|F(s)\|_\infty$$
$$= \sqrt{\epsilon_n}\gamma < \gamma \qquad (2.42)$$

we already know that the Riccati equation (in the unknown S)

$$0 = SA + A'S + \gamma^{-2}SBB'S + C_n'C_n \qquad (2.43)$$

admits an unique symmetric, positive semidefinite solution S_n such that matrix $A_n := A + \gamma^{-2}BB'S_n$ is stable. Additionally, the sequence $S_n \geq 0$, $n = 1, 2, \cdots$ is nondecreasing because as it can be verified

$$0 = (S_{n+1} - S_n)A_n + A_n'(S_{n+1} - S_n) +$$
$$+ \gamma^{-2}(S_{n+1} - S_n)BB'(S_{n+1} - S_n) + (\epsilon_{n+1} - \epsilon_n)C'C$$

Once again, due to (2.42), the Riccati equation (in the unknown P)

$$0 = PA' + AP + \gamma^{-2}PC_n'C_nP + BB' \qquad (2.44)$$

also admits an unique symmetric, positive semidefinite and stabilizing solution P_n which in fact is positive definite since A stable and (A, B) reachable yield

$$P_n \geq \int_0^\infty e^{At} BB' e^{A't} dt > 0$$

Using the same reasoning adopted before we can show that the sequence $P_n > 0$, $n = 1, 2, \cdots$ is nondecreasing as well. Hence the sequence of positive matrices $\bar{S}_n :=$ $\gamma^2 P_n^{-1}$ is nonincreasing and satisfy the Riccati equation (2.43). Moreover, defining matrix $\bar{A}_n := A + \gamma^{-2} BB' \bar{S}_n$ we have from (2.44) with $P = P_n$

$$\begin{aligned} P_n \bar{A}'_n P_n^{-1} &= P_n (A + BB' P_n^{-1})' P_n^{-1} \\ &= -(A + \gamma^{-2} P_n C'_n C_n) \end{aligned}$$

implying that $-\bar{A}_n$ is stable. Finally taking into account that S_n and $\bar{S}_n$ solve the Riccati equation (2.43) we get

$$0 = (\bar{S}_n - S_n) A_n + A'_n (\bar{S}_n - S_n) + \gamma^{-2} (\bar{S}_n - S_n) BB' (\bar{S}_n - S_n)$$

which shows from the stability of matrix A_n that $S_n \leq \bar{S}_n$. This last inequality together with the fact that the sequence $S_n \geq 0$, $n = 1, 2, \cdots$ is nondecreasing and the sequence $\bar{S}_n > 0$, $n = 1, 2, \cdots$ is nonincreasing allow the conclusion that both sequences converge to some matrices such that

$$0 \leq S_\infty := \lim_{n \to \infty} S_n \leq \lim_{n \to \infty} \bar{S}_n := \bar{S}_\infty$$

At this point, it remains to prove that $\bar{S}_\infty$ is positive definite even though some eigenvalues of matrix $\bar{A}_\infty$ lie on the imaginary axis. Indeed, assume by contradiction that $\bar{S}_\infty \geq 0$. Since it solves the Riccati equation

$$0 = \bar{S}_\infty A + A' \bar{S}_\infty + \gamma^{-2} \bar{S}_\infty BB' \bar{S}_\infty + C'C$$

from Lemma C.1 and the *PBH* test there exits $x \neq 0$ such that $Ax = \lambda x$, $B' \bar{S}_\infty x = 0$ and $Cx = 0$, that is $\bar{A}_\infty x = \lambda x$. However, being A stable, this is impossible since, as proved before, all eigenvalues of matrix $\bar{A}_\infty$ are located on the right part (including the imaginary axis) of the complex plane.   □

Example 2.20 Consider the rational function $F(s)$ defined in Example 2.19. We notice that for this transfer function the pair (A, B) is reachable and the pair (A, C) is observable. Taking $\gamma = 5/3$, matrices S_∞ and $\bar{S}_\infty$ (recall the proof of Theorem 2.16) are found to be

$$S_\infty = \bar{S}_\infty = \begin{bmatrix} 55.56 & 5.56 \\ 5.56 & 5.56 \end{bmatrix}$$

It is interesting to see that, in this case, the eigenvalues of matrices $A_\infty = A + \gamma^{-2} BB' S_\infty = \bar{A}_\infty$ are $\pm 2.83j$. That is, the corresponding Riccati equation solution $\bar{S}_\infty$ is no more stabilizing but still, for $\gamma = 5/3 = \|F(s)\|_\infty$, there exists a symmetric and positive definite matrix satisfying the Riccati inequality (2.40). One of such matrices is exactly $\bar{S}_\infty$.   □

Example 2.21 Consider the rational function $F(s)$

$$F(s) = \frac{1}{s+2} = \frac{s+1}{s^2 + 3s + 2} = \left[\begin{array}{cc|c} 0 & 1 & 0 \\ -2 & -3 & 1 \\ \hline 1 & 1 & 0 \end{array} \right]$$

so that $\|F(s)\|_\infty = 1/2$. It is apparent that the pair (A, B) is reachable but the pair (A, C) is not observable. With $\gamma = 0.5$ the following matrices appearing in the proof of Theorem 2.16 have been calculated

$$S_\infty = \begin{bmatrix} 0.50 & 0.50 \\ 0.50 & 0.50 \end{bmatrix} \leq \begin{bmatrix} 0.50 & 0.50 \\ 0.50 & 1.00 \end{bmatrix} = \bar{S}_\infty$$

In this case, matrix S_∞ is positive semidefinite and the eigenvalues of A_∞ are 0 and -1. On the other hand, matrix $\bar{S}_\infty$ is positive definite and the eigenvalues of $\bar{A}_\infty$ are 0 and 1. By construction, matrix $\bar{S}_\infty$ satisfies the Riccati inequality (2.40). □

An important role in the development of the discussion in Chapter 5 will be played by a few linear operators acting on the spaces previously defined.

Definition 2.32 (The Laurent operator with symbol F) *Let $F(s) \in RL_\infty$. The map $\Lambda_F : RL_2 \to RL_2$ defined as*

$$\Lambda_F : G(s) \mapsto \Lambda_F G(s) := F(s)G(s)$$

is called the Laurent operator with symbol $F(s)$. □

Notice that the operator Λ_F is obviously linear and, thanks to Theorem 2.12, $\|\Lambda_F\| = \|F(s)\|_\infty$ so that Λ_F is bounded.

Definition 2.33 (Orthogonal stable and antistable projections) *Let $G(s)$ be a generic element of RL_2 such that $G(s) := G_s(s) + G_a(s)$ with $G_s(s) \in RH_2$ and $G_a(s) \in RH_2^\perp$. The map $\Pi_s : RL_2 \to RH_2$, defined by*

$$\Pi_s : G(s) \mapsto \Pi_s G(s) := G_s(s)$$

is called the stable (orthogonal) projection , whereas the map $\Pi_a : RL_2 \to RH_2^\perp$, defined by

$$\Pi_a : G(s) \mapsto \Pi_a G(s) := G_a(s)$$

is called the antistable (orthogonal) projection. □

It is immediate to check that Π_s and Π_a are actually linear operators.

Definition 2.34 (The Hankel operator with symbol F) *Let $F(s) \in RL_\infty$. The map $\Gamma_F : RH_2^\perp \to RH_2$ defined by*

$$\Gamma_F : G(s) \mapsto \Gamma_F G(s) := \Pi_s \Lambda_F G(s)$$

is called the Hankel operator with symbol F. □

Remark 2.23 It should be evident from the above definition that the Hankel operator is the result of the composition of the Laurent operator with the stable projection, i.e. $\Gamma_F = \Pi_s \Lambda_F$. Hence, Γ_F is indeed an operator. □

Remark 2.24 Sometimes the Hankel operator is defined as mapping RH_2 to $RH_2^\perp$ in such a way that $\Gamma_F = \Pi_a \Lambda_F$. This should not be surprising. Actually, given the existing isomorphism between RH_2 and $RH_2^\perp$, the deep understanding of the basic facts of the underlying theory is certainly not blurred by the presence of these different (although equivalent) definitions. □

Remark 2.25 Thanks to Definition 2.34, it turns out that only the strictly proper and stable part of $F(s)$ contributes to $\Gamma_F G(s)$. As a matter of fact, letting $F(s) = F_\infty + F_s(s) + F_a(s)$, where $F_\infty = \lim_{s \to \infty} F(s)$, $F_s(s) \in RH_2$, $F_a(s) \in RH_2^\perp$, it follows

$$\Gamma_F G(s) = \Pi_s \Lambda_F G(s)$$
$$= \Pi_s F(s) G(s)$$
$$= \Pi_s \left[F_\infty + F_s(s) + F_a(s) \right] G(s)$$
$$= \Gamma_{F_s} G(s)$$

Hence, $\Gamma_F = \Gamma_{F_s}$.   □

Example 2.22 A way to compute the components F_∞, $F_a(s)$ and $F_s(s)$ of $F(s) \in RL_\infty$, calls for a (minimal) realization $\Sigma(A, B, C, D)$ of it. Let n_s (resp. n_a) be the number of eigenvalues of A with negative (resp. positive) real parts and define $\mathcal{X}_s$ (resp. $\mathcal{X}_a$) as the n_s dimensional (n_a dimensional) subspace generated by n_s (resp. n_a) linear independent generalized eigenvectors of A associated with the eigenvalues with negative (resp. positive) real part. Such subspaces are obviously complementary and can be identified by the image of suitable full rank matrices X_s and X_a, namely $\mathcal{X}_s = \text{Im}[X_s]$ and $\mathcal{X}_a = \text{Im}[X_a]$. Then, matrix $[X_s\ X_a]$ is invertible and, letting $T := [X_s\ X_a]^{-1}$, it results

$$\hat{A} := TAT^{-1} := \begin{bmatrix} A_s & 0 \\ 0 & A_a \end{bmatrix}$$

where A_s is stable and A_a antistable. Now, let

$$\hat{B} := TB := \begin{bmatrix} B_s \\ B_a \end{bmatrix} , \quad \hat{C} := CT^{-1} := \begin{bmatrix} C_s & C_a \end{bmatrix}$$

so that

$$F(s) = \hat{C}(sI - \hat{A})^{-1}\hat{B} + D$$
$$= \begin{bmatrix} C_s & C_a \end{bmatrix} \begin{bmatrix} sI - A_s & 0 \\ 0 & sI - A_a \end{bmatrix}^{-1} \begin{bmatrix} B_s \\ B_a \end{bmatrix} + D$$
$$= C_s(sI - A_s)^{-1}B_s + C_a(sI - A_a)^{-1}B_a + D$$
$$:= F_s(s) + F_a(s) + F_\infty$$

Specifically, consider the function

$$F(s) = \Sigma(A, B, C, D) = \frac{s^2 + s + 1}{s^2 - 1} \in RL_\infty$$

with

$$A = \begin{bmatrix} 0 & 1 \\ 1 & 0 \end{bmatrix} , \quad B = \begin{bmatrix} 0 \\ 1 \end{bmatrix} , \quad C = \begin{bmatrix} 2 & 1 \end{bmatrix} , \quad D = 1$$

It follows

$$X_s = \begin{bmatrix} 1 \\ -1 \end{bmatrix} , \quad X_a = \begin{bmatrix} 1 \\ 1 \end{bmatrix}$$

and $A_s = -1$, $A_a = 1$, $B_s = -1/2$, $B_a = 1/2$, $C_s = 1$, $C_a = 3$ so that $F_\infty = 1$, $F_s(s) = -1/2(s+1)$ and $F_a(s) = 3/2(s-1)$. Let now

$$G(s) := \sum_{i=1}^{k} \frac{\alpha_i}{(s - p_i)^{n_i}} , \quad Re(p_i) > 0$$

It turns out that $F(s)G(s) = G(s) - G(s)/2(s+1) + 3G(s)/2(s-1)$ so that

$$\Gamma_F G(s) = \Pi_s[F(s)G(s)] = -\frac{G(-1)}{2(s+1)}$$

On the other hand

$$F_s(s)G(s) = -\frac{G(-1)}{2(s+1)} + \sum_{i=1}^{k} \frac{\beta_i}{(s-p_i)^{n_i}} \ , \ \ Re(p_i) > 0$$

where β_i are suitable scalars. Hence

$$\Gamma_{F_s} G(s) = \Pi_s[F_s(s)G(s)] = -\frac{G(-1)}{2(s+1)}$$

$\square$

The adjoint Laurent and Hankel operators are now defined in the two lemmas below according to Definition 2.22.

Lemma 2.26 (The adjoint Laurent operator with symbol F) *The adjoint Laurent operator with symbol F is the Laurent operator with symbol $F^\sim$, i.e.*

$$\Lambda_F^* = \Lambda_{F^\sim}$$

Proof Let $G_1(s)$ and $G_2(s)$ be two generic elements of RL_2 and $F(s) \in RL_\infty$. Then, by recalling Definition 2.22,

$$< G_1(s), \Lambda_F^* G_2(s) > \ = \ < \Lambda_F G_1(s), G_2(s) >$$
$$= \frac{1}{2\pi} \int_{-\infty}^{\infty} \text{trace}[G_1^\sim(j\omega)F^\sim(j\omega)G_2(j\omega)d\omega$$
$$= \ < G_1(s), F^\sim(s)G_2(s) >$$
$$= \ < G_1(s), \Lambda_{F^\sim} G_2(s) >$$

Being $G_1(s)$ arbitrary, from this expression it follows

$$\Lambda_F^* G_2(s) = \Lambda_{F^\sim} G_2(s)$$

so that $\Lambda_F^* = \Lambda_{F^\sim}$, since $G_2(s)$ is arbitrary too. $\square$

Lemma 2.27 (The adjoint Hankel operator with symbol F) *The adjoint Hankel operator with symbol F is the operator $\Gamma_F^* := \Pi_a \Lambda_F^* = \Pi_a \Lambda_{F^\sim}$*

Proof Let $G_1(s) \in RH_2^\perp$, $H(s) \in RH_2$ and $F(s) \in RL_\infty$. Preliminarily observe that

$$< G(s), \Gamma_F^* H(s) > \ = \ < \Gamma_F G(s), H(s) >$$
$$= \ < \Pi_s \Lambda_F G(s), H(s) >$$
$$= \ < \Pi_s F(s)G(s), H(s) > = \ < F(s)G(s), H(s) >$$

since $F(s)G(s) = \Pi_s F(s)G(s) + \Pi_a F(s)G(s)$ and being $H(s) \in RH_2$, it turns out that $< \Pi_a F(s)G(s), H(s) > = 0$ (recall Theorem 2.11). Analogously, $F^\sim(s)H(s) = \Pi_a F^\sim(s)H(s) + \Pi_s F^\sim(s)H(s)$, so that Theorem 2.11 implies $< G(s), \Pi_s F^\sim(s)H(s) > = 0$, since $G(s) \in RH_2^\perp$. Taking in mind this fact, it follows

$$< F(s)G(s), H(s) > = \ < G(s), F^\sim(s)H(s) > = \ < G(s), \Pi_a F^\sim(s)H(s) >$$

so that

$$< G(s), \Gamma_F^* H(s) > = < G(s), \Pi_a F^\sim(s) H(s) > = < G(s), \Pi_a \Lambda_{F^\sim} H(s) >$$

Since $G(s)$ is arbitrary it follows that $\Gamma_F^* H(s) = \Pi_a \Lambda_{F^\sim} H(s)$. Finally, being $H(s)$ arbitrary, it results $\Gamma_F^* = \Pi_a \Lambda_{F^\sim} = \Pi_a \Lambda_F^*$.   □

Interestingly, one can associate with a generic Hankel operator, defined in the frequency domain, a suitable function of time. Precisely, consider the operator Γ_F and assume, without any loss of generality, (recall Remark 2.25), that

$$F(s) := \left[\begin{array}{c|c} A & B \\ \hline C & 0 \end{array} \right]$$

with A stable and $\Sigma(A, B, C, 0)$ in minimal form with order n. Then, the function

$$f(t) := \left\{ \begin{array}{ll} 0 & t \leq 0 \\ Ce^{At}B & t > 0 \end{array} \right.$$

is the inverse Laplace transform of $F(s)$. The map

$$\Gamma_f : RL_2(-\infty \ 0] \rightarrow RL_2[0 \ \infty)$$

defined by

$$\Gamma_f : u(t) \mapsto \Gamma_f u(t) := y(t) := \left\{ \begin{array}{ll} 0 & t < 0 \\ Ce^{At} \int_{-\infty}^0 e^{-A\tau} Bu(\tau) d\tau & t \geq 0 \end{array} \right.$$

is easily shown to be an operator, which is legitimated to be considered as the time domain counterpart of the operator Γ_F. Actually, the Laplace transform of y equals the stable projection of the product of $F(s)$ with the Laplace transform of u, i.e.

$$y_L = (\Gamma_f u)_L = \Gamma_F u_L \tag{2.45}$$

It is also useful to characterize further the operator Γ_f as follows. Consider the system $\Sigma(A, B, C, 0)$ and assume that its initial state at $t = -\infty$ is zero, i.e. $x(-\infty) = 0$. Now apply to the system an input $u(\cdot)$ which is different from zero only for nonpositive time instants. Hence,

$$x(0) = \int_{-\infty}^0 e^{-A\tau} Bu d\tau$$

Moreover, consider the free output y, defined for nonnegative time instants, due to the above defined initial state at $t = 0$. It turns out that the operator Γ_f maps the input u ($t \leq 0$) to the output y ($t \geq 0$), through $x(0)$. As such, Γ_f can be viewed as the composition of two operators $\Psi_r : RL_2(-\infty \ 0] \rightarrow R^n$ and $\Psi_o : R^n \rightarrow RL_2[0 \ \infty)$, defined by

$$\Psi_r : u(t) \mapsto \Psi_r u(t) := x := \int_{-\infty}^0 e^{-A\tau} Bu d\tau$$

$$\Psi_o : x \mapsto \Psi_o x := y(t) := \left\{ \begin{array}{ll} 0 & t < 0 \\ Ce^{At} x & t \geq 0 \end{array} \right.$$

Actually, it results $\Gamma_f = \Psi_o \Psi_r$. The observability *operator* Ψ_o is obviously *injective* (the unique element of R^n which is mapped in zero is the zero element), thanks to the assumed observability condition. Slightly less obvious is to recognize that the operator Ψ_r is *surjective*. In fact, thanks to the reachability condition, any element of R^n is the result of the transformation of a suitable element of $RL_2(\infty\ 0]$. To see this, let $x \in R^n$ be a fixed element . The input

$$u(t) := \begin{cases} 0 & t > 0 \\ B'e^{-A't}\left[\int_{-\infty}^0 e^{-A\sigma}BB'e^{-A'\sigma}d\sigma\right]^{-1}x, & t \leq 0 \end{cases}$$

is such that $\Psi_r u = x$. Actually, recalling the discussion in Remark 2.24, it follows

$$\int_{-\infty}^0 e^{-A\sigma}BB'e^{-A'\sigma}d\sigma = \int_0^\infty e^{A\sigma}BB'e^{A'\sigma}d\sigma = P_r$$

where P_r is the unique solution of the Lyapunov equation (in the unknown P) $PA' + AP + BB' = 0$. Moreover, such solution is positive definite (for this consider points i) and ii) of the "dual" version of Lemma C.1, i.e. when the pair (A, C) is replaced by the pair (A', B')). Hence u is well defined. Moreover, since A is stable, $u \in RL_2(-\infty\ 0]$. Finally,

$$\Psi_r u = \int_{-\infty}^0 e^{-A\tau}BB'e^{-A'\tau}d\tau\left[\int_{-\infty}^0 e^{-A\sigma}BB'e^{-A'\sigma}d\sigma\right]^{-1}x = x$$

By resorting to Definition 2.22 it is easy to determine the adjoint operators of Ψ_r and Ψ_o. It results that $\Psi_r^* : R^n \to RL_2(-\infty\ 0]$ and $\Psi_o^* : RL_2[0\ \infty) \to R^n$ are defined by

$$\Psi_r^* : x \mapsto \Psi_r^*x := u(t) := \begin{cases} B'e^{-A't}x & t \leq 0 \\ 0 & t > 0 \end{cases}$$

$$\Psi_o^* : y(t) \mapsto \Psi_o^*y(t) := x := \int_0^\infty e^{A'\sigma}C'yd\sigma$$

respectively. By exploiting the fact that $\Gamma_f = \Psi_o\Psi_r$ and taking into account the above expressions, it follows, in view of Theorem 2.9, that $\Gamma_f^* : RL_2[0\ \infty) \to RL_2(-\infty\ 0]$ is defined by

$$\Gamma_f^* : y(t) \mapsto \Gamma_f^*y(t) := u(t) := \begin{cases} B'e^{A't}\int_0^\infty e^{A'\tau}C'yd\tau & t \leq 0 \\ 0 & t > 0 \end{cases}$$

This operator is legitimated to be the time domain counterpart of the operator Γ_F^*. Actually, the Laplace transform of u equals the antistable projection of the product of $F^\sim(s)$ with the Laplace transform of y, i.e.

$$u_L = (\Gamma_f^*y)_L = \Gamma_F^*y_L \tag{2.46}$$

The operators Ψ_r and Ψ_o have rank equal to n, i.e. equal to the dimension of the minimal realization $\Sigma(A, B, C, 0)$. Moreover, since Ψ_r is surjective and Ψ_o is injective, it turns out that $\Gamma_f = \Psi_o\Psi_r$ and, obviously, Γ_F have rank equal to n. Therefore the self adjoint operator $\Gamma_F^*\Gamma_F$ has rank n and, thanks to Remark 2.14 it admits eigenvalues.

Remark 2.26 The operator $\Gamma_F^*\Gamma_F$ has a zero eigenvalue. Actually, write $F(s) \in RH_2$ as $F(s) = N(s)/\psi(s)$, where $N(s)$ is a polynomial matrix and $\psi(s)$ is the least common multiple of all denominators of the elements of $F(s)$. Moreover, let $H(s) \in RH_2^\perp$ be such that

$$G(s) := \psi(s)H(s) \in RH_2^\perp$$

Then,

$$\Gamma_F G(s) = \Pi_s F(s)G(s)$$
$$= \Pi_s \frac{N(s)}{\psi(s)}\psi(s)H(s)$$
$$= \Pi_s N(s)H(s) = 0$$

since $N(s)H(s) \in RH_2^\perp$. Consequently, $\Gamma_F^*\Gamma_F G(s) = 0 = 0G(s)$ so that $\lambda = 0$ is an eigenvalue. $\qquad\square$

Example 2.23 Let $F(s) = 1/(s+1)$. Taken $G(s) = 1/(s-1)$, it turns out that

$$\Gamma_F G(s) = \Pi_s F(s)G(s) = \frac{G(-1)}{s+1} = -\frac{1}{2(s+1)}$$

On the other hand,

$$\Gamma_F^*\Gamma_F G(s) = \Gamma_F^* \left[-\frac{1}{2(s+1)} \right] = \Pi_a F^\sim(s)\left[-\frac{1}{2(s+1)} \right] = \frac{1}{4}G(s)$$

so that $\lambda = 1/4$ is an eigenvalue and $G(s)$ an eigenvector of $\Gamma_F^*\Gamma_F$. $\qquad\square$

The operator $\Gamma_F^*\Gamma_F$ enjoys the interesting properties stated in the following result, for the proof of which the reader is referred to specialized texts.

Theorem 2.17 Let $F(s) \in RL_\infty$. The eigenvalues of $\Gamma_F^*\Gamma_F$ are real and nonnegative. The greatest of them is $\|\Gamma_F\|^2$.

Example 2.24 Consider the function $F(s) = 1/(s+1)$ already introduced in Example 2.23. It is easy to check that the equation $\Gamma_F^*\Gamma_F G(s) = \lambda G(s)$, with $G(s) \in RH_2^\perp$ and $G(s) \neq 0$, is satisfied only for $\lambda = 0$ or $\lambda = 0.25$. Hence, $\Gamma_F^*\Gamma_F$ has two distinct eigenvalues, which are real and nonnegative. The norm of $\Gamma_F^*\Gamma_F$ is equal to 0.25 and $\|\Gamma_F\| = 0.5$ (recall Theorem 2.9, point 2)). $\qquad\square$

The operators $\Psi_r\Psi_r^*$ and $\Psi_o^*\Psi_o$ map the space R^n into itself and, as such, can be represented by suitable (real) n-dimensional matrices. Actually, as for the former, it follows

$$x_r := \Psi_r\Psi_r^*x_0$$
$$= \int_{-\infty}^0 e^{-A\tau}BB'e^{-A'\tau}d\tau x_0$$
$$= \int_0^\infty e^{A\tau}BB'e^{A'\tau}d\tau x_0 = P_r x_0$$

where P_r again represents the unique solution of the Lyapunov equation in the unknown P (recall Lemma C.1)

$$0 = AP + PA' + BB' \tag{2.47}$$

Such a solution is also positive definite. Since the equation $\Psi_r \Psi_r^* x_0 = P_r x_0$ holds for any x_0, it results

$$\Psi_r \Psi_r^* = P_r \tag{2.48}$$

As for the operator $\Psi_o^* \Psi_o$ it follows that, for any x_0,

$$x_f := \Psi_o^* \Psi_o x_0 = \int_o^\infty e^{A'\tau} C' C e^{A\tau} d\tau x_0 = P_o x_0$$

where, thanks to what has been shown in Remark 2.20, matrix P_o is the unique solution of the Lyapunov equation in the unknown P

$$0 = PA + A'P + C'C \tag{2.49}$$

Such a solution is positive definite (recall points i) and iii) of Lemma C.1). Hence, since x_0 is arbitrary,

$$\Psi_o^* \Psi_o = P_o \tag{2.50}$$

The link previously expressed between Γ_F and Γ_f allows one to precisely state the following important result, which also provides a procedure for the computation of the eigenvalues of $\Gamma_F^* \Gamma_F$ and hence the norm of Γ_F.

Theorem 2.18 *The operator $\Gamma_F^* \Gamma_F$ and the matrix $P_r P_o$ share the same nonzero eigenvalues.*

Proof Let $\lambda \neq 0$ be an eigenvalue of $\Gamma_F^* \Gamma_F$. In view of eqs. (2.45),(2.46) it is easy to check that, if $0 \neq u_L \in RH_2^\perp$ is an associate eigenvector, then its inverse Laplace transform, $u \neq 0$, is an eigenvector of $\Gamma_f^* \Gamma_f$, associated with the same eigenvalue. Moreover, in view of Theorem 2.9, it is $\Gamma_f^* = \Psi_r^* \Psi_o^*$ so that

$$\Psi_r^* \Psi_o^* \Psi_o \Psi_r u = \lambda u \tag{2.51}$$

Premultiplying both sides of this equation by Ψ_r and defining $x_0 := \Psi_r u$, one obtains (recall eqs. (2.48),(2.50))

$$P_r P_o x_0 = \lambda x_0 \tag{2.52}$$

Notice that $x_0 \neq 0$. Actually, if not, then $\Psi_r u = 0$ and, from eq. (2.51), $\lambda u = 0$. Hence $\lambda \neq 0$ would imply $u = 0$, a contradiction.

Conversely, if $\lambda \neq 0$ is an eigenvalue of $P_r P_o$, then there exists $x_0 \neq 0$ such that eq. (2.52) holds. Premultiplying this expression by $\Psi_r^* P_o$, defining $u(t) := \Psi_r^* P_o x_0$, and recalling eqs. (2.48),(2.50), eq. (2.51) is derived. It is left to show that $u \neq 0$. If it were not so, eqs. (2.52),(2.48) would imply $\lambda x_0 = 0$, a contradiction. Hence u is an eigenvector of $\Gamma_f^* \Gamma_f$ associated with λ and its Laplace transform is an eigenvector of $\Gamma_F^* \Gamma_F$, associated with the same eigenvalue. $\square$

Remark 2.27 (Computation of the norm of the Hankel operator) As already said, Theorem 2.18 provides a procedure for the computation of $\|\Gamma_F\|$. It can be summarized in the following steps, with reference to the case in which $F(s) \in RH_2$. Such situation can be always be matched by using the arguments discussed in Remark 2.20.

1) Determine a minimal realization $\Sigma(A, B, C, 0)$ of $F(s)$.

2) Solve the two Lyapunov equations (2.47),(2.49).

3) Compute the greatest eigenvalue λ_M^2 of $P_r P_o$

4) $\|\Gamma_F\| = \lambda_M$

Example 2.25 Take again the function $F(s) = 1/(s + 1)$ considered in Examples 2.23, 2.24. Being $F(s) = \Sigma(-1, 1, 1, 0)$, eqs. (2.47),(2.49) have solutions $P_r = P_o = 0.5$, so that $P_r P_o = 0.25$ and $\|\Gamma_F\| = 0.5$.   □

At the end of the present section, it is presented an important result, the so called *Nehari theorem*, which sets at the basis of an operatorial technique for the solution of the control problem in the RH_∞ context, treated in Chapter 5. The proof of this result is not simple and therefore is not reported here.

Theorem 2.19 (Nehari theorem) *Let $F(s) \in RL_\infty$. There exists $X^o(s) \in RH_\infty$ such that*

$$\inf_{X(s) \in RH_\infty} \|F(s) - X(s)\|_\infty = \|F(s) - X^o(s)\|_\infty = \|\Gamma_{F^\sim}\|$$

The Nehari theorem concerns the problem of finding a stable function $X(s)$ which approximate an assigned function of RL_∞ by minimizing the *distance* between $X(s)$ and $F(s)$. Such a distance is defined as the RL_∞ norm of the difference $F(s) - X(s)$. The theorem comes out with two important conclusions: first, the existence of a function $X^o(s) \in RH_\infty$ which represents the best approximation of $F(s)$ and, second, that the minimal distance is given by the norm of the Hankel operator with symbol $F^\sim$. It should be apparent, consistently with such an interpretation, that if $F(s)$ is stable, the minimal distance is zero, since one can choose $X^o(s) = F(s)$. In general, the optimal approximation is not unique. A way to compute an optimal approximation will be presented in Chapter 5, in the case where $F(s)$ is a scalar.

Example 2.26 Consider the function $F(s) = \gamma/(s - \beta)$ with $\beta > 0$ and $\gamma > 0$. Then, $F^\sim(s) = -\gamma/(s + \beta)$. Based on Remark 2.27 one finds that $\|\Gamma_{F^\sim}\| = |\gamma/2\beta|$.   □

2.9   Notes and references

More about the material of this chapter can be found in many places. Restricting the attention to system and control theory point of view (which is the one mostly pursued in the present book), it is worth quoting the following sources with reference to the various sections. Sections 2.3 and 2.4: the books of Kailath [29], Maciejowski [43] and Vidyasagar [60]. Section 2.5: besides the books [29], [60], the papers by MacFarlane and Karcanias [42] and Kouvaritakis and MacFarlane [33], [34]. Section 2.6: the book of Lawson and Hanson [38]. Section 2.7: any text of functional analysis, for instance that of Rudin [55]. Section 2.8: the book of Francis [19], the paper by Boyd et al. [9] on the computation of the H_∞ norm and the book of Power [52] for the Nehari theorem.

Chapter 3

Feedback Systems Stability

3.1 Introduction

One of the most significant problems of linear control theory is no doubt that of characterizing the set of all controllers which stabilize a given system. The present chapter is devoted to this problem, which is also conceptually linked with the results presented in the forthcoming Chapters 4 and 5.

In order to avoid any possible misunderstanding, it is well advisable to place in the right context the problem covered in the present treatment. Therefore, consider the system shown in fig. 3.1 where Σ_1 is a given time invariant and finite dimensional linear system and Σ_2 is a controller which receives informations from the system through the output variable y only and can drive the system through the control variable u. Its duty is rendering the feedback connected system in fig. 3.1 (asymptotically) stable. An obvious *necessary* condition for a stabilizing regulator to exist is *stabilizability and detectability* of system Σ_1. This condition is also sufficient if Σ_2 can be chosen in the class $\mathcal{C}_L$ of linear time invariant finite dimensional systems. This is precisely the class to which the present discussion will be limited.

The actual determination of an element of $\mathcal{C}_{LS}(\Sigma_1)$, the subset of $\mathcal{C}_L$ constituted by the stabilizing controllers for Σ_1, can be worked out by resorting to a number of classical techniques. Among them, it is worth recalling the optimal *linear quadratic Gaussian* control or the *pole assignment* technique. Greatly more difficult is rather the problem of individuating the whole set $\mathcal{C}_{LS}(\Sigma_1)$, in the sense of establishing a significant correspondence between an element of the set and a suitable *free parameter*.

In Section 3.4 it will be shown how the set $\mathcal{C}_{LS}(\Sigma_1)$ can be parametrized as a

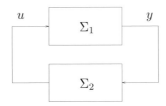

Figure 3.1: The feedback connection of two systems

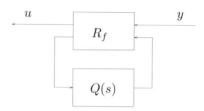

Figure 3.2: The set $\mathcal{C}_{LS}(\Sigma_1)$

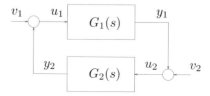

Figure 3.3: The system Σ - feedback connection of $G_1(s)$ and $G_2(s)$

function of a free parameter $Q(s)$. Specifically, it will be proved that: i) each element of $\mathcal{C}_{LS}(\Sigma_1)$ can be represented accordingly to the block scheme in fig. 3.2, where R_f is a system which is determined from the given system Σ_1 once for ever, whereas $Q(s)$ is a linear system belonging to a suitable subset of RH_∞; ii) chosen an element, one obtains an element of $\mathcal{C}_{LS}(\Sigma_1)$. In order to present the relevant results, some preliminary facts are now introduced. In particular, Section 3.2 is devoted to discuss the basic relationships between internal and *BIBO* (external) stability of a feedback system, whereas in Section 3.3 the important concept of *double coprime factorization* is introduced.

3.2   Internal and external stability

Consider system Σ depicted in fig. 3.3 where each one of the two blocks with transfer functions $G_1(s)$ and $G_2(s)$ is a detectable and stabilizable system (otherwise Σ should never be stable). The aim of this section is presenting some simple results which establish precise connections between the (internal) stability of Σ and the (external) stability of some transfer functions defined in the block-scheme of fig. 3.3. Notice that the (stable) dynamic matrix of the unreachable and/or unobservable parts of the two subsystems which constitute Σ do not affect anyone of the transfer functions which can be defined in the scheme of fig. 3.3. For this reason reference will be made, without loss of generality, to minimal realizations of both $G_1(s)$ and $G_2(s)$, i.e.

$$G_i(s) := C_i(sI - A_i)^{-1}B_i + D_i , \quad i = 1, 2$$

with

$$\dot{x}_i = A_i x_i + B_i u_i , \quad i = 1, 2$$
$$y_i = C_i x_i + D_i u_i , \quad i = 1, 2$$
$$u_1 = v_1 + y_2$$

$$u_2 = v_2 + y_1$$

where, for $i = 1, 2$, the pair (A_i, B_i) is reachable and the pair (A_i, C_i) is observable. Obviously, the above equations make sense only provided that the matrix $\Delta_{21} := (I - D_2 D_1)$ (or, equivalently, matrix $\Delta_{12} := (I - D_1 D_2)$) is nonsingular. Letting $x := [x_1' \ x_2']' \in R^n$, $u := [u_1' \ u_2']'$, $v := [v_1' \ v_2']'$, a realization of Σ is given by

$$\dot{x} = Ax + Bv$$
$$u = Cx + Dv$$

with

$$A := \begin{bmatrix} A_1 + B_1 \Delta_{21}^{-1} D_2 C_1 & B_1 \Delta_{21}^{-1} C_2 \\ B_2 \Delta_{12}^{-1} C_1 & A_2 + B_2 \Delta_{12}^{-1} D_1 C_2 \end{bmatrix}$$

$$B := \begin{bmatrix} B_1 \Delta_{21}^{-1} & B_1 \Delta_{21}^{-1} D_2 \\ B_2 \Delta_{12}^{-1} D_1 & B_2 \Delta_{12}^{-1} \end{bmatrix}$$

$$C := \begin{bmatrix} \Delta_{21}^{-1} D_2 C_1 & \Delta_{21}^{-1} C_2 \\ \Delta_{12}^{-1} C_1 & \Delta_{12}^{-1} D_1 C_2 \end{bmatrix}$$

$$D := \begin{bmatrix} \Delta_{21}^{-1} & \Delta_{21}^{-1} D_2 \\ \Delta_{12}^{-1} D_1 & \Delta_{12}^{-1} \end{bmatrix}$$

having exploited Lemma B.9. Letting $T(u, v; s) := C(sI - A)^{-1}B + D$ denote the transfer function from the input v to the output u, the following result can be proved.

Theorem 3.1 *The system Σ in fig. 3.3 is internally stable if and only if $T(u, v; s) \in RH_\infty$.*

Proof If matrix A is stable, then, obviously, $T(u, v; s) \in RH_\infty$. In order to prove the converse statement it will be shown that all the eigenvalues of A coincide with the poles of $T(u, v; s)$, or, in other words that the four matrices (A, B, C, D) constitute a minimal realization of $T(u, v; s)$. The proof is worked out by exploiting the *PBH* test (see Lemmas D.1 and D.3).

Recall that the pair (A, C) is observable if and only if

$$\text{rank}[P_C(\lambda)] := \text{rank} \begin{bmatrix} \begin{bmatrix} \lambda I - A \\ C \end{bmatrix} \end{bmatrix} = n , \quad \forall \lambda$$

Moreover, the matrix

$$T_o := \begin{bmatrix} I & 0 & B_1 & 0 \\ 0 & I & 0 & B_2 \\ 0 & 0 & I & -\Delta_{21}^{-1} D_2 \Delta_{12} \\ 0 & 0 & -\Delta_{12}^{-1} D_1 \Delta_{21} & I \end{bmatrix}$$

is nonsingular since the Schur formula for the computation of the determinant of a block matrix (see Lemma B.13) gives

$$\det \begin{bmatrix} \begin{bmatrix} I & -\Delta_{21}^{-1} D_2 \Delta_{12} \\ -\Delta_{12}^{-1} D_1 \Delta_{21} & I \end{bmatrix} \end{bmatrix} = \det[I - \Delta_{12}^{-1} D_1 \Delta_{21} \Delta_{21}^{-1} D_2 \Delta_{12}] = \det[\Delta_{12}]$$

Thus,

$$\mathrm{rank}[P_C(\lambda)] = \mathrm{rank}[T_o P_C(\lambda)]$$

$$= \mathrm{rank}\left[\left[\begin{array}{cc} \lambda I - A_1 & 0 \\ 0 & \lambda I - A_2 \\ 0 & C_2 \\ C_1 & 0 \end{array}\right]\right] = n, \quad \forall \lambda$$

since, for $i = 1, 2$, the pair (A_i, C_i) is observable. In a similar way the pair (A, B) is proved to be reachable. To this aim, let

$$T_r := \left[\begin{array}{cccc} I & 0 & 0 & 0 \\ 0 & I & 0 & 0 \\ 0 & -C_2 & -\Delta_{21} D_2 \Delta_{12}^{-1} & I \\ -C_1 & 0 & I & -\Delta_{12} D_1 \Delta_{21}^{-1} \end{array}\right]$$

Simple computations show that (recall also Lemma B.9)

$$P_B(\lambda) T_r = \left[\begin{array}{cc} \lambda I - A & -B \end{array}\right] T_r$$

$$= \left[\begin{array}{cccc} \lambda I - A_1 & 0 & 0 & -B_1 \\ 0 & \lambda I - A_2 & -B_2 & 0 \end{array}\right]$$

In view of the *PBH* test the pair (A, B) is reachable if and only if $\mathrm{rank}[P_B(\lambda)] = \mathrm{rank}[T_r P_B(\lambda)] = n, \forall \lambda$. This condition holds thanks to the reachability of the pair (A_i, B_i), $i = 1, 2$. □

Example 3.1 Let $G_1(s) = \bar{G}_1(s) := -1/(s - 1)$ and $G_2(s) = (s - 1)/(s + 1)$. The dynamic matrix of a particular realization of system Σ is

$$A = \left[\begin{array}{cc} 0 & -2 \\ -1 & -1 \end{array}\right]$$

the eigenvalues of which are $\lambda_1 = -2$ and $\lambda_2 = 1$, so that Σ is not internally stable. Consistently, the matrix

$$T(u, v; s) = \left[\begin{array}{cc} (s + 1)/(s + 2) & (s - 1)/(s + 2) \\ -(s + 1)/(s - 1)(s + 2) & (s + 1)/(s + 2) \end{array}\right]$$

does not belong to RH_∞. On the contrary, if $G_1(s) = \widetilde{G}_1(s) := 1/(s + 2)$, the dynamic matrix of a particular realization of Σ is

$$A = \left[\begin{array}{cc} -1 & -2 \\ 1 & -1 \end{array}\right]$$

whose eigenvalues are $\lambda_{1,2} = -1 \pm j\sqrt{2}$, so that the system is internally stable. Consistently, the matrix

$$T(u, v; s) = \frac{1}{s^2 + 2s + 3} \left[\begin{array}{cc} (s + 1)(s + 2) & (s - 1)(s + 2) \\ (s + 1) & (s + 1)(s + 2) \end{array}\right]$$

is an element of RH_∞. □

This result can be specialized to the case where one of the two transfer functions $G_1(s)$ or $G_2(s)$ is stable, leading to the following theorem where $T_{ij}(u, v; s)$, $i = 1, 2$, $j = 1, 2$ denotes the transfer function from v_j to u_i.

Theorem 3.2 *Let $G_2(s)$ be stable. Then the system Σ in fig. 3.3 is internally stable if and only if $T_{21}(u, v; s) \in RH_\infty$.*

Proof In view of Theorem 3.1, the present theorem is proved once it is shown that stability of $T_{21}(u, v; s)$ implies stability of $T_{11}(u, v; s)$, $T_{12}(u, v; s)$, $T_{22}(u, v; s)$. By exploiting Lemma B.9 one has

$$I + G_2(s)T_{21}(u, v; s) = I + G_2(s)G_1(s)[I - G_2(s)G_1(s)]^{-1}$$
$$= [I - G_2(s)G_1(s)]^{-1} = T_{11}(u, v; s)$$

so that, if $T_{21}(u, v; s) \in RH_\infty$, also $T_{11}(u, v; s) \in RH_\infty$. Similarly, by exploiting once more Lemma B.9,

$$I + T_{21}(u, v; s)G_2(s) = I + [I - G_1(s)G_2(s)]^{-1}G_1(s)G_2(s)$$
$$= [I - G_1(s)G_2(s)]^{-1} = T_{22}(u, v; s)$$

so that, again, if $T_{21}(u, v; s) \in RH_\infty$, $T_{22}(u, v; s) \in RH_\infty$, as well. Finally, being $T_{12}(u, v; s) = T_{11}(u, v; s)G_2(s)$ then stability of $T_{11}(u, v; s)$ entails stability of $T_{12}(u, v; s)$. □

Example 3.2 Again consider the functions $G_1(s)$ and $G_2(s)$ defined in Example 3.1. Being $G_2(s)$ stable, the internal stability of Σ can be tested by checking the stability of $T_{21}(u, v; s)$ only. When $G_1(s) = \bar{G}_1(s)$, $T_{21}(u, v; s)$ is not stable, whereas $T_{21}(u, v; s) \in RH_\infty$ when $G_1(s) = \widetilde{G}_1(s)$, consistently with Σ being stable only in the second case. □

A further specialization of the above results can be found when both transfer functions $G_1(s)$ and $G_2(s)$ are stable.

Theorem 3.3 *Assume that $G_i(s) \in RH_\infty$, $i = 1, 2$. Then the system Σ depicted in fig. 3.3 is internally stable if and only if*

$$\det[I - G_1(s)G_2(s)] \neq 0 , \quad Re(s) \geq 0$$

Proof Preliminarily notice that the transfer function of the system

$$\Sigma_a := \left[\begin{array}{cc|c} A_1 & B_1C_2 & B_1D_2 \\ 0 & A_2 & B_2 \\ \hline -C_1 & -D_1C_2 & \Delta_{12} \end{array} \right]$$

is $I - G_1(s)G_2(s)$. By exploiting Lemma B.9 and recalling Definition 2.14 it is easy to check that the dynamic matrix of Σ_a^{-1} is precisely A. Moreover, any common root of the polynomial $\pi_{zt}(s)$ of the transmission zeros and $\pi_p(s)$ of the poles of system $I - G_1(s)G_2(s)$ must necessarily lie in the open left half plane, since such a system is stable.

Now suppose that $\det[I - G_1(s)G_2(s)] \neq 0$, $Re(s) \geq 0$, so that, in view of Remark 2.7,

$$\frac{\pi_{zt}(s)}{\pi_p(s)} = \det[I - G_1(s)G_2(s)] \neq 0 , \quad Re(s) \geq 0$$

which implies that all transmission zeros of Σ_a have negative real part. The stability of Σ_a (notice that A_i is stable, $i = 1, 2$, since $G_i(s) \in RH_\infty$, $i = 1, 2$, and the associated realizations are minimal) entails that any invariant zero of it which is not a transmission zero must lie in the open left half plane. Hence, thanks to Theorem

2.7, all the eigenvalues of Σ_a^{-1} have negative real part, matrix A is stable and in turn system Σ is internally stable.

Conversely, if system Σ is internally stable, then matrix A is stable and system Σ_a^{-1} is internally stable, so that, in view of Theorem 2.7, all invariant zeros of system Σ_a are in the open left half plane. Since the set of transmission zeros is contained in the set of invariant zeros (Theorem 2.4), all the transmission zeros of system Σ_a have negative real part. Then, from Remark 2.7, it follows

$$\det[I - G_1(s)G_2(s)] \neq 0 , \quad Re(s) \geq 0$$

$\square$

Example 3.3 Consider the functions $G_1(s) = \widetilde{G}_1(s)$ and $G_2(s)$ defined in Example 3.1. Both of them are stable and system Σ is internally stable if and only if $\det[I - G_1(s)G_2(s)]$ is not zero in the closed right half plane. Actually, it is

$$\det[I - \widetilde{G}_1(s)G_2(s)] = \frac{s^2 + 2s + 3}{(s+1)(s+2)}$$

the zeros of which have negative real part, consistently with the (already established) internal stability of system Σ.

$\square$

3.3 Double coprime factorizations

The definitions of right and left coprimeness for rational matrices which, together with some related properties, have been presented in Section 2.4, are now exploited in order to introduce an intermediate result on the way of presenting the so called parametrization of all stabilizing controllers in the next section.

Theorem 3.4 *Let $G(s)$ be any proper rational matrix. Then there exist eight matrices $M(s)$, $N(s)$, $\hat{M}(s)$, $\hat{N}(s)$, $X(s)$, $Y(s)$, $\hat{X}(s)$, $\hat{Y}(s)$, all belonging to RH_∞, such that*

a)

$$G(s) = N(s)M^{-1}(s) = \hat{M}^{-1}(s)\hat{N}(s)$$

b)

$$\begin{bmatrix} \hat{X}(s) & -\hat{Y}(s) \\ -\hat{N}(s) & \hat{M}(s) \end{bmatrix} \begin{bmatrix} M(s) & Y(s) \\ N(s) & X(s) \end{bmatrix} = I$$

The matrices $M(s)$, $N(s)$, $\hat{M}(s)$, $\hat{N}(s)$ constitute a double coprime factorization of $G(s)$.

Proof First notice that, if the statement of the theorem is correct, the matrices $M(s)$ and $N(s)$ are right coprime in view of Lemma 2.5, while $\hat{M}(s)$ and $\hat{N}(s)$ are left coprime thanks to the "left version" of the same lemma.

Now let $G(s) = \Sigma(A, B, C, D)$ and assume that the triple (A, B, C) is stabilizable and detectable. Further, let F and H be any two matrices such that $(A + BF)$ and $(A + HC)$ are stable and set $\bar{C} := C + DF$, $\bar{B} := B + HD$. Define

$$M(s) := \left[\begin{array}{c|c} A + BF & B \\ \hline F & I \end{array} \right] \quad , \quad N(s) := \left[\begin{array}{c|c} A + BF & B \\ \hline \bar{C} & D \end{array} \right] \qquad (3.1)$$

$$X(s) := \left[\begin{array}{c|c} A + BF & H \\ \hline -\bar{C} & I \end{array}\right] \quad , \quad Y(s) := \left[\begin{array}{c|c} A + BF & H \\ \hline -F & 0 \end{array}\right] \qquad (3.2)$$

$$\hat{M}(s) := \left[\begin{array}{c|c} A + HC & H \\ \hline C & I \end{array}\right] \quad , \quad \hat{N}(s) := \left[\begin{array}{c|c} A + HC & B \\ \hline C & D \end{array}\right] \qquad (3.3)$$

$$\hat{X}(s) := \left[\begin{array}{c|c} A + HC & B \\ \hline -F & I \end{array}\right] \quad , \quad \hat{Y}(s) := \left[\begin{array}{c|c} A + HC & H \\ \hline -F & 0 \end{array}\right] \qquad (3.4)$$

Point a) It is obvious that the four matrices $M(s)$, $N(s)$, $\hat{M}(s)$, $\hat{N}(s)$ belong to RH_∞. Moreover, $M(s)$ and $\hat{M}(s)$ have inverses since

$$\lim_{s\to\infty} M(s) = I , \quad \lim_{s\to\infty} \hat{M}(s) = I$$

Now consider system Σ_1 (with input v and outputs y and u) defined by

$$\dot{x} = (A + BF)x + Bv$$
$$y = \bar{C}x + Dv$$
$$u = Fx + v$$

Thus, Σ_1 is nothing but system $G(s)$ after the control law $u = Fx + v$ has been implemented. It follows

$$u_{L0} = M(s)v_L$$
$$y_{L0} = N(s)v_L$$

so that $y_{L0} = G(s)u_{L0} = N(s)M^{-1}(s)u_{L0}$. Hence

$$G(s) = N(s)M^{-1}(s) \qquad (3.5)$$

Consider system Σ_2 (with inputs u and y and output η) defined by

$$\dot{\vartheta} = A\vartheta + Bu + H\eta$$
$$\eta = C\vartheta + Du - y$$

Thus, Σ_2 is nothing but a state observer for the system with transfer function $G(s)$, so that η is the output observation error, namely

$$\eta = C(\vartheta - x)$$

which is well known not to depend on u, since $\dot{\vartheta} - \dot{x} = (A + HC)(\vartheta - x)$. Therefore,

$$\eta_{L0} = [C(sI - (A + HC))^{-1}\bar{B} + D]u_L -$$
$$-[C(sI - (A + HC))^{-1}H + I]y_{L0}$$
$$= \hat{N}(s)u_L - \hat{M}(s)y_{L0}$$
$$= [\hat{N}(s) - \hat{M}(s)G(s)]u_L = 0$$

which implies

$$G(s) = \hat{M}^{-1}(s)\hat{N}(s) \qquad (3.6)$$

Point *a)* is then proved in view of eqs. (3.5),(3.6).

Point b) Consider system Σ_3 (with inputs u and y and output γ) defined by

$$\dot{\mu} = (A + HC)\mu + \bar{B}u - Hy$$
$$\gamma = u - F\mu$$

so that $\gamma_{L0} = \hat{X}(s)u_L - \hat{Y}(s)y_L$. Then in the series connection $\Sigma_3\Sigma_1$ it is

$$\gamma_{L0} = \hat{X}(s)u_{L0} - \hat{Y}(s)y_{L0} = [\hat{X}(s)M(s) - \hat{Y}(s)N(s)]v_L$$

However, $\dot{\mu} - \dot{x} = (A + HC)(\mu - x)$ and $\gamma = v - F(\mu - x)$ which implies that the transfer function from v to γ is I, that is

$$\hat{X}(s)M(s) - \hat{Y}(s)N(s) = I \tag{3.7}$$

Now consider system Σ_4 (with input w and outputs y and u) defined by

$$\dot{z} = (A + BF)z + Hw$$
$$y = \bar{C}z - w$$
$$u = Fz$$

so that $y_{L0} = -X(s)w_L$ and $u_{L0} = -Y(s)w_L$. Then in the series connection $\Sigma_2\Sigma_4$ it is

$$\eta_{L0} = \hat{N}(s)u_{L0} - \hat{M}(s)y_{L0} = [-\hat{N}(s)Y(s) + \hat{M}(s)X(s)]w_L$$

However, $\dot{z} - \dot{\vartheta} = (A + HC)(z - \vartheta)$ and $\eta = w - C(z - \vartheta)$ which implies that the transfer function from w to η is I, that is

$$- \hat{N}(s)Y(s) + \hat{M}(s)X(s) = I \tag{3.8}$$

Finally, in the series connection $\Sigma_3\Sigma_4$ it is

$$\gamma_{L0} = \hat{X}(s)u_{L0} - \hat{Y}(s)y_{L0} = [-\hat{X}(s)Y(s) + \hat{Y}(s)X(s)]w_L$$

However it is $\dot{z} - \dot{\mu} = (A + HC)(z - \mu)$ and $\gamma = F(z - \mu)$ which implies that the transfer function from w to γ is 0,

$$- \hat{X}(s)Y(s) + \hat{Y}(s)X(s) = 0 \tag{3.9}$$

From eqs. (3.5),(3.6) it follows that $\hat{M}(s)N(s) - \hat{N}(s)M(s) = 0$ which, together with eqs. (3.7)-(3.9), proves point *b)*.   □

Remark 3.1 The proof of Theorem 3.4 outlines the fact that, corresponding to a given function $G(s)$, there exists an *infinite* number of distinct double coprime factorizations in RH_∞ (recall that matrices F and H can be chosen in an almost arbitrary way).   □

Example 3.4 Consider the rational matrix

$$G(s) = \begin{bmatrix} (s^2 + s - 1)/(s^3 - s) & s/(s^2 - 1) \\ (2s^2 - 1)/(s^3 - s) & s^2/(s^2 - 1) \end{bmatrix}$$

which admits the quadruple

$$A = \begin{bmatrix} 0 & 1 & 0 \\ 1 & 0 & 0 \\ 0 & 0 & 0 \end{bmatrix}, \quad B = \begin{bmatrix} 0 & 1 \\ 1 & 0 \\ 1 & 0 \end{bmatrix}$$

$$C = \begin{bmatrix} 1 & 0 & 1 \\ 0 & 1 & 1 \end{bmatrix}, \quad D = \begin{bmatrix} 0 & 0 \\ 0 & 1 \end{bmatrix}$$

as a minimal realization. Chosen

$$F := \begin{bmatrix} 0 & 0 & -1 \\ -1 & -2 & 0 \end{bmatrix}, \quad H := \begin{bmatrix} -4 & 0 \\ -4 & 0 \\ 1 & 0 \end{bmatrix}$$

it is easy to check the stability of matrices $A + BF$ and $A + HC$ and then computing the eight matrices referred to in Theorem 3.4. It results

$$M(s) = \begin{bmatrix} s/(s+1) & 0 \\ -s(2s+1)/(s+1)(s^2+s+1) & (s^2-1)/(s^2+s+1) \end{bmatrix}$$

$$N(s) = \begin{bmatrix} (s^2+1)/(s+1)(s^2+s+1) & s/(s^2+s+1) \\ 1/(s^2+s+1) & s^2/(s^2+s+1) \end{bmatrix}$$

$$X(s) = \begin{bmatrix} (s^3+5s^2+s-5)/(s+1)(s^2+s+1) & 0 \\ -(9s+5)/(s^2+s+1) & 1 \end{bmatrix}$$

$$Y(s) = \begin{bmatrix} 1/(s+1) & 0 \\ -(12s^2+26s+13)/(s+1)(s^2+s+1) & 0 \end{bmatrix}$$

$$\hat{M}(s) = \begin{bmatrix} s(s-1)/(s+1)^2 & 0 \\ -(3s+1)/(s+1)^2 & 1 \end{bmatrix}$$

$$\hat{N}(s) = \begin{bmatrix} (s^2+s-1)/(s+1)^3 & s^2/(s+1)^3 \\ (2s^2+3s)/(s+1)^3 & (s^3+3s^2+s)/(s+1)^3 \end{bmatrix}$$

$$\hat{X}(s) = \begin{bmatrix} (s^3+4s^2+7s+5)/(s+1)^3 & s/(s+1)^3 \\ (2s^2-5s-13)/(s+1)^3 & (s^3+4s^2-4s-1)/(s+1)^3 \end{bmatrix}$$

$$\hat{Y}(s) = \begin{bmatrix} (s-1)/(s+1)^2 & 0 \\ -12s/(s+1)^2 & 0 \end{bmatrix}$$

These matrices verify Theorem 3.4. □

Theorem 3.4 allows one stating an *equivalence* condition between internal and external stability for a feedback system as the one depicted in fig. 3.3 which is different from the condition presented in Theorem 3.1. As done in the preceding section, it is assumed that a minimal realization of the two systems $G_1(s)$ and $G_2(s)$ is considered and their feedback connection is well defined (the matrix $(I - D_1 D_2)$ is nonsingular, $D_i := \lim_{s \to \infty} G_i(s), i = 1, 2$). For $i = 1, 2$, let $N_i(s)$ and $M_i(s)$ be elements of RH_∞ such that $G_i(s) = N_i(s)M_i^{-1}(s)$ with $N_i(s)$ and $M_i(s)$ right coprime. Then the system in fig. 3.3 can be represented as in fig. 3.4 as well. With reference to this figure let $z := [z_1' \ z_2']'$, $u := [u_1' \ u_2']'$ and $v := [v_1' \ v_2']'$. Then the following result holds.

Theorem 3.5 *The system in fig. 3.4 is internally stable if and only if the transfer function $T(z, v; s)$ from the input v to the output z is stable.*

Proof In view of Theorem 3.1, the proof consists in showing that the matrix $T(z, v; s)$ is stable if and only if the transfer function $T(u, v; s)$ from the input v to the output u is stable.

Sufficiency With reference to fig. 3.4 it is

$$u_{L0} = T(z, v; s)v_L$$

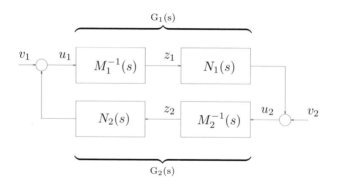

Figure 3.4: Feedback connection of two systems

$$= \begin{bmatrix} 0 & N_2(s) \\ N_1(s) & 0 \end{bmatrix} z_{L0} + v_L$$

$$= \left[\begin{bmatrix} 0 & N_2(s) \\ N_1(s) & 0 \end{bmatrix} T(z, v; s) + I \right] v_L$$

Therefore, if $T(z, v; s) \in RH_\infty$ also $T(u, v; s) \in RH_\infty$, since, for $i = 1, 2$, $N_i(s) \in RH_\infty$.

Necessity Being $N_i(s)$ and $M_i(s)$, $i = 1, 2$, right coprime, there exist matrices $X_i(s)$ and $Y_i(s)$, $i = 1, 2$, which are elements of RH_∞ and such that

$$\begin{bmatrix} X_1(s) & 0 \\ 0 & X_2(s) \end{bmatrix} \begin{bmatrix} M_1(s) & 0 \\ 0 & M_2(s) \end{bmatrix} +$$
$$+ \begin{bmatrix} Y_1(s) & 0 \\ 0 & Y_2(s) \end{bmatrix} \begin{bmatrix} N_1(s) & 0 \\ 0 & N_2(s) \end{bmatrix} = I$$

so that

$$\begin{bmatrix} X_1(s) & 0 \\ 0 & X_2(s) \end{bmatrix} \begin{bmatrix} M_1(s) & 0 \\ 0 & M_2(s) \end{bmatrix} z_{L0} +$$
$$+ \begin{bmatrix} Y_1(s) & 0 \\ 0 & Y_2(s) \end{bmatrix} \begin{bmatrix} N_1(s) & 0 \\ 0 & N_2(s) \end{bmatrix} z_{L0} = z_{L0}$$

from which it follows

$$\begin{bmatrix} X_1(s) & 0 \\ 0 & X_2(s) \end{bmatrix} u_{L0} + \begin{bmatrix} 0 & Y_1(s) \\ Y_2(s) & 0 \end{bmatrix} (u_{L0} - v_L) =$$
$$= \left[\begin{bmatrix} X_1(s) & Y_1(s) \\ Y_2(s) & X_2(s) \end{bmatrix} T(u, v; s) - \begin{bmatrix} 0 & Y_1(s) \\ Y_2(s) & 0 \end{bmatrix} \right] v_L$$
$$= T(z, v; s) v_L$$

Being $X_i(s)$, $Y_i(s)$, $i = 1, 2$, and $T(u, v; s)$ all elements of RH_∞, also $H(s)$ must be such.   □

Thanks to Theorem 3.5 it is possible to prove a result useful for the discussion in the next section. It clarifies the circumstances under which a system with transfer function $G_1(s)$ is internally stabilized by a system with transfer function $G_2(s)$ when they are feedback connected as shown in fig. 3.4, where, for $i = 1, 2$, matrices $N_i(s)$ and $M_i(s)$ possess the above mentioned properties.

Theorem 3.6 *With reference to fig. 3.4, system $G_2(s)$ internally stabilizes system $G_1(s)$, that is, the resulting system is internally stable, if and only if $G^{-1}(s) \in RH_\infty$, where*

$$G(s) := \begin{bmatrix} M_1(s) & N_2(s) \\ N_1(s) & M_2(s) \end{bmatrix}$$

Proof Preliminarily observe that matrix $G(s)$ is nonsingular, so that the statement makes sense. Indeed, one can write

$$\begin{bmatrix} M_1(s) & N_2(s) \\ N_1(s) & M_2(s) \end{bmatrix} = \begin{bmatrix} I & G_2(s) \\ G_1(s) & I \end{bmatrix} \begin{bmatrix} M_1(s) & 0 \\ 0 & M_2(s) \end{bmatrix}$$

For the first matrix on the right hand side of this equation it is

$$\lim_{s \to \infty} \det \left[\begin{bmatrix} I & G_2(s) \\ G_1(s) & I \end{bmatrix} \right] = \lim_{s \to \infty} \det[I - G_1(s)G_2(s)]$$
$$= \det[I - D_1 D_2]$$

having exploited the Schur formula for the determinant of a block matrix (see Lemma B.13). Thus, matrix $G(s)$ is nonsingular since it is the product of two nonsingular matrices (recall that, for $i = 1, 2$, $M_i^{-1}(s)$ exists).

By taking into account fig. 3.4 one has

$$\begin{bmatrix} M_1(s) & -N_2(s) \\ -N_1(s) & M_2(s) \end{bmatrix} z_{L0} = \begin{bmatrix} M_1(s)z_{1L0} - N_2(s)z_{2L0} \\ -N_1(s)z_{1L0} + M_2(s)z_{2L0} \end{bmatrix}$$
$$= \begin{bmatrix} u_{1L0} + v_{1L} - u_{1L0} \\ v_{2L} - u_{2L0} + u_{2L0} \end{bmatrix} = v_L$$

so that, thanks to Theorem 3.5, $G_2(s)$ internally stabilizes $G_1(s)$ if and only if

$$T(z, v; s) := \begin{bmatrix} M_1(s) & -N_2(s) \\ -N_1(s) & M_2(s) \end{bmatrix}^{-1} \in RH_\infty$$

Finally, $T(z, v; s) \in RH_\infty$ if and only if $G^{-1}(s) \in RH_\infty$. In fact,

$$T(z, v; s) = \begin{bmatrix} I & 0 \\ 0 & -I \end{bmatrix} G^{-1}(s) \begin{bmatrix} I & 0 \\ 0 & -I \end{bmatrix}$$

$\square$

Example 3.5 Consider the functions $G_1(s)$ and $G_2(s)$ defined in Example 3.1. Being $G_2(s)$ stable one can set (see the proof of Theorem 3.4)

$$N_2(s) = \frac{s-1}{s+1}, \quad M_2(s) = 1$$

whereas for $G_1(s) = \bar{G}_1(s)$, the choice $F = -3$ entails (see the proof of Theorem 3.4)

$$N_1(s) = \bar{N}_1(s) = -\frac{1}{s+2}, \quad M_1(s) = \bar{M}_1(s) = \frac{s-1}{s+2}$$

Finally, being $\widetilde{G}_1(s)$ stable, for $G_1(s) = \widetilde{G}_1(s)$, one can set

$$N_1(s) = \widetilde{N}_1(s) = \frac{1}{s+2}, \quad M_1(s) = \widetilde{M}_1(s) = 1$$

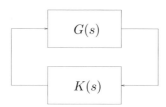

Figure 3.5: Feedback connection of two systems

Thus,

$$\bar{G}^{-1}(s) = \left[\begin{array}{cc} (s-1)/(s+2) & (s-1)/(s+1) \\ -1/(s+2) & 1 \end{array} \right]^{-1}$$

$$= \left[\begin{array}{cc} (s+1)/(s-1) & -1 \\ (s+1)/(s^2+s-2) & (s+1)/(s+2) \end{array} \right]$$

which does not belong to RH_∞, consistently with $\bar{G}_1(s)$ not being stabilized by $G_2(s)$, whereas

$$\widetilde{G}^{-1}(s) = \left[\begin{array}{cc} 1 & (s-1)/(s+1) \\ 1/(s+2) & 1 \end{array} \right]^{-1}$$

$$= \left[\begin{array}{cc} (s^2+3s+2)/(s^2+2s+3) & -(s^2+s-2)/(s^2+2s+3) \\ -(s+1)/(s^2+2s+3) & (s^2+3s+2)/(s^2+2s+3) \end{array} \right]$$

belongs to RH_∞, consistently with $\widetilde{G}_1(s)$ being stabilized by $G_2(s)$.   □

3.4   The set of stabilizing controllers

The so called parametrization of *all* stabilizing controllers is presented in this section. This result is particularly significant as it allows to completely characterize the set of all linear, finite dimensional and time invariant controllers $K(s)$ which stabilize a given system with transfer function $G(s)$, if connected to it according to the block-scheme of fig. 3.5. It will be shown that the system in fig. 3.5 is internally stable if and only if the controller transfer function can be given a very simple form in terms of a *free parameter* which, besides ensuring the well-posedness of the feedback connection, has to comply with a single constraint only, namely being an element of the space RH_∞.

Obviously, there exist stabilizing controllers for system $G(s) := C(sI-A)^{-1}B+D$ only provided that the pair (A,B) is stabilizable and the pair (A,C) is detectable. Therefore, such an unavoidable assumptions will be done throughout the subsequent discussion.

Consistently with Theorem 3.4, let the eight matrices $N(s)$, $M(s)$, $\hat{N}(s)$, $\hat{M}(s)$, $X(s)$, $Y(s)$, $\hat{X}(s)$, $\hat{Y}(s)$ specify a double coprime factorization in RH_∞ of the transfer function $G(s)$. These matrices are derived according to eqs. (3.1)-(3.4) and are such that

$$G(s) = N(s)M^{-1}(s) = \hat{M}^{-1}(s)\hat{N}(s)$$

and

$$\begin{bmatrix} \hat{X}(s) & -\hat{Y}(s) \\ -\hat{N}(s) & \hat{M}(s) \end{bmatrix} \begin{bmatrix} M(s) & Y(s) \\ N(s) & X(s) \end{bmatrix} = I \tag{3.10}$$

It is now possible to state the following result.

Theorem 3.7 (Youla parametrization) *The set of all rational and proper transfer functions $K(s)$ which internally stabilize $G(s)$ is defined by controllers of the form*

$$\begin{aligned} K(s) &= [Y(s) - M(s)Q(s)][X(s) - N(s)Q(s)]^{-1} \\ &= [\hat{X}(s) - Q(s)\hat{N}(s)]^{-1}[\hat{Y}(s) - Q(s)\hat{M}(s)] \end{aligned} \tag{3.11}$$

where the matrix $Q(s)$ is any element of RH_∞ such that

$$\det[I - D \lim_{s \to \infty} Q(s)] \neq 0 \tag{3.12}$$

Proof Preliminarily notice that the existence of the inverse matrices in eq. (3.11) is ensured by eqs. (3.1)-(3.4) and (3.12) (also recall Lemma B.8).

For any matrix $Q(s)$, eq. (3.10) implies that

$$\begin{aligned} I &= \begin{bmatrix} I & Q(s) \\ 0 & I \end{bmatrix} \begin{bmatrix} I & -Q(s) \\ 0 & I \end{bmatrix} \\ &= \begin{bmatrix} I & Q(s) \\ 0 & I \end{bmatrix} \begin{bmatrix} \hat{X}(s) & -\hat{Y}(s) \\ -\hat{N}(s) & \hat{M}(s) \end{bmatrix} \cdot \\ &\quad \cdot \begin{bmatrix} M(s) & Y(s) \\ N(s) & X(s) \end{bmatrix} \begin{bmatrix} I & -Q(s) \\ 0 & I \end{bmatrix} \\ &= \begin{bmatrix} \hat{X}(s) - Q(s)\hat{N}(s) & -[\hat{Y}(s) - Q(s)\hat{M}(s)] \\ -\hat{N}(s) & \hat{M}(s) \end{bmatrix} \cdot \\ &\quad \cdot \begin{bmatrix} M(s) & -M(s)Q(s) + Y(s) \\ N(s) & -N(s)Q(s) + X(s) \end{bmatrix} \end{aligned} \tag{3.13}$$

The (1,2) block of the product of the two last matrices in eq. (3.13) is zero, i.e.

$$\begin{aligned} [\hat{X}(s) - Q(s)\hat{N}(s)][-M(s)Q(s) + Y(s)] - \\ -[\hat{Y}(s) - Q(s)\hat{M}(s)][-N(s)Q(s) + X(s)] = 0 \end{aligned} \tag{3.14}$$

so that the second equality sign in eq. (3.11) is proved.

Now the system with transfer function

$$K(s) := [Y(s) - M(s)Q(s)][X(s) - N(s)Q(s)]^{-1}$$

where $Q(s) \in RH_\infty$ satisfies condition (3.12), is proved to internally stabilize $G(s)$. To this aim, let $U(s) := Y(s) - M(s)Q(s)$, $V(s) := X(s) - N(s)Q(s)$, $\hat{U}(s) := \hat{Y}(s) - Q(s)\hat{M}(s)$, $\hat{V}(s) := \hat{X}(s) - Q(s)\hat{N}(s)$, so that $K(s) = U(s)V^{-1}(s)$. From eq. (3.10) it follows

$$\begin{bmatrix} \hat{V}(s) & -\hat{U}(s) \\ -\hat{N}(s) & \hat{M}(s) \end{bmatrix} \begin{bmatrix} M(s) & U(s) \\ N(s) & V(s) \end{bmatrix} = I$$

that is

$$\begin{bmatrix} M(s) & U(s) \\ N(s) & V(s) \end{bmatrix}^{-1} = \begin{bmatrix} \hat{V}(s) & -\hat{U}(s) \\ -\hat{N}(s) & \hat{M}(s) \end{bmatrix}$$

The matrix on the right hand side of this last equation belongs to RH_∞ (each block in it is an element of RH_∞), so it does the inverse on the left hand side. Hence, thanks to Theorem 3.6, the system in fig. 3.5 is internally stable.

Conversely, assume that $K(s)$ stabilizes $G(s)$. In order that the feedback system in fig. 3.5 is well defined, matrix $K(s)$ must satisfy the condition

$$\det \left[I - D\bar{K} \right] \neq 0 \tag{3.15}$$

where $\bar{K} := \lim_{s \to \infty} K(s)$. Let $U(s)$ and $V(s)$ be two right coprime elements of RH_∞ such that $K(s) = U(s)V^{-1}(s)$ and consider the identity

$$
\begin{bmatrix} \hat{X}(s) & -\hat{Y}(s) \\ -\hat{N}(s) & \hat{M}(s) \end{bmatrix} \begin{bmatrix} M(s) & U(s) \\ N(s) & V(s) \end{bmatrix} =
$$
$$
= \begin{bmatrix} I & \hat{X}(s)U(s) - \hat{Y}(s)V(s) \\ 0 & -\hat{N}(s)U(s) + \hat{M}(s)V(s) \end{bmatrix} \tag{3.16}
$$

which follows from eq. (3.10). The two matrices on the left hand side of eq. (3.16) have inverses in RH_∞, the first one thanks to eq. (3.10), the second one thanks to Theorem 3.6. Hence also the matrix on the right hand side has inverse in RH_∞, so that $P^{-1}(s) := [-\hat{N}(s)U(s) + \hat{M}(s)V(s)]^{-1} \in RH_\infty$. Therefore the matrix $Q(s) := -[\hat{X}(s)U(s) - \hat{Y}(s)V(s)]P^{-1}(s)$ belongs to RH_∞, too. Now notice that from eqs. (3.1)-(3.4) it follows

$$I - D \lim_{s \to \infty} Q(s) = I + D \lim_{s \to \infty} \left[[\hat{X}(s)U(s) - \hat{Y}(s)V(s)] \cdot \right.$$
$$\left. \cdot [-\hat{N}(s)U(s) + \hat{M}(s)V(s)]^{-1} \right]$$
$$= I + D \lim_{s \to \infty} \left[[\hat{X}(s)U(s)V^{-1}(s) - \hat{Y}(s)]V(s)V^{-1}(s) \cdot \right.$$
$$\left. \cdot [-\hat{N}(s)U(s)V^{-1}(s) + \hat{M}(s)]^{-1} \right]$$
$$= I + D\bar{K}(I - D\bar{K})^{-1}$$
$$= (I - D\bar{K})^{-1}$$

where the last equality sign is due to Lemma B.9. Hence $Q(s)$ satisfies condition (3.12) in view of eq. (3.15).

From eq. (3.16) one obtains

$$
\begin{bmatrix} M(s) & Y(s) \\ N(s) & X(s) \end{bmatrix} \begin{bmatrix} \hat{X}(s) & -\hat{Y}(s) \\ -\hat{N}(s) & \hat{M}(s) \end{bmatrix} \begin{bmatrix} M(s) & U(s) \\ N(s) & V(s) \end{bmatrix} =
$$
$$
= \begin{bmatrix} M(s) & Y(s) \\ N(s) & X(s) \end{bmatrix} \begin{bmatrix} I & -Q(s)P(s) \\ 0 & P(s) \end{bmatrix}
$$

which, in view of eq. (3.10), becomes

$$
\begin{bmatrix} M(s) & U(s) \\ N(s) & V(s) \end{bmatrix} = \begin{bmatrix} M(s) & Y(s) \\ N(s) & X(s) \end{bmatrix} \begin{bmatrix} I & -Q(s)P(s) \\ 0 & P(s) \end{bmatrix}
$$

This last equation implies

$$U(s) = [Y(s) - M(s)Q(s)]P(s)$$
$$V(s) = [X(s) - N(s)Q(s)]P(s)$$

so that

$$K(s) = U(s)V^{-1}(s)$$
$$= [Y(s) - M(s)Q(s)][X(s) - N(s)Q(s)]^{-1}$$

according to eq. (3.11).   □

The controller $K_0(s)$ corresponding to the somehow most natural choice of $Q(s)$, namely $Q(s) = 0$, is usually referred to as the *central* controller. In view of eq. (3.11) such a controller is given by

$$K_0(s) = Y(s)X^{-1}(s) = \hat{X}^{-1}(s)\hat{Y}(s)$$

and therefore only depends on the particular double coprime factorization of $G(s)$ which has been selected. Further, in spite of corresponding to the *simplest* matrix $Q(s) \in RH_\infty$ which satisfies condition (3.12), the central controller is *not*, in general, the *lowest* order stabilizing controller.

Example 3.6 Let $G(s) = 1/s$ and choose $F = H = -1$, thus obtaining $N(s) = \hat{N}(s) = 1/(s+1)$, $M(s) = \hat{M}(s) = s/(s+1)$, $X(s) = \hat{X}(s) = (s+2)/(s+1)$, $Y(s) = \hat{Y}(s) = -1/(s+1)$. From eq. (3.11) it follows

$$K(s) = -\frac{1 + sQ(s)}{s + 2 - Q(s)}$$

so that, corresponding to the choice $Q(s) = 0$ it is $K_0(s) = -1/(s+2)$, while, if $Q(s) = 1$, it is $K(s) = -1$ (note that the first controller, the central one, is a dynamic system of order 1, while the second one is purely algebraic). Vice-versa, by making use of the expressions for $Q(s)$ given in the proof of Theorem 3.7, one gets

$$Q(s) = -\frac{V(s) + (s+2)U(s)}{sV(s) - U(s)}$$

where $U(s)$ and $V(s)$ are such that $K(s) = U(s)V^{-1}(s)$. As an example, if $K(s) = -1/(s+1)$, then $Q(s) = 1/(s^2 + s + 1)$.   □

The set of all stabilizing controllers can be further analyzed in order to enlighten interesting connections with other results of linear control theory. A preliminary result is presented in the forthcoming lemma, where reference is made to a double coprime factorization of $G(s)$ satisfying eq. (3.10).

Lemma 3.1 *The set of proper rational transfer functions which internally stabilize $G(s)$ is given by*

$$K(s) = K_0(s) - \hat{X}^{-1}(s)Q(s)[I - X^{-1}(s)N(s)Q(s)]^{-1}X^{-1}(s)$$

where $K_0(s) = Y(s)X^{-1}(s) = \hat{X}^{-1}(s)\hat{Y}(s)$ and $Q(s) \in RH_\infty$ is such that

$$\det[I - D \lim_{s \to \infty} Q(s)] \neq 0$$

Proof By exploiting the results in Lemma B.9 and the expression of a generic stabilizing controller (Theorem 3.7), one obtains

$$K(s) = [Y(s) - M(s)Q(s)][X(s) - N(s)Q(s)]^{-1}$$
$$= [Y(s) - M(s)Q(s)]\{[I - N(s)Q(s)X^{-1}(s)]X(s)\}^{-1}$$

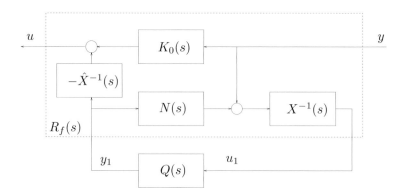

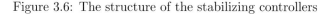

Figure 3.6: The structure of the stabilizing controllers

$$
\begin{aligned}
&= [Y(s) - M(s)Q(s)]X^{-1}(s)[I - N(s)Q(s)X^{-1}(s)]^{-1} \\
&= [K_0(s) - M(s)Q(s)X^{-1}(s)]\{I + N(s)Q(s) \cdot \\
&\quad \cdot [I - X^{-1}(s)N(s)Q(s)]^{-1}X^{-1}(s)\} \\
&= K_0(s) + K_0(s)N(s)Q(s)[I - X^{-1}(s)N(s)Q(s)]^{-1} \cdot \\
&\quad \cdot X^{-1}(s) - M(s)Q(s)X^{-1}(s) \cdot \\
&\quad \cdot \{I + N(s)Q(s)[I - X^{-1}(s)N(s)Q(s)]^{-1}X^{-1}(s)\} \\
&= K_0(s) + K_0(s)N(s)Q(s)[I - X^{-1}(s)N(s)Q(s)]^{-1} \cdot \\
&\quad \cdot X^{-1}(s) - M(s)Q(s)X^{-1}(s)[I - N(s)Q(s)X^{-1}(s)]^{-1} \\
&= K_0(s) + K_0(s)N(s)Q(s)[I - X^{-1}(s)N(s)Q(s)]^{-1} \cdot \\
&\quad \cdot X^{-1}(s) - M(s)Q(s)[I - X^{-1}(s)N(s)Q(s)]^{-1}X^{-1}(s) \\
&= K_0(s) + [K_0(s)N(s) - M(s)]Q(s) \cdot \\
&\quad \cdot [I - X^{-1}(s)N(s)Q(s)]^{-1}X^{-1}(s)
\end{aligned}
$$

Thanks to eq. (3.10) it is

$$
\begin{aligned}
K_0(s)N(s) - M(s) &= \hat{X}^{-1}(s)\hat{Y}(s)N(s) - M(s) \\
&= \hat{X}^{-1}(s)[\hat{Y}(s)N(s) - \hat{X}(s)M(s)] \\
&= \hat{X}^{-1}(s)
\end{aligned}
$$

so that $K(s)$ is given by

$$
K(s) = K_0(s) - \hat{X}^{-1}(s)Q(s)[I - X^{-1}(s)N(s)Q(s)]^{-1}X^{-1}(s)
$$

and the lemma is proved. $\square$

The form of $K(s)$ as given by Lemma 3.1 shows that the generic stabilizing controller is constituted by two subsystems in feedback connection. The first subsystem, denoted with $R_f(s)$ in fig. 3.6, only depends on the particular double coprime factorization which has been selected for $G(s)$, while the second one is simply constituted by the system with transfer function $Q(s)$. In fact, it is straightforward to check that the transfer function from y to u of the controller depicted in fig. 3.6 is precisely $K(s)$. In view of this fact it is particularly meaningful to look for a realization of the subsystem $R_f(s)$, since the generic stabilizing controller *must* result from the feedback connection of it with a stable system $Q(s)$ which is only constrained to actually produce a well defined overall system (condition (3.12)).

With reference to fig. 3.6, the transfer function of system $R_f(s)$ from the two inputs y and y_1 to the two outputs u and u_1 is

$$R_f(s) = \begin{bmatrix} K_0(s) & -\hat{X}^{-1}(s) \\ X^{-1}(s) & X^{-1}(s)N(s) \end{bmatrix}$$

Preliminarily a realization is presented for each one of the transfer functions appearing in $R_f(s)$. The proof of Theorem 3.4 (as for the form of $N(s)$, $X(s)$, $Y(s)$, $\hat{X}(s)$) and Section 2.5 (as for the form of the inverse system) are expedient to such an operation. Letting $\bar{C} := C + DF$, $\bar{B} := B + HD$ and $\bar{A} := A + BF + H\bar{C}$, one obtains

$$X^{-1}(s)N(s) = \left[\begin{array}{cc|c} A+BF & 0 & B \\ H\bar{C} & \bar{A} & HD \\ \hline \bar{C} & \bar{C} & D \end{array} \right]$$

It is easy to ascertain that a lower order realization can be found by performing a change of variables which put into evidence an unobservable part and is defined by the matrix

$$T_1 := \begin{bmatrix} I & I \\ 0 & I \end{bmatrix}$$

namely,

$$X^{-1}(s)N(s) = \left[\begin{array}{c|c} \bar{A} & \bar{B} \\ \hline \bar{C} & D \end{array} \right]$$

In a similar way one gets

$$K_0(s) = Y(s)X^{-1}(s) = \left[\begin{array}{cc|c} \bar{A} & 0 & H \\ H\bar{C} & A+BF & H \\ \hline 0 & -F & 0 \end{array} \right]$$

simplified to

$$K_0(s) = \left[\begin{array}{c|c} \bar{A} & H \\ \hline -F & 0 \end{array} \right]$$

through the change of variables defined by the matrix

$$T_2 := \begin{bmatrix} -I & I \\ 0 & I \end{bmatrix}$$

which put into evidence an unreachable part. Finally, one has

$$X^{-1}(s) = \left[\begin{array}{c|c} \bar{A} & H \\ \hline \bar{C} & I \end{array} \right] \quad , \quad \hat{X}^{-1}(s) = \left[\begin{array}{c|c} \bar{A} & \bar{B} \\ \hline F & I \end{array} \right]$$

Thus a realization for system $R_f(s)$ is

$$R_f(s) = \left[\begin{array}{c|cc} \bar{A} & H & \bar{B} \\ \hline -F & 0 & -I \\ \bar{C} & I & D \end{array} \right] = \left[\begin{array}{c|cc} \bar{A} & -H & -\bar{B} \\ \hline F & 0 & -I \\ -\bar{C} & I & D \end{array} \right]$$

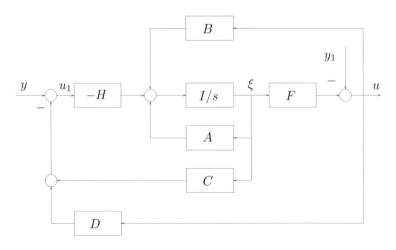

Figure 3.7: A realization of $R_f(s)$

This particular realization corresponds to the block-scheme depicted in fig. 3.7. This scheme clearly shows that when $y_1 = 0$, that is when $Q(s) = 0$ or, in other words, when the central controller has been adopted, the stabilizing controller generates the control variable u through the law $u = F\xi$, ξ being the state of an observer with dynamic matrix $A + HC$. Thus the central controller coincides with the controller designed via the well known *pole placement* technique and any stabilizing controller for $G(s)$ can be said to be "built around" a state observer.

Finally, the eigenvalues of the resulting control system, that is of the system in fig. 3.5 with $K(s)$ given by the block-scheme of fig. 3.6 where $R_f(s)$ is specified by the block-diagram of fig. 3.7, are those of matrices $A + BF$ and $A + HC$ together with those of the dynamic matrix of $Q(s) := \Sigma(A_q, B_q, C_q, D_q)$. In fact, denoting by x and x_q the state variables of $G(s)$ and $Q(s)$, respectively, the choice $x_t := [x' - \xi'\ x_q'\ x']'$ as state variables of resulting system, yields the (closed loop) dynamic matrix

$$A_t = \begin{bmatrix} A + HC & 0 & 0 \\ B_q C & A_q & 0 \\ -B(F + D_q C) & -B C_q & A + BF \end{bmatrix}$$

3.5   Notes and references

The line of reasoning pursued in this chapter and the derivation of some results presented herein are inspired by the books of Francis [19] and Maciejowski [43]. The result on the parametrization of stabilizing controllers can be found in the pioneering paper by Youla et al. [63].

Chapter 4

RH_2 Control

4.1   Introduction

This chapter presents the most significant results concerning the control problem in the RH_2 context: it simply consists in minimizing the RH_2 norm of a transfer function. As it will be apparent in the sequel, a number of connections can be established between this problem (more precisely, between the subproblems which actually constitute its frame) and the most celebrated set of results in optimal filtering and control problems, the well known *Linear Quadratic Gaussian* theory. Approaching these problems within the RH_2 setting gives a somehow more complete picture of the structure of the results (see the forthcoming Remarks 4.3, 4.10, 4.19).

Throughout this chapter reference will be made to a controlled system described by the following equations

$$\dot{x} = Ax + B_1 w + B_2 u \tag{4.1}$$
$$z = C_1 x + D_{11} w + D_{12} u \tag{4.2}$$
$$y = C_2 x + D_{21} w + D_{22} u \tag{4.3}$$

The *measured output* variable y is the input to the controller, which is constrained to be a finite dimensional, linear, time invariant system, while the *control* variable u is its output. Therefore, the controller takes on the form

$$\dot{\xi} = F\xi + Gy \tag{4.4}$$
$$u = H\xi + Ey \tag{4.5}$$

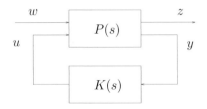

Figure 4.1: The standard control system

The situation is summarized in the block-scheme of fig. 4.1, where $P(s)$ and $K(s)$ denote the transfer functions of the controlled system and the controller, respectively. More precisely,

$$P(s) := \left[\begin{array}{c|cc} A & B_1 & B_2 \\ \hline C_1 & D_{11} & D_{12} \\ C_2 & D_{21} & D_{22} \end{array} \right]$$

$$K(s) := \left[\begin{array}{c|c} F & G \\ \hline H & E \end{array} \right]$$

It is apparent that the feedback connection of system (4.1)-(4.3) with system (4.4),(4.5) is well defined only if the controller is such that

$$\det[I - ED_{22}] \neq 0 \tag{4.6}$$

that is only if the *algebraic loop* deriving from its implementation can be solved.

The structure of system (4.1)-(4.3) is typical of the control problems stated within the RH_2 and RH_∞ context. Indeed, it is sufficiently general so as to encompass the most meaningful situations. A fairly comprehensive discussion on this aspect is presented at the beginning of Chapter 5 to adequately motivate the significance of the results there presented for the RH_∞ control problems.

The *input variable w* collects all exogenous signals which can be viewed as *disturbances* acting on the control system, while the *output performance variable z* is expedient to specify an index of the performances to be attained by the control system.

With reference to the control system depicted in fig. 4.1, the main objective is minimizing (with respect to $K(s)$) the RH_2 norm of the transfer function $T(z, w; s)$ from the input w to the output z.

If $T(z, w; s)$ has to belong to RH_2, then it must necessarily be *strictly proper* and *stable*. The first requirement is equivalent to the condition

$$D_{11} + D_{12}(I - ED_{22})^{-1}ED_{21} = 0 \tag{4.7}$$

As for the second requirement, it is no doubt satisfied if the internal stability of the control system is ensured, that is if the controller (4.4),(4.5) internally stabilizes the controlled system (4.1)-(4.3), which simply amounts to asking for the stability of the dynamic matrix of the resulting system, namely

$$Re(\lambda_i(A_F)) < 0 \ , \ \forall i \tag{4.8}$$

where

$$A_F := \left[\begin{array}{cc} A + B_2(I - ED_{22})^{-1}EC_2 & B_2(I - ED_{22})^{-1}H \\ G[I + D_{22}(I - ED_{22})^{-1}E]C_2 & F + GD_{22}(I - ED_{22})^{-1}H \end{array} \right] \tag{4.9}$$

In view of the discussion above the notion of *admissible* controller is introduced in the following definition.

Definition 4.1 (Admissible controller in RH_2) *A controller $K(s)$ is said to be admissible in RH_2 for $P(s)$ if conditions (4.6)-(4.9) are verified.* □

Three problems will be discussed in the forthcoming Sections 4.2 - 4.4: (i) the *full information* problem; (ii) the *output estimation* problem; (iii) the *partial information* problem. Each of these problems refers to a particular structure of the controlled system $P(s)$ and exhibits deep and significant mutual connections. Indeed, the last one presents an interesting separation property and will be dealt with by exploiting the solutions relevant to the first two, which, in turn, are characterized by strong duality properties.

The main result concerning each one of the above problems provides the answer to three important questions: (i) the *existence* of the *optimal* controller; (ii) the actual *form* of such a controller; (iii) the *parametrization* of *suboptimal* controllers. The material will be presented according to the scheme formally stated in Problem 4.1 which refers to the feedback connected system shown in fig. 4.1 and to the set $\mathcal{F}_{2\gamma}$ of controllers which are admissible in RH_2 for $P(s)$ and such that $\|T(z,w;s)\|_2$ is bounded by a given positive scalar γ.

Problem 4.1 (Standard problem in RH_2) *Find*

  a) *The minimum value (if any) of $\|T(z,w;s)\|_2$ attainable by a controller $K(s)$ which is admissible in RH_2 for $P(s)$.*

  b) *An admissible controller which minimizes $\|T(z,w;s)\|_2$.*

  c) *A set of controllers $\mathcal{F}_{2\gamma r} \subseteq \mathcal{F}_{2\gamma}$ whose elements generate the whole set of functions $T(z,w;s)$ which are generated by the elements of $\mathcal{F}_{2\gamma}$.*

Remark 4.1 An obvious necessary condition for the existence of a stabilizing controller (and therefore for the existence of an admissible controller in RH_2 for $P(s)$) is the stabilizability of the pair (A, B_2) and the detectability of the pair (A, C_2). The statement of Problem 4.1 makes sense only if both properties actually hold true. □

In the forthcoming sections the parametrization of the controllers in the family $\mathcal{F}_{2\gamma}$ will be presented in specific remarks which follow the main theorems concerning the solution of Problem 4.1. Hence, the issues relative to the family $\mathcal{F}_{2\gamma}$ and those relative to Problem 4.1 will be treated separately.

Problem 4.1 will be tackled in the subsequent sections by assuming $D_{11} = 0$ in eq. (4.2). Actually, this assumption does not cause any loss of generality. In fact, assume that the set of matrices E which satisfy eqs. (4.6),(4.7) is not empty (otherwise the problem would not admit any solution in RH_2) and let $\hat{E}$ be one of such matrices. It is straightforward to check that the situation $D_{11} = 0$ is recovered after the output feedback

$$u = \hat{E}y + v$$

has been implemented.

4.2   The full information problem

In this section Problem 4.1 is considered by assuming that the output signal y of system (4.1)-(4.3) is constituted by the state vector x and the disturbance vector w. Moreover, no direct influence of the input signals w and u exists on the outputs z and y, respectively. As a consequence, the controlled plant, whose transfer function from

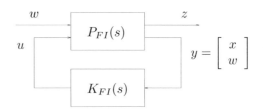

Figure 4.2: The full information problem

$[w'\ u']'$ to $[z'\ y']'$ will be denoted by $P_{FI}(s)$, is described by

$$\dot{x} = Ax + B_1 w + B_2 u \tag{4.10}$$

$$z = C_1 x + D_{12} u \tag{4.11}$$

$$y = [y'_1\ y'_2]' \tag{4.12}$$

$$y_1 = x \tag{4.13}$$

$$y_2 = w \tag{4.14}$$

Furthermore, the following assumptions will be made.

Assumption 4.1 *The pair (A, B_2) is stabilizable and no eigenvalue of the unobservable part of the pair $[(A - B_2 D'_{12} C_1), (I - D_{12} D'_{12}) C_1]$ lies on the imaginary axis.*

Assumption 4.2 $D'_{12} D_{12} = I$.

In the theorem below reference is made to the block-scheme of fig. 4.2 where $K_{FI}(s)$ is a generic controller admissible in RH_2 for $P_{FI}(s)$. In this diagram $T(z, w; s)$ is the transfer function from w to z.

Theorem 4.1 (Full information) *Consider Problem 4.1 relative to system (4.10)-(4.14). Then, under Assumptions 4.1, 4.2, it has the solution*

 a)

$$\min \|T(z, w; s)\|_2 = \|P_c(s) B_1\|_2 = \sqrt{\text{trace}[B'_1 P_2 B_1]}$$

 b)

$$K^o_{FI}(s) = \left[\begin{array}{c|cc} \emptyset & \emptyset & \emptyset \\ \hline \emptyset & F_2 & 0 \end{array} \right]$$

 c) The set $\mathcal{F}_{2\gamma r}$ of the controllers $K_{FIr}(s)$ is defined by the diagram of fig. 4.3, where $Q(s) := \Sigma(A_q, B_q, C_q, 0)$ with A_q stable and $\|Q(s)\|_2^2 < \gamma^2 - \|P_c(s) B_1\|_2^2$.

In the three points above, γ is a positive scalar such that $\gamma > \|P_c(s) B_1\|_2$ and

$$F_2 := -B'_2 P_2 - D'_{12} C_1 \tag{4.15}$$

$$P_c(s) := \left[\begin{array}{c|c} A_c - B_2 B'_2 P_2 & I \\ \hline C_{1c} - D_{12} B'_2 P_2 & 0 \end{array} \right] \tag{4.16}$$

$$A_c := A - B_2 D'_{12} C_1, \quad C_{1c} := (I - D_{12} D'_{12}) C_1 \tag{4.17}$$

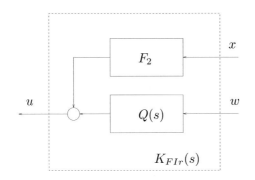

Figure 4.3: The set $\mathcal{F}_{2\gamma r}$ of the controllers $K_{FIr}(s)$

where P_2 is the symmetric, positive semidefinite and stabilizing solution of the Riccati equation (in the unknown P)

$$0 = PA_c + A'_c P - PB_2 B'_2 P + C'_{1c} C_{1c} \tag{4.18}$$

i.e. such that matrix A_{cc} defined by

$$A_{cc} := A_c - B_2 B'_2 P_2 = A + B_2 F_2 \tag{4.19}$$

is stable.

Proof First observe that the necessary condition for the problem at hand to make sense (recall Remark 4.1) is satisfied. Indeed, being measurable the state of the system, detectability of the pair (A, C_2) trivially holds, while, on the other hand, stabilizability of the pair (A, B_2) is guaranteed by Assumption 4.1.

Points a) and b) Notice that the pair (A, B_2) is stabilizable if and only if the pair (A_c, B_2) is such (recall that state feedback does not modify the stabilizability property). Therefore, Assumptions 4.1 and 4.2 together with Lemma C.4 guarantee the existence of the symmetric, positive semidefinite and stabilizing solution P_2 of eq. (4.18), so that the matrix A_{cc} defined by eqs. (4.19),(4.15) is stable.

Furthermore, let

$$v := u - F_2 x \tag{4.20}$$

Equation (4.20) apparently defines the control law

$$u := v + F_2 x \tag{4.21}$$

From eqs. (4.10),(4.11),(4.21) it follows

$$z_{L0} = P_c(s)B_1 w_L + U(s)v_L \tag{4.22}$$

where

$$U(s) := \Sigma(A_c - B_2 B'_2 P_2, B_2, C_{1c} - D_{12} B'_2 P_2, D_{12}) \tag{4.23}$$

The variable v defined by eq. (4.20) is one of the output of the system

$$P_v(s) := \begin{cases} \dot{x} = Ax + B_1 w + B_2 u \\ v = -F_2 x + u \\ y = [y'_1\ y'_2]' , \quad y_1 = x , \quad y_2 = w \end{cases}$$

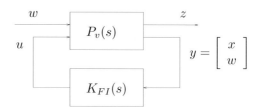

Figure 4.4: The equivalent full information problem

In order to evaluate the transfer function $T(z, w; s)$ (relevant to the scheme in fig. 4.2) by means of eq. (4.22), it is convenient to make reference to the block-diagram in fig. 4.4 and to the related transfer function $T(v, w; s)$ from the input w to the output v. From eqs. (4.10)-(4.14) and the definition of $P_v(s)$ it follows that the systems in fig. 4.2 and 4.4 are well defined for all $K_{FI}(s)$, since condition (4.6) is satisfied in both cases. Therefore, it is possible to write

$$T(z, w; s) = P_c(s)B_1 + U(s)T(v, w; s) \qquad (4.24)$$

From eqs. (4.10)-(4.14) and the definition of $P_v(s)$ it also follows that the system in fig. 4.4 is stable if and only if the system in fig. 4.2 is such. Moreover, letting $K_{FI}(\infty) := [E_1\ E_2]$, it is $T(z, w; \infty) = D_{12}E_2$ and $T(v, w; \infty) = E_2$. Thus, $D_{12}E_2 = 0$ if and only if $E_2 = 0$ since, thanks to Assumption 4.2, the matrix D_{12} has rank equal to the number of its columns. Therefore, the function $T(z, w; s)$ is strictly proper if and only if the function $T(v, w; s)$ is such. In conclusion, it can be claimed that the controller $K_{FI}(s)$ is admissible in RH_2 for $P_{FI}(s)$ if and only if it is such also for $P_v(s)$.

From Assumption 4.2 it follows that $(I - D_{12}D'_{12})D_{12} = 0$, so that $C'_{1c}D_{12} = 0$. This fact, together with Lemma C.5 implies that the function $U(s)$ defined by eq. (4.23) is inner and $U^\sim(s)P_c(s) \in RH_2^\perp$.

In view of eq. (4.24), it follows that

$$\|T(z, w; s)\|_2^2 = \|P_c(s)B_1\|_2^2 + \|U(s)T(v, w; s)\|_2^2 +$$
$$+2 < U(s)T(v, w; s), P_c(s)B_1 > \qquad (4.25)$$

for any controller $K_{FI}(s)$ admissible in RH_2 for $P_{FI}(s)$. Being $U(s)$ an inner function, one gets

$$\|U(s)T(v, w; s)\|_2^2 = < U(s)T(v, w; s), U(s)T(v, w; s) >$$
$$= < T(v, w; s), T(v, w; s) >= \|T(v, w; s)\|_2^2$$

while, being $U^\sim(s)P_c(s) \in RH_2^\perp$, it is

$$< U(s)T(v, w; s), P_c(s)B_1 >=< T(v, w; s), U^\sim(s)P_c(s)B_1 >= 0$$

since $T(v, w; s) \in RH_2$ (recall Theorem 2.11). Then, eq. (4.25) becomes

$$\|T(z, w; s)\|_2^2 = \|P_c(s)B_1\|_2^2 + \|T(v, w; s)\|_2^2 \qquad (4.26)$$

so that

$$\min_{K_{FI}(s)} \|T(z, w; s)\|_2^2 = \|P_c(s)B_1\|_2^2 + \min_{K_{FI}(s)} \|T(v, w; s)\|_2^2$$

The minimum is by sure attained if $\|T(v, w; s)\|_2^2 = 0$, that is if $v = 0$. This is the case if

$$u = F_2 x \tag{4.27}$$

Now observe that P_2, being a solution of eq. (4.18), is also a positive semidefinite solution of the Lyapunov equation (in the unknown P)

$$0 = PA_{cc} + A'_{cc}P + P_2B_2B'_2P_2 + C'_{1c}C_{1c}$$

where A_{cc} is defined by eq. (4.19). Matrix A_{cc} is the dynamic matrix of the system (4.10)-(4.14),(4.27), namely

$$\dot{x} = A_{cc}x + B_1 w \tag{4.28}$$
$$z = (C_{1c} - D_{12}B'_2P_2)x \tag{4.29}$$

The norm of the (optimal) transfer function of system (4.28), (4.29) may be computed by exploiting what has been presented in Remark 2.20. To this aim, notice that from Assumption 4.2 and $C'_{1c}D_{12} = 0$ it follows

$$(C_{1c} - D_{12}B'_2P_2)'(C_{1c} - D_{12}B'_2P_2) = P_2B_2B'_2P_2 + C'_{1c}C_{1c}$$

and recall that P_2 solves the above Lyapunov equation. Therefore, $\|T(z, w; s)\|_2^2 = \text{trace}[B'_1 P_2 B_1]$. Points $a)$ and $b)$ are thus proved.

Point c) Let a generic controller $K_{FI}(s) \in \mathcal{F}_{2\gamma}$ be described by the equations

$$\dot{\xi} = L\xi + M_1 x + M_2 w$$
$$u = N\xi + O_1 x + O_2 w$$

If v is given by eq. (4.20), then the same arguments exploited for proving points $a)$ and $b)$ lead to eq. (4.26), so that

$$\|T(v, w; s)\|_2^2 < \gamma^2 - \|P_c(s)B_1\|_2^2$$

Recall that if $K_{FI}(s)$ is admissible in RH_2 for $P_{FI}(s)$, then it is also admissible in RH_2 for $P_v(s)$. Therefore, $T(v, w; s)$ can be written as

$$T(v, w; s) := Q(s) := \Sigma(A_q, B_q, C_q, 0)$$

with A_q stable. A realization of $Q(s)$ can be easily derived by recalling the definition of $P_v(s)$ and the above given expression for $K_{FI}(s)$. It results

$$\dot{\vartheta} = L\vartheta + M_1\sigma + M_2 w$$
$$\dot{\sigma} = B_2 N\vartheta + (A + B_2 O_1)\sigma + (B_1 + B_2 O_2)w$$
$$v = N\vartheta + (O_1 - F_2)\sigma + O_2 w$$

The controller $K_{FIr}(s)$ defined by these equations and $u = F_2 x + v$ is now shown to belong to the set $\mathcal{F}_{2\gamma r}$. Obviously, it possesses the structure of the system in fig. 4.3. Moreover, thanks to eqs. (4.10),(4.11), the equations of the system resulting from the feedback connection of $P_{FI}(s)$ with $K_{FIr}(s)$ are, letting $\varepsilon := \sigma - x$,

$$\dot{x} = Ax + B_1 w + B_2 u$$
$$\dot{\vartheta} = L\vartheta + M_1 x + M_2 w + M_1 \varepsilon$$
$$\dot{\varepsilon} = (A + B_2 F_2)\varepsilon$$
$$u = N\vartheta + O_1 x + (O_1 - F_2)\varepsilon + O_2 w$$
$$z = C_1 x + D_{12} u$$

The comparison of these equations with those relevant to the feedback connection of system $P_{FI}(s)$ with the controller $K_{FI}(s)$ allows one to derive the following two conclusions: first, the transfer functions from w to z are equal in the two cases; second, the system having $K_{FIr}(s)$ as a controller is stable because the system having $K_{FI}(s)$ as a controller is such and matrix $A + B_2 F_2$ is stable.

Vice versa, if $K_{FIr}(s)$ belongs to the set described in fig. 4.3, then it is $u = F_2 x + v$ where v is the output of the system $T(v, w; s) = \Sigma(A_q, B_q, C_q, 0)$, with A_q stable and $\|T(v, w; s)\|_2^2 < \gamma^2 - \|P_c(s)B_1\|_2^2$. The same arguments exploited in proving points $a)$ and $b)$ imply that $K_{FIr}(s)$ is admissible in RH_2 for $P_{FI}(s)$ and, thanks to eq. (4.26), $\|T(z, w; s)\|_2^2 < \gamma^2$.   □

Example 4.1 Consider system (4.10)-(4.14) with

$$A = \begin{bmatrix} 0 & 1 \\ 0 & 0 \end{bmatrix}, \quad B_1 = \begin{bmatrix} 0 \\ 2 \end{bmatrix}, \quad B_2 = \begin{bmatrix} 0 \\ 1 \end{bmatrix}$$

$$C_1 = \begin{bmatrix} \alpha & 1 \end{bmatrix}, \quad D_{12} = 1$$

where $\alpha \neq 0$ and let $u = u_w + u_x$. The classic way of designing a controller which reduces the effect of the disturbance w on the output z is trying to zeroing the transfer function from w to z by means of a suitable controller $K_{DC}(s)$ with input w and output u_w which performs a *direct compensation*. In the problem at hand such a transfer function vanishes (namely, z does not depend on w) if

$$K_{DC}(s) = -\frac{2(s + \alpha)}{s^2 + s + \alpha}$$

However, it is apparent that the resulting control system can not be stabilized (by means of a control law $u = u_w + u_x$ which makes u_x to depend on x in a suitable manner) if $K_{DC}(s)$ is not stable by itself. Hence, if $\alpha > 0$, a perfect direct compensation of the disturbance can be achieved. On the contrary, if $\alpha < 0$, this is no more possible. By resorting to Theorem 4.1 one obtains, correspondingly to $Q(s) = 0$, $u = -\alpha x_1 - x_2$ and $T(z, w; s) = 0$, when $\alpha > 0$, while, when $\alpha < 0$, one gets $u = \alpha x_1 - \sqrt{1 - 4\alpha} x_2$ and

$$T(z, w; s) = 2\frac{2\alpha + (1 - \sqrt{1 - 4\alpha})s}{s^2 + \sqrt{1 - 4\alpha}s - \alpha}$$

                                                                                                □

Remark 4.2 The structure of the controllers $K_{FIr}(s)$ which are admissible in RH_2 for $P_{FI}(s)$ and defined by the block-scheme in fig. 4.3 allows one to easily conclude that the eigenvalues of the resulting control system are those of matrix $A + B_2 F_2$ and those of matrix A_q.   □

Remark 4.3 (Parametrization of the set $\mathcal{F}_{2\gamma}$) Observe that the set $\mathcal{F}_{2\gamma r}$ is a *proper subset* of the set $\mathcal{F}_{2\gamma}$. In fact, consider a generic controller $K_{FI}(s)$ admissible in RH_2 for $P_{FI}(s)$ which is purely algebraic and makes the control variables to depend on the state variables only. Namely, let

$$u = \Lambda x \tag{4.30}$$

with $\Lambda \neq F_2$. Such a controller certainly exists because of a continuity argument, since the controller $K_{FI}^o(s)$ actually has the form given in eq. (4.30).

Within the set $\mathcal{F}_{2\gamma r}$ the only element with the form given in eq. (4.30) is $K_{FI}^o(s)$, so that it is necessary to resort to a dynamic controller in the set $\mathcal{F}_{2\gamma r}$ in order to generate the same transfer function $T(z, w; s)$. As an example, for the system $P_{FI}(s)$ given by

$$\dot{x} = w + u$$
$$z = \begin{bmatrix} 1 \\ 0 \end{bmatrix} x + \begin{bmatrix} 0 \\ 1 \end{bmatrix} u$$
$$y = \begin{bmatrix} x \\ w \end{bmatrix}$$

the controller (admissible in RH_2 for such a $P_{FI}(s)$) described by

$$u = -3x \qquad (4.31)$$

has a "corresponding element" in the set $\mathcal{F}_{2\gamma r}$ given by (recall what has been shown in the proof of Theorem 4.1, point c))

$$\dot{\sigma} = -3\sigma + w \qquad (4.32)$$
$$u = -x - 2\sigma \qquad (4.33)$$

Indeed, by exploiting eq. (4.31), it follows

$$T(z, w; s) = \frac{1}{s+3} \begin{bmatrix} 1 \\ -3 \end{bmatrix}$$

which coincides with the expression deriving from eqs. (4.32),(4.33).

A *parametrization* of the set $\mathcal{F}_{2\gamma}$ is now presented. Consider the system $P_F(s)$ which is obtained from system (4.10)-(4.14) after the control law (4.21) has been implemented, namely the system

$$P_F(s) := \begin{bmatrix} P_{F11}(s) & P_{F12}(s) \\ P_{F21}(s) & P_{F22}(s) \end{bmatrix}$$

$$= \left[\begin{array}{c|cc} A_{cc} & B_1 & B_2 \\ \hline C_{1F} & 0 & D_{12} \\ \begin{bmatrix} I \\ 0 \end{bmatrix} & \begin{bmatrix} 0 \\ I \end{bmatrix} & \begin{bmatrix} 0 \\ 0 \end{bmatrix} \end{array} \right]$$

where eqs. (4.19) and (4.15) have been taken into account and

$$C_{1F} := C_{1c} - D_{12} B_2' P_2$$

The set of controllers $K_F(s)$ which stabilize $P_F(s)$ apparently coincides with the set of controllers which stabilize $P_{F22}(s)$, the latter being the system

$$P_{F22}(s) = \left[\begin{array}{c|c} A_{cc} & B_2 \\ \hline I & 0 \\ 0 & 0 \end{array} \right]$$

consistently to what has been previously defined. In view of Theorem 3.7, such a set can be expressed as

$$K_F(s) = [Y(s) - M(s)\Theta(s)][X(s) - N(s)\Theta(s)]^{-1}$$

with $\Theta(s) \in RH_\infty$. Notice that condition (3.12) is no doubt satisfied in the present context, since $P_{F22}(s)$ is strictly proper. From the proof of Theorem 3.4 it follows that, being the matrix A_{cc} stable, the choice

$$Y(s) = 0 , \quad M(s) = I , \quad X(s) = I , \quad N(s) = P_{F22}(s)$$

is admissible, so that

$$K_F(s) = -\Theta(s)[I - P_{F22}(s)\Theta(s)]^{-1} \tag{4.34}$$

In the closed loop situation it is

$$y_{L0} = P_{F21}(s)w_L - P_{F22}(s)\Theta(s)[I - P_{F22}(s)\Theta(s)]^{-1}y_{L0}$$

By exploiting Lemma B.9, this equation can be written as

$$y_{L0} = [I - P_{F22}(s)\Theta(s)]P_{F21}(s)w_L$$

which, in turn, can be utilized to obtain

$$z_{L0} = P_{F11}(s)w_L - P_{F12}(s)\Theta(s)[I - P_{F22}(s)\Theta(s)]^{-1}y_{L0}$$
$$= [P_{F11}(s) - P_{F12}(s)\Theta(s)P_{F21}(s)]w_L$$

so that, when the controller described by

$$u_{L0} = F_2 x_{L0} + K_F(s)y_{L0}$$

is adopted, the closed loop transfer function from w to z is

$$T_\Theta(z, w; s) = P_{F11}(s) - P_{F12}(s)\Theta(s)P_{F21}(s)$$

On the other hand, if a controller $K_{FIr}(s)$ of the set $\mathcal{F}_{2\gamma r}$ is adopted, namely, a controller described by

$$u_{L0} = F_2 x_{L0} + Q(s)w_L$$

with (recall point (c) in the statement of Theorem 4.1) $Q(s) \in RH_2$, and $\|Q(s)\|_2^2 < \gamma^2 - \|P_c(s)B_1\|_2^2 = \gamma^2 - \|P_{F11}(s)\|_2^2$, one gets

$$z_{L0} = [P_{F11}(s) + P_{F12}(s)Q(s)]w_L$$

Therefore, the set of the $K_{FIr}(s)$ generates the set of transfer functions from w to z

$$T_Q(z, w; s) = P_{F11}(s) + P_{F12}(s)Q(s)$$

By equating the transfer functions $T_Q(z, w; s)$ and $T_\Theta(z, w; s)$ it is possible to characterize the set of functions $\Theta(s)$ associated with controllers $K_F(s)$ admissible in RH_2 for $P_F(s)$ and such that $\|T_\Theta(z, w; s)\|_2 < \gamma$. In so doing one obtains

$$P_{F12}(s)[\Theta(s)P_{F21}(s) + Q(s)] = 0$$

Letting $\Phi(s) := (sI - A_{cc})^{-1}$, from Assumption 4.2 it follows that the rank of $P_{F12}(s) = C_{1F}\Phi(s)B_2 + D_{12}$ equals the number of its columns, so that the above written equation is equivalent to

$$\Theta(s)P_{F21}(s) = -Q(s) \tag{4.35}$$

A particular solution of this equation is

$$\bar{\Theta}(s) = Q(s)[0 \ -I]$$

since

$$P_{F21}(s) = \begin{bmatrix} \Phi(s)B_1 \\ I \end{bmatrix}$$

Thus, the general solution of eq. (4.35) is

$$\Theta_Q(s) = \bar{\Theta}(s) + \hat{\Theta}(s)$$

where $\hat{\Theta}(s)$ denotes any solution in RH_∞ of the homogeneous equation

$$\Theta(s)P_{F21}(s) = 0$$

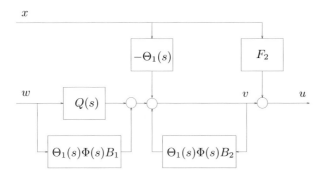

Figure 4.5: The generic admissible controller for $P_{FI}(s)$

Letting $\Theta(s) := [\Theta_1(s)\ \Theta_2(s)]$, from the last equation it follows that $\Theta_2(s) = -\Theta_1(s)\Phi(s)B_1$, so that

$$\hat{\Theta}(s) = \Theta_1(s)[I - \Phi(s)B_1]$$

belongs to RH_∞ if and only if $\Theta_1(s) \in RH_\infty$, since $\Phi(s) \in RH_\infty$, being A_{cc} a stable matrix.

Therefore, the set of functions $\Theta_Q(s)$ which generate controllers $K_F(s)$ admissible in RH_2 for $P_F(s)$ and such that $\|T_\Theta(z,w;s)\|_2 < \gamma$ is defined by

$$\Theta_Q(s) = [\Theta_1(s)\ \ -[Q(s) + \Theta_1(s)\Phi(s)B_1]]$$

$$\Theta_1(s) \in RH_\infty,\ Q(s) \in RH_2,\ \|Q(s)\|_2^2 < \gamma^2 - \|P_{F11}(s)\|_2^2$$

By exploiting Lemma B.9, from eq. (4.34) it follows

$$K_F(s) = -[I - \Theta_Q(s)P_{F22}(s)]^{-1}\Theta_Q(s)$$
$$= [I - \Theta_1(s)\Phi(s)B_2]^{-1} \cdot$$
$$\cdot [-\Theta_1(s)\ [Q(s) + \Theta_1(s)\Phi(s)B_1]] \tag{4.36}$$

since $\Theta_Q(s)P_{F22}(s) = \Theta_1(s)\Phi(s)B_2$. In conclusion, by recalling eq. (4.21), the generic controller $K_{FI}(s)$ in the set $\mathcal{F}_{2\gamma}$ is described by

$$K_{FI}(s) = [I - \Theta_1(s)\Phi(s)B_2]^{-1} \cdot$$
$$\cdot [-\Theta_1(s)\ [Q(s) + \Theta_1(s)\Phi(s)B_1]] + [F_2\ 0]$$

$$\Theta_1(s) \in RH_\infty,\ Q(s) \in RH_2,\ \|Q(s)\|_2^2 < \gamma^2 - \|P_{F11}(s)\|_2^2$$

Such a set is depicted in fig. 4.5. In the controller shown in this figure it is

$$v_{L0} = \Theta_1(s)\Phi(s)[-\Phi^{-1}(s)x_L + B_2v_{L0} + B_1w_L] + Q(s)w_L$$
$$= -\Theta_1(s)\Phi(s)x(0) + Q(s)w_L$$

having taken into account eqs. (4.10) and (4.21). Therefore, the effect of the parameter $\Theta_1(s)$ (which is responsible of the difference between the controllers in $\mathcal{F}_{2\gamma}$ and those in $\mathcal{F}_{2\gamma r}$) amounts to a term which only depends on the *initial state* of system (4.10)-(4.14).

With reference to the example previously considered it is easy to verify that the algebraic controller characterized by $u = -3x$ corresponds to the choice

$$Q(s) = -\frac{2}{s+3}\ ,\quad \Theta_1(s) = 2\frac{s+1}{s+3}$$

□

Remark 4.4 (Optimal control problems) The control problem to which Theorem 4.1 can be applied is strictly related to the *linear quadratic deterministic (LQ)* or *stochastic (LQS)* control problem with *measurable* state. This connection can be made explicit by exploiting what has been presented in Remark 2.20.

LQ Problem Consider the *n-th* order system

$$\dot{x} = Ax + B\bar{u} \tag{4.37}$$

$$x(0) = x_0 \tag{4.38}$$

and the cost functional

$$J_1 := \int_0^\infty \left[x'(t)Qx(t) + 2x'(t)S\bar{u}(t) + \bar{u}'(t)R\bar{u}(t) \right] dt \tag{4.39}$$

where

$$\begin{bmatrix} Q & S \\ S' & R \end{bmatrix} := L = L' \geq 0 , \quad R > 0 \tag{4.40}$$

Observe that $L \geq 0$ and $R > 0$ imply that

$$\hat{Q} := Q - SR^{-1}S' \geq 0$$

since $\hat{Q} = Z'LZ$ with $Z' := [I \ -SR^{-1}]$. Let $C_{11} \in R^{n \times n}$ be a factorization of $\hat{Q}$, so that

$$C_{11}'C_{11} = \hat{Q} \tag{4.41}$$

and define

$$C_1 := \begin{bmatrix} C_{11} \\ R^{-1/2}S' \end{bmatrix} , \quad D_{12} := \begin{bmatrix} 0 \\ I \end{bmatrix} \tag{4.42}$$

$$u := R^{1/2}\bar{u} , \quad z := C_1x + D_{12}u \tag{4.43}$$

It is easy to verify that it is

$$J_1 = \int_0^\infty z'(t)z(t)dt$$

On the other hand, the state free motion can always be interpreted as the forced motion caused by an impulsive input acting on the system through a suitable input matrix. Therefore, if eqs. (4.41)-(4.43) are taken into account, system (4.37),(4.38) can be described by

$$\dot{x} = Ax + B_1w + B_2u \tag{4.44}$$
$$z = C_1x + D_{12}u \tag{4.45}$$

with $w := \delta(t)$, $B_1 := x_0$, $B_2 := BR^{-1/2}$ and $x(0) = 0$.

The optimal control problem at hand (*LQ* problem) consists in finding a controller of the form

$$\dot{\xi} = F\xi + Gx \tag{4.46}$$
$$u = H\xi + Nx \tag{4.47}$$

such that the system (4.44)-(4.47) is stable and the performance index J_1 is minimized. Notice that the feedback connection of any controller of the form (4.46),(4.47) to the system (4.44),(4.45) is well defined and the relevant transfer function $T(z, w; s)$ from w to z is strictly proper. Therefore, if system (4.44)-(4.47) is stable, then the controller (4.46),(4.47) is admissible in RH_2 for system (4.44),(4.45). Further, in view of Remark 2.20, any controller (4.46),(4.47) which is admissible in RH_2 for system (4.44),(4.45) is such that $J_1 = \|T(z, w; s)\|_2^2$.

If the pair (A, B) is stabilizable and no eigenvalue of the unobservable part of the pair $(A - BR^{-1}S', C_{11})$ lies on the imaginary axis, Assumptions 4.1 and 4.2 are satisfied. In fact, from one hand, Assumption 4.2 is readily ascertained to hold in view of eqs. (4.41)-(4.43). On the other hand, by performing the required substitutions, it is still straightforward to check that the unobservable part of the pair $[(A - B_2 D'_{12} C_1), (I - D_{12} D'_{12}) C_1]$ coincides with the unobservable part of the pair $[(A - B_2 R^{-1/2} S'), C_{11}]$. Finally, observe that the pair (A, B_2) is stabilizable if and only if the pair (A, B) is such (recall the definition of B_2).

If stated in terms of the system (4.44),(4.45) with the additional output equation $y = [x'\ w']'$, the control problem addressed to by Theorem 4.1 is solved by a controller of the form (4.46),(4.47). Therefore, the controller defined at point (b) of Theorem 4.1 constitutes the solution of the LQ control problem, too.

However, observe that within the context of the classical optimal control theory the linear quadratic problem is stated without requiring the stability of the resulting control system. Thus the assumptions which are necessary to guarantee the existence of the solution in that context (stability of the observable and unreachable part of system $\Sigma(A - BR^{-1}S', B, C_{11}, 0)$) are *weaker* than those required within the RH_2 context. As an example, consider system (4.37) and the performance index (4.39) with

$$A = \begin{bmatrix} \alpha & 1 \\ 0 & 0 \end{bmatrix}, \quad B = \begin{bmatrix} 0 \\ 1 \end{bmatrix}, \quad Q = \begin{bmatrix} 0 & 0 \\ 0 & 1 \end{bmatrix}, \quad R = 1, \quad S = 0$$

where $\alpha = 1$ or $\alpha = -1$. Within the framework of the classical optimal control theory the solution of this problem is given by the control law $u^o_{CL}(x) = -[0\ 1]x$ to which there corresponds the value $J^o_{1CL}(x(0)) = x_2^2(0)$. The resulting control system is stable when $\alpha = -1$ and unstable when $\alpha = 1$. In the RH_2 context the optimal control law is $u^o_{RH_2}(x) = -[4\ 3]x$ when $\alpha = 1$. Correspondingly, the value of the performance index is $J^o_{1RH_2}(x(0)) = 8x_1^2(0) + 8x_1(0)x_2(0) + 3x_2^2(0)$. When $\alpha = -1$ the same control law as in the classical setting is found. In the RH_2 framework the resulting control system is stable in both cases. Finally, notice that $J^o_{1CL}(x(0)) \leq J^o_{1RH_2}(x(0)), \forall x(0)$.

LQS Problem Assume that the controlled system is

$$\dot{x} = Ax + B_1 w + B\bar{u}$$

where w is a zero mean white noise with identity intensity. Let the pair (A, B) be stabilizable and consider either the cost functional

$$J_2 := \lim_{t \to \infty} E\left[x'(t)Qx(t) + 2x'(t)S\bar{u}(t) + \bar{u}'(t)R\bar{u}(t)\right]$$

or the cost functional

$$J_3 := \lim_{T \to \infty} E\left[\frac{1}{T} \int_0^T [x'(t)Qx(t) + 2x'(t)S\bar{u}(t) + \bar{u}'(t)R\bar{u}(t)]dt\right]$$

where matrices Q, R, S satisfy eq. (4.40). From eqs. (4.41)-(4.43) it follows

$$J_2 = \lim_{t \to \infty} E[z'(t)z(t)]$$

$$J_3 = \lim_{T \to \infty} E\left[\frac{1}{T} \int_0^T z'(t)z(t)dt\right]$$

and the controlled system is described by

$$\dot{x} = Ax + B_1 w + B_2 u \tag{4.48}$$

$$z = C_1 x + D_{12} u \tag{4.49}$$

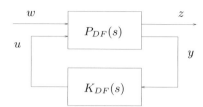

Figure 4.6: The disturbance feedforward problem

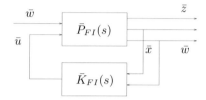

Figure 4.7: The auxiliary full information problem

The optimal control problem at hand (LQS problem) consists in finding a controller of the form given in eqs. (4.46),(4.47) such that stability of the resulting system is ensured and J_2 or J_3 is minimized when it is connected to system (4.48),(4.49). The same arguments exploited for the LQ problem lead to $J_2 = J_3 = \|T(z, w; s)\|_2^2$. If system (4.48),(4.49) verifies the same assumptions as system (4.44),(4.45), then the solution of the LQS problem is again the one specified under point (b) of Theorem 4.1.   □

Remark 4.5 (Disturbance feedforward) Here reference is made to the block-scheme of fig. 4.6, where $P_{DF}(s)$ is described by the equations

$$\dot{x} = Ax + B_1 w + B_2 u \tag{4.50}$$
$$z = C_1 x + D_{12} u \tag{4.51}$$
$$y = C_2 x + w \tag{4.52}$$

while $K_{DF}(s)$ is a generic controller admissible in RH_2 for $P_{DF}(s)$.

It is assumed that the pair (A, B_2) is stabilizable and no eigenvalue of the unobservable part of the pair $[(A - B_2 D'_{12}C_1), (I - D_{12}D'_{12})C_1]$ lies on the imaginary axis. Moreover the matrix $A - B_1 C_2$ is supposed to be stable and $D'_{12}D_{12} = I$. Observe that stability of $A - B_1 C_2$ implies detectability of the pair (A, C_2) which, together with the stabilizability of the pair (A, B_2), guarantees the fulfillment of the necessary condition reported in Remark 4.1.

Now, notice that system $P_{DF}(s)$ is equal to system $\hat{P}(s)$ defined in Lemma E.2. Moreover, let

$$\bar{P}_{FI}(s) := \left[\begin{array}{c|cc} A & B_1 & B_2 \\ \hline C_1 & 0 & D_{12} \\ I & 0 & 0 \\ 0 & I & 0 \end{array} \right]$$

Observe that: (i) System $\bar{P}_{FI}(s)$ is equal to system $\bar{P}(s)$ defined in Lemma E.2; (ii) System $\bar{P}_{FI}(s)$ is equal to system $P_{FI}(s)$ relative to which the full control problem has been stated.

Lemma E.2 (which can be exploited, thanks to stability of matrix $A - B_1 C_2$) ensures that the controller $\bar{K}_{FI}(s)$ connected to system $\bar{P}_{FI}(s)$ as shown in fig. 4.7, stabilizes system

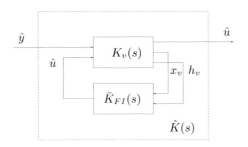

Figure 4.8: The structure of the controller $\hat{K}(s)$ in terms of $\bar{K}_{FI}(s)$

$\bar{P}_{FI}(s)$ if and only if the controller $\hat{K}(s)$, defined in the block-scheme of fig. 4.8, where

$$K_v(s) := \left[\begin{array}{c|cc} A - B_1C_2 & B_1 & B_2 \\ \hline 0 & 0 & I \\ I & 0 & 0 \\ -C_2 & I & 0 \end{array} \right]$$

stabilizes $\hat{P}(s)$ to which it is connected according to the scheme shown in fig. 4.9, this latter scheme being identical to the one depicted in fig. 4.6. Moreover, the transfer function from $\bar{w}$ to $\bar{z}$ in fig. 4.7 and the transfer function from $\hat{w}$ to $\hat{z}$ in fig. 4.9 are equal. Therefore, the solution of Problem 4.1 relative to system $P_{DF}(s)$ (fig. 4.6) can be found by solving the same problem relative to system $\bar{P}_{FI}(s)$ (fig. 4.7) via Theorem 4.1. Notice that such a theorem can be exploited because the assumptions made for system $P_{DF}(s)$ imply the fulfillment of Assumptions 4.1 and 4.2. Thus

a) $\min \|T(z, w; s)\|_2 = \|P_c(s)B_1\|_2 = \sqrt{\operatorname{trace}[B_1' P_2 B_1]}$;

b) The optimal controller is given by

$$K^o_{DF}(s) = \left[\begin{array}{c|c} A - B_1C_2 + B_2F_2 & B_1 \\ \hline F_2 & 0 \end{array} \right]$$

c) The set $\mathcal{F}_{2\gamma r}$ of the controllers $K_{DFr}(s)$ is defined by the block-scheme of fig. 4.10 where

$$N_2(s) := \left[\begin{array}{c|cc} A - B_1C_2 + B_2F_2 & B_1 & B_2 \\ \hline F_2 & 0 & I \\ -C_2 & I & 0 \end{array} \right]$$

$Q(s) := \Sigma(A_q, B_q, C_q, 0)$, $\|Q(s)\|_2^2 < \gamma^2 - \|P_c(s)B_1\|_2^2$ and A_q is a stable matrix.

In the three points above γ is a positive scalar such that $\gamma > \|P_c(s)B_1\|_2$ and reference has been made to eqs. (4.15)-(4.20). The problem at hand is referred to as the *disturbance feedforward* problem in view of the following discussion.

Preliminarily, observe that, if in eq. (4.52) $C_2 = 0$, then $y = w$ so that the disturbance w can be measured (*direct compensation*). However, in such a case the controller $K_{DF}(s)$ is connected to the controlled system $P_{DF}(s)$ in an *open* rather than *closed* loop configuration (see also fig. 4.6), so that the stability assumption of the matrix $A - B_1C_2$ simply reduces to the stability assumption of the controlled system, which is obviously necessary to ensure the stability of the resulting control system.

If, on the contrary, $C_2 \neq 0$, it is still possible to get the disturbance w in terms of the variables u and y. In fact, from eqs. (4.50),(4.52) it follows

$$y_{L0} = P_{21}(s)w_L + P_{22}(s)u_L \tag{4.53}$$

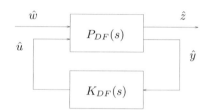

Figure 4.9: The equivalent disturbance feedforward problem

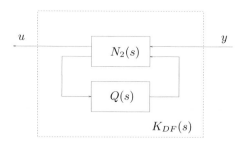

Figure 4.10: The set $\mathcal{F}_{2\gamma r}$ of the controllers $K_{DFr}(s)$

where

$$P_{21}(s) := \left[\begin{array}{c|c} A & B_1 \\ \hline C_2 & I \end{array}\right] \;, \quad P_{22} := \left[\begin{array}{c|c} A & B_2 \\ \hline C_2 & 0 \end{array}\right]$$

System $P_{21}(s)$ is invertible (recall what has been presented in Section 2.5) so that from eq. (4.53) one gets

$$w_L = P_{21}^{-1}(s)[y_{L0} - P_{22}(s)u_L]$$

with

$$P_{21}^{-1}(s) = \left[\begin{array}{c|c} A - B_1 C_2 & B_1 \\ \hline -C_2 & I \end{array}\right]$$

Now define a precompensator $K_{FIDF}(s)$ by means of the equations

$$\dot{\xi} = (A - B_1 C_2)\xi + B_2 u + B_1 y \tag{4.54}$$

$$\hat{w} = -C_2\xi + y \tag{4.55}$$

Comparing eqs. (4.50),(4.52) with eq. (4.54) leads to the conclusion that, letting $\varepsilon := \xi - x$,

$$\dot{\varepsilon} = (A - B_1 C_2)\varepsilon \tag{4.56}$$

Thus, $x_{L0} = \xi_{L0}$ and, from eqs. (4.52),(4.55), also $w_L = \hat{w}_{L0}$, so that, also when $C_2 \neq 0$, it is still possible to think to w as being measurable by resorting to a suitable dynamical system (4.54),(4.55) (*indirect compensation*).

Therefore, one can consider Problem 4.1 relative to system (4.50)-(4.52), rather than to system (4.50),(4.51) with the additional equation

$$\hat{y} := \left[\begin{array}{c} x \\ w \end{array}\right] \tag{4.57}$$

The new problem has the very structure of a full information problem. Thus, the controller $K_{FI}(s)$ which solves Problem 4.1 relative to system (4.50),(4.51),(4.57) must be connected

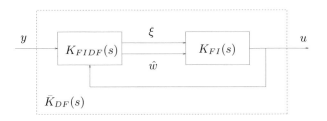

Figure 4.11: $K_{DF}(s)$ in terms of the precompensator $K_{FIDF}(s)$

to system (4.54),(4.55) in order to obtain the controller $K_{DF}(s)$ (see also fig. 4.11). The resulting system includes an unreachable part (recall eq. (4.56)) which is nevertheless stable, thanks to the stated assumptions. Moreover, it is easy to check that whenever the controller solving Problem 4.1 relative to system (4.50),(4.51),(4.57) belongs to the set of controllers $K_{FIr}(s)$, then the controller which is obtained by substituting $K_{FI}(s)$ with $K_{FIr}(s)$ in fig. 4.11 belongs to the set of controllers $K_{DFr}(s)$. In other words, the procedure just now presented for solving Problem 4.1 leads to the results previously found.

Finally, notice that $\mathcal{F}_{2\gamma r} = \mathcal{F}_{2\gamma}$. In fact, by recalling what has been said at the end of Remark 4.3, resorting to the controller $K_{FI}(s)$ rather than to the (simpler) controller $K_{FIr}(s)$ implies adding the term $\Theta_1(s)\Phi(s)\xi(0)$ to the control variable u_{L0}. Such a term depends on the initial state of the precompensator $K_{FIDF}(s)$ only, so that the transfer function from y to u is independent of the parameter $\Theta_1(s)$. Alternatively, from eqs. (4.21),(4.54),(4.55) one gets

$$\xi_{L0} = \Phi(s)(B_2 v_{L0} + B_1 \hat{w}_{L0})$$

where $\Phi(s) := (sI - A - B_2 F_2)^{-1}$. Hence, by recalling eq. (4.36),

$$v_{L0} = [I - \Theta_1(s)\Phi(s)B_2]^{-1}\{-\Theta_1(s)\Phi(s)(B_2 v_{L0} + B_1 \hat{w}_{L0}) +$$
$$+[Q(s) + \Theta_1(s)\Phi(s)B_1]\hat{w}_{L0}\}$$
$$= [I - \Theta_1(s)\Phi(s)B_2]^{-1}[Q(s)\hat{w}_{L0} - \Theta_1(s)\Phi(s)B_2 v_{L0}]$$

Therefore, it follows

$$v_{L0} = Q(s)\hat{w}_{L0}$$

so that, in view of eq. (4.21),

$$u_{L0} = F_2 \xi_{L0} + Q(s)\hat{w}_{L0}$$

which, thanks to eqs. (4.54),(4.55), implies that the transfer function of the generic controller $K_{DF}(s)$ is independent of $\Theta_1(s)$.   □

Example 4.2 Consider system (4.50)-(4.52) with

$$A = \begin{bmatrix} 0 & 1 \\ 0 & 0 \end{bmatrix}, \quad B_1 = \begin{bmatrix} 0 \\ 2 \end{bmatrix}, \quad B_2 = \begin{bmatrix} 0 \\ 1 \end{bmatrix}$$

$$C_1 = \begin{bmatrix} 1 & \alpha \end{bmatrix}, \quad C_2 = \begin{bmatrix} 1 & 1 \end{bmatrix}, \quad D_{12} = 1$$

where $\alpha \neq 0$. The classical synthesis procedure for a controller which reduces the influence of the disturbance w on the output z consist in trying to make zero the transfer function from w to z by introducing a controller $K_{IC}(s)$ with input y and output u (*indirect compensation*). In the problem at hand such a transfer function is zero (hence z does not depend on w) if

$$K_{IC}(s) = -\frac{2(\alpha s + 1)}{s^2 + (\alpha + 2)s + 3}$$

However, it is easy to verify that the resulting system is stable when $\alpha > 0$, while it is not stable when $\alpha < 0$. Thus, in this second case it is no more possible to perform a (perfect) indirect compensation of the disturbance w. On the contrary, by resorting to Remark 4.5, one gets $K_{DF}^o(s) = K_{IC}(s)$ when $\alpha > 0$ and

$$K_{DF}^o(s) = -2\frac{1-\alpha s}{s^2 + (2-\alpha)s + 3}$$

when $\alpha < 0$.   □

Remark 4.6 Assumption 4.2 is restrictive though standard in control problems. It implies that matrix D_{12} has rank equal to the number of its columns. Whenever such an assumption is not verified one has to tackle the so called *singular* problem the solution of which (if it exists) calls for a theoretical development far beyond the scope of this book.

On the contrary, notice that the condition $D_{12}'D_{12} = I$ can be replaced by $D_{12}'D_{12} = R$, with $R > 0$. Indeed, the here presented case can be easily recovered by a suitable redefinition of the control variable, according to eq. (4.43).

Finally, it is often set $D_{12}'C_1 = 0$. This simplifying *orthogonality* assumption, besides greatly reducing the notational burden, corresponds to the absence of the *cross* term $x'(t)S\bar{u}(t)$ in the performance index of the optimal control problems dealt with in Remark 4.4.   □

Remark 4.7 The Riccati equation (4.18) coincides with the one encountered within the (classical) *Optimal Regulator* problem defined in terms of the system

$$\dot{x} = A_c x + B_2 u$$

and the performance index

$$J := \int_0^\infty [x'C_{1c}'C_{1c}x + u'u]dt$$

                                                                                                    □

Remark 4.8 Under Assumption 4.2, the fact that the no eigenvalue of the unobservable part of the pair $[(A - B_2 D_{12}'C_1), (I - D_{12}D_{12}')C_1]$ lies on the imaginary axis (Assumption 4.1) is equivalent to the subsystem of $P_{FI}(s)$ corresponding to the transfer function $P_{FI12}(s)$ from the input u and the output z (namely system $\Sigma(A, B_2, C_1, D_{12})$) not to have invariant zeros with $Re(s) = 0$. In fact, in view of the material in Section 2.5 and recalling that the dimension of u is not greater than the dimension of z, if it exists λ with $Re(\lambda) = 0$ such that

$$(\lambda I - A)x - B_2 u = 0$$

$$C_1 x + D_{12}u = 0$$

with $[x'\ u']' \neq 0$, then from the identity $D_{12}'D_{12} = I$ it follows

$$u = -D_{12}'C_1 x$$

and hence also

$$(A - B_2 D_{12}'C_1)x = \lambda x$$

$$(I - D_{12}D_{12}')C_1 x = 0$$

Being $x \neq 0$, since, otherwise, also $u = 0$, this violates the assumption that no eigenvalue of the unobservable part of the pair $[(A - B_2 D_{12}'C_1), (I - D_{12}D_{12}')C_1]$ lies on the imaginary axis (recall Lemma D.1, part (a)). By going the other way on, if the last two equations hold corresponding to $x \neq 0$, the proof of the above mentioned equivalence can be carried out by letting $u := -D_{12}'C_1 x$.   □

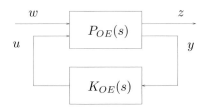

Figure 4.12: The output estimation problem

4.3   The output estimation problem

The problem of optimally observing a linear combination of the state variables is faced in this section by following an approach which relies on RH_2 techniques.

Consider the system

$$\dot{x} = Ax + B_1 w + B_2 u \tag{4.58}$$

$$z = C_1 x + u \tag{4.59}$$

$$y = C_2 x + D_{21} w \tag{4.60}$$

and let $P_{OE}(s)$ denote its transfer function. With reference to the block-scheme in fig. 4.12, suppose that $K_{OE}(s)$ is a RH_2 admissible controller for $P_{OE}(s)$ which makes "small" $\|T(z, w; s)\|_2$ (the norm of the transfer function from w to z). Then, the variable u provides a "good" estimate of the linear combination $-C_1 x$ of the state variables. Indeed, should $T(z, w; s) = 0$ (namely, $z_{L0} = 0$) then it would follows $u_{L0} = -C_1 x_{L0}$. In such a case the signal u apparently constitutes the best possible estimate of $-C_1 x$ (when x(0)=0). The following assumptions are now introduced.

Assumption 4.3 *The pair (A, C_2) is detectable and no eigenvalue of the unreachable part of the pair $[(A - B_1 D'_{21} C_2), B_1(I - D'_{21} D_{21})]$ lies on the imaginary axis.*

Assumption 4.4 $D_{21} D'_{21} = I.$

Assumption 4.5 $A - B_2 C_1$ *is stable.*

Under these assumptions it is possible to state the following theorem.

Theorem 4.2 (Output estimation) *Consider Problem 4.1 relative to system (4.58)-(4.60). Then, under Assumptions 4.3 - 4.5, it has the solution*

a)

$$\min\|T(z, w; s)\|_2 = \|C_1 P_f(s)\|_2 = \sqrt{\operatorname{trace}[C_1 \Pi_2 C'_1]}$$

b)

$$K^o_{OE}(s) = \left[\begin{array}{c|c} A_f - B_2 C_1 - \Pi_2 C'_2 C_2 & L_2 \\ \hline C_1 & 0 \end{array} \right]$$

c) The set $\mathcal{F}_{2\gamma r}$ of controllers $K_{OEr}(s)$ is defined by the block-scheme in fig. 4.13, where

$$M_2(s) := \left[\begin{array}{c|cc} A_f - B_2 C_1 - \Pi_2 C'_2 C_2 & L_2 & -B_2 \\ \hline C_1 & 0 & I \\ C_2 & I & 0 \end{array} \right]$$

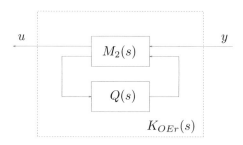

Figure 4.13: The set $\mathcal{F}_{2\gamma r}$ of the controllers $K_{OEr}(s)$

with $Q(s) := \Sigma(A_q, B_q, C_q, 0)$, $\|Q(s)\|_2^2 < \gamma^2 - \|C_1 P_f(s)\|_2^2$ and A_q stable.

In the three preceding points γ is a positive scalar such that $\gamma > \|C_1 P_f(s)\|_2$, while

$$L_2 := -\Pi_2 C_2' - B_1 D_{21}' \tag{4.61}$$

$$P_f(s) := \left[\begin{array}{c|c} A_f - \Pi_2 C_2' C_2 & B_{1f} - \Pi_2 C_2' D_{21} \\ \hline I & 0 \end{array}\right] \tag{4.62}$$

$$A_f := A - B_1 D_{21}' C_2 \ , \quad B_{1f} := B_1(I - D_{21}' D_{21}) \tag{4.63}$$

where Π_2 is the symmetric, positive definite and stabilizing solution of the Riccati equation (in the unknown Π)

$$0 = \Pi A_f' + A_f \Pi - \Pi C_2' C_2 \Pi + B_{1f} B_{1f}' \tag{4.64}$$

that is such that matrix A_{fc} defined by

$$A_{fc} := A_f - \Pi_2 C_2' C_2 = A + L_2 C_2 \tag{4.65}$$

is stable.

Proof Preliminarily, notice that Assumption 4.5 implies that the pair (A, B_2) is stabilizable. This fact, together with the assumed detectability of the pair (A, C_2), makes the necessary condition in Remark 4.1 satisfied.

Now consider system $\hat{P}_{OE}(s)$ obtained from system $P_{OE}(s)$ by transposition. From eqs. (4.58)-(4.60) it follows that it is described by

$$\dot{\xi} = F\xi + G_1\zeta + G_2\eta$$
$$\omega = H_1\xi + E\eta$$
$$\varphi = H_2\xi + \zeta$$

where

$$F := A' \ , \quad E := D_{21}' \tag{4.66}$$

and, for $i = 1, 2$,

$$G_i := C_i' \ , \quad H_i := B_i' \tag{4.67}$$

Therefore, system $\hat{P}_{OE}(s)$ possesses the very same structure of system $P_{DF}(s)$ considered in Remark 4.5. Thus, Problem 4.1 relative to system $\hat{P}_{OE}(s)$ is solved as it is mentioned in such a remark (relatively to $P_{OE}(s)$). Indeed, the assumption required for system $P_{OE}(s)$ are precisely Assumptions 4.3 - 4.5. Thanks to Lemma E.1, the results concerning system $P_{OE}(s)$ can be derived by transposition of those relevant to system $\hat{P}_{OE}(s)$, provided that eqs. (4.66),(4.67) are taken into account. $\square$

Example 4.3 Consider system (4.58)-(4.60) with

$$A = \begin{bmatrix} 0 & 1 \\ 0 & 0 \end{bmatrix}, \quad B_1 = \begin{bmatrix} 1 & 0 \\ 1 & 0 \end{bmatrix}, \quad B_2 = \begin{bmatrix} 0 \\ 1 \end{bmatrix}$$

$$C_1 = \begin{bmatrix} 1 & 1 \end{bmatrix}, \quad C_2 = \begin{bmatrix} 1 & 0 \end{bmatrix}, \quad D_{21} = \begin{bmatrix} 0 & 1 \end{bmatrix}$$

One obtains $L_2 = [-\sqrt{3} \ -1]'$ and $\|C_1 P_f(s)\|_2^2 = 4.46$. Taken $\gamma^2 = 9$ and

$$Q(s) = \left[\begin{array}{cc|c} -1 & 0 & 1 \\ 0 & -2 & 1 \\ \hline 1 & 1 & 0 \end{array} \right]$$

it is $\|T(z, w; s)\|_2^2 = 5.88 < \gamma^2$, consistently with $1.42 = \|Q(s)\|_2^2 < \gamma^2 - 4.46$. □

Remark 4.9 The structure of a generic controller $K_{OE}(s)$ admissible in RH_2 for $P_{OE}(s)$ as defined by the block-scheme of fig. 4.13 allows one to easily verify that the eigenvalues of the resulting control system are those of matrices $A + L_2 C_2$, $A - B_2 C_1$ and A_q. In fact, letting x_m and x_q be the state variables of systems $M_2(s)$ and $Q(s)$, respectively, and choosing the state vector of the resulting control system as $x_t := [x' + x'_m \ x'_q \ x']'$, its dynamic matrix is

$$A_t = \begin{bmatrix} A + L_2 C_2 & 0 & 0 \\ B_q C_2 & A_q & 0 \\ B_2 C_1 & B_2 C_q & A - B_2 C_1 \end{bmatrix}$$

the eigenvalues of which are precisely those above mentioned.

Observe that the order of the resulting system is $2n + n_q$, where n is the order of system (4.58)-(4.60) and n_q is the order of system $Q(s)$. Moreover, when $Q(s) = 0$ it results $T(z, w; s) = C_1(sI - A_{fc})^{-1}(B_1 + L_2 D_{21})$, so that the transfer function from w to z does not depend on B_2 (recall that, in view of eqs. (4.61),(4.63),(4.64), L_2 is independent of B_2). □

Remark 4.10 (Parametrization of the set $\mathcal{F}_{2\gamma}$) Notice that $\mathcal{F}_{2\gamma} = \mathcal{F}_{2\gamma r}$. In fact, as shown in the proof of Theorem 4.2, the RH_2 admissible controllers for $P_{OE}(s)$ can be obtained by transposing those which are admissible for $\hat{P}_{OE}(s)$, this last system possessing the structure of system $P_{DF}(s)$ considered in Remark 4.5. Having proved that $\mathcal{F}_{2\gamma} = \mathcal{F}_{2\gamma r}$ relative to $P_{DF}(s)$, the same conclusion must hold for $P_{OE}(s)$. □

Remark 4.11 (Optimal filtering) The control problem relative to system $P_{OE}(s)$ can be interpreted as an *optimal filtering* problem relative to the *n-th* order system

$$\dot{x} = Ax + \zeta_1 + B_2 u \tag{4.68}$$
$$\bar{y} = Cx + \zeta_2 \tag{4.69}$$

where $\zeta := [\zeta_1' \ \zeta_2']'$ is a zero mean, Gaussian white noise with intensity

$$W := \begin{bmatrix} W_{11} & W_{12} \\ W_{12}' & W_{22} \end{bmatrix}, \quad W_{22} > 0 \tag{4.70}$$

If an asymptotic estimate of a generic linear combination Sx of the system state has to be found, then one of the two functionals

$$J_4 := \lim_{t \to \infty} E\left[[Sx(t) - u(t)]'[Sx(t) - u(t)] \right]$$

and

$$J_5 := \lim_{T \to \infty} E\left[\frac{1}{T} \int_0^T [Sx(t) - u(t)]'[Sx(t) - u(t)]dt \right]$$

may conveniently be associated with the system. Notice that $\hat{W} := W_{11} - W_{12}W_{22}^{-1}W_{12}' \geq 0$, since $W \geq 0$ and $W_{22} > 0$. Indeed, it is $\hat{W} = Z'WZ$, with

$$Z := \begin{bmatrix} I \\ -W_{22}^{-1}W_{12}' \end{bmatrix}$$

Let the $n \times n$ matrix B_{11} be a factorization of $\hat{W}$, so that

$$B_{11}B_{11}' = \hat{W} \tag{4.71}$$

and define

$$D_{21} := [0 \ \ I] \ , \quad B_1 := [B_{11} \ \ W_{12}W_{22}^{-1/2}] \tag{4.72}$$

$$y := W_{22}^{-1/2}\bar{y} \ , \quad C_2 := W_{22}^{-1/2}C \tag{4.73}$$

$$z := C_1 x + u \ , \quad C_1 := -S \tag{4.74}$$

Then system (4.68)-(4.70) can be rewritten as

$$\dot{x} = Ax + B_1 w + B_2 u \tag{4.75}$$

$$z = C_1 x + u \tag{4.76}$$

$$y = C_2 x + D_{21}w \tag{4.77}$$

where w is the zero mean Gaussian white noise with identity intensity which satisfies the equation

$$\begin{bmatrix} \zeta_1 \\ \zeta_2 \end{bmatrix} = \begin{bmatrix} B_{11} & W_{12}W_{22}^{-1/2} \\ 0 & W_{22}^{1/2} \end{bmatrix} w \tag{4.78}$$

Observe that

$$B_1 B_1' = W_{11} \ , \quad B_1 D_{21}' = W_{12}W_{22}^{-1/2} \ , \quad D_{21}D_{21}' = I \tag{4.79}$$

It is easy to verify that

$$J_4 = \lim_{t \to \infty} E[z'(t)z(t)]$$

and

$$J_5 = \lim_{T \to \infty} E\left[\frac{1}{T}\int_0^T z'(t)z(t)dt\right]$$

The problem at hand consists in finding a controller of the form

$$\dot{\xi} = F\xi + Gy \tag{4.80}$$

$$u = H\xi + Ny \tag{4.81}$$

such that the control system (4.75)-(4.81) is stable and either the criterion J_4 or J_5 is minimized. Observe that the feedback connection of any controller of the above form with system (4.75)-(4.79) is always well defined, while the transfer function $T(z, w; s)$ from w to z relevant to such a connection is strictly proper if and only if $ND_{21} = 0$, that is, in view of eq. (4.72), if and only if $N = 0$. Therefore, if system (4.75)-(4.81) is stable and $N = 0$, then the controller (4.80),(4.81) is RH_2 admissible for system (4.75)-(4.79). Further, from Remark 2.20, any RH_2 admissible controller for system (4.75)-(4.79) is such that

$$J_4 = J_5 = \|T(z, w; s)\|_2^2$$

If the pair (A, C) is detectable and no eigenvalue of the unreachable part of the pair $[(A - W_{12}W_{22}^{-1}C_1), B_{11}]$ lies on the imaginary axis, then Assumptions 4.3 and 4.4 are verified. In fact, eqs. (4.71)-(4.74),(4.79) are readily seen to imply the fulfillment of Assumption 4.4, while, by performing the required substitutions, it is easy to ascertain that the unreachable part of the pair $[(A - W_{12}W_{22}^{-1}C_1), B_{11}]$ coincides with the unreachable part of the pair

$[(A - B_1 D_{21}' C_2), B_1(I - D_{21}' D_{21})]$. As for detectability of the pair (A, C_2), it is equivalent to the detectability of the pair (A, C), thanks to eq. (4.73).

Finally, if matrix $A + B_2 S$ is stable, so that also Assumption 4.5 is satisfied, then Theorem 4.2 (point (b)) ensures that the controller which solves the underlying problem relative to system (4.75)-(4.79) has the form (4.80),(4.81) with $N = 0$. Precisely, the controller is

$$\dot{\xi} = (A - B_2 C_1 + L_2 C_2)\xi + L_2 y \tag{4.82}$$
$$u = C_1 \xi \tag{4.83}$$

and provides a solution of the considered filtering problem. Here L_2 is given by eqs. (4.61),(4.63) and (4.64). However, the filtering problem for system (4.68)-(4.70) could have been more classically tackled via Kalman theory, yielding the filter

$$\dot{\xi}_K = (A + L_K C)\xi_K - L_K \bar{y} \tag{4.84}$$
$$u_K = S \xi_K \tag{4.85}$$

where u_K is the optimal estimate of Sx. In eq. (4.84) it is

$$L_K = -(\Pi_K C' + W_{12}) W_{22}^{-1}$$

and Π_K is the symmetric, positive definite and *minimal* solution of the Riccati equation (in the unknown Π)

$$0 = \Pi(A - W_{12} W_{22}^{-1} C)' + (A - W_{12} W_{22}^{-1} C)\Pi - \Pi C' W_{22}^{-1} C\Pi + \hat{W}$$

where, again, $\hat{W} = W_{11} - W_{12} W_{22}^{-1} W_{12}'$. Notice that this equation coincides with eq. (4.64) once the substitutions (4.71)-(4.73) have been performed.

The assumption which guarantees the existence of Π_K (namely, the stability of the unobservable but reachable part of system $\Sigma(A - W_{12} W_{22}^{-1} C, \hat{W}^{1/2}, C, 0)$), is weaker than those assuring the existence of the stabilizing (i.e. *maximal*) solution of eq. (4.64). Hence, in general, Π_K might exist and Π_2 not exist; moreover it can also happen that $\Pi_K \neq \Pi_2$ so that the RH_2 and Kalman filtering problems may substantially differ one from the other.

However, if the Kalman filter is required to be stable, then $\Pi_2 = \Pi_K$ and $L_2 = L_K W_{22}$. Despite of being the stable Kalman filter a device different from the RH_2 controller, the transfer function from the input noise ζ to the estimation error $S(\xi_K - x)$ (system (4.68),(4.69),(4.84) and (4.85)) coincides with the transfer function from the input noise w to the estimation error $z = C_1(x + \xi)$ (system (4.75)-(4.77),(4.82) and (4.83)). On the contrary, the two devices do coincide in the particular case where $B_2 = 0$, $W_{12} = 0$ and A stable. $\square$

Example 4.4 Consider system (4.68)-(4.70) with

$$A = \begin{bmatrix} 0 & 1 \\ -1 & 0 \end{bmatrix}, \quad B_2 = \begin{bmatrix} 0 \\ -1 \end{bmatrix}, \quad C = \begin{bmatrix} 1 & 0 \end{bmatrix}$$

$$W_{11} = \begin{bmatrix} 1 & 0 \\ 0 & 1 \end{bmatrix}, \quad W_{12} = \begin{bmatrix} 0 \\ 0 \end{bmatrix}, \quad W_{22} = 1$$

The Kalman filter relative to the linear combination $\eta := Sx$ with $S = [1 \ 1]$ is described by the equations

$$\dot{\xi}_K = (A + L_K C)\xi_K + B_2 u - L_K \bar{y}$$
$$u_K = S\xi$$

where $L_K = -[1.35 \ 0.41]$ and $(A + L_K C)$ is stable since the relevant Riccati equation admits a unique symmetric and positive definite solution which is also stabilizing.

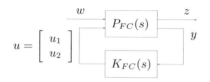

$$u = \begin{bmatrix} u_1 \\ u_2 \end{bmatrix}$$

Figure 4.14: The full control problem

If, on the contrary, Theorem 4.2 has to be exploited according to Remark 4.11, the controller (corresponding to the choice $Q(s) = 0$) is obtained

$$\dot{\xi} = (A - B_2 C_1 + L_2 C_2)\xi + L_2 y$$
$$u = C_1 \xi$$

where $C_1 = -S$, $C_2 = C$, $y = \bar{y}$, $L_2 = L_K$. Thus the optimal estimate of η is $-C_1\xi$.   □

Remark 4.12 (Full control) Here reference is made to the block-scheme of fig. 4.14 where the control vector u is partitioned into two components u_1 and u_2. Assume that the first one of them acts in a direct way on the state derivative only, while the second one directly affects the performance output only. More precisely, the considered system $P_{FC}(s)$ is described by the equations

$$\dot{x} = Ax + B_1 w + [I\ 0]u \tag{4.86}$$
$$z = C_1 x + [0\ I]u \tag{4.87}$$
$$y = C_2 x + D_{21} w \tag{4.88}$$
$$u = [u_1'\ u_2']' \tag{4.89}$$

Further, let the pair (A, C_2) be detectable, $D_{21}D_{21}' = I$ and assume that no eigenvalue of the unreachable part of the pair $[(A - B_1 D_{21}'C_2), B_1(I - D_{21}'D_{21})]$ lies on the imaginary axis.

First, observe that it makes sense dealing with Problem 4.1 relative to the system above since the necessary condition in Remark 4.1 is verified. Indeed, stabilizability of the pair (A, B_2) is guaranteed by the form of matrix B_2, while detectability of the pair (A, C_2) holds by assumption. Now consider system $\hat{P}_{FC}(s) := P_{FC}'(s)$. From eqs. (4.86)-(4.89) it follows that $\hat{P}_{FC}(s)$ is described by

$$\dot{\xi} = F\xi + G_1\zeta + G_2\eta$$
$$\omega = H_1\xi + E\eta$$
$$\varphi = \begin{bmatrix} I \\ 0 \end{bmatrix}\xi + \begin{bmatrix} 0 \\ I \end{bmatrix}\zeta$$

where

$$F := A' , \quad E := D_{21}' , \quad H_1 := B_1'$$

and, for $i = 1, 2$,

$$G_i := C_i'$$

Therefore, system $\hat{P}_{FC}(s)$ possesses the same structure as system $P_{FI}(s)$ which has been considered in Section 4.2. The assumptions on system $P_{FC}(s)$ make system $\hat{P}_{FC}(s)$ to satisfy Assumptions 4.1, 4.2, so that the results concerning the solution of Problem 4.1 relative to system $P_{FC}(s)$ can be derived, thanks to Lemma E.1, by transposing those concerning system $\hat{P}_{FC}(s)$ which, in turn, coincide with those supplied by Theorem 4.1 for the full information problem. Thus one obtains

a)

$$\min \|T(z, w; s)\|_2 = \|C_1 P_f(s)\|_2 = \sqrt{\text{trace}[C_1\Pi_2 C_1']}$$

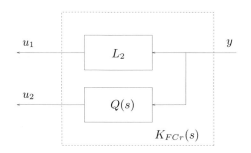

Figure 4.15: The set $\mathcal{F}_{2\gamma r}$ of the controllers $K_{FCr}(s)$

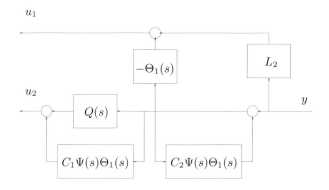

Figure 4.16: The generic admissible controller for $P_{FC}(s)$

b)

$$K_{FC}^o(s) = \begin{bmatrix} \emptyset & \emptyset \\ \hline \emptyset & L_2 \\ \emptyset & 0 \end{bmatrix}$$

c) The set $\mathcal{F}_{2\gamma r}$ of the controllers $K_{FCr}(s)$ is defined in the block-scheme of fig. 4.15, where $Q(s) := \Sigma(A_q, B_q, C_q, 0)$, with the matrix A_q stable and $\|Q(s)\|_2^2 < \gamma^2 - \|C_1 P_f(s)\|_2^2$.

In the three points above $\gamma > \|C_1 P_f(s)\|_2$ and reference has been made to eqs. (4.61)-(4.65).

Finally, the set $\mathcal{F}_{2\gamma}$ of the RH_2 admissible controllers for system $P_{FC}(s)$ can easily be found by exploiting (through transposition) the content of Remark 4.3 (which refers to system $P_{FI}(s)$). Therefore, this set is described by the block-scheme of fig. 4.16 (which has been obtained by "transposing" fig. 4.5), where $\Theta_1(s) \in RH_\infty$ and $\Psi(s) := (sI - A_{fc})^{-1}$. $\square$

Remark 4.13 Assumption 4.4 is somehow restrictive though customary in estimation theory. It implies that matrix D_{21} has rank equal to the number of its rows. Should this assumption not be verified one would have to face a *singular* problem the solution of which, if any, requires a discussion far beyond the scope of this book.

On the contrary, observe that the condition $D_{21} D_{21}' = I$ can be substituted, without troubles, by $D_{21} D_{21}' = R$ with $R > 0$. Indeed, the present derivation can be exploited by redefining the output variable as shown in eq. (4.73).

Finally, the *orthogonality* assumption $D_{21} B_1' = 0$ is often made. Besides making simpler the notation, such an assumption implies that the noises ζ_1 and ζ_2, introduced in Remark 4.11 when dealing with filtering problems, are uncorrelated ($W_{12} = 0$). $\square$

Remark 4.14 The Riccati equation (4.64) is the one encountered in deriving the Kalman filter for the system $\dot{x} = Ax + w_1$, $y = C_2x + w_2$, where the zero mean Gaussian white noise $w := [w_1' \ w_2']'$ has intensity

$$W := \begin{bmatrix} B_1 B_1' & B_1 D_{21}' \\ D_{21} B_1' & I \end{bmatrix}$$

$\square$

Remark 4.15 Under Assumption 4.4, the eigenvalues of the unreachable part of the pair $[(A - B_1 D_{21}' C_2), B_1(I - D_{21}' D_{21})]$ do not lie on the imaginary axis (Assumption 4.3) if and only if the invariant zeros of the subsystem of $P_{OE}(s)$ corresponding to the transfer function $P_{OE21}(s)$ from the input w to the output y, namely the system $\Sigma(A, B_1, C_2, D_{21})$, all have real part different from zero. In fact, from Section 2.5 and by recalling that the number of components of the disturbance w is not smaller than the number of components of the output y, if a scalar λ with $Re(\lambda) = 0$ exists such that

$$(\lambda I - A')x - C_2'y = 0$$
$$B_1'x + D_{21}'y = 0$$

with $[x' \ y']' \neq 0$, then, from the identity $D_{21} D_{21}' = I$ it follows

$$y = -D_{21} B_1' x$$

so that

$$(A' - C_2' D_{21} B_1')x = \lambda x$$
$$(I - D_{21}' D_{21})B_1' x = 0$$

Being $x \neq 0$, since, otherwise, also $y = 0$, these two equations would violate the assumption that no eigenvalue of the unreachable part of the pair $[(A - B_1 D_{21}' C_2), B_1(I_D 21' D_{21})]$ lies on the imaginary axis (recall Lemma D.3, point (a)). On the contrary, if the two last relations hold true for a certain $x \neq 0$, then letting $y := -D_{21} B_1' x$ and proceeding in the reverse way the conclusion straightforwardly follows. $\square$

Remark 4.16 The above results can be generalized to the fairly frequent case in which the output variable y explicitly depends on the control variable u, that is when eq. (4.60) is substituted by

$$y = C_2x + D_{12}w + D_{22}u \tag{4.90}$$

Indeed, letting

$$\bar{y} := y - D_{22}u = C_2x + D_{12}w \tag{4.91}$$

Problem 4.1 relative to system $\bar{P}_{OE}(s)$ (described by eqs. (4.58),(4.59), (4.91)) admits the solution presented in Theorem 4.2. If $\bar{K}_{OE}(s)$ is a RH_2 admissible controller for $\bar{P}_{OE}(s)$, then the controller $K_{OE}(s)$ defined in the block-scheme of fig. 4.17 is apparently a RH_2 admissible controller for system $P_{OE}(s)$ (described by eqs. (4.58),(4.59),(4.90)), only provided that it is well defined. This is certainly the case since the set of controllers $\bar{K}_{OE}(s)$ is constituted by strictly proper systems (see Theorem 4.2). $\square$

4.4   The partial information problem

In the control problem considered in the present section only a partial information on the system state is available to the controller. Therefore, the controlled system is described by

$$\dot{x} = Ax + B_1w + B_2u \tag{4.92}$$
$$z = C_1x + D_{12}u \tag{4.93}$$
$$y = C_2x + D_{21}w \tag{4.94}$$

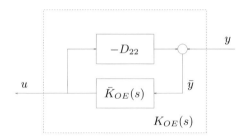

Figure 4.17: The controller $K_{OE}(s)$ when y directly depends on u

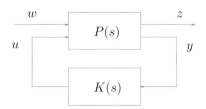

Figure 4.18: The partial information problem

and its transfer function is denoted by $P(s)$. Further, the following assumptions are done.

Assumption 4.6 *The pair* (A, B_2) *is stabilizable and the pair* (A, C_2) *is detectable.*

Assumption 4.7 $D'_{12}D_{12} = I$.

Assumption 4.8 *The eigenvalues of the unobservable and unreachable part of the pairs* $[(A - B_2 D'_{12}C_1)(I - D_{12}D'_{12})C_1]$ *and* $[(A - B_1 D'_{21}C_2), B_1(I - D'_{21}D_{21})]$ *respectively, do not lie on the imaginary axis.*

Assumption 4.9 $D_{21}D'_{21} = I$.

In the forthcoming theorem reference is made to the block-scheme of fig. 4.18 where $K(s)$ denotes a generic RH_2 admissible controller for $P(s)$ and $T(z, w; s)$ is the transfer function from w to z.

Theorem 4.3 (Partial information) *Consider system (4.92)-(4.94). Then, under Assumptions 4.6 - 4.9, Problem 4.1 has the following solution.*

a)

$$
\begin{aligned}
\min \|T(z, w; s)\|_2^2 &= \|P_c(s)B_1\|_2^2 + \|F_2 P_f(s)\|_2^2 \\
&= \|P_c(s)L_2\|_2^2 + \|C_1 P_f(s)\|_2^2 \\
&:= (\gamma^o)^2
\end{aligned}
$$

b)

$$
K^o(s) := \left[\begin{array}{c|c} A + B_2 F_2 + L_2 C_2 & -L_2 \\ \hline F_2 & 0 \end{array} \right]
$$

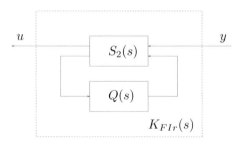

Figure 4.19: The set $\mathcal{F}_{2\gamma r}$ of the controllers $K_{FIr}(s)$

c) The set $\mathcal{F}_{2\gamma r}$ of the controller $K_r(s)$ is defined by the block-scheme of fig. 4.19 where

$$S_2(s) := \left[\begin{array}{c|cc} A + B_2 F_2 + L_2 C_2 & -L_2 & B_2 \\ \hline F_2 & 0 & I \\ -C_2 & I & 0 \end{array} \right]$$

$Q(s) := \Sigma(A_q, B_q, C_q, 0)$, the matrix A_q is stable and $\|Q(s)\|_2^2 < \gamma^2 - (\gamma^o)^2$.

In the three points above reference has been made to eqs. (4.15)-(4.18) and (4.61)-(4.64), while γ is a positive scalar such that $\gamma > \gamma^o$.

Proof First observe that Assumption 4.6 coincides with the necessary condition for the problem at hand to make sense (see Remark 4.1). Then notice that the above assumptions guarantee, thanks to Lemma C.3, the existence of the solutions P_2 and Π_2 of the Riccati equations (4.18) and (4.64) endowed with the relevant properties. Repeat now the first part of the proof of Theorem 4.1 (eqs. (4.20)-(4.26)) by making reference to fig. 4.18 and 4.19 rather than to fig. 4.2 and 4.3, respectively, and defining the system $P_v(s)$ in fig. 4.4 as

$$P_v(s) := \left[\begin{array}{c|cc} A & B_1 & B_2 \\ \hline -F_2 & 0 & I \\ C_2 & D_{21} & 0 \end{array} \right]$$

Then, one gets

$$\|T(z, w; s)\|_2^2 = \|P_c(s)B_1\|_2^2 + \|T(v, w; s)\|_2^2$$

On the other hand, system $P_v(s)$ has the same structure as system $P_{OE}(s)$ (the system considered in Section 4.3) and equals it if one let $-F_2 = C_1$. It is easy to ascertain that under Assumptions 4.6 - 4.9, Assumptions 4.3 - 4.5 are verified for system $P_v(s)$. This is straightforward as for the first two of them, while Assumption 4.5, namely stability of the matrix $A - B_2 C_1 = A + B_2 F_2$, follows from P_2 being the stabilizing solution of the Riccati equation (recall eq. (4.19)). Since the set of RH_2 admissible controllers for $P(s)$ coincides with the set of RH_2 admissible controllers for $P_v(s)$, it is possible to minimize $\|T(v, w; s)\|_2$ by resorting to Theorem 4.2. Thus the first equality sign in point a) and points b) and c) follow.

As for the second equality sign in point a), notice that system $\hat{P}(s) := P'(s)$ has the same structure as $P(s)$. Therefore, the solution of Problem 4.1 relative to system $\hat{P}(s)$ is fully described by the statement of Theorem 4.3 apart, for the moment being, from the second part of point a). The relevant results can be utilized also for system $P(s)$, thanks to Lemma E.1. In particular, if a *cap* sign marks the items concerning system $\hat{P}(s)$ which correspond to those introduced for system $P(s)$, in view of point (a) of Theorem 4.3, first equality sign, one obtains

$$\min \|\hat{T}(\hat{z}, \hat{w}; s)\|_2^2 = \|\hat{P}_c(s)\hat{B}_1\|_2^2 + \|\hat{F}_2\hat{P}_f(s)\|_2^2$$
$$= \|P_c(s)L_2\|_2^2 + \|C_1 P_f(s)\|_2^2$$

since $\hat{P}'_c(s) = P_f(s)$, $\hat{B}'_1 = C_1$, $\hat{P}'_f(s) = P_c(s)$ and $\hat{F}'_2 = L_2$, as it can be verified. □

Example 4.5 Consider system (4.92)-(4.94) with

$$A = \begin{bmatrix} 0 & 1 \\ 0 & 0 \end{bmatrix}, \quad B_1 = \begin{bmatrix} 1 & 0 \\ 1 & 0 \end{bmatrix}, \quad B_2 = \begin{bmatrix} 0 \\ 1 \end{bmatrix}, \quad D_{12} = \begin{bmatrix} 0 \\ 1 \end{bmatrix}$$

$$C_1 = \begin{bmatrix} 1 & 0 \\ 0 & 0 \end{bmatrix}, \quad C_2 = \begin{bmatrix} 1 & 0 \end{bmatrix}, \quad D_{21} = \begin{bmatrix} 0 & 1 \end{bmatrix}$$

One obtains $F_2 = -[1 \ \sqrt{2}]$, $L_2 = -[\sqrt{3} \ 1]'$ and

$$(\gamma^o)^2 := \|P_c(s)B_1\|_2^2 + \|F_2 P_f(s)\|_2^2 = 10.85$$

Taken $\gamma = 16$ and

$$Q(s) = \left[\begin{array}{cc|c} -1 & 0 & 1 \\ 0 & -3 & 1 \\ \hline 1 & 1 & 0 \end{array} \right]$$

it is $\|T(z, w; s)\|_2^2 = 12.02 < \gamma^2$, consistently with $1.17 = \|Q(s)\|_2^2 < \gamma^2 - (\gamma^o)^2$. □

Example 4.6 Consider system (4.92)-(4.94) with

$$A = \begin{bmatrix} 0 & 1 \\ 0 & 0 \end{bmatrix}, \quad B_1 = B_2 = \begin{bmatrix} 0 \\ 1 \end{bmatrix}, \quad D_{12} = D_{21} = 1$$

$$C_1 = \begin{bmatrix} 1 & \beta \end{bmatrix}, \quad C_2 = \begin{bmatrix} 1 & \alpha \end{bmatrix}$$

where $\alpha \neq 0$ and $\beta \neq 0$. It is easy to verify that the controller which makes the control variable u to depend on the output variable y through the transfer function

$$K_{IC}(s) := -\frac{\beta s + 1}{s^2 + (\alpha + \beta)s + 2}$$

performs the *indirect perfect compensation* of the disturbance w, since the transfer function from w to z is zero. However, the resulting system is stable only for $\alpha \geq 0$ and $\beta \geq 0$. Therefore, this kind of solution is no more feasible for all other values of the pair (α, β).

On the contrary, by applying Theorem 4.3 one obtains (corresponding to the choice $Q(s) = 0$) $K^o(s) = K_{IC}(s)$ when $\alpha > 0$ and $\beta > 0$, otherwise

$$K^o(s) = \frac{(2\alpha + \beta)s - 1}{s^2 - (\alpha + \beta)s + 2\alpha^2 + 2\alpha\beta + 2} , \quad \alpha < 0, \ \beta < 0$$

$$K^o(s) = \frac{\beta s - 1}{s^2 + (\alpha - \beta)s + 2} , \quad \alpha > 0, \ \beta < 0$$

$$K^o(s) = \frac{(2\alpha - \beta)s - 1}{s^2 + (\beta - \alpha)s + 2\alpha^2 - 2\alpha\beta + 2} , \quad \alpha < 0, \ \beta > 0$$

□

Remark 4.17 The structure of a generic RH_2 admissible controller for $P(s)$ (as defined by the block-scheme of fig. 4.19) allows checking that the eigenvalues of the resulting control system are those of the matrices $A + B_2F_2$, $A + L_2C_2$ and A_q. In fact, letting x_s and x_q denote the state variables of systems $S_2(s)$ and $Q(s)$, respectively, the dynamic matrix of the resulting system (with state $x_t := [x' - x_s'\ x_q'\ x']'$) is

$$A_t = \begin{bmatrix} A + L_2C_2 & 0 & 0 \\ B_qC_2 & A_q & 0 \\ B_2F_2 & B_2C_q & A + B_2F_2 \end{bmatrix}$$

□

Remark 4.18 The optimal controller given in point *b*) of Theorem 4.3 may be interpreted as the result of a synthesis procedure made up of two independent steps. The first one consists in solving the full information problem (dealt with in Section 4.2) yielding matrix F_2. The second step tackles the output estimation problem (dealt with in Section 4.3) relative to the linear state combination F_2x, that is relative to system (4.58)-(4.60) with $C_1 = -F_2$.   □

Remark 4.19 (Parametrization of the set $\mathcal{F}_{2\gamma}$) Notice that $\mathcal{F}_{2\gamma} = \mathcal{F}_{2\gamma r}$. In fact, as it was done in the proof of Theorem 4.3, recall that the set of RH_2 admissible controllers for $P(s)$ coincides with the set of RH_2 admissible controllers for $P_v(s)$. However, the latter system has the same structure as system $P_{OE}(s)$ (considered in Section 4.3), so that, in view of Remark 4.10, the above claim is correct.   □

Remark 4.20 The control problem addressed to in Theorem 4.3 can be viewed as an optimal *linear quadratic stochastic* problem with *unmeasurable* state. In fact, consider the *n-th* order system

$$\dot{x} = Ax + B\bar{u} + \zeta_1 \qquad (4.95)$$
$$\bar{y} = Cx + \zeta_2 \qquad (4.96)$$

where $\zeta := [\zeta_1'\ \zeta_2']'$ is a zero mean white Gaussian noise with intensity

$$W := \begin{bmatrix} W_{11} & W_{12} \\ W_{12}' & W_{22} \end{bmatrix} , \quad W_{22} > 0 \qquad (4.97)$$

Also consider the cost functionals

$$J_6 := \lim_{t \to \infty} E[x'(t)Qx(t) + 2x'(t)S\bar{u}(t) + \bar{u}'(t)R\bar{u}(t)]$$

and

$$J_7 := \lim_{T \to \infty} E\left[\frac{1}{T} \int_0^T [x'(t)Qx(t) + 2x'(t)S\bar{u}(t) + \bar{u}'(t)R\bar{u}(t)]dt \right]$$

where

$$\begin{bmatrix} Q & S \\ S' & R \end{bmatrix} := L = L' \geq 0 , \quad R > 0$$

Letting

$$\hat{Q} := Q - SR^{-1}S' , \quad \hat{W} := W_{11} - W_{12}W_{22}^{-1}W_{12}'$$

which are positive semidefinite (recall Remarks 4.4 and 4.11) and

$$C_{11}'C_{11} := \hat{Q} , \ C_{11} \in R^{n \times n} , \quad B_{11}B_{11}' := \hat{W}, \ B_{11} \in R^{n \times n}$$
$$B_2 := BR^{-1/2} , \quad B_1 := [B_{11}\ W_{12}W_{22}^{-1/2}]$$
$$D_{21} := [0\ I] , \quad C_2 := W_{22}^{-1/2}C$$
$$C_1 := \begin{bmatrix} C_{11} \\ R^{-1/2}S' \end{bmatrix} , \quad D_{12} := \begin{bmatrix} 0 \\ I \end{bmatrix} , \quad u := R^{1/2}\bar{u}$$
$$y := W_{22}^{-1/2}\bar{y} , \quad z := C_1x + D_{12}u$$

system (4.95)-(4.97) can be rewritten as

$$\dot{x} = Ax + B_1 w + B_2 u \tag{4.98}$$
$$z = C_1 x + D_{12} \tag{4.99}$$
$$y = C_2 x + D_{21} \tag{4.100}$$

where w is a zero mean white Gaussian noise with identity intensity and dimension $n + p$ which satisfies the equation

$$\zeta = \begin{bmatrix} B_{11} & W_{12}W_{22}^{-1/2} \\ 0 & W_{22}^{1/2} \end{bmatrix} w$$

It is also easy to verify that

$$J_6 := \lim_{t \to \infty} E[z'(t)z(t)]$$

and

$$J_7 := \lim_{T \to \infty} E\left[\frac{1}{T}\int_0^T z'(t)z(t)dt\right]$$

The problem under consideration is finding a controller of the form

$$\dot{\xi} = F\xi + Gy \tag{4.101}$$
$$u = H\xi + Ny \tag{4.102}$$

such that system (4.98)-(4.102) is stable and J_6 or J_7 is minimized. Notice that the feedback connection of any controller described by eqs. (4.101),(4.102) with system (4.98)-(4.100) is well defined and the relevant transfer function from w to z is strictly proper if and only if $ND_{21} = 0$, that is if and only if $N = 0$ (recall the definition of D_{21}). Therefore, if system (4.98)-(4.102) is stable and $N = 0$, then the controller (4.101),(4.102) is RH_2 admissible for system (4.98)-(4.100). In view of Remark 2.20 any RH_2 admissible controller for system (4.98)-(4.100) is such that

$$J_6 = J_7 = \|T(z, w; s)\|_2^2$$

Assume that neither the eigenvalues of the unobservable part of the pair $[(A - BR^{-1}S'), C_{11}]$ nor those of the unreachable part of the pair $[(A - W_{12}W_{22}^{-1}C), B_{11}]$ lie on the imaginary axis. Moreover, assume that the pair (A, C) is detectable and the pair (A, B) is stabilizable. With the same kind of reasoning developed in Remark 4.11 it is easy to see that these assumptions are equivalent to Assumptions 4.6 and 4.8, while Assumptions 4.7 and 4.9 are satisfied because of the definition of D_{12} and D_{21}.

Therefore, Theorem 4.3 can be applied to system (4.98)-(4.100) and supplies the optimal controller which is described (with reference to system (4.95)-(4.97)) by

$$\dot{\xi} = A\xi + B\bar{u} + L(C\xi - \bar{y}) \tag{4.103}$$
$$\bar{u} = \bar{H}\xi \tag{4.104}$$

if the relevant substitutions have been done. In eqs. (4.103),(4.104) it is $L := -(\Pi_2 C' + W_{12})W_{22}^{-1}$, $\bar{H} := -R^{-1}(B'P_2 + S')$, P_2 and Π_2 being the symmetric, positive semidefinite and stabilizing solutions of the Riccati equations (in the unknown P and Π, respectively)

$$0 = PA_c + A_c'P - PBR^{-1}B'P + Q - SR^{-1}S'$$
$$0 = \Pi A_f' + A_f\Pi - \Pi C'W_{22}^{-1}C\Pi + W_{11} - W_{12}W_{22}^{-1}W_{12}'$$

with $A_c := A - BR^{-1}S'$, $A_f := A - W_{12}W_{22}^{-1}C$. Notice that in the controller (4.103),(4.104) it is $N = 0$.

Equation (4.103) is the equation of the stable (see Remark 4.11) Kalman filter for system (4.95)-(4.97). Thus, the controller (4.103),(4.104) can be considered as a Kalman filter on the state of which the control law has been implemented (eq. (4.104)) which is optimal and stabilizing for the Linear Quadratic problem with infinite horizon defined on system (4.95) with $\zeta_1 \equiv 0$ and performance index

$$\bar{J}_7 := \int_0^\infty [x'(t)Qx(t) + 2x'(t)S\bar{u}(t) + \bar{u}'(t)R\bar{u}(t)]dt$$

The criterion $\bar{J}_7$ is the "deterministic version" of the functional J_7. This structural *separation* of the solution (optimal filtering and regulator problem) is outlined also by the spectrum of the dynamic matrix of the resulting system. In fact, letting $e := \xi - x$, from eqs. (4.95)-(4.97),(4.103),(4.104) it follows

$$\dot{e} = (A + LC)e - \zeta_1 - L\zeta_2$$
$$\dot{x} = (A + B\bar{H})x + B\bar{H}e + \zeta_1$$

The problem of minimizing either the functional J_6 or the functional J_7 under the unique constraint expressed by eqs. (4.95)-(4.97) is known as the *linear quadratic Gaussian (LQG)* problem. The solution of such a problem (whenever it exists) is specified by eqs. (4.103),(4.104), where, however, the two matrices Π_2 and P_2 which determine L and $\bar{H}$, respectively, may not be the stabilizing solutions of the relevant Riccati equations. Consequently, system (4.95)-(4.97), (4.103),(4.104) may be unstable. This outcome is consistent with the absence of any stability requirement put forth by the classical *LQG* theory. As a matter of fact, the assumptions required by such a theory (stability of the unreachable but observable part of system $\Sigma(A - BR^{-1}S', B, C_{11}, 0)$ and of the unobservable but reachable part of system $\Sigma(A - W_{12}W_{22}^{-1}C, B_{11}, C, 0)$) are weaker than those required within the RH_2 context.   □

Remark 4.21 The solution of the output estimation problem (see Section 4.2) can be derived as an application of Theorem 4.3 to a particular case. In fact, it suffices to set $D_{12} = I$ in Assumptions 4.6 - 4.9 in order to conclude that Assumptions 4.3 - 4.5 are satisfied. In particular, it results $C_{1c} = 0$ so that the (unique) symmetric, positive semidefinite and stabilizing solution of the Riccati equation (4.18) is $P_2 = 0$. In such a context, the conclusions of Theorem 4.3 are immediately redrawn to those of Theorem 4.2. In a similar way, the solution of the disturbance feedforward problem dealt with in Remark 4.5 can be obtained by solving the partial information problem in the particular case $D_{21} = I$. Indeed, letting $D_{21} = I$ in Assumptions 4.6 - 4.9, it is easy to verify that the assumptions made in Remark 4.5 are satisfied. In particular, it results $B_{1f} = 0$ and hence $\Pi_2 = 0$, so that the conclusions of Theorem 4.3 coincide with those illustrated in Remark 4.5.   □

Remark 4.22 The contents of Remark 4.6 (as for matrix D_{12}) and of Remark 4.13 (as for matrix D_{21}) apply with no changes to the problem dealt with in the present section.   □

Remark 4.23 In view of Remarks 4.8 and 4.15 it can be said that, under Assumptions 4.7 and 4.9, Assumption 4.8 amounts to requiring that the two subsystems of $P(s)$ having transfer functions $P_{12}(s)$ (that is, system $\Sigma(A, B_2, C_1, D_{12})$ with input u and output z) and $P_{21}(s)$ (that is, system $\Sigma(A, B_1, C_2, D_{21})$ with input w and output y), respectively, do not have zeros on the imaginary axis.   □

Remark 4.24 The discussion in Remark 4.16 can be applied with no changes to the present context. Therefore, the results presented in Theorem 4.3 can be extended with no difficulty to encompass the case in which eq. (4.94) is replaced by

$$y = C_2 x + D_{21}w + D_{22}u$$

□

4.5 Notes and references

The material of this section mostly relies on the paper by Doyle et al. [17]. However, the assumptions under which Theorems 4.1, 4.2, 4.3 have been stated are more general than those adopted in such a paper. Moreover the results concerning the parametrization of the admissible controllers have been modified so as to encompass the remark put forward by Mita et al. [44]. Further insight on the connections existing between the LQG and RH_2 problems was given by Kucera [35].

Chapter 5

RH_∞ Control

5.1 Introduction

The control problem for linear time-invariant systems has been classically tackled in the frequency domain. A typical, though not completely general, context is the one shown in fig. 5.1, where $G(s)$ and $K(s)$ are the transfer functions of the *process to be controlled* (possibly including actuators and sensors) and of the *controller* to be synthesized. The *disturbances* d_c, d_u, d_r act on the controlled variable c, on the process input variable u_p and on the feedback path, respectively. Finally, c^o is the *opposite* of the *set point*.

As well known, the aim in designing $K(s)$ is, loosely speaking, guaranteeing the *stability* of the control system and achieving *satisfactory performances*. Usually, such performances are evaluated in terms of the behavior of suitable variables of interest to be specified according to the problem at hand and must be attained in spite of the disturbances acting on the system and inaccurate knowledge of the process model.

In general, the philosophy underlying the adopted synthesis procedure strongly affects the result: for instance, having either ignored or taken into account the *inaccurate* knowledge of the process model makes the controller quite different. Moreover, the design procedure significantly depends on the adopted description of the uncertainty: thus, subdividing the relevant discussion (and consequently this section) into three parts, appears to be a fairly natural strategy. First, the design problem is faced in *nominal conditions*, that is the process is supposed perfectly known. Second, some

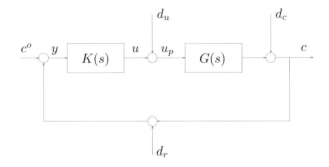

Figure 5.1: A typical control system

typical situations in which the process model is not precisely known are shown to be conveniently described by inserting a suitable perturbation $\Delta(s)$ into the block scheme of fig. 5.1. The topology of this scheme does therefore reflects the particular *nature* of the considered uncertainty. Third, the design problem is tackled in a *robust way*, namely it is stated within various uncertainty scenarios, modeled in accordance with the previous discussion.

5.1.1   Nominal design

The process under control is assumed to be perfectly described by its *nominal* transfer function $G_n(s)$ so that in the block scheme of fig. 5.1 $G(s) = G_n(s)$. The transfer functions which are suited to evaluate the effects of the disturbances and hence can be exploited to express meaningful performance requirements are the following:

$$S_n(s) = [I - G_n(s)K(s)]^{-1} \tag{5.1}$$

$$T_n(s) = G_n(s)K(s)[I - G_n(s)K(s)]^{-1} \tag{5.2}$$

$$V_n(s) = K(s)[I - G_n(s)K(s)]^{-1} \tag{5.3}$$

The function $S_n(s)$, usually referred to as *sensitivity* function, describes the effect of the disturbance d_c on the controlled variable c and the effect of the signal c^o on the controller input variable y. The function $T_n(s)$, which is readily recognized to equal $S_n(s) - I$, is, for such a reason, referred to as *complementary sensitivity* function and accounts for the effect of c^o or d_r on the controlled variable. Finally, the function $V_n(s)$ is responsible of the effect of c^o, d_r or d_c on the process input: thus, it will be referred to as *input sensitivity* function.

   A good solution to the design problem spontaneously calls for making small, in some suitable sense to be specified, the effects of the disturbances on the variables of interest. Thus, the desire of *making small* the above introduced transfer functions naturally arises.

   In the scalar case (that is the case where all the relevant variables are scalar), the desire of *making small* a transfer function $\varphi(s)$ is consistent with the request that the *absolute value* of $\varphi(j\omega)$ be, for each frequency ω, smaller than a given (possibly frequency dependent) quantity, namely $| \varphi(j\omega) | < \vartheta(\omega), \forall \omega$. For this to make sense, it is necessary that no poles of $\varphi(s)$ lie on the imaginary axis: in the here considered framework this requirement is naturally fulfilled since, in view of the unavoidable stability constraint, all the transfer functions of interest must belong to RH_∞. The natural extension of this philosophy to the multivariable case leads to asking that an inequality of the above type be verified by the *maximum singular value* of $\varphi(j\omega)$, namely

$$\bar\sigma[\varphi(j\omega)] < \vartheta(\omega) , \quad \forall \omega \tag{5.4}$$

   Let $W(s) \in RH_\infty$ be any matrix such that $W^{-1}(s) \in RH_\infty$ and $\bar\sigma[W^{-1}(j\omega)] = \vartheta(\omega)$. Then, the inequality (5.4) is no doubt satisfied if

$$\bar\sigma[W(j\omega)\varphi(j\omega)] < 1 , \quad \forall \omega$$

that is if

$$\|W(s)\varphi(s)\|_\infty < 1 \tag{5.5}$$

since, in view of Lemma 2.21,

$$\bar\sigma[\varphi(j\omega)] = \bar\sigma[W^{-1}(j\omega)W(j\omega)\varphi(j\omega)]$$
$$\leq \bar\sigma[W(j\omega)\varphi(j\omega)]\bar\sigma[W^{-1}(j\omega)]$$

The function $W(s)$ is usually referred to as *shaping function*.

For the control system of fig. 5.1 the more classical request is the *sensitivity performance* one, which, consistently with eq. (5.5), can be expressed as

$$\| W_1(s)S_n(s) \|_\infty < 1 \tag{5.6}$$

Not at all less meaningful are the requests related to the *complementary sensitivity performance* and *input sensitivity performance* which can be expressed as

$$\| W_2(s)T_n(s) \|_\infty < 1 \tag{5.7}$$

and

$$\| W_3(s)V_n(s) \|_\infty < 1 \tag{5.8}$$

respectively.

Equally of interest are the requests calling for the *simultaneous* satisfaction of more than one of the inequalities (5.6)-(5.8). As an example, a common request involves both the sensitivity and the complementary sensitivity performance by asking for the fulfillment of

$$\left\| \begin{bmatrix} W_1(s)S_n(s) \\ W_2(s)T_n(s) \end{bmatrix} \right\|_\infty < 1 \tag{5.9}$$

since (recall Definition 2.24 and Lemmas 2.17 and 2.23) if (5.9) holds, then both (5.6) and (5.7) hold. It should be apparent that a multiple goal as the one expressed by eq. (5.9) must not ignore the intrinsic constraints existing among the involved functions. A wise selection of the shaping functions $W_i(s)$ to be associated with the performance functions $S_n(s)$, $T_n(s)$, $V_n(s)$ is therefore mandatory.

With reference to the particular case of eq. (5.9), the selection of the shaping functions must reflect the identity $S_n(s) - T_n(s) = I$. In general, $\bar{\sigma}[S_n(j\omega)]$ is asked to be small at *low* frequencies in order to endove the control system with good capabilities of tracking the set point, whose bandwidth is usually limited from above. This can be achieved by selecting a shaping function $W_1(s)$ such that $\bar{\sigma}[W_1^{-1}(j\omega)]$ is small at low frequencies and equal to 1 at higher frequencies. On the contrary, $\bar{\sigma}[T_n(j\omega)]$ is requested to be small at *high* frequencies in order to effectively counteract, for instance, the disturbances in the feedback path, as their spectra are usually located at high frequency. Consistently, the shaping function $W_2(s)$ can be selected so as to have $\bar{\sigma}[W_2^{-1}(j\omega)]$ equal to 1 at low frequencies and as small as possible at higher frequencies.

5.1.2 Uncertainty description

It is often more realistic to assume that the process model belongs to some specified set $\mathcal{G}$ rather than being perfectly known. Moreover, the so called nominal model $G_n(s)$ is usually taken as an element of the set $\mathcal{G}$ and therefore viewed as a *first order approximation* of the *true model* $G(s)$. Consistently, a description of the set $\mathcal{G}$ can be performed by *parametrizing* it by means of a transfer function $\Delta(s)$ belonging to a suitable set $\mathcal{D}_\alpha$: the *perturbations* which $G_n(s)$ may undergo are then defined by the adopted parametrization and the structure of the set $\mathcal{D}_\alpha$.

Here reference will be made to *unstructured* perturbations only, that is to perturbations which are qualified only in terms of their *amplitude* as specified by the set

$$\mathcal{D}_\alpha := \{\Delta(s) \mid \Delta(s) \in RH_\infty \,, \ \|\Delta(s)\|_\infty < \alpha\} \tag{5.10}$$

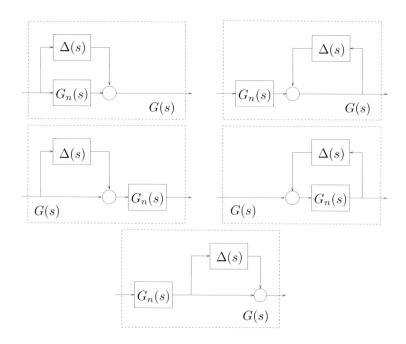

Figure 5.2: Different uncertainty models

Some particularly meaningful examples of parametrization of $\mathcal{G}$ are presented in the following equations, where the parameter $\Delta(s)$ is any element of the set $\mathcal{D}_\alpha$ defined by eq. (5.10):

$$\mathcal{G} := \{G(s) \mid G(s) = G_n(s) + \Delta(s)\} \tag{5.11}$$
$$\mathcal{G} := \{G(s) \mid G(s) = G_n(s)[I + \Delta(s)]\} \tag{5.12}$$
$$\mathcal{G} := \{G(s) \mid G(s) = [I + \Delta(s)]G_n(s)\} \tag{5.13}$$
$$\mathcal{G} := \{G(s) \mid G(s) = [I - \Delta(s)]^{-1}G_n(s)\} \tag{5.14}$$
$$\mathcal{G} := \{G(s) \mid G(s) = [I - G_n(s)\Delta(s)]^{-1}G_n(s)\} \tag{5.15}$$

It is easy to verify that each one of the sets (5.11)-(5.15) is suited to describe meaningful types of uncertainties in a fairly natural way (see also fig. 5.2 where a block-scheme version of eqs. (5.11)-(5.15) is presented). The set (5.11) may model an uncertain location of right half plane zeros of $G(s)$ (as an example: $G_n(s) = (s-2)/(s+1)(s+2)$, $\Delta(s) = \varepsilon/(s+1)(s+2)$). The sets (5.12) and (5.13) may suitably account for neglected high frequency poles as well as right half plane zeros (as an example: $\Delta(s) = -\varepsilon s/(1 + \varepsilon s)$ or $\Delta(s) = -2/(1 + s)$, so that $1 + \Delta(s) = 1/(1 + \varepsilon s)$ and $1 + \Delta(s) = (s-1)/(1+s)$, respectively). These sets can obviously be exploited in describing the model uncertainties of both actuators and sensors as well. The set (5.14) can easily model neglected right half plane poles (as an example: $\Delta(s) = 10/(1 + s)$, so that $[1 - \Delta(s)]^{-1} = (1 + s)/(s - 9)$). Finally, the set (5.15) can easily account for the uncertain location of a right half plane pole (as an example: $\Delta(s) = \varepsilon$, $G_n(s) = 1/(s - 1)$, so that $G(s) = 1/(s - 1 - \varepsilon)$).

5.1.3 Robust design

The design problem in an uncertain environment consists in selecting a controller $K(s)$ which ensures stability as well as satisfactory performances not only in nominal conditions (e.g., $G(s) = G_n(s)$), but also when the plant undergoes *finite perturbations*. As for the basic stability requirement, a controller $K(s)$ is said to guarantee *robust stability* if, given a set $\mathcal{D}_\alpha$, the control system is stable for each $G(s) \in \mathcal{G}$. In a similar way, a controller $K(s)$ is said to guarantee *robust performances* if, given a set $\mathcal{D}_\alpha$, the control system satisfies some specified performance requirements (like those defined through eqs. (5.6)-(5.9)) for each $G(s) \in \mathcal{G}$.

Within this framework a natural question arises, namely whether a control system which has been designed in nominal conditions can, for a given set $\mathcal{G}$ and some finite α, guarantee robust performances and/or stability relative to the set $\mathcal{D}_\alpha$. Whenever possible, the answer to such a question is supplied by the so called procedures for the *robustness analysis* of a control system.

The same approach adopted for the nominal design problem can be exploited for the *robust design* problem, provided that reference is made to sensitivity functions defined in terms of $G(s)$ rather than of $G_n(s)$, namely

$$S(s) = [I - G(s)K(s)]^{-1} \tag{5.16}$$
$$T(s) = G(s)K(s)[I - G(s)K(s)]^{-1} \tag{5.17}$$
$$V(s) = K(s)[I - G(s)K(s)]^{-1} \tag{5.18}$$

Accordingly, the *robust sensitivity performance*, the *robust complementary sensitivity performance* and the *robust control sensitivity performance* are guaranteed if, given the sets $\mathcal{G}$ and $\mathcal{D}_\alpha$, the control system is stable for all $G(s) \in \mathcal{G}$ and

$$\|W_1(s)S(s)\|_\infty < 1 , \quad \forall G(s) \in \mathcal{G} \tag{5.19}$$
$$\|W_2(s)T(s)\|_\infty < 1 , \quad \forall G(s) \in \mathcal{G} \tag{5.20}$$
$$\|W_3(s)V(s)\|_\infty < 1 , \quad \forall G(s) \in \mathcal{G} \tag{5.21}$$

respectively.

Not differently from the nominal design framework, it is possible to call for the simultaneous matching of two or even all the inequalities (5.19)-(5.21). Thus, for instance, a design problem could be stated requiring that

$$\left\| \begin{bmatrix} W_1(s)S(s) \\ W_2(s)T(s) \end{bmatrix} \right\|_\infty < 1 , \quad \forall G(s) \in \mathcal{G} \tag{5.22}$$

Finally, under some circumstances, the controller might be required to guarantee robust stability together with satisfactory performances (as specified by the inequalities presented in this section) in nominal conditions only (*nominal design with robust stability*).

The next section is devoted to showing how the design problems introduced in this section can all be reduced to a unique *standard problem* in the RH_∞ context which is completely defined in terms of $G_n(s)$ and $\mathcal{G}$ only.

5.2 The standard problem

Consider the block-structure depicted in fig. 5.3 where $P(s)$ is the so called *augmented*

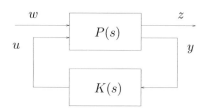

Figure 5.3: The standard 2-block configuration

system and $K(s)$ is the *controller* to be designed. It is now shown that the controller $K(s)$ in fig. 5.1 which solves one of the design problems defined in Section 5.1 is the same controller that in fig. 5.3 guarantees stability and the boundedness of the RH_∞ norm of the transfer function $T(z; w, s)$ from w to z. The interest in reformulating the original design problem in terms of the block structure of fig. 5.3 lies on the fact that the augmented plant $P(s)$ depends only on the nominal plant $G_n(s)$, on the particular set $\mathcal{G}$ of the given perturbations and on the performances requested to the control system.

The procedure at the basis of such a reformulation, i.e. the definition of the system $P(s)$ and signals w and z, comes up to be very simple when dealing with a design problem in nominal conditions, henceforth referred to as *nominal design*. Such procedure is now presented at the light of simple, but illustrative, examples, which also serve as preliminaries in the cases not specifically considered herein.

Nominal design: sensitivity performance   With reference to fig. 5.1 and fig. 5.3, define $z := W_1(s)y$ and $w := c^o$, or $z := W_1(s)y$ and $w := d_r$, or $z := W_1(s)c$ and $w := d_c$. Then, in the three cases,

$$P(s) = \begin{bmatrix} W_1(s) & W_1(s)G_n(s) \\ I & G_n(s) \end{bmatrix}$$

Nominal design: complementary sensitivity performance   With reference to fig. 5.1 and fig. 5.3, define $z := W_2(s)c$ and $w := d_r$ or $w := c^o$. Then, in both cases,

$$P(s) = \begin{bmatrix} 0 & W_2(s)G_n(s) \\ I & G_n(s) \end{bmatrix}$$

Nominal design: input sensitivity performance   With reference to fig. 5.1 and fig. 5.3, define $z := W_3(s)u_p$ and $w := d_r$ or $w := c^o$, or $w := d_c$. Then, in all the three cases,

$$P(s) = \begin{bmatrix} 0 & W_3(s) \\ I & G_n(s) \end{bmatrix}$$

Nominal design: joint sensitivity and complementary sensitivity performance   With reference to fig. 5.1 and fig. 5.3, define $z := [y'W_1(s)' \quad c'W_2(s)']'$ and $w := c^o$ or $w := d_r$. Then, in both cases,

$$P(s) = \begin{bmatrix} W_1(s) & W_1(s)G_n(s) \\ 0 & W_2(s)G_n(s) \\ I & G_n(s) \end{bmatrix}$$

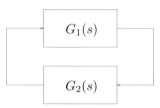

Figure 5.4: Feedback connection of two systems

Restating in terms of the block-structure of fig. 5.3 the synthesis problems in uncertain conditions requires a preliminary result (usually called *small gain theorem*) which refers to the feedback connection of fig. 5.4.

Theorem 5.1 *Let $G_1(s) \in RH_\infty$ be an assigned $p \times m$ transfer function and $G_2(s) \in RH_\infty$ an arbitrary $m \times p$ transfer function with $\|G_2\|_\infty < \alpha \neq 0$. Then*

i) *The feedback connected system of fig. 5.4 is stable for any $G_2(s)$ if $\|G_1(s)\|_\infty \leq \alpha^{-1}$*

ii) *If $\|G_1(s)\|_\infty > \alpha^{-1}$, there exists a transfer function $G_2(s)$ which destabilizes the feedback connected system of fig. 5.4.*

Proof If $\|G_1(s)\|_\infty \leq \alpha^{-1}$ then, recalling Lemma 2.21, Remark 2.13 and Remark 2.16, it follows that $\bar{\sigma}[G_1(s) \, G_2(s)] < 1$, $\forall Re(s) \geq 0$. Hence, Lemma 2.18 entails that all eigenvalues of $G_1(s)G_2(s)$ have modulus less than one in the closed right half plane, so that $\det[I - G_1(s)G_2(s)] \neq 0$, $\forall Re(s) \geq 0$. In view of Theorem 3.3 the conclusion is drawn that system in fig. 5.4 is (internally) stable.

On the contrary, suppose that $\|G_1(s)\|_\infty = \alpha^{-1}(1 + \epsilon) := \rho^{-1}, \epsilon > 0$ and consider the case $m \leq p$. Write a singular value decomposition of $G_1(j\omega)$ as $G_1(j\omega) = U(j\omega)\Sigma(j\omega)V^\sim(j\omega)$ with $\Sigma(j\omega) = [S(j\omega)' \; 0]'$, where $S(j\omega)$ is a square and m-dimensional matrix. Letting $G_2(j\omega) := \rho V(j\omega)TU^\sim(j\omega)$, where $T = [I \; 0]$ is a $m \times p$ dimensional matrix, it follows that

$$\det[I - G_1(j\omega)G_2(j\omega)] = \det\left[I - \rho U(j\omega)\begin{bmatrix} S(j\omega) & 0 \\ 0 & 0 \end{bmatrix}U^\sim(j\omega)\right]$$

$$= \det\left[\begin{bmatrix} I - \rho S(j\omega) & 0 \\ 0 & I \end{bmatrix}\right]$$

$$= \det[I - \rho S(j\omega)]$$

Since $\|G_1(s)\|_\infty = \rho^{-1}$, there exists a frequency $\bar{\omega}$ such that $\lim_{\omega \to \bar{\omega}} \bar{\sigma}[G_1(j\omega)] = \rho^{-1}$. Then, recalling that $S(j\omega) = \text{diag}\{\sigma_i[G_1(j\omega)]\}$, it follows that one of the nonzero entries of $\rho S(j\omega)$ tends to one so that $\lim_{\omega \to \bar{\omega}} \det[I - G_1(j\omega)G_2(j\omega)] = 0$. In view of Theorem 3.3 the system in fig. 5.4 is unstable. The case $m \geq p$ can be dealt with in a similar way. ☐

Remark 5.1 (Stability of an uncertain matrix) Theorem 5.1 allows one to give a simple answer to a peculiar problem which concerns the stability of an uncertain linear system. Consider a system described by

$$\dot{x} = (A + \Delta_A)x$$

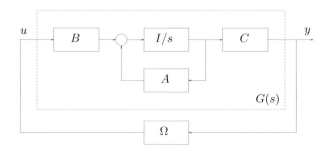

Figure 5.5: An uncertain system in feedback configuration

where A is stable and Δ_A represents the uncertainty which is assumed to belong to a set of perturbations specified by

$$\Delta_A = B\Omega C \ , \ \ \|\Omega\| \le \gamma^{-1}$$

where B and C are two assigned matrices and γ is an assigned positive scalar. Notice that it is possible, through a suitable choice of matrices B and C, to effectively describe various situations, for instance the case in which only a parameter of the system dynamic matrix is really uncertain. Looking at fig. 5.5 it is apparent that the uncertain system can be viewed as a closed loop system obtained by performing the control law $u = \Omega y$ on the system with transfer function $G(s) = C(sI - A)^{-1}B$. Then, it can be concluded that, if $\|G(s)\|_\infty < \gamma$, the stability is guaranteed for any Ω such that $\|\Omega\| \le \gamma^{-1}$.

Thanks to Theorem 2.14, the condition $\|G(s)\|_\infty < \gamma$ is equivalent to the existence of a symmetric positive semidefinite stabilizing solution S_s of the Riccati equation (in the unknown S)

$$0 = SA + A'S + \gamma^{-2}SBB'S + C'C$$

$\square$

Remark 5.2 (Covariance bound) Consider the stochastic system described by

$$\dot{x} = (A + \Delta_A)x + B_1 w$$

where A is stable, w is a zero-mean white noise with identity intensity. Similarly to Remark 5.1, the perturbation Δ_A is assumed here to be described by

$$\Delta_A = B_2\Omega_2 C \ , \ \ \|\Omega_2\| \le \gamma^{-1}$$

where B_2 and C are specified matrices and γ a positive scalar. It is also assumed that the system is stable for any perturbation Δ_A of the given form. Hence, it makes sense to tackle the problem of finding a meaningful (i.e. "small") upper bound of the asymptotic covariance matrix $X_a(\Omega_2)$ of the system state.

Let $\beta > 0$ be fixed and assume for the moment that there exists a symmetric positive semidefinite and stabilizing solution $P_s(\beta)$ to the algebraic Riccati equation (in the unknown P)

$$0 = PA' + AP + \gamma^{-2}PC'CP + B_2B_2' + \beta^{-2}B_1B_1'$$

It is now proved that $\beta^2 P_s(\beta)$ is a solution of the problem stated above. Recall that the asymptotic state covariance $X_a(\Omega_2)$ is the (unique) solution of the Lyapunov equation (in the unknown X)

$$0 = (A + B_2\Omega_2 C)X + X(A + B_2\Omega_2 C)' + B_1B_1'$$

This fact can be readily verified by exploiting arguments similar to those used in Remark 2.20 for the computation of J_2.

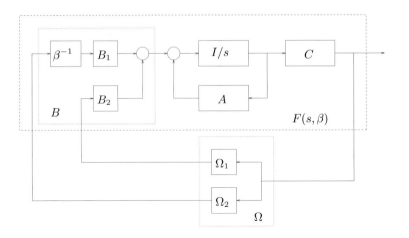

Figure 5.6: The covariance bound problem

Now, let $S(\Omega_2, \beta) := \beta^2 P_s(\beta) - X_a(\Omega_2)$, and subtract the Lyapunov equation from the Riccati one. Then, $S(\Omega_2, \beta)$ satisfies the following equation (in the unknown S)

$$0 = (A + B_2\Omega_2 C)S + S(A + B_2\Omega_2 C)' +$$
$$+ (\beta\gamma^{-1}P_sC' - \gamma\beta B_2\Omega_2)(\beta\gamma^{-1}P_sC' - \gamma\beta B_2\Omega_2)' +$$
$$+ \beta^2 B_2(I - \gamma\Omega_2\Omega_2')B_2'$$

Being $(A + B_2\Omega_2 C)$ stable, Lemma C.1 leads to the conclusion that $S(\Omega_2, \beta) \geq 0$, i.e.

$$\beta^2 P_s(\beta) \geq X_a(\Omega_2) , \quad \forall \Omega_2 , \quad \|\Omega_2\|_\infty \leq \gamma^{-1}$$

This inequality raises the interest in determining the value of β to which the "minimum" value of the bound of $X_a(\Omega_2)$ corresponds. Let $\mathcal{B}$ be the set of β's for which the solution $P_s(\beta)$ of the Riccati equation actually exists. Observe that the existence of a solution $P_s(\beta)$ entails that the system shown in fig. 5.6 is stable for any $\Omega := [\Omega_1' \ \Omega_2']', \|\Omega\| \leq \gamma^{-1}$. This is simply verified in view of Theorems 5.1, 2.13 and Remark 2.22 since $\|F(s, \beta)\|_\infty < \gamma$, with $F(s, \beta) := C(sI - A)^{-1}B(\beta)$, and $B(\beta) := [\beta^{-1}B_1 \ B_2]$. Hence, $\forall \beta \in \mathcal{B}$ it results $\|F(s, \beta)\|_\infty < \gamma$. In order to show that $\mathcal{B} \neq \emptyset$, it is sufficient to observe that $\|F(s, \beta)\|_\infty$ is a monotonic nonincreasing function of β and

$$\lim_{\beta \to 0} \|F(s, \beta)\|_\infty = \infty , \quad \lim_{\beta \to \infty} \|F(s, \beta)\|_\infty = \|C(sI - A)^{-1}B_2\|_\infty < \gamma$$

where the last inequality follows from the assumed stability of $A + B_2\Omega_2 C, \|\Omega_2\| \leq \gamma^{-1}$.

In conclusion, there exists β_1 such that $\|F(s, \beta_1)\|_\infty = \gamma$ and $\mathcal{B} = (\beta_1, \infty)$. The problem of finding a meaningfully optimal value of β (namely, a β° to which a "small" bound $\beta^{\circ^2} P_s(\beta^\circ)$ of $X_a(\Omega_2)$ corresponds) can easily be faced since it can be proven that for $\beta \in \mathcal{B}$

$$\frac{d^2}{d\beta^2}[\beta^2 P_s(\beta)] \geq 0$$

This entails that, for instance, trace$[\beta^2 P_s(\beta)]$ is a convex function of β so that the problem

$$\inf_{\beta \in \mathcal{B}} \text{trace}[\beta^2 P_s(\beta)]$$

actually admits a solution.   □

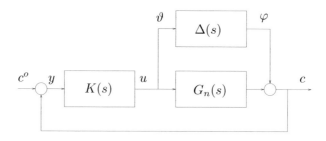

Figure 5.7: A control system with additive perturbations

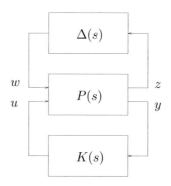

Figure 5.8: The standard 3-block configuration

Uncertain design: robust stability Assume that one of the parametrizations of $\mathcal{G}$ in eq. (5.11)-(5.15) has been adopted to account for the uncertain knowledge of the plant. It is then obvious how important would be the choice of a controller $K(s)$ which guarantees closed loop stability for an assigned set of perturbations, i.e. for an assigned value of the scalar variable α (recall the definition of $\mathcal{D}_\alpha$, eq. (5.10)). Along these lines the control problem that spontaneously arises is a *robust stability* design problem. Such a problem is easily reformulated in that of determining (if any) a controller $K(s)$ that, with reference to the scheme of fig. 5.3, guarantees stability and is such that the RH_∞ norm of the transfer function $T(z, w; s)$ is less than a suitable scalar β. The augmented plant $P(s)$ depends on the choice of the particular set $\mathcal{G}$, as now shown for the sets in eqs. (5.11) and (5.13). Dealing with the remaining cases is simple at the light of these examples. Considering the set $\mathcal{G}$ given by eq. (5.11) (*additive perturbations*) is equivalent to considering the control system of fig. 5.7. Then, define $z := \vartheta$ and $w := \varphi$ and observe that the control system of fig. 5.8 is completely equivalent to that of fig. 5.7 if

$$P(s) = \begin{bmatrix} 0 & I \\ I & G_n(s) \end{bmatrix}$$

In view of Theorem 5.1, $K(s)$ guarantees stability for any $\Delta(s) \in \mathcal{D}_\alpha$ if and only if $\|T(z, w; s)\|_\infty < \beta := \alpha^{-1}$ in the system of fig. 5.3.

Analogously, consider the set $\mathcal{G}$ given by eq. (5.13) (*multiplicative perturbations*) and observe that this corresponds to considering the scheme in fig. 5.9, which, in

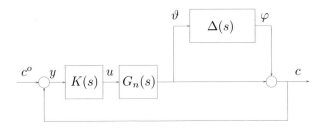

Figure 5.9: A control system with multiplicative perturbations

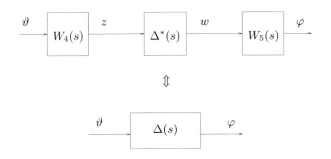

Figure 5.10: A less primitive description of uncertainty

turn, is completely equivalent to that of fig. 5.8 if $z := \vartheta$, $w := \varphi$ and

$$P(s) = \begin{bmatrix} 0 & G_n(s) \\ I & G_n(s) \end{bmatrix}$$

Hence, stability for any $\Delta(s) \in \mathcal{D}_\alpha$ is guaranteed if and only if, in the system of fig. 5.3, $\|T(z, w; s)\|_\infty < \beta := \alpha^{-1}$.

Remark 5.3 At the light of what has been now shown, a less primitive class of uncertainty than the one defined in eq. (5.10) can readily be introduced by substituting the set $\mathcal{D}_\alpha$ with

$$\mathcal{D}_\alpha^* := \{\Delta(s) \mid \Delta(s) = W_5(s)\Delta^*(s)W_4(s) , \quad \Delta^*(s) \in RH_\infty , \quad \|\Delta^*(s)\|_\infty < \alpha\} \quad (5.23)$$

where $W_4(s)$ and $W_5(s)$ are assigned elements of RH_∞. Without any loss of generality, one can assume $\|W_5(s)\|_\infty = 1$.

The two shaping functions $W_4(s)$ and $W_5(s)$ allows one to more finely qualify the perturbations which affect the control loop, in terms of both their harmonic components (nature of the functions) and their very structure (configuration of the matrices). With reference to the set $\mathcal{D}_\alpha^*$ given in eq. (5.23), it is possible to substitute $\Delta^*(s)$ for $\Delta(s)$ in fig. 5.8, provided that $z := W_4(s)\vartheta$ and $w := W_5^{-1}(s)\varphi$ (recall fig. 5.7, 5.9, 5.10).

In conclusion, the controller $K(s)$ is stabilizing for any $\Delta(s) \in \mathcal{D}_\alpha^*$ if and only if $\|T(z, w; s)\|_\infty < \beta := \alpha^{-1}$ in the system of fig. 5.3, where (recall fig. 5.7, 5.9) the augmented plant $P(s)$ is given, in the two considered cases, by

$$P(s) = \begin{bmatrix} 0 & W_4(s) \\ W_5(s) & G_n(s) \end{bmatrix}$$

and

$$P(s) = \begin{bmatrix} 0 & W_4(s)G_n(s) \\ W_5(s) & G_n(s) \end{bmatrix}$$

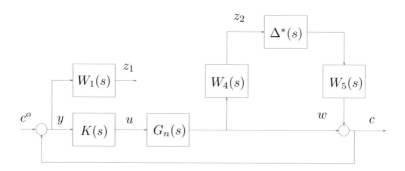

Figure 5.11: The control system with multiplicative perturbations

respectively.   □

Remark 5.4 Assume that one of the (single performance) nominal design problem previously introduced has been performed and let $K(s)$ be the resulting controller. In view of Theorem 5.1 such a controller also guarantees the robust stability relatively to suitable sets $\mathcal{G}$ (of the type of those defined in eqs. (5.11)-(5.15)) and $\mathcal{D}_\alpha^*$ (of the type of that defined in eq. (5.23)). Actually, the controller that satisfies the constraint $\|T(z, w; s)\|_\infty < \beta$ in fig. 5.3 also guarantees the stability of the control system for any perturbation $\Delta^*(s)$, $\|\Delta^*(s)\|_\infty < \beta^{-1}$ acting between the signals z and w. Hence stability is ensured for any $\Delta(s) \in \mathcal{D}_\alpha^*, \alpha < \beta^{-1}$ with $W_5(s) = I$ and $W_4(s)$ depending on the shaping function which is possibly introduced in the formulation of the performance objective (recall eqs. (5.6)-(5.8)).

   As an example, consider a controller that solves the nominal design problem in terms of the input sensitivity performance. It also guarantees the robust stability with respect to the perturbations in the set (5.11) with $W_4(s) = W_3(s)$. On the contrary, a controller that solves the nominal design problem for the sensitivity (complementary sensitivity) performance guarantees the robust stability with respect to the perturbations in the set (5.14) (resp. (5.13)) with $W_4(s) = W_1(s)$ (resp. $W_4(s) = W_2(s)$).   □

Remark 5.5 If the shaping function $W_4(s)$ is suitably scaled, one can assume $\alpha = 1$ in eq. (5.23) so that for given $W_4(s)$ and $W_5(s), \|W_5(s)\|_\infty = 1$, the reference set is

$$\mathcal{D}^* := \{\Delta(s) \mid \Delta(s) = W_5(s)\Delta^*(s)W_4(s) \ , \ \Delta^*(s) \in RH_\infty \ , \ \|\Delta^*(s)\|_\infty < 1\} \qquad (5.24)$$

                                                                                                      □

Uncertain design: robust stability and nominal performances Consider the control system depicted in fig. 5.1 and assume that a description of the uncertainty, based on two sets $\mathcal{G}$ and $\mathcal{D}^*$, has been selected (recall eqs. (5.11)-(5.15) and (5.24)). The design problem is that of determining a controller $K(s)$ which guarantees the robust stability and the fulfillment, in nominal conditions, of preassigned performance requirements (of the type of those specified in eqs. (5.6)-(5.9)).

   For example, if one is interested in the sensitivity performance (eq. (5.6)), being the set $\mathcal{G}$ specified by eq. (5.13), the problem is stated by requiring the determination of $K(s)$ such that (see fig. 5.11) $\|W_1(s)S_n(s)\|_\infty < 1$ and $\|W_4(s)T_n(s)W_5(s)\|_\infty < 1$. Recalling now what has been said about eq. (5.9), Lemma 2.21 and the fact that $\|W_5(s)\|_\infty = 1$, it is apparent that the design specifications are met with if

$$\left\| \begin{bmatrix} W_1(s)S_n(s) \\ W_4(s)T_n(s) \end{bmatrix} \right\|_\infty < 1$$

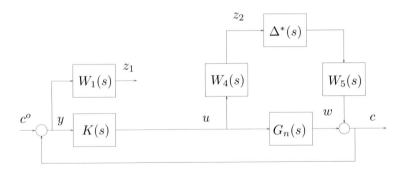

Figure 5.12: The control system with additive perturbations

i.e. if $\|T(z, w; s)\|_\infty < 1$ in the block-scheme of fig. 5.3, with

$$z := \left[\begin{array}{c} z_1 \\ z_2 \end{array} \right], \quad P(s) := \left[\begin{array}{cc} W_1(s) & W_1(s)G_n(s) \\ 0 & W_4(s)G_n(s) \\ I & G_n(s) \end{array} \right]$$

On the contrary, if the set $\mathcal{G}$ is the one defined in eq. (5.11), the sensitivity performance requirement is met with if (see fig. 5.12)

$$\left\| \left[\begin{array}{c} W_1(s)S_n(s) \\ W_4(s)V_n(s) \end{array} \right] \right\|_\infty < 1$$

i.e. if $\|T(z, w; s)\|_\infty < 1$ in the block-scheme of fig. 5.3, with

$$z := \left[\begin{array}{c} z_1 \\ z_2 \end{array} \right], \quad P(s) := \left[\begin{array}{cc} W_1(s) & W_1(s)G_n(s) \\ 0 & W_4(s) \\ I & G_n(s) \end{array} \right]$$

The statement of the design problems aimed at contemporarily achieving robust stability and robust sensitivity performances requires the result presented in the following lemma.

Lemma 5.1 *Let $X(s)$ and $Y(s)$ be assigned elements of RL_∞ and $\Psi(s)$ a generic element of RL_∞ such that $\|\Psi(s)\|_\infty < 1$. If*

$$\sup_\omega \{ \|X(j\omega)\| + \|Y(j\omega)\| \} < 1 \tag{5.25}$$

then

$$\sup_\omega \|X(j\omega)\| < 1 \tag{5.26}$$

$$\sup_\omega \|Y(j\omega)[I - \Psi(j\omega)X(j\omega)]^{-1}\| < 1 \tag{5.27}$$

Proof Eq. (5.26) derives directly from eq. (5.25). If $X(s) = 0$, eq. (5.27) follows trivially from eq. (5.25). Hence, let $X(s) \neq 0$. Remark 2.13 and Lemmas 2.21 and 2.18 (point 3), entail that, for each ω,

$$\|Y(j\omega)[I - \Psi(j\omega)X(j\omega)]^{-1}\| \leq \|Y(j\omega)\| \cdot \|[I - \Psi(j\omega)X(j\omega)]^{-1}\|$$

$$= \frac{\|Y(j\omega)\|}{\underline{\sigma}[I - \Psi(j\omega)X(j\omega)]} \tag{5.28}$$

But, for each ω,

$$
\begin{aligned}
\underline{\sigma}[I - \Psi(j\omega)X(j\omega)] &\geq \underline{\sigma}(I) - \bar{\sigma}[-\Psi(j\omega)X(j\omega)] \\
&\geq 1 - \bar{\sigma}[-\Psi(j\omega)X(j\omega)] \\
&\geq 1 - \bar{\sigma}[\Psi(j\omega)]\bar{\sigma}[X(j\omega)] \\
&\geq 1 - \bar{\sigma}[X(j\omega)] \\
&\geq 1 - \|X(j\omega)\|
\end{aligned}
\tag{5.29}
$$

The above relations have been written by exploiting Lemma 2.22 (first inequality), Lemma 2.21 (second inequality) and the assumption on the norm of $\Psi(j\omega)$ (last inequality). From eqs. (5.28) and (5.29) it follows that, for each ω,

$$
\|Y(j\omega)[I - \Psi(j\omega)X(j\omega)]^{-1}\| \leq \frac{\|Y(j\omega)\|}{1 - \|X(j\omega)\|}
$$

Thanks to eq. (5.25), the right hand side of this inequality is less than one, so that eq. (5.27) is proven.   □

Uncertain design: robust sensitivity performances First consider the case where the uncertainty is described by means of the sets $\mathcal{G}$ and $\mathcal{D}^*$ defined in eqs. (5.13) and (5.24), respectively (notice also fig. 5.11). Then,

$$
\begin{aligned}
S(s) &= [I - G(s)K(s)]^{-1} \\
&= [I - (I + W_5(s)\Delta^*(s)W_4(s))G_n(s)K(s)]^{-1} \\
&= [I - G_n(s)K(s) - W_5(s)\Delta^*(s)W_4(s)G_n(s)K(s)]^{-1} \\
&= \{[I - W_5(s)\Delta^*(s)W_4(s)G_n(s)K(s)(I - G_n(s)K(s))^{-1}] \cdot \\
&\quad \cdot [I - G_n(s)K(s)]\}^{-1} \\
&= S_n(s)[I - W_5(s)\Delta^*(s)W_4(s)T_n(s)]^{-1}
\end{aligned}
$$

Thus the robust sensitivity performance can be expressed as

$$
\|W_1(s)S_n(s)[I - W_5(s)\Delta^*(s)W_4(s)T_n(s)]^{-1}\|_\infty < 1
$$

$$
\forall \Delta^*(s) \in RH_\infty , \quad \|\Delta^*(s)\|_\infty < 1
\tag{5.30}
$$

whereas the robust stability requirement is stated as

$$
\|W_4(s)T_n(s)W_5(s)\|_\infty < 1
\tag{5.31}
$$

It is easy to verify that eqs. (5.30) and (5.31) are satisfied if

$$
\left\| \begin{bmatrix} W_1(s)S_n(s) \\ W_4(s)T_n(s) \end{bmatrix} \right\|_\infty < \frac{\sqrt{2}}{2}
\tag{5.32}
$$

Actually, letting $X(s) := W_4(s)T_n(s)W_5(s)$ and $Y(s) := W_1(s)S_n(s)$, eq. (5.32) implies that (recall that $\|W_5(s)\|_\infty = 1$)

$$
\left\| \begin{bmatrix} Y(s) \\ X(s) \end{bmatrix} \right\|_\infty^2 < \frac{1}{2}
$$

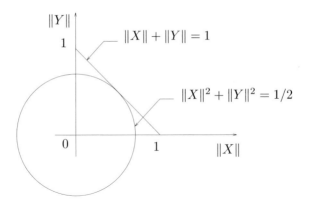

Figure 5.13: Sensitivity and complementary sensitivity constraints

and this is equivalent to

$$\sup_{\omega} \left\{ \, \|X(j\omega)\|^2 + \|Y(j\omega)\|^2 \, \right\} < \frac{1}{2}$$

This last relation implies (see fig. 5.13)

$$\sup_{\omega} \left\{ \, \|X(j\omega)\| + \|Y(j\omega)\| \, \right\} < 1$$

so that, thanks to Lemma 5.1, one can conclude about the correctness of what is claimed above. Notice that the simultaneous request of both robust stability and robust performance has lowered, not really surprisingly, the bound on the value of the norm.

Let now consider the case where the uncertainty is described by means of the sets $\mathcal{G}$ and $\mathcal{D}^*$ defined in eqs. (5.11) and (5.24), respectively (consider also fig. 5.12). Then, in strict analogy to what has been done in the previous case,

$$S(s) = S_n(s)[I - W_5(s)\Delta^*(s)W_4(s)V_n(s)]^{-1}$$

Thus, the robust sensitivity performance can be expressed in the following way

$$\|W_1(s)S_n(s)[I - W_5(s)\Delta^*(s)W_4(s)V_n(s)]^{-1}\|_\infty < 1,$$

$$\forall \Delta^*(s) \in RH_\infty , \quad \|\Delta^*(s)\|_\infty < 1$$

whereas the robust stability requirement is formulated as

$$\|W_4(s)V_n(s)W_5(s)\|_\infty < 1$$

The previously presented arguments can be exploited again to conclude that the goal of the design problem is achieved if the controller $K(s)$ is such that

$$\left\| \begin{bmatrix} W_1(s)S_n(s) \\ W_4(s)V_n(s) \end{bmatrix} \right\|_\infty < \frac{\sqrt{2}}{2}$$

The foregoing discussion has shown that a number of meaningful control problems can be treated in a unified fashion. As a matter of fact, it has been shown that they are all amenable to the problem of synthesizing a controller $K(s)$ which stabilizes the control system in fig. 5.3 and is such that the RH_∞ norm of the transfer function $T(z, w; s)$ from the input w to the output z is less than one. However, if such objective is not achievable, it makes sense to look for a controller that, besides stabilizing the control system, is such that $\|T(z, w; s)\|_\infty$ is minimized.

The so resulting design problem is apparently well posed only if the augmented plant $P(s)$ is stabilizable: given this, its precise statement is presented in the remaining part of this section, whose development closely follows the one presented in Section 4.1 within the RH_2 framework.

First of all, it is convenient to introduce the following time domain description of the process under control (augmented plant), which will be frequently quoted in the development of the present chapter.

$$\dot{x} = Ax + B_1 w + B_2 u \tag{5.33}$$

$$z = C_1 x + D_{11} w + D_{12} u \tag{5.34}$$

$$y = C_2 x + D_{21} w + D_{22} u \tag{5.35}$$

The controller is constrained to be a finite dimensional, time invariant, linear system, described by:

$$\dot{\xi} = F\xi + Gy \tag{5.36}$$

$$u = H\xi + Ey \tag{5.37}$$

Hence,

$$P(s) = \left[\begin{array}{c|cc} A & B_1 & B_2 \\ \hline C_1 & D_{11} & D_{12} \\ C_2 & D_{21} & D_{22} \end{array} \right] \quad , \quad K(s) = \left[\begin{array}{c|c} F & G \\ \hline H & E \end{array} \right]$$

Of course, the feedback connection of system (5.33)-(5.35) with system (5.36),(5.37) must be well defined. For such a condition to be verified it is necessary that

$$\det[I - ED_{22}] \neq 0 \tag{5.38}$$

so that the algebraic loop which is created by the insertion of the controller is automatically solvable. With reference to the block structure of fig. 5.3, the primary scope is that of determining $K(s)$ in such a way that the RH_∞ norm of the transfer function $T(z, w; s)$ is less than a specified positive value γ.

Notice that $T(z, w; s) \in RH_\infty$ entails that such a function is stable: this objective is obviously satisfied if the internal stability of the closed loop system is ensured, i.e. if $K(s)$ in (5.36),(5.37) internally stabilizes system (5.33)-(5.35). This is equivalent to requiring the stability of the dynamic matrix of the closed-loop system, i.e.

$$Re(\lambda_i(A_F)) < 0 , \quad \forall i \tag{5.39}$$

$$A_F := \left[\begin{array}{cc} A + B_2(I - ED_{22})^{-1}EC_2 & B_2(I - ED_{22})^{-1}H \\ G[I + D_{22}(I - ED_{22})^{-1}E]C_2 & F + GD_{22}(I - ED_{22})^{-1}H \end{array} \right] \tag{5.40}$$

In view of the above considerations, it is useful to precisely formalize the concept of *admissible controller*.

Definition 5.1 (Admissible controller in RH_∞) *A controller $K(s)$ is said to be admissible in RH_∞ for $P(s)$ if conditions (5.38)-(5.40,) are verified.* □

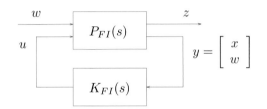

Figure 5.14: The full information problem

In Sections 5.3, 5.5 the attention will be focused on three main problems, each of them associated with a particular structure of the system $P(s)$: they are referred to as the *full information* problem, the *output estimation* problem and the *partial information* problem. More in detail, the last problem will be tackled by exploiting the solutions of the former ones, which, in turn, are strictly related each to the other by structural relations: the complete picture put into sharp relief important duality and separation properties.

The main result relevant to these problems will concern the solution of two precise points, following the scheme formally presented in Problem 5.1 below: the existence of a controller such that $\|T(z,w;s)\|_\infty < \gamma$ and the parametrization of such controllers. Problem 5.1 refers to the feedback configuration of fig. 5.3 and to the set $\mathcal{F}_{\infty\gamma}$ which represents the family of all admissible controllers in RH_∞ for $P(s)$ such that $\|T(z,w;s)\|_\infty < \gamma$.

Problem 5.1 (Standard problem in RH_∞) *Let a positive scalar γ be fixed.*

a) *Find a necessary and sufficient condition for the existence of a controller $K(s)$ which is admissible in RH_∞ for $P(s)$ and such that $\|T(z,w;s)\|_\infty < \gamma$.*

b) *Find a family of controllers $\mathcal{F}_{\infty\gamma r} \subseteq \mathcal{F}_{\infty\gamma}$ whose elements generate the whole set of functions $T(z,w;s)$ which are generated by the elements of $\mathcal{F}_{\infty\gamma}$.*

Remark 5.6 An obvious necessary condition for the existence of a stabilizing controller (and therefore for the existence of an admissible controller in RH_∞ for $P(s)$) is the stabilizability of the pair (A, B_2) and the detectability of the pair (A, C_2). The statement of Problem 5.1 makes sense only if both properties actually hold true.   □

In the forthcoming sections the parametrization of the controllers in the family $\mathcal{F}_{\infty\gamma}$ for all the considered cases will be presented. Such parametrization will be the subject of specified remarks which follow the main theorems concerning the solution of Problem 5.1. In other words, the issues relative to the family $\mathcal{F}_{\infty\gamma}$ and those relative to Problem 5.1 are treated separately. Actually, a unified treatment would further weight down the solution of Problem 5.1.

5.3   The full information problem

As in Section 4.2, the output signal y is constituted by the state variable x and the disturbance vector w relevant to the controlled plant. Therefore the plant is described by the equations

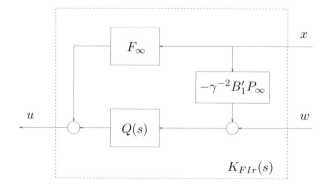

Figure 5.15: The set $\mathcal{F}_{\infty\gamma r}$ of the controllers $K_{FIr}(s)$

$$\dot{x} = Ax + B_1 w + B_2 u \tag{5.41}$$
$$z = C_1 x + D_{12} u \tag{5.42}$$
$$y = [y_1'\ y_2']' \tag{5.43}$$
$$y_1 = x \tag{5.44}$$
$$y_2 = w \tag{5.45}$$

while the transfer function from $[w'\ u']'$ to $[z'\ y']'$ will be denoted with $P_{FI}(s)$. A complete answer to Problem 5.1 can be given under the following assumptions.

Assumption 5.1 *The pair* $[(A - B_2 D_{12}' C_1), (I - D_{12} D_{12}') C_1]$ *is detectable and the pair* (A, B_2) *is stabilizable.*

Assumption 5.2 $D_{12}' D_{12} = I$.

The main result of this section makes reference to the block-diagram of fig. 5.14, where $K_{FI}(s)$ is any controller RH_∞ admissible for $P_{FI}(s)$ (recall Definition 5.1). In this diagram the transfer function from w to z is, as usual, denoted by $T(z, w; s)$.

Theorem 5.2 (Full information) *Consider Problem 5.1 relative to system (5.41)-(5.45). Then, under Assumptions 5.1, 5.2, it has the solution*

a) *The existence of a symmetric, positive semidefinite and stabilizing solution* P_∞ *of the Riccati equation (in the unknown P)*

$$0 = PA_c + A_c'P - P(B_2 B_2' - \gamma^{-2} B_1 B_1')P + C_{1c}'C_{1c} \tag{5.46}$$

i.e., such that the matrix A_{cc} given by

$$A_{cc} := A_c - B_2 B_2' P_\infty + \gamma^{-2} B_1 B_1' P_\infty \tag{5.47}$$

is stable. In eqs. (5.46),(5.47)

$$A_c := A - B_2 D_{12}' C_1 \ , \quad C_{1c} := (I - D_{12} D_{12}') C_1 \tag{5.48}$$

b) *The set $\mathcal{F}_{\infty\gamma r}$ of the controllers $K_{FIr}(s)$ is defined by the diagram of fig. 5.15, where*

$$F_\infty := -B_2' P_\infty - D_{12}' C_1 \tag{5.49}$$

and $Q(s) := \Sigma(A_q, B_q, C_q, D_q)$ with A_q stable and $\|Q(s)\|_\infty < \gamma$.

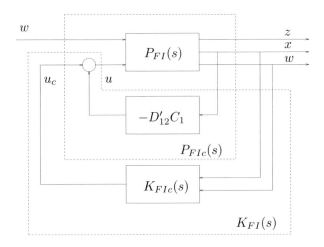

Figure 5.16: The modified full information problem

Proof First observe that the necessary condition for the problem at hand to make sense (recall Remark 5.6) is satisfied. Indeed, being measurable the state of the system, detectability of the pair (A, C_2) trivially holds, while, on the other hand, stabilizability of the pair (A, B_2) is guaranteed by Assumption 5.1.

Let now

$$u_c := u + D_{12}'C_1 x \tag{5.50}$$

which simply amounts to defining a control law as shown in fig. 5.16. The resulting system $P_{FIc}(s)$ is therefore described by

$$\dot{x} = A_c x + B_1 w + B_2 u_c \tag{5.51}$$
$$z = C_{1c} x + D_{12} u_c \tag{5.52}$$
$$y = [y_1' \; y_2']' \tag{5.53}$$
$$y_1 = x \tag{5.54}$$
$$y_2 = w \tag{5.55}$$

Being measurable the state, solving Problem 5.1 relative to system $P_{FIc}(s)$ is equivalent to solving the same problem relative to system $P_{FI}(s)$. As for point (b) in particular, the solution relative to system $P_{FI}(s)$ follows from that relative to system $P_{FIc}(s)$ by recalling eq. (5.50).

The proof is organized into three main parts. In part (i) it will be shown that the observability of the pair (A_c, C_{1c}) can be assumed without any loss of generality. Then, in part (ii), the necessity of point $a)$ will be proved. Finally, part (iii) is devoted to proving sufficiency of point $a)$ and point $b)$.

Part (i) With reference to fig. 5.16, denote by $T(z, w; s)$ the transfer function from w to z and suppose that a controller $K_{FIc}(s)$ admissible in RH_∞ exists such that $\|T(z, w; s)\|_\infty < \gamma$.

Now it will be shown that there is no loss of generality in assuming the pair (A_c, C_{1c}) observable rather than simply detectable. To this aim, let this pair already

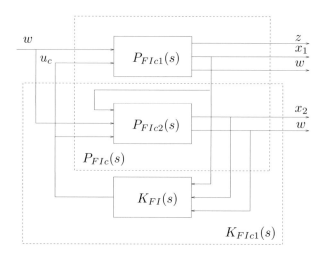

Figure 5.17: The canonical decomposition of the modified problem

be in Kalman's canonical observability form, namely

$$A_c = \begin{bmatrix} A_{c1} & 0 \\ A_{c2} & A_{c3} \end{bmatrix} , \quad C_{1c} = \begin{bmatrix} C_{1c1} & 0 \end{bmatrix}$$

with the pair (A_{c1}, C_{1c1}) observable, and, accordingly, decompose the matrices B_1 and B_2 as

$$B_1 = \begin{bmatrix} B_{11} \\ B_{12} \end{bmatrix} , \quad B_2 = \begin{bmatrix} B_{21} \\ B_{22} \end{bmatrix}$$

Finally, consider the state equations for $K_{FIc}(s)$

$$\dot{\xi} = F\xi + G_1 x_1 + G_2 x_2 + G_3 w \tag{5.56}$$
$$u_c = H\xi + E_1 x_1 + E_2 x_2 + E_3 w \tag{5.57}$$

where x_1 and x_2 are the components of the state vector $x = [x_1'\ x_2']'$ of system $P_{FIc}(s)$ in Kalman's canonical form. The overall system can then be viewed (see also fig. 5.17) as resulting from the feedback connection of the subsystem $P_{FIc1}(s)$

$$\dot{x}_1 = A_{c1} x_1 + B_{11} w + B_{21} u_c \tag{5.58}$$
$$z = C_{1c1} x_1 + D_{12} u_c \tag{5.59}$$

with the controller $K_{FIc1}(s)$ which, in turn, is constituted by $K_{FIc}(s)$ (see eqs. (5.56),(5.57)) and the subsystem $P_{FIc2}(s)$

$$\dot{x}_2 = A_{c2} x_1 + A_{c3} x_2 + B_{12} w + B_{22} u_c$$

Indeed, by substituting eq. (5.57) for u_c, the controller $K_{FIc1}(s)$ for the system $P_{FIc1}(s)$ (which is defined by eqs. (5.58),(5.59)) turns out to be described by

$$\dot{\xi} = F\xi + G_2 x_2 + G_1 x_1 + G_3 w$$
$$\dot{x}_2 = B_{22} H\xi + (A_{c3} + B_{22} E_2) x_2 + (A_{c2} + B_{22} E_1) x_1$$
$$\qquad + (B_{12} + B_{22} E_3) w$$
$$u_c = H\xi + E_2 x_2 + E_1 x_1 + E_3 w$$

Obviously, the controller $K_{FIc1}(s)$ is admissible in RH_∞ for system $P_{FIc1}(s)$ since $K_{FIc}(s)$ is such for $P_{FIc}(s)$.

Finally, observe that (the simple check is left to the reader) if $P_{1\infty}$ is the symmetric, positive semidefinite and stabilizing solution of the Riccati equation (in the unknown P_1)

$$0 = P_1 A_{c1} + A'_{c1} P_1 - P_1(B_{21}B'_{21} - \gamma^{-2}B_{11}B'_{11})P_1 + C'_{1c1}C_{1c1} \tag{5.60}$$

i.e., such that $A_{cc1} := A_{c1} - (B_{21}B'_{21} - \gamma^{-2}B_{11}B'_{11})P_{1\infty}$ is stable, then

$$P_\infty := \begin{bmatrix} P_{1\infty} & 0 \\ 0 & 0 \end{bmatrix} \tag{5.61}$$

is a symmetric and positive semidefinite solution of eq. (5.46). Moreover, such a solution is stabilizing since the eigenvalues of the matrix A_{cc}, defined by eqs. (5.47),(5.48), are those of the matrix A_{c3} (which is stable by assumption, because the pair (A_c, C_{1c}) is detectable) together with those of the matrix A_{cc1}.

Having proved that the existence of a controller $K_{FIc}(s)$ admissible in RH_∞ for system $P_{FIc}(s)$ and such that $\|T(z, w; s)\|_\infty < \gamma$ implies the existence of a controller $K_{FIc1}(s)$ admissible in RH_∞ for system $P_{FIc1}(s)$ and such that $\|T(z, w; s)\|_\infty < \gamma$, the necessity of the existence of the solution of the Riccati equation (5.46) is proved, thanks to eq. (5.61), once the necessity of the existence of the solution of eq. (5.60) has been ascertained. It is therefore possible to assume the pair (A_c, C_{1c}) to be observable from the very beginning.

Part (ii) With reference to system $P_{FIc}(s)$ (recall eqs. (5.51),(5.55)), consider the control law

$$u_c := v - B'_2 P_2 x \tag{5.62}$$

where P_2 is the symmetric, positive semidefinite and stabilizing solution of the Riccati equation (in the unknown P)

$$0 = P A_c + A'_c P - P B_2 B'_2 P + C'_{1c}C_{1c} \tag{5.63}$$

Indeed, such a solution exists (recall Lemma C.4) because the pair (A_c, C_{1c}) is observable and the pair (A_c, B_2) is stabilizable. This latter claim derives from the stabilizability assumption on the pair (A, B_2) which is equivalent to that of the pair (A_c, B_2), in view of the form of A_c. Moreover, $P_2 > 0$ due to the observability property.

From eqs. (5.51),(5.52) and (5.62) it follows

$$z_{L0} = P_c(s)B_1 w_L + U(s)v_L \tag{5.64}$$

where

$$P_c(s) := \Sigma(A_c - B_2 B'_2 P_2, I, C_{1c} - D_{12}B'_2 P_2, 0)$$
$$U(s) := \Sigma(A_c - B_2 B'_2 P_2, B_2, C_{1c} - D_{12}B'_2 P_2, D_{12})$$

It is now shown that the quadruple $(A_c, B_2, C_{1c}, D_{12})$ verifies the assumptions which are required by Lemma C.5.

Indeed, from one side it is $D'_{12}D_{12} = I$ because of Assumption 5.2, while, on the other side, from the same assumption and eq. (5.48) it is $C'_{1c}D_{12} = 0$. Moreover, P_2 is the symmetric, positive definite and stabilizing solution of eq. (5.63). Thus, from Lemma C.5 it follows that system

$$F(s) := \begin{bmatrix} U(s) & U^\perp(s) \end{bmatrix} \tag{5.65}$$

is square and inner and it results

$$B_1'P_c^\sim(s)F(s) = \left[\begin{array}{c|cc} A_c - B_2B_2'P_2 & B_2 & -P_2^{-1}C_{1c}'D_{12}^\perp \\ \hline B_1'P_2 & 0 & 0 \end{array}\right] \qquad (5.66)$$

where

$$U^\perp(s) := \Sigma(A_c - B_2B_2'P_2, -P_2^{-1}C_{1c}'D_{12}^\perp, C_{1c} - D_{12}B_2'P_2, D_{12}^\perp)$$

and $D_{12}^\perp$ is such that the matrix $[D_{12}\ D_{12}^\perp]$ is orthogonal, i.e., such that $D_{12}^{\perp\prime}D_{12}^\perp = I$ and $D_{12}^{\perp\prime}D_{12} = 0$. Consider now the vector space

$$\mathcal{Q} := \left\{ q = \left[\begin{array}{c} q_1 \\ q_2 \end{array}\right], q_1 \in RH_2^\perp, q_2 \in RL_2 \right\}$$

(which is a subspace of RL_2) and the operator $\Xi : \mathcal{Q} \to RH_2$ defined by (recall Definition 2.33)

$$\Xi : \left[\begin{array}{c} q_1 \\ q_2 \end{array}\right] \mapsto \Pi_s B_1'P_c^\sim(s)F(s)\left[\begin{array}{c} q_1 \\ q_2 \end{array}\right] := w \qquad (5.67)$$

Now it will be shown that the operator $\Xi^* : RH_2 \to \mathcal{Q}$ defined by (again recall Definition 2.33)

$$\Xi^* : w \mapsto \left[\begin{array}{c} \Pi_a U^\sim(s) \\ U^{\perp\sim}(s) \end{array}\right] P_c(s)B_1 w := q \qquad (5.68)$$

is, consistently with the adopted symbol, the adjoint of the operator Ξ.

In fact, from $< w, \Xi q >=< \Xi^* w, q >, \forall q \in \mathcal{Q}, \forall w \in RH_2$, it follows that

$$\begin{aligned}
< w, \Xi q > &= < w, \Pi_s B_1'P_c^\sim(s)U(s)q_1 > + < w, \Pi_s B_1'P_c^\sim(s)U^\perp(s)q_2 > \\
&= < w, B_1'P_c^\sim(s)U(s)q_1 > + < w, B_1'P_c^\sim(s)U^\perp(s)q_2 > \\
&= < U^\sim(s)P_c(s)B_1 w, q_1 > + < U^{\perp\sim}(s)P_c(s)B_1 w, q_2 > \\
&= < \Pi_a U^\sim(s)P_c(s)B_1 w, q_1 > + < U^{\perp\sim}(s)P_c(s)B_1 w, q_2 > \\
&= < \left[\begin{array}{c} \Pi_a U^\sim(s) \\ U^{\perp\sim}(s) \end{array}\right] P_c(s)B_1 w, \left[\begin{array}{c} q_1 \\ q_2 \end{array}\right] > \\
&= < \Xi^* w, q >
\end{aligned}$$

In the above equations the two identities $< \Pi_a \alpha, q_1 >=< \alpha, q_1 >$ and $< w, \beta >=< w, \Pi_s \beta >$, which hold whenever $q_1 \in RH_2^\perp$ and $w \in RH_2$, have been exploited now and then. Since $F(s)$ is square and inner, then $F^\sim(s)F(s) = F(s)F^\sim(s) = I$ and therefore, recalling Definition 2.30,

$$\|z\|_2 = \|F^\sim(s)z_L\|_2 \qquad (5.69)$$

Now suppose that a controller exists which is admissible in RH_∞ for $P_{FIc}(s)$ and such that $\|T(z,w;s)\|_\infty < \gamma$. Therefore, with reference to the block-scheme of fig. 5.16, it is

$$\sup_{\substack{w \in RH_2 \\ \|w\|_2=1}} \|z\|_2^2 < \gamma^2$$

from which it follows

$$\sup_{\substack{w \in RH_2 \\ \|w\|_2=1}} \inf_{u_c \in RH_2} \|z\|_2^2 < \gamma^2$$

As a matter of fact, this inequality is fulfilled when u_c, rather than being suitably chosen in RH_2, is the output of the controller $K_{FIc}(s)$ (observe that such an output belongs to RH_2, because the controller is admissible and $w \in RH_2$).

By recalling eq. (5.62) and $x \in RH_2$, it then follows

$$\sup_{\substack{w \in RH_2 \\ \|w\|_2=1}} \inf_{v \in RH_2} \|z\|_2^2 < \gamma^2 \qquad (5.70)$$

Eqs. (5.70),(5.69),(5.65) and (5.64) imply

$$\gamma^2 > \sup_{\substack{w \in RH_2 \\ \|w\|_2=1}} \inf_{v \in RH_2} \|z\|_2^2$$

$$> \sup_{\substack{w \in RH_2 \\ \|w\|_2=1}} \inf_{v \in RH_2} \left\| \begin{bmatrix} U^\sim(s)P_c(s)B_1w + U^\sim(s)U(s)v \\ U^{\perp\sim}(s)P_c(s)B_1w + U^{\perp\sim}(s)U(s)v \end{bmatrix} \right\|_2^2$$

$$> \sup_{\substack{w \in RH_2 \\ \|w\|_2=1}} \inf_{v \in RH_2} \left\| \begin{bmatrix} U^\sim(s)P_c(s)B_1w + v \\ U^{\perp\sim}(s)P_c(s)B_1w \end{bmatrix} \right\|_2^2 \qquad (5.71)$$

The identities $U^\sim(s)U(s) = I$ and $U^{\perp\sim}(s)U(s) = 0$ which are consequences of $F(s)$ being inner, have been taken into account in writing down eq. (5.71). Moreover, by defining

$$\xi := v + \Pi_s U^\sim(s)P_c(s)B_1w$$

and noticing that $\xi \in RH_2$ and

$$\Pi_s U^\sim(s)P_c(s)B_1w + \Pi_a U^\sim(s)P_c(s)B_1w = U^\sim(s)P_c(s)B_1w$$

from eqs. (5.71) and (5.68) it results

$$\gamma^2 > \sup_{\substack{w \in RH_2 \\ \|w\|_2=1}} \inf_{\xi \in RH_2} \|z\|_2^2$$

$$> \sup_{\substack{w \in RH_2 \\ \|w\|_2=1}} \inf_{\xi \in RH_2} \left\| \begin{bmatrix} \Pi_a U^\sim(s)P_c(s)B_1w + \xi \\ U^{\perp\sim}(s)P_c(s)B_1w \end{bmatrix} \right\|_2^2$$

$$> \sup_{\substack{w \in RH_2 \\ \|w\|_2=1}} \inf_{\xi \in RH_2} \left\| \Xi^*w + \begin{bmatrix} \xi \\ 0 \end{bmatrix} \right\|_2^2 \qquad (5.72)$$

But

$$\left\| \Xi^*w + \begin{bmatrix} \xi \\ 0 \end{bmatrix} \right\|_2^2 = \|\Xi^*w\|_2^2 + \|\xi\|_2^2 + 2 < \xi, \Pi_a U^\sim(s)P_c(s)B_1w >$$

$$= \|\Xi^*w\|_2^2 + \|\xi\|_2^2 \qquad (5.73)$$

since $\xi \in RH_2$, while $\Pi_a U^\sim(s)P_c(s)B_1w \in RH_2^\perp$. Therefore, from eqs. (5.72) and (5.73) it follows

$$\gamma^2 > \sup_{\substack{w \in RH_2 \\ \|w\|_2=1}} \inf_{\xi \in RH_2} \|z\|_2^2$$

$$> \sup_{\substack{w \in RH_2 \\ \|w\|_2=1}} \inf_{\xi \in RH_2} \{\|\Xi^*w\|_2^2 + \|\xi\|_2^2\}$$

$$> \sup_{\substack{w \in RH_2 \\ \|w\|_2=1}} \|\Xi^* w\|_2^2$$

$$> \|\Xi^*\|^2$$

$$> \|\Xi\|^2$$

The last equality sign follows from Theorem 2.9. Then, by recalling Remark 2.13, it is

$$\sup_{\substack{q \in Q \\ \|q\|_2=1}} \|\Xi q\|_2 < \gamma \tag{5.74}$$

By exploiting Lemma G.2, applied to the operator Ξ defined in eq. (5.67) and to the system specified in eq. (5.66), from eq. (5.74) the conclusion can be drawn that a symmetric, positive semidefinite and stabilizing solution W exists for the Riccati equation in the unknown P (recall that $D_{12}^\perp D_{12}^{\perp\prime} = I - D_{12}D_{12}'$)

$$0 = P(A_c - B_2 B_2' P_2) + (A_c - B_2 B_2' P_2)'P +$$
$$+ \gamma^{-2} P P_2^{-1} C_{1c}' C_{1c} P_2^{-1} P + P_2 B_1 B_1' P_2 \tag{5.75}$$

Lemma G.2 also implies that $r_s(W L_c) < \gamma^{-2}$, where L_c solves the Lyapunov equation (in the unknown L)

$$0 = L(A_c - B_2 B_2' P_2)' + (A_c - B_2 B_2' P_2)L +$$
$$+ P_2^{-1} C_{1c}' C_{1c} P_2^{-1} + B_2 B_2' \tag{5.76}$$

By recalling that P_2 solves eq. (5.63), it is easy to verify that P_2^{-1} satisfies the Lyapunov equation (5.76) so that it actually coincides with L_c, as such an equation admits a unique solution, thanks to Lemma C.1 (all the eigenvalues of the matrix $A_c - B_2 B_2' P_2$ have negative real parts). Therefore, $L_c = P_2^{-1}$ and $r_s(W P_2^{-1}) < \gamma^2$, that is (recall Lemma B.11)

$$\gamma^2 P_2 - W > 0 \tag{5.77}$$

The Hamiltonian matrix Z_W associated with the Riccati equation (5.75) is

$$Z_W = \begin{bmatrix} A_c - B_2 B_2' P_2 & \gamma^{-2} P_2^{-1} C_{1c}' C_{1c} P_2^{-1} \\ -P_2 B_1 B_1' P_2 & -(A_c - B_2 B_2' P_2)' \end{bmatrix}$$

By letting

$$T := \begin{bmatrix} -\gamma^2 I & P_2^{-1} \\ -\gamma^2 P_2 & 0 \end{bmatrix}$$

and recalling that P_2 solves eq. (5.63), it is easy to verify that

$$Z_\infty := T Z_W T^{-1} = \begin{bmatrix} A_c & \gamma^{-2} B_1 B_1' - B_2 B_2' \\ -C_{1c}' C_{1c} & -A_c' \end{bmatrix} \tag{5.78}$$

The matrix Z_∞ is the Hamiltonian matrix associated with the Riccati equation (5.46). Now, remember that, being W a stabilizing solution of eq. (5.75), it is

$$Z_W = \begin{bmatrix} I \\ W \end{bmatrix} = \begin{bmatrix} I \\ W \end{bmatrix} V$$

where V is a stable matrix. By taking into account eq. (5.78) it then follows

$$Z_\infty T \begin{bmatrix} I \\ W \end{bmatrix} = T Z_W \begin{bmatrix} I \\ W \end{bmatrix} = T \begin{bmatrix} I \\ W \end{bmatrix} V$$

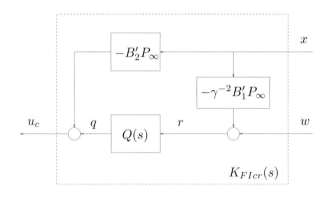

Figure 5.18: The set $\mathcal{F}_{\infty\gamma cr}$ of the controllers $K_{FIcr}(s)$

so that $\mathrm{Im}[T[I\ W]']$ is the Z_∞-invariant subspace associated with the stable eigenvalues of such a matrix. Finally, being

$$T\begin{bmatrix} I \\ W \end{bmatrix} = \begin{bmatrix} -\gamma^2 I + P_2^{-1}W \\ -\gamma^2 P_2 \end{bmatrix}$$

from Lemma C.3 it can be concluded that

$$\gamma^2 P_2(\gamma^2 I - P_2^{-1}W)^{-1} = \gamma^2 P_2(\gamma^2 P_2 - W)^{-1}P_2$$

is the (unique) symmetric and stabilizing solution P_∞ of the Riccati equation (5.46). In view of eq. (5.77) such a solution is positive definite: hence P_∞ possesses all the properties listed in the statement of the theorem and the necessity of point a) is proved.

Part (iii) Here reference is made to fig. 5.16 and the symmetric, positive semidefinite and stabilizing solution P_∞ of eq. (5.46) is supposed to exist. Denote with $\mathcal{F}_{\infty\gamma cr}$ the set (described in fig. 5.18) of controllers $K_{FIcr}(s)$ for the system $P_{FIc}(s)$. By comparing fig. 5.14, 5.15 with fig. 5.16, 5.18 and recalling eqs. (5.49),(5.50) it is apparent that proving that the controllers $K_{FIr}(s)$ of the set $\mathcal{F}_{\infty\gamma r}$ are admissible in RH_∞ for $P_{FI}(s)$ and $\|T(z,w;s)\|_\infty < \gamma$ is completely equivalent to proving that the controllers $K_{FIcr}(s)$ of the set $\mathcal{F}_{\infty\gamma cr}$ are admissible in RH_∞ for $P_{FIc}(s)$ and $\|T(z,w;s)\|_\infty < \gamma$. Moreover, the set $\mathcal{F}_{\infty\gamma cr}$ solves point (b) of Problem 5.1 for system $P_{FIc}(s)$ if and only if the set $\mathcal{F}_{\infty\gamma r}$ solves point b) of the same problem for system $P_{FI}(s)$. As far as the sufficiency of point (a) and point b) are concerned, it is therefore correct considering Problem 5.1 relative to the system $P_{FIc}(s)$ and the set $\mathcal{F}_{\infty\gamma cr}$. Consequently, it will first be shown that if a controller $K_{FIcr}(s)$ belongs to $\mathcal{F}_{\infty\gamma cr}$, then it is admissible in RH_∞ for $P_{FIc}(s)$ and $\|T(z,w;s)\|_\infty < \gamma$ (part (iii.1)). Second, it will be proved that if a controller admissible in RH_∞ for $P_{FIc}(s)$ exists and $\|T(z,w;s)\|_\infty < \gamma$, then there exists in the set $\mathcal{F}_{\infty\gamma cr}$ a controller which generates the same transfer function $T(z,w;s)$ (part (iii.2)).

Part (iii.1) Consider the block-scheme of fig. 5.18. If the system $Q(s)$ with realization $Q(s) := \Sigma(A_q, B_q, C_q, D_q)$ is stable and such that $\|Q(s)\|_\infty < \gamma$, $r := w - \gamma^{-2}B_1'P_\infty x$, $q := u_c + B_2'P_\infty x$, then (recall eqs. (5.51)-(5.55)) the resulting system (that is the system of fig. 5.16 with $K_{FIc}(s)$ given by the block-scheme in fig. 5.18) with transfer function $T(z,w;s)$ is described by $T(z,w;s) = \Sigma(A_z, B_z, C_z, D_z)$,

where

$$A_z = \begin{bmatrix} A_c - B_2 B_2' P_\infty - \gamma^{-2} B_2 D_q B_1' P_\infty & B_2 C_q \\ -\gamma^{-2} B_q B_1' P_\infty & A_q \end{bmatrix} \tag{5.79}$$

$$B_z = \begin{bmatrix} B_1 + B_2 D_q \\ B_q \end{bmatrix} \tag{5.80}$$

$$C_z = \begin{bmatrix} C_{1c} - D_{12} B_2' P_\infty - \gamma^{-2} D_{12} D_q B_1' P_\infty & D_{12} C_q \end{bmatrix} \tag{5.81}$$

$$D_z = D_{12} D_q \tag{5.82}$$

The system (5.79)-(5.82) coincides with the one referred to in Lemma E.3 and illustrated in fig. E.5. The conditions under which such a lemma can be applied are verified, so that it can be deduced that matrix A_z is stable and $\|T(z, w; s)\|_\infty < \gamma$.

  $Part(iii.2)$ Suppose that there exists a controller $K_{FIc}(s)$ which is admissible in RH_∞ for $P_{FIc}(s)$ and such that $\|T(z, w; s)\|_\infty < \gamma$. Let this controller be described by the equations

$$\dot{\xi} = L\xi + M_1 w + M_2 x \tag{5.83}$$

$$u_c = N\xi + O_1 w + O_2 x \tag{5.84}$$

  Now define the variable $q := u_c + B_2' P_\infty x$ and let $w := r + \gamma^{-2} B_1' P_\infty x$. By putting together eqs. (5.83),(5.84) with eqs. (5.51)-(5.55) it is straightforward to verify that a possible realization of the transfer function $Q(s)$ from r to q is given by

$$\dot{\vartheta} = L\vartheta + (M_2 + \gamma^{-2} M_1 B_1' P_\infty)\sigma + M_1 r \tag{5.85}$$

$$\dot{\sigma} = B_2 N\vartheta + (A_c + B_2 O_2 + \gamma^{-2} B_1 B_1' P_\infty + $$
$$+\gamma^{-2} B_2 O_1 B_1' P_\infty)\sigma + (B_1 + B_2 O_1)r \tag{5.86}$$

$$q = N\vartheta + (O_2 + B_2' P_\infty + \gamma^{-2} O_1 B_1' P_\infty)\sigma + O_1 r \tag{5.87}$$

The controller $K_{FIcr}(s)$ described by these equations and by

$$u_c := -B_2' P_\infty x + q \tag{5.88}$$

$$r := w - \gamma^{-2} B_1' P_\infty x \tag{5.89}$$

is now shown to be an element of the set $\mathcal{F}_{\infty\gamma cr}$. The structure of such a controller apparently coincides with the one in fig. 5.18. Moreover, from eqs. (5.51)-(5.55), (5.85)-(5.89), it follows that, letting $\varepsilon := \sigma - x$, the system resulting from the feedback connection of $P_{FIc}(s)$ with $K_{FIcr}(s)$ is described by the equations

$$\dot{x} = A_c x + B_1 w + B_2 u_c$$
$$\dot{\vartheta} = L\vartheta + (M_2 + \gamma^{-2} M_1 B_1' P_\infty)\varepsilon + M_2 x + M_1 w$$
$$\dot{\varepsilon} = (A_c - B_2 B_2' P_\infty + \gamma^{-2} B_1 B_1' P_\infty)\varepsilon$$
$$u_c = N\vartheta + O_2 x + O_1 w + (O_2 + B_2' P_\infty + \gamma^{-2} O_1 B_1' P_\infty)\varepsilon$$
$$z = C_{1c} x + D_{12} u_c$$

so that a comparison of these last equations with those relevant to the feedback connection of $P_{FIc}(s)$ with $K_{FIc}(s)$ (eqs. (5.51)-(5.55) and (5.83),(5.84)) leads to the following conclusions: First, the transfer functions from w to z are the same in both cases; second, the stability of the system which adopts $K_{FIcr}(s)$ is entailed by the stability of the system which utilizes $K_{FIc}(s)$, since matrix $A_c - B_2 B_2' P_\infty + \gamma^{-2} B_1 B_1' P_\infty$ is stable.

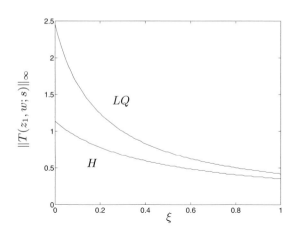

Figure 5.19: The RH_∞ performances of the controllers LQ and H

It is only left to be proved that any controller possessing the structure shown in fig. 5.18, admissible in RH_∞ for $P_{FIc}(s)$ and such that $\|T(z, w; s)\|_\infty < \gamma$ belongs to the set $\mathcal{F}_{\infty\gamma cr}$ or, in other words, that $Q(s) := (A_q, B_q, C_q, D_q)$ is stable (A_q is stable) and $\|Q(s)\|_\infty < \gamma$. These conclusions are immediately drawn by exploiting Lemma E.3.   □

Example 5.1 Consider system (5.41)-(5.45) with

$$A = \begin{bmatrix} 0 & 1 \\ -1 & -1 \end{bmatrix} , \quad B_1 = \begin{bmatrix} 0 \\ 1 \end{bmatrix} , \quad B_2 = \begin{bmatrix} 0 \\ 1 \end{bmatrix} ,$$

$$C_1 = \begin{bmatrix} C_{11} \\ C_{12} \end{bmatrix} = \begin{bmatrix} 0 & 1 \\ 0 & 0 \end{bmatrix} , \quad D_{12} = \begin{bmatrix} 0 \\ 1 \end{bmatrix}$$

A stabilizing controller is sought which makes small the effects of the disturbance w over the first component z_1 of the output z. This goal can be attained by looking for the controller which is admissible in RH_∞ for the given system and minimizes $\|T(z, w; s)\|_\infty$. Notice that, in so doing, some kind of uncertainty in the knowledge of the system dynamics can be taken into account as well. It is in fact apparent (recall fig. 5.8), that perturbations of the matrix A amounting to $\Delta_A = B_1\Omega C_{11}$ are effectively counteracted (under the stability point of view) by such a controller.

Eq. (5.46) admits a symmetric, positive semidefinite and stabilizing solution for $\gamma \geq 0.71 := \gamma_M$, while this does not happen when $\gamma \leq 0.705 := \gamma_m$, so that the minimal value attainable by $\|T(z, w; s)\|_\infty$ belongs to the interval $(\gamma_m, \gamma_M]$.

For $\gamma = \gamma_M$ it results $F_\infty = -[0 \ 0.89]$. The performances of the resulting controller, labeled with H, are compared with those, labeled with LQ, relative to a controller designed within the RH_2 framework with $\Omega = 0$ ($F_2 = -[0 \ 0.41]$). The quantities $\|T(z_1, w; s)\|_\infty$ and $\|T(z_1, w; s)\|_2$ are plotted against the damping factor $\xi := (1 - \Omega)/2$ in fig. 5.19 and 5.20 respectively. It is fairly apparent the more satisfactory behavior of the H controller.   □

Remark 5.7 The proof of Theorem 5.2 presents a result which deserves some interest per se. More precisely, the existence of a symmetric, positive semidefinite and stabilizing solution of eq. (5.46), that is such that the matrix

$$A_{cc} := A_c - B_2 B_2' P_\infty + \gamma^{-2} B_1 B_1' P_\infty$$

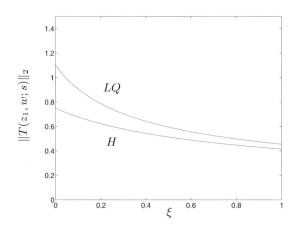

Figure 5.20: The RH_2 performances of the controllers LQ and H

is stable, also entails the stability of the matrix

$$A_{c\infty} := A_c - B_2 B_2' P_\infty = A + B_2 F_\infty$$

which is the dynamic matrix of the control system resulting from the choice $Q(s) = 0$ in the controller scheme of fig. 5.15. The correctness of this result is implicitly guaranteed by the proof of Theorem 5.2 but can be verified with a different argument too. The matrix $P_\infty = P_\infty' \geq 0$, being the stabilizing solution of eq. (5.46), is also the solution of the Lyapunov equation (in the unknown P)

$$0 = P A_{c\infty} + A_{c\infty}' P + W$$

where

$$W := P_\infty (B_2 B_2' + \gamma^{-2} B_1 B_1') P_\infty + C_{1c}' C_{1c}$$

If the pair $(A_{c\infty}, W)$ is detectable, then stability of matrix $A_{c\infty}$ follows from point (ii) of Lemma C.1. Detectability is readily proved by contradiction. Indeed, if $A_{c\infty} x = \lambda x, Re(\lambda) \geq 0, W x = 0, x \neq 0$, then $B_1' P_\infty x = 0$ and, consequently, $A_{cc} x = A_{c\infty} x = \lambda x$, so that λ would be an eigenvalue, with nonnegative real part, of A_{cc}, which is stable by assumption.   □

The assumption that the pair $[(A - B_2 D_{12}' C_1), (I - D_{12} D_{12}') C_1]$ is detectable is not exploited in the proof of part (b) and of sufficiency of part (a) of Theorem 5.2. Moreover, the existence of the stabilizing solution of eq. (5.46) implies, thanks to Remark 5.7, the stabilizability of the pair (A, B_2). It is therefore possible to state the following corollary which can be viewed as a useful side-product of the result presented in Theorem 5.2.

Corollary 5.1 *Suppose that Assumption 5.2 holds. If, for a given positive scalar γ, there exists the symmetric, positive semidefinite and stabilizing solution of the Riccati equation (5.46), then there exists a controller $K_{FI}(s)$ which is admissible in RH_∞ for $P_{FI}(s)$ and such that $\|T(z, w; s)\|_\infty < \gamma$. Moreover, the set of controllers defined by point (b) in the statement of Theorem 5.2 constitutes the set $\mathcal{F}_{\infty \gamma r}$.*

Remark 5.8 It is worth noticing that when $\gamma \to \infty$ the RH_∞ controller of fig. 5.15 (with $Q(s) = 0$) tends to the RH_2 controller of fig. 4.3 (with $Q(s) = 0$).   □

Remark 5.9 (Parametrization of the set $\mathcal{F}_{\infty\gamma}$) A parametrization of the set $\mathcal{F}_{\infty\gamma}$ will be presented now. To this aim, consider the system $P_F(s)$ which results from system (5.41)-(5.45) after the control law

$$u = F_\infty x + v \tag{5.90}$$

has been implemented, where F_∞ is defined by eq. (5.49), namely the system

$$P_F(s) = \begin{bmatrix} P_{F11}(s) & P_{F12}(s) \\ P_{F21}(s) & P_{F22}(s) \end{bmatrix}$$

$$= \left[\begin{array}{c|cc} A_{c\infty} & B_1 & B_2 \\ \hline C_{1\infty} & 0 & D_{12} \\ \begin{bmatrix} I \\ 0 \end{bmatrix} & \begin{bmatrix} 0 \\ I \end{bmatrix} & \begin{bmatrix} 0 \\ 0 \end{bmatrix} \end{array} \right]$$

where

$$A_{c\infty} := A + B_2 F_\infty , \quad C_{1\infty} := C_{1c} - D_{12} B_2' P_\infty$$

having taken into account eq. (5.48). The set of controllers $K_F(s)$ which stabilize $P_F(s)$ obviously coincides with the set of controllers which stabilize $P_{F22}(s)$. By mimicking the discussion in Remark 4.3 (recall that, thanks to Remark 5.7, matrix $A_{c\infty}$ is stable) and adopting the same kind of notation, it follows

$$[I + \gamma^{-2} Q(s) B_1' P_\infty \Phi(s) B_2]^{-1} Q(s)[I - \gamma^{-2} B_1' P_\infty \Phi(s) B_1] = -\Theta(s) P_{F21}(s)$$

where $\Phi(s) := (sI - A_{c\infty})^{-1}, \Theta(s) \in RH_\infty, Q(s) \in RH_\infty$. Letting

$$\Delta(s) := I + \gamma^{-2} Q(s) B_1' P_\infty \Phi(s) B_2$$

the last equation can be rewritten as

$$\Delta^{-1}(s) Q(s)[-\gamma^{-2} B_1' P_\infty \ \ I] P_{F21}(s) = -\Theta(s) P_{F21}(s)$$

since

$$P_{F21}(s) = \begin{bmatrix} \Phi(s) B_1 \\ I \end{bmatrix}$$

Assuming, for the moment being, that $\Delta^{-1}(s) \in RH_\infty$, it is easy to verify that a particular solution in RH_∞ of such an equation is

$$\bar{\Theta}(s) = \Delta^{-1}(s) Q(s)[\gamma^{-2} B_1' P_\infty \ \ - I]$$

The general solution in RH_∞ is therefore

$$\Theta_Q(s) = \bar{\Theta}(s) + \hat{\Theta}(s)$$

where $\hat{\Theta}(s)$ is any solution in RH_∞ of the homogeneous equation

$$\Theta(s) P_{F21}(s) = 0$$

Letting $\Theta(s) := [\Delta^{-1}(s)\Theta_1(s) \ \ \Theta_2(s)]$, this last equation implies that $\Theta_2(s)$ is given by $\Theta_2(s) = -\Delta^{-1}(s)\Theta_1(s)\Phi(s) B_1$ and thus

$$\hat{\Theta}(s) = \Delta^{-1}(s)\Theta_1(s)[I \ \ - \Phi(s) B_1]$$

Notice that $\hat{\Theta}(s)$ is an element of RH_∞ if and only if $\Theta_1(s)$ is such, since $\Delta^{-1}(s)$ has been assumed to belong to RH_∞ and $\Phi(s) \in RH_\infty$ being $A_{c\infty}$ stable. Therefore, the set of functions $\Theta_Q(s)$ giving rise to controllers $K_F(s)$ which are admissible in RH_∞ for $P_F(s)$ and such that $\|T(z, w; s)\|_\infty < \gamma$ is defined by

$$\Theta_Q(s) = \Delta^{-1}(s)\Lambda(s)$$

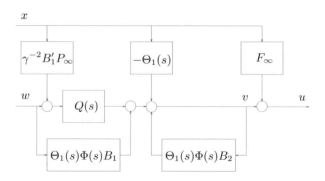

Figure 5.21: The generic admissible controller for $P_{FI}(s)$

where

$$\Lambda(s) := \begin{bmatrix} \Theta_1(s) + \gamma^{-2}Q(s)B_1'P_\infty & -[Q(s) + \Theta_1(s)\Phi(s)B_1] \end{bmatrix}$$
$$\Theta_1(s) \in RH_\infty, \ Q(s) \in RH_\infty, \ \|Q(s)\|_\infty < \gamma$$

Thus, the transfer function (from $[x' \ w']'$ to v) of a generic controller which is admissible in RH_∞ for $P_F(s)$ is (again, recall what has been presented in Remark 4.3)

$$\begin{aligned}
K_F(s) &= -[I - \Theta_Q(s)P_{F22}(s)]^{-1}\Theta_Q(s) \\
&= -[I - \Delta^{-1}(s)\Lambda(s)P_{F22}(s)]^{-1}\Delta^{-1}(s)\Lambda(s) \\
&= -[\Delta(s) - \Lambda(s)P_{F22}(s)]^{-1}\Lambda(s) \\
&= [I - \Theta_1(s)\Phi(s)B_2]^{-1} \cdot \\
&\quad \cdot \begin{bmatrix} -\gamma^{-2}Q(s)B_1'P_\infty - \Theta_1(s) & Q(s) + \Theta_1(s)\Phi(s)B_1 \end{bmatrix}
\end{aligned}$$

By exploiting this equation, the generic controller which is admissible in RH_∞ for $P_{FI}(s)$ can be represented by the block-scheme depicted in fig. 5.21. Such a scheme shows that

$$\begin{aligned}
v_{L0} &= \Theta_1(s)\Phi(s)[-\Phi^{-1}(s)x_L + B_2v_{L0} + B_1w_L] + \\
&\quad +Q(s)(w_L - \gamma^{-2}B_1'P_\infty x_L) \\
&= -\Theta_1(s)\Phi(s)x(0) + Q(s)(w_L - \gamma^{-2}B_1'P_\infty x_L)
\end{aligned}$$

having taken into consideration eqs. (5.41) and (5.90). Therefore, the effect of the parameter $\Theta_1(s)$ (which is responsible for the difference between the elements of the set $\mathcal{F}_{\infty\gamma}$ and those of the set $\mathcal{F}_{\infty\gamma r}$) on the control variable u_L amounts to a term which depends on the initial conditions of system (5.41)-(5.45), only.

It remains to verify that $\Delta^{-1}(s) \in RH_\infty$. First, recall that P_∞, being the stabilizing solution of eq. (5.46), is the stabilizing solution of the Riccati equation (in the unknown P)

$$0 = PA_{c\infty} + A_{c\infty}'P + PB_2B_2'P + C_{1c}'C_{1c} + \gamma^{-2}PB_1B_1'P$$

as well. In writing down this equation reference has been made to eqs. (5.48) and (5.49). This implies, thanks to Theorem 2.13, that system

$$\Upsilon(s) := \left[\begin{array}{c|c} A_{c\infty} & B_2 \\ \hline C_{1c} & 0 \\ \gamma^{-1}B'P_\infty & 0 \end{array} \right]$$

is such that $\|\Upsilon(s)\|_\infty < 1$, which, in turn, entails (recall Lemma 2.23, Definition 2.23 and Remark 2.16) that

$$\|\gamma^{-1}B_1'P_\infty\Phi(s)B_2\|_\infty < 1$$

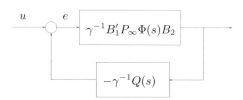

Figure 5.22: Proving stability of $\Delta^{-1}(s)$

Being $\|Q(s)\|_\infty < \gamma$, the system shown in fig. 5.22 is internally stable thanks to Theorem 5.1. The transfer function from u to e of such a system is precisely $\Delta^{-1}(s)$.   □

Remark 5.10 (Linear quadratic!differential game) The control problem whose solution has been presented in the preceding theorem may be viewed as a *linear quadratic deterministic differential game (LQDG)* with *measurable* state.

LQDG Problem   Consider the *n-th* order system with initial state $x(0) = x_0$

$$\dot{x} = Ax + \bar{B}_1 \bar{w} + \bar{B}_2 \bar{u} \tag{5.91}$$

together with the cost functional

$$J_1 := \int_0^\infty \left\{ \begin{bmatrix} x'(t) & \bar{u}'(t) \end{bmatrix} L \begin{bmatrix} x(t) \\ \bar{u}(t) \end{bmatrix} - \gamma^2 \bar{w}'(t) R_1 \bar{w}(t) \right\} dt \tag{5.92}$$

where

$$L := \begin{bmatrix} Q & S \\ S' & R_2 \end{bmatrix} \geq 0 \,, \quad R_2 > 0 \,, \quad R_1 > 0$$

Notice that the sign assumptions on L and R_2 imply that

$$\hat{Q} := Q - SR_2^{-1}S' \geq 0 \tag{5.93}$$

since $\hat{Q} = Z'LZ$ with $Z' = [I \; -SR_2^{-1}]$. Let $C_{11} \in R^{n \times n}$ be a *factorization* of $\hat{Q}$, so that

$$C'_{11}C_{11} = \hat{Q} \tag{5.94}$$

and define

$$C_1 := \begin{bmatrix} C_{11} \\ R_2^{-\frac{1}{2}} S' \end{bmatrix} \,, \quad D_{12} := \begin{bmatrix} 0 \\ I \end{bmatrix} \tag{5.95}$$

$$u := R_2^{\frac{1}{2}} \bar{u} \,, \quad z := C_1 x + D_{12} u \,, \quad w := R_1^{\frac{1}{2}} \bar{w} \tag{5.96}$$

$$B_2 := \bar{B}_2 R_2^{-\frac{1}{2}} \,, \quad B_1 := \bar{B}_1 R_1^{-\frac{1}{2}} \tag{5.97}$$

In view of eqs. (5.93), (5.97), eqs. (5.91)-(5.92) become

$$\dot{x} = Ax + B_1 w + B_2 u \tag{5.98}$$

$$J_1 = \int_0^\infty [z'(t)z(t) - \gamma^2 w'(t)w(t)]dt \tag{5.99}$$

The solution of the differential game consists in finding the controller which stabilizes the resulting system and generates the control $u \in RH_2$ so as to *minimize* the functional J_1 corresponding to the *worst* input (disturbance) $w \in RH_2$, that is corresponding to the input

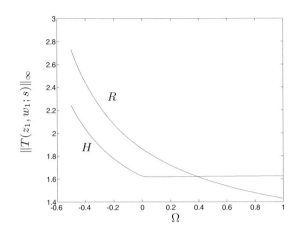

Figure 5.24: The performances of the controllers R and H

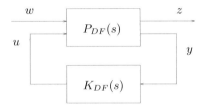

Figure 5.25: The disturbance feedforward problem

For $\Omega = 0$ the Riccati equation (5.46) relevant to this system admits the positive semidefinite and stabilizing solution whenever $\gamma \geq 1.62$, while such a solution does not exist if $\gamma \leq 1.61$.

Taken $\gamma = 1.62$, it results $F_\infty = -[367.68 \quad 227.58]$. The corresponding controller will be denoted with the label H.

By proceeding along the same lines as in Remark 5.11, an output z_2 is defined as

$$z_2 := \begin{bmatrix} 0 & 1 \end{bmatrix} x$$

and an input w_2 is introduced, acting on the system through the matrix $[1\ 1]'$.

Corresponding to $\Omega = 0$, the eq. (5.46) relevant to the new system characterized by the disturbance $w := [w_1\ w_2]'$ and the performance output $z := [z_1\ z_2]'$, admits a positive semidefinite and stabilizing solution for $\gamma \geq 2.46$, while such a solution does not exist if $\gamma \leq 2.45$. Taken $\gamma = 2.46$ it results $F_{\infty r} = -[333.04 \quad 286.98]$. The controller adopting such a $F_{\infty r}$ will be denoted with the label R. The graphs of $\|T(z_1, w_1; s)\|_\infty$ corresponding to the adoption of the two controllers R and H are shown, as functions of Ω, in fig. 5.24. The controller R apparently behaves better for high values of Ω.   □

Remark 5.12 (Disturbance feedforward) A problem similar to the one discussed in Remark 4.4 can be dealt with in the RH_∞ context as well. Consider the system depicted in fig. 5.25 where $P_{DF}(s)$ is described by

$$\dot{x} = Ax + B_1 w + B_2 u$$
$$z = C_1 x + D_{12} u$$
$$y = C_2 x + w$$

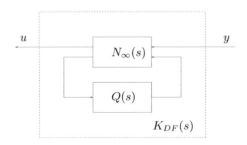

Figure 5.26: The set $\mathcal{F}_{\infty\gamma r}$ of the controllers $K_{DFr}(s)$

while $K_{DF}(s)$ is any controller which is admissible in RH_∞ for $P_{DF}(s)$.

Assume that the pair (A, B_2) is stabilizable, the pair $[(A - B_2 D'_{12}C_1), (I - D_{12}D'_{12})C_1]$ is detectable, the matrix $A - B_1C_2$ is stable and $D'_{12}D_{12} = I$.

By proceeding as done in the RH_2 setting (in particular, observe that Lemma E.2 can be applied in the present RH_∞ context as well) it is possible to directly give the solution of Problem 5.1 relative to system $P_{DF}(s)$

a) Existence of the symmetric, positive semidefinite and stabilizing solution P_∞ of the Riccati equation (5.46);

b) The set $\mathcal{F}_{\infty\gamma r}$ of the controllers $K_{DFr}(s)$ is defined by the block-scheme of fig. 5.26 where

$$N_\infty(s) := \left[\begin{array}{c|cc} A - B_1C_2 + B_2F_\infty & B_1 & B_2 \\ \hline F_\infty & 0 & I \\ -C_2 - \gamma^{-2}B'_1P_\infty & I & 0 \end{array}\right]$$

$Q(s) := \Sigma(A_q, B_q, C_q, D_q)$, with A_q stable, $\|Q(s)\|_\infty < \gamma$ and F_∞ given by eq. (5.49).

Finally, notice that $\mathcal{F}_{\infty\gamma r} = \mathcal{F}_{\infty\gamma}$. In fact, what has been said concerning the problem in the RH_2 setting can be applied with no modifications in the present framework, coming up to the same conclusions. □

Example 5.3 Consider the system

$$\dot{x} = (A + \Delta_A)x + B_1w + B_2u$$
$$z = C_1x + D_{12}u$$
$$y = C_2x + w$$

where

$$A = \left[\begin{array}{cc} 0 & 1 \\ -1 & -1 \end{array}\right], \quad \Delta_A = L\Omega M, \ B_1 = B_2 = L = \left[\begin{array}{c} 0 \\ 1 \end{array}\right]$$

$$C_1 = C_2 = \left[\begin{array}{cc} 1 & 0 \end{array}\right], \quad M = \left[\begin{array}{cc} 0 & 1 \end{array}\right], \quad D_{12} = 1$$

The only uncertain system parameter is Ω: its nominal value is 0 and corresponds to a damping factor $\xi := (1 - \Omega)/2 = 0.5$. A stabilizing controller is sought which makes the RH_∞ norm of the transfer function from w to z small. If the perturbation Ω is ignored ($\Omega = 0$), it is possible to design a controller, denoted by the label IC and with input y and output u, which achieves a *perfect indirect compensation* of the disturbance, that is a controller such that $T(z, w; s) = 0$. More in detail, the transfer function of this controller is

$$K_{IC}(s) = -\frac{1}{s^2 + s + 3}$$

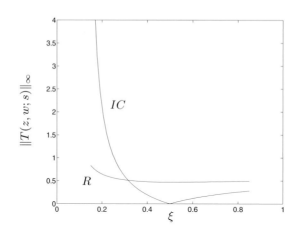

Figure 5.27: The performances of the controllers IC and R

On the contrary, by exploiting what has been presented in Remark 5.11, a controller (denoted with the label R) can be designed which can account for Ω not being 0 in a more effective way. Notice that being $L = B_1$ it is useless introducing a further disturbance input, while being $M \neq C_1$ it is mandatory adding a new performance variable $z_a = Mx$ to the formerly existing one z. Thus, the resulting problem has the same structure as those considered in Remark 5.12.

The Riccati equation relevant to the so restated problem does not admit the positive semidefinite and stabilizing solution when $\gamma \leq 0.7$, while such a solution exists if $\gamma \geq 0.71$. Associated with this last value of γ it results $F_\infty = -[1 \ \ 0.89]$.

In fig. 5.27 the plots of $\|T(z, w; s)\|_\infty$ corresponding to the above controllers are shown. The controller R apparently behaves better for small values of the actual damping factor; moreover, its performance is somehow less sensitive (in terms of the RH_∞ norm) to the variations of such a parameter. On the contrary, the controller IC is to be preferred, as it should be expected, whenever the parameter Ω is rather precisely known. $\square$

Remark 5.13 The content of Remark 4.6 concerning the rank of D_{12} and the condition $D'_{12}D_{12} = I$ applies to the RH_∞ setting with no changes. It is also apparent that the above given expressions get much simpler under the *orthogonality* assumption $D'_{12}C_1 = 0$ $\square$

Remark 5.14 Under Assumption 5.2, the detectability of the pair $[(A - B_2D'_{12}C_1), (I - D_{12}D'_{12})C_1]$ (Assumption 5.1) is equivalent to asking the subsystem of $P_{FI}(s)$ corresponding to the transfer function $P_{FI12}(s)$ between the input u and the output z (that is, system $\Sigma(A, B_2, C_1, D_{12})$ not to have zeros in $Re(s) \geq 0$. This claim can easily be checked by means of the same arguments exploited in Remark 4.8. $\square$

5.4   The output estimation problem

In this section the problem of *observing* linear combinations of the state variables is dealt with in the RH_∞ context. The system considered here is described by the state equations

$$\dot{x} = Ax + B_1w + B_2u \tag{5.102}$$

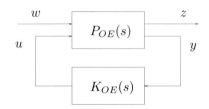

Figure 5.28: The output estimation problem

$$z = C_1 x + u \tag{5.103}$$
$$y = C_2 x + D_{21} w \tag{5.104}$$

As it has been done in Section 4.3, $P_{OE}(s)$ denotes the transfer function of this system, while $T(z, w; s)$ is the transfer function from w to z in the block-scheme of fig. 5.28. The structure of system $P_{OE}(s)$ clearly indicates that solving Problem 5.1 for such a system can be really viewed as finding a "good" estimate (in the RH_∞ sense) of the linear combination $-C_1 x$ of the state variables. Indeed, should the controller $K_{OE}(s)$ be capable of zeroing $T(z, w; s)$ so that $z_{L0} = 0$, then, apparently, $u_{L0} = -C_1 x_{L0}$ would represent the best possible estimate of that linear combination.

The statement of the result below requires the following assumptions.

Assumption 5.3 *The pair $[(A - B_1 D'_{21} C_2), B_1 (I - D'_{21} D_{21})]$ is stabilizable and the pair (A, C_2) is detectable.*

Assumption 5.4 $D_{21} D'_{21} = I$.

Assumption 5.5 $A - B_2 C_1$ *is stable.*

It is now possible to prove the next theorem which makes reference to fig. 5.28.

Theorem 5.3 (Output estimation) *Consider Problem 5.1 relative to system (5.102)-(5.104). Then, under Assumptions 5.3 - 5.5, it has the solution*

a) *The existence of the symmetric, positive semidefinite and stabilizing solution Π_∞ of the Riccati equation (in the unknown Π)*

$$0 = \Pi A'_f + A_f \Pi - \Pi (C'_2 C_2 - \gamma^{-2} C'_1 C_1) \Pi + B_{1f} B'_{1f} \tag{5.105}$$

i.e., such that the matrix A_{fc} given by

$$A_{fc} := A_f - \Pi_\infty (C'_2 C_2 - \gamma^{-2} C'_1 C_1) \tag{5.106}$$

is stable. In eqs. (5.105), (5.106)

$$A_f := A - B_1 D'_{21} C_2, \quad B_{1f} := B_1 (I - D'_{21} D_{21}) \tag{5.107}$$

b) *The set $\mathcal{F}_{\infty\gamma r}$ of the controllers $K_{OEr}(s)$ is defined by the diagram of fig. 5.29, where*

$$M_\infty(s) := \left[\begin{array}{c|cc} A_f - \Pi_\infty C'_2 C_2 - B_2 C_1 & L_\infty & -B_2 - \gamma^{-2} \Pi_\infty C'_1 \\ \hline C_1 & 0 & I \\ C_2 & I & 0 \end{array} \right]$$

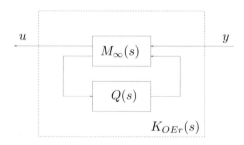

Figure 5.29: The set $\mathcal{F}_{\infty\gamma r}$ of the controllers $K_{OEr}(s)$

$Q(s) := \Sigma(A_q, B_q, C_q, D_q)$ with A_q stable and $\|Q(s)\|_\infty < \gamma$, having defined

$$L_\infty := -\Pi_\infty C_2' - B_1 D_{21}' \tag{5.108}$$

Proof First observe that Assumption 5.5 implies that the pair (A, B_2) is stabilizable. This fact, together with the assumed detectability of the pair (A, C_2), makes the necessary condition of Remark 5.6 satisfied.

Now consider the system $\hat{P}_{OE}(s) := P_{OE}'(s)$. Then, from eqs. (5.102)-(5.104) it follows that such a system is given by

$$\dot{\xi} = F\xi + G_1\zeta + G_2\eta$$
$$\omega = H_1\xi + E\eta$$
$$\varphi = H_2\xi + \zeta$$

with

$$F := A' , \quad E := D_{21}' \tag{5.109}$$

and, for $i = 1, 2$,

$$G_i := C_i' , \quad H_i := B_i' \tag{5.110}$$

System $\hat{P}_{OE}(s)$ possesses the structure of system $P_{DF}(s)$ which has been introduced in Remark 5.12. Therefore, Problem 5.1 relative to system $\hat{P}_{OE}(s)$ is solved in the very same way, since the requirements there are satisfied by Assumptions 5.3 - 5.5. Thanks to Lemma E.1 the results concerning system P_{OE} follow by transposition of those concerning system $\hat{P}_{OE}(s)$, provided that eqs. (5.109) and (5.110) are taken into account. $\square$

Example 5.4 This example is aimed at pointing out an important difference existing between the solution of the estimation problem carried out in the RH_2 context and that obtained in the RH_∞ one. Thus, consider system (5.102)-(5.104) with

$$A = \begin{bmatrix} 0 & 1 \\ 0 & 0 \end{bmatrix} , \quad B_1 = I , \quad B_2 = \begin{bmatrix} 0 \\ 1 \end{bmatrix} , \quad C_2 = \begin{bmatrix} 1 & 0 \end{bmatrix}$$

$$D_{21} = \begin{bmatrix} 0 & 1 \end{bmatrix} , \quad C_1 = \begin{bmatrix} 1 & \alpha \end{bmatrix} , \quad \alpha > 0$$

Assumptions 5.3 - 5.5 (and therefore also Assumptions 4.3 - 4.5) are satisfied for each value of α. The RH_2 problem calls for the determination of the matrix Π_2 which is the stabilizing solution of the Riccati equation (4.64). Such a matrix is independent of C_1, so that also the matrix L_2 is such (recall Theorem 4.2). As a consequence of this fact and the structure of system $M_2(s)$, the transfer function from the input w to the state estimation error $\varepsilon := x + x_m$ does not depend on C_1 (recall that x_m, the state of system $M_2(s)$, is the *opposite* of the estimate of the state of system $P_{OE}(s)$, since the relevant linear combination of the state variables is $-C_1 x$). Indeed, for any $Q(s)$, it results

$$T((x + x_m), w; s) = [sI - (A + L_2 C_2)]^{-1}(L_2 D_{21} + B_1)$$

This property in no more verified in the RH_∞ context. In fact, matrix Π_∞, which is the stabilizing solution of the Riccati equation (5.105), and, consequently, matrix L_∞ too, depends on C_1 (see Theorem 5.3). This fact, together with the structure of system $M_\infty(s)$, implies that the transfer function from the input w to the state estimation error $\varepsilon := x + x_m$, where x_m is the state variable of system $M_\infty(s)$, is given by

$$T((x + x_m), w; s) = [sI - (A + L_\infty C_2)]^{-1}(L_\infty D_{21} + B_1)$$

when $Q(s) = 0$. This function, though formally identical to that resulting in the RH_2-context, depends on C_1, that is from the actual linear combination of the state of $P_{OE}(s)$ which one is willing to estimate.

With reference to the considered system, for instance, if α is set equal to 10 or equal to 100, one obtains, corresponding to $\gamma = 130$, $L_\infty = -[1.00 \quad 1.00]'$ and $L_\infty = -[6.81 \quad 8.17]'$, respectively. □

Remark 5.15 (Parametrization of the set $\mathcal{F}_{\infty\gamma}$) Observe that $\mathcal{F}_{\infty\gamma} = \mathcal{F}_{\infty\gamma r}$. In fact, as in the proof of Theorem 5.3, the RH_∞ admissible controllers for $P_{OE}(s)$ can be obtained by transposition of those which are RH_∞ admissible for $\hat{P}_{OE}(s)$, this last system possessing the structure of system $P_{DF}(s)$ (see Remark 5.12). Therefore, being $\mathcal{F}_{\infty\gamma} = \mathcal{F}_{\infty\gamma r}$ for system $P_{DF}(s)$, the same conclusion must hold for system $P_{OE}(s)$. □

Remark 5.16 (Optimal state filtering) The control problem associated with system $P_{OE}(s)$ can be viewed as an (optimal) state filtering problem in RH_∞ for the stable system described by

$$\dot{x} = Ax + \bar{B}_1 w_1$$
$$y = C_2 x + w_2$$

The signals w_1 and w_2 belong to the set

$$\mathcal{W} := \{w_\alpha \mid w_\alpha \in RL_2[0, \infty) \, , \, \|w_\alpha\|_2 \le \alpha\}$$

A filter is sought for a given linear combination Sx of the state variables such that, denoting with u the filter output, it results

$$J = \min_{u \in RL_2[0,\infty)} \sup_{\substack{w_1 \in \mathcal{W} \\ w_2 \in \mathcal{W}}} \|u - Sx\|_2$$

Define $w := [w_1' \; w_2']'$, $B_1 := [\bar{B}_1 \; 0]$, $D_{21} := [0 \; I]$, $C_1 := -S$, $z := C_1 x + u$. Then the problem can conveniently be stated by considering the system described by the equations

$$\dot{x} = Ax + B_1 w$$
$$z = C_1 x + u$$
$$y = C_2 x + D_{21} w$$

and seeking for the controller-filter such that the RH_∞ norm of the transfer function from w to z is less than a given positive scalar γ. Indeed, observing that $\|w\|_2 \le \alpha\sqrt{2} := \beta$, it results

$$\sup_{\substack{w_1 \in \mathcal{W} \\ w_2 \in \mathcal{W}}} \|z\|_2 \le \sup_{\substack{w \in RH_2 \\ \|w\|_2 \le \beta}} \|z\|_2 = \beta \sup_{\substack{w \in RH_2 \\ w \ne 0}} \frac{\|z\|_2}{\|w\|_2}$$

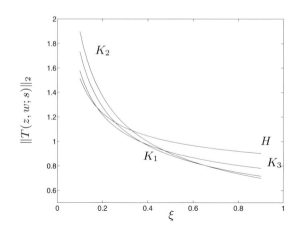

Figure 5.30: The RH_2 norm of the four filters

so that $J \leq \beta \|T(z, w; s)\|_\infty$. Thus, the smaller the value of γ, the tighter would be the upper bound of the maximum attainable value of J. What has previously been shown in Theorem 5.3 can now be applied to the so recast problem, since the underlying system possesses the very same structure of system $P_{OE}(s)$ and Assumptions 5.3 - 5.5 are verified (recall the definition of D_{21}, the stability of matrix A and notice that $B_1 D'_{21} = 0$ and $B_2 = 0$).   □

Example 5.5 Consider the system

$$\dot{x} = Ax + \bar{B}_1 w_1$$
$$y = C_2 x + w_2$$

where w_1 and w_2 are noises the features of which are to be specified later on, while

$$A = \begin{bmatrix} 0 & 1 \\ -1 & -1+\Omega \end{bmatrix} , \quad \bar{B}_1 = \begin{bmatrix} 0 \\ 1 \end{bmatrix} , \quad C_2 = \begin{bmatrix} 1 & 0 \end{bmatrix}$$

The parameter Ω accounts for the uncertain knowledge of the system damping factor ξ. Indeed, the characteristic polynomial of A is $\psi(s) = s^2 + s(1 - \Omega) + 1 = s^2 + 2\xi\omega_n s + \omega_n^2$, with damping factor $\xi = (1 - \Omega)/2$ and natural frequency $\omega_n = 1$.

The performances of four different filters for the whole state vector are now compared. Three of them, denoted with K_1, K_2 and K_3, respectively, are standard Kalman's filters which have been designed under the assumption that w_1 and w_2 are uncorrelated, zero-mean, Gaussian white noises. In the considered cases, their intensity are $W_1 = W_2 = 1$ for the K_1 filter, $W_1 = 0.5$, $W_2 = 1$ for the K_2 filter, $W_1 = 1$, $W_2 = 0.5$ for the K_3 filter.

On the contrary, the fourth filter, denoted with the label H, is designed accordingly to what has been presented in Remark 5.16. The adopted value for the scalar γ is 1.001 (notice that for $\gamma \leq 1$ the positive semidefinite and stabilizing solution of the relevant Riccati equation does not exist).

The design of the four filters is carried out in nominal conditions ($\Omega = 0 \Leftrightarrow \xi = .5$), while their performances are evaluated by connecting them to the system perturbed in correspondence with various values of ξ.

Two performance criteria have been adopted. Both of them make reference to the transfer function $T(z, w; s)$, where $w := [w_1 \ w_2]'$ and z is the state estimation error. The criteria are the RH_2 and RH_∞ norms of $T(z, w; s)$, respectively: their plots against ξ are shown in fig. 5.30 and 5.31.

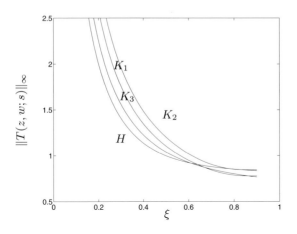

Figure 5.31: The RH_∞ norm of the four filters

The two graphs show that the filter designed within the RH_∞ framework is somehow less sensitive with respect to ξ and entails lower values of the norm for small values of the damping factor, namely close to instability. $\qquad \square$

Remark 5.17 Analogously to what has been done in Remark 5.7 with reference to the Riccati equation (5.46), the existence of the symmetric, positive semidefinite and stabilizing solution of eq. (5.105), i. e. , such that

$$A_{fc} := A_f - \Pi_\infty(C_2'C_2 - \gamma^{-2}C_1'C_1)$$

is stable, also implies the stability of

$$A_{f\infty} := A_f - \Pi_\infty C_2'C_2 = A + L_\infty C_2$$

which is the dynamic matrix of the RH_∞ state observer discussed in Remark 5.16 when $Q(s) = 0$. The proof of this claim is identical to the one presented in Remark 5.7. $\qquad \square$

It is easy to verify that the proof of point (b) and of the sufficiency part of point (a) of Theorem 5.3 does not exploit the assumption that the pair $[(A - B_1 D_{21}'C_2), B_1(I - D_{21}'D_{21})]$ is stabilizable. Moreover, the existence of the stabilizing solution of eq. (5.105) implies, thanks to Remark 5.17, the detectability of the pair (A, C_2). Therefore, it is possible to state the following corollary which has to be viewed as a useful side-product of Theorem 5.3.

Corollary 5.2 *Suppose that Assumptions 5.4 and 5.5 hold. If, for a given positive scalar γ, there exists the symmetric, positive semidefinite and stabilizing solution of the Riccati equation (5.105), then there exists a controller $K_{OE}(s)$ which is admissible in RH_∞ for $P_{OE}(s)$ and such that $\|T(z, w; s)\|_\infty < \gamma$. Moreover, the set of controllers defined by point (b) in the statement of Theorem 5.3 constitutes the set $\mathcal{F}_{\infty\gamma r}$.*

Remark 5.18 It is worth noticing that, also in the output estimation case, when $\gamma \to \infty$ the RH_∞ controller of fig. 5.29 (with $Q(s) = 0$) tends to the RH_2 controller of fig. 4.13 (with $Q(s) = 0$). $\qquad \square$

Remark 5.19 (Robust parametric filtering) An interesting role is played by the Riccati equation (5.105) also when the problem at hand is filtering the state of a system affected by *parametric* uncertainty and contemporarily looking for a *meaningful* (namely, *small*) upper bound to the filtering error. More precisely, consider the *n-th* order stochastic system

$$\dot{x} = (A + \Delta_A)x + B_1 w$$
$$y = Cx + v$$

with the matrix A stable and the pair (A, B_1) reachable. Moreover, the inputs w and v are zero mean, uncorrelated white noises with identity intensity. The perturbation Δ_A is defined by

$$\Delta_A := B_2 \Omega C_1 \ , \quad \|\Omega\| \le \gamma^{-1}$$

where the matrices B_2, C_1 and the positive scalar γ are given. The problem of *robust parametric filtering with cost Q* consists in finding a stable state filter of the form

$$\dot{\xi} = A_f \xi + K_f y$$

and a positive semidefinite matrix Q which provides an upper bound for $X_{\varepsilon a}(\Omega)$, the asymptotic covariance matrix of the estimation error

$$\varepsilon := \xi - x$$

correspondingly to *any* perturbation Δ_A of the above specified form. In other words, it is required that

$$X_{\varepsilon a}(\Omega) := \lim_{t \to \infty} E[\varepsilon(t)\varepsilon'(t)] \le Q \ , \quad \forall \ \|\Omega\| \le \gamma^{-1}$$

Let $C_2(\beta) := \beta C$, $B(\beta) := [\beta^{-1} B_1 \quad B_2]$. Then it will be proved that, for each $\beta > 0$ belonging to the set of the $\beta's$ for which the Riccati equation (in the unknown Π)

$$0 = A\Pi + \Pi A' + \Pi(\gamma^{-2} C_1' C_1 - C_2'(\beta) C_2(\beta))\Pi + B(\beta)B'(\beta) \tag{5.111}$$

admits the stabilizing solution $\Pi_\infty(\beta) = \Pi_\infty'(\beta) \ge 0$, the matrices

$$A_{f\beta} := A + \Pi_\infty(\beta)(\gamma^{-2} C_1' C_1 - \beta^2 C' C) \tag{5.112}$$
$$K_{f\beta} := \beta^2 \Pi_\infty(\beta) C' \ , \quad Q_\beta := \beta^2 \Pi_\infty(\beta) \tag{5.113}$$

define a solution of the above stated problem.

Remark 5.2 is exploited to prove this claim. In fact, let $\eta := [\varepsilon' \ \xi']'$ and $\varphi := [w' \ v']'$, then the connection of the system with the filter is described by the equation

$$\dot{\eta} = (F + \Delta_F)\eta + G\varphi$$

where

$$F := \begin{bmatrix} A - K_f C & A_f + K_f C - A \\ -K_f C & A_f + K_f C \end{bmatrix} , \quad G := \begin{bmatrix} -B_1 & K_f \\ 0 & K_f \end{bmatrix}$$

$$\Delta_F := \bar{B}\Omega\bar{C} := \begin{bmatrix} B_2 \\ 0 \end{bmatrix} \Omega \begin{bmatrix} C_1 & -C_1 \end{bmatrix}$$

Of course, matrix F is stable since matrices A and A_f are both stable. Corresponding to the given matrices K_f and A_f, assume that there exists the stabilizing solution $\Theta_\infty(\beta) = \Theta_\infty'(\beta) \ge 0$ of the Riccati equation (in the unknown Θ)

$$0 = \Theta F' + F\Theta + \gamma^{-2}\Theta\bar{C}'\bar{C}\Theta + \bar{B}\bar{B}' + \beta^{-2}GG' \tag{5.114}$$

Then, in view of Remark 5.2 it follows

$$X_{\eta a}(\Omega) := \lim_{t \to \infty} E[\eta(t)\eta'(t)] \le \beta^2 \Theta_\infty(\beta) \ , \quad \forall \ \|\Omega\| \le \gamma^{-1}$$

so that, by partitioning matrix Θ consistently with the structure of F, that is letting

$$\Theta := \begin{bmatrix} \Theta_{11} & \Theta_{12} \\ \Theta_{21} & \Theta_{22} \end{bmatrix}$$

it results

$$X_{\varepsilon a}(\Omega) \le \beta^2 \Theta_{\infty 11}(\beta) , \quad \forall \, \|\Omega\| \le \gamma^{-1}$$

Further, if the matrices A_f and K_f are given by eqs. (5.112) and (5.113), then it is straightforward to verify that

$$\bar{\Theta}(\beta) := \begin{bmatrix} \Pi_\infty(\beta) & 0 \\ 0 & S_\infty(\beta) - \Pi_\infty(\beta) \end{bmatrix}$$

satisfies eq. (5.114), provided that the stabilizing solution $S_\infty(\beta) = S'_\infty(\beta) \ge 0$ exists of the Riccati equation (in the unknown S)

$$0 = AS + SA' + \gamma^{-2} SC'_1 C_1 S + B(\beta)B'(\beta) \tag{5.115}$$

Notice that $\bar{\Theta}(\beta)$ is a solution of the Lyapunov equation (in the unknown Θ)

$$0 = \Theta F' + F\Theta + \gamma^{-2}\bar{\Theta}(\beta)\bar{C}'\bar{C}\bar{\Theta}(\beta) + \bar{B}\bar{B}' + \beta^{-2}GG'$$

Hence, from Lemma C.1, the stability of matrix F entails that $\bar{\Theta}(\beta) \ge 0$. Moreover,

$$F + \gamma^{-2}\bar{\Theta}(\beta)\bar{C}'\bar{C} = \begin{bmatrix} A_{f\beta} & 0 \\ \star & A + \gamma^{-2}S_\infty(\beta)C'_1 C_1 \end{bmatrix}$$

where the symbol "$\star$" denotes a matrix of no interest. Hence, $\bar{\Theta}(\beta)$ is also stabilizing, so that $\bar{\Theta}(\beta) = \Theta_\infty(\beta)$. Observe that, in view of Remark 5.2, it is

$$X_{xa}(\Omega) := \lim_{t \to \infty} E[x(t)x'(t)] \le \beta^2 S_\infty(\beta) , \quad \forall \, \|\Omega\| \le \gamma^{-1}$$

As for the existence of the solutions $\Pi_\infty(\beta)$ and $S_\infty(\beta)$, first notice that the arguments exploited in Remark 5.2 can be applied here, thus allowing one to make reference to a nonempty set $\mathcal{B}$ of β's corresponding to which a solution $S_\infty(\beta)$ of eq. (5.115) exists.

Second, observe that for $\beta \in \mathcal{B}$, $S_\infty(\beta)$ is nonsingular because the pair (A, B_1) is reachable. In fact, consider eq. (5.115) (which is solved by $S_\infty(\beta)$) and suppose that $S_\infty(\beta)x = 0$, $x \ne 0$. If both sides of eq. (5.115) are premultiplied by x' and postmultiplied by x, then, in view of the definition of $B(\beta)$, one can conclude that $B'_1 x = 0$ and $S_\infty(\beta)A'x = 0$. If both sides of eq. (5.115) are now premultiplied by $x'A$ and postmultiplied by $A'x$, one can conclude that $B'_1 A'x = 0$ and $S_\infty(\beta)(A')^2 x = 0$. By iterating these operations it follows

$$x' \begin{bmatrix} B_1 & AB_1 & A^2 B_1 & \cdots & A^{n-1}B_1 \end{bmatrix} = 0$$

which contradicts the reachability of the pair (A, B_1).

Since $S_\infty(\beta)$ is nonsingular, its inverse satisfies the Riccati equation (in the unknown S^{-1}) which results from premultiplying and postmultiplying by S^{-1} eq. (5.115), namely

$$0 = S^{-1}A + A'S^{-1} + \gamma^{-2}C'_1 C_1 + S^{-1}B(\beta)B'(\beta)S^{-1} \tag{5.116}$$

Third, recall that for $\beta \in \mathcal{B}$, $A + \gamma^{-2}S_\infty(\beta)C'_1 C_1$ is stable, so that if both sides of eq. (5.115) with $S = S_\infty(\beta)$ are premultiplied by $S_\infty^{-1}(\beta)$ one gets

$$S_\infty^{-1}(\beta)[A + \gamma^{-2}S_\infty(\beta)C'_1 C_1]S_\infty(\beta) = -[A + B(\beta)B'(\beta)S_\infty^{-1}(\beta)]'$$

from which it follows that matrix

$$A_s := -[A + B(\beta)B'(\beta)S_\infty^{-1}(\beta)]$$

is stable. Then the Riccati equation (in the unknown V)

$$0 = V[A + B(\beta)B'(\beta)S_\infty^{-1}(\beta)] + [A + B(\beta)B'(\beta)S_\infty^{-1}(\beta)]'V +$$
$$+VB(\beta)B'(\beta)V - C_2'(\beta)C_2(\beta) \tag{5.117}$$

which can equivalently be written as

$$0 = VA_s + A_s'V - VB(\beta)B'(\beta)V + C_2'(\beta)C_2(\beta)$$

admits the stabilizing solution $V_\infty(\beta) = V_\infty'(\beta) \geq 0$ since matrix A_s is stable (recall Lemma C.4). Therefore, $\Lambda_\infty(\beta) := S_\infty^{-1}(\beta) + V_\infty(\beta)$ is positive definite because S_∞^{-1} is positive definite. By summing up eq. (5.117) to eq. (5.116) one can conclude that $\Lambda_\infty(\beta)$ solves the Riccati equation (in the unknown Λ)

$$0 = \Lambda A + A'\Lambda + \Lambda B(\beta)B'(\beta)\Lambda + \gamma^{-2}C_1'C_1 - C_2'(\beta)C_2(\beta) \tag{5.118}$$

Thus, its inverse (which actually exists because $\Lambda_\infty(\beta) > 0$) solves the equation which derives from eq. (5.118) after both sides have been premultiplied and postmultiplied by Λ^{-1}. Such an equation coincides with eq. (5.111) which therefore admits a symmetric, positive definite solution given by

$$\Pi_\infty(\beta) := \Lambda_\infty^{-1}(\beta) = [S_\infty^{-1}(\beta) + V_\infty(\beta)]^{-1}$$

It is only left to be proved that such a solution is the stabilizing one. By exploiting the fact that $\Pi_\infty(\beta)$ solves eq. (5.111), it results

$$A_s - B(\beta)B'(\beta)V_\infty(\beta) = -A - B(\beta)B'(\beta)\Pi_\infty^{-1}(\beta)$$
$$= -[A\Pi_\infty(\beta) + B(\beta)B'(\beta)]\Pi_\infty^{-1}(\beta)$$
$$= \Pi_\infty(\beta)[A' + (\gamma^{-2}C_1'C_1 - C_2'(\beta)C_2(\beta)) \cdot$$
$$\cdot \Pi_\infty(\beta)]\Pi_\infty^{-1}(\beta)$$

so that $A' + (\gamma^{-2}C_1'C_1 - C_2'(\beta)C_2(\beta))\Pi_\infty(\beta)$ is stable because $A_s - B(\beta)B'(\beta)V_\infty(\beta)$ is stable (recall that $V_\infty(\beta)$ is the stabilizing solution).

In order to tighten as much as possible the so obtained bound for $X_{\varepsilon a}(\Omega)$, the best value for the parameter β should be found: this task may be performed by means of a suitable (iterative) searching method. $\qquad\square$

Example 5.6 Consider the system described in Example 5.5. A filter, labeled with the symbol R, is to be designed according to what has been presented in Remark 5.19. More precisely, letting w_1 and w_2 be zero mean white noises with identity intensity, a filter is sought which supplies an as small as possible upper bound to the trace of the asymptotic estimation error covariance matrix. This goal corresponds to minimizing trace$[\beta^2\Pi_\infty(\beta)]$ with respect to the parameter β, where $\Pi_\infty(\beta)$ is the stabilizing solution of eq. (5.111). A suitable computer routine gives $\beta^\circ = 1.13$. The performance of the corresponding filter are compared with those of the Kalman's filter K_1 and observer H which have been designed in Example 5.5. The same performance criteria (namely, RH_2 and RH_∞ norms of the transfer function from $w := [w_1 \; w_2]$) are adopted. The more satisfactory behavior of the robust (with respect to parameter variations) filter for small values of the system damping factor is put into evidence in fig. 5.32 and 5.33. $\qquad\square$

Remark 5.20 (Full control) The problem considered in Remark 4.12 can be dealt with also within the RH_∞ context. The problem refers to the system $P_{FC}(s)$ depicted in fig. 5.34 and described by the equations

$$\dot{x} = Ax + B_1 w + [I \; 0]u$$
$$z = C_1 x + [0 \; I]u$$
$$y = C_2 x + D_{21}w$$
$$u = [u_1' \; u_2']'$$

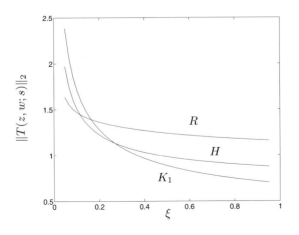

Figure 5.32: The RH_2 norm of the three filters

In fig. 5.34 $K_{FC}(s)$ is a generic controller admissible in RH_∞ for $P_{FC}(s)$. Assume that the pair (A, C_2) is detectable, the pair $[(A - B_1 D'_{21} C_2), B_1(I - D'_{21} D_{21})]$ is stabilizable and $D_{21} D'_{21} = I$. By exploiting also in the present framework what has been utilized in the RH_2 context (in particular, note that Lemma E.1 can be applied both in the RH_2 and RH_∞ settings), it is possible to directly claim that Problem 5.1 relative to system $P_{FC}(s)$ has the solution

a) Existence of the symmetric, positive semidefinite and stabilizing solution Π_∞ of the Riccati equation (5.105);

b) The set $\mathcal{F}_{\infty\gamma r}$ of the controllers $K_{FCr}(s)$ is defined by the block-scheme of fig. 5.35 where $Q(s) := \Sigma(A_q, B_q, C_q, D_q)$, with A stable and $\|Q(s)\|_\infty < \gamma$.

Finally, the set $\mathcal{F}_{\infty\gamma}$ of the controllers which are RH_∞ admissible for $P_{FC}(s)$ can be easily obtained by exploiting (via transposition) what has been shown in Remark 5.9 with reference to system $P_{FI}(s)$. The block-scheme which defines such a set is shown in fig. 5.36 (note that it is the "transpose" of fig. 5.21) where $\Theta_1(s) \in RH_\infty$, $\Psi(s) := (sI - A - L_\infty C_2)^{-1}$.   □

Remark 5.21 The comments on the matrix D_{21} presented in Remark 4.13 for the RH_2 setting can be done with no changes in the present context too with reference to Assumption 5.4. Further, many of the above given formulas greatly simplify if the additional orthogonality assumption $D_{21} B'_1 = 0$ is done.   □

Remark 5.22 Under Assumption 5.4, the stabilizability of the pair $[(A - B_1 D'_{21} C_2), B_1(I - D'_{21} D_{21})]$ (Assumption 5.3) is equivalent to requiring that the subsystem of $P_{OE}(s)$ corresponding to the transfer function P_{OE21} between the input w and the output y, that is system $\Sigma(A, B_1, C_2, D_{21})$, does not have zeros in the closed right half plane. The proof of this claim exploits the same arguments adopted in proving an analogous result in Remark 4.15.   □

Remark 5.23 The discussion in Remark 4.16 concerning a possible direct dependence of the output variable y on the control variable u applies to the RH_∞ setting as well. In fact, suppose that eq. (5.104) is substituted by

$$y = C_2 x + D_{21} w + D_{22} u \qquad (5.119)$$

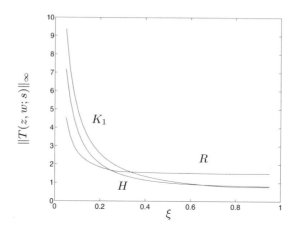

Figure 5.33: The RH_∞ norm of the three filters

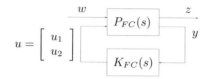

Figure 5.34: The full control problem

Letting

$$\bar{y} := y - D_{22}u = C_2x + D_{21}w \tag{5.120}$$

the solution to Problem 5.1 relative to the system $\bar{P}_{OE}(s)$, namely to the system described by eqs. (5.102), (5.103), (5.120), is supplied by Theorem 5.3. If $\bar{K}_{OE}(s)$ is a RH_∞ admissible controller for $\bar{P}_{OE}(s)$, then the controller $K_{OE}(s)$ defined in the block scheme of fig. 5.37 is apparently RH_∞ admissible for $P_{OE}(s)$, provided that it is well defined, that is, provided that matrix $I + D_q D_{22}$ is nonsingular. □

5.5 The partial information problem

The partial information problem in RH_∞ is discussed now. As in Section 4.4, only the output variable y can be measured: consistently, the system under control is described by

$$\dot{x} = Ax + B_1w + B_2u \tag{5.121}$$

$$z = C_1x + D_{12}u \tag{5.122}$$

$$y = C_2x + D_{21}w \tag{5.123}$$

while its transfer function is denoted by $P(s)$. Notice that system (5.121)-(5.123) is less general than system (5.33)-(5.35) however, the latter can be redrawn to the former by exploiting Remark 5.24. The solution of Problem 5.1 relative to system (5.121)-(5.123) will be presented under the following assumptions.

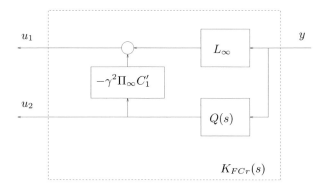

Figure 5.35: The set $\mathcal{F}_{\infty\gamma r}$ of the controllers $K_{FCr}(s)$

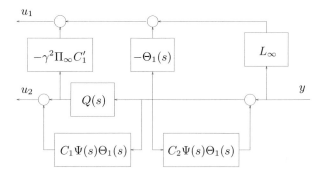

Figure 5.36: The generic admissible controller for $P_{FC}(s)$

Assumption 5.6 *The pair $[(A - B_2 D'_{12} C_1), (I - D_{12} D'_{12}) C_1]$ is detectable and the pair $[(A - B_1 D'_{21} C_2), B_1 (I - D'_{21} D_{21})]$ is stabilizable.*

Assumption 5.7 *The pair (A, B_2) is stabilizable and the pair (A, C_2) is detectable.*

Assumption 5.8 $D'_{12} D_{12} = I.$

Assumption 5.9 $D_{21} D'_{21} = I.$

The result in Theorem 5.4 below makes reference to the block-scheme of fig. 5.38 relative to which the transfer function from w to z is denoted by $T(z, w; s)$, while $K(s)$ is a generic RH_∞ admissible controller for $P(s)$.

Theorem 5.4 (Partial information problem) *Consider Problem 5.1 relative to system (5.121)-(5.123). Then, under Assumptions 5.6-5.9, it has the solution*

a1) Existence of the symmetric, positive semidefinite and stabilizing solution P_∞ of the Riccati equation (in the unknown P)

$$0 = PA_c + A'_c P - P(B_2 B'_2 - \gamma^{-2} B_1 B'_1) P + C'_{1c} C_{1c} \qquad (5.124)$$

that is, such that

$$A_{cc} := A_c - B_2 B'_2 P_\infty + \gamma^{-2} B_1 B'_1 P_\infty \qquad (5.125)$$

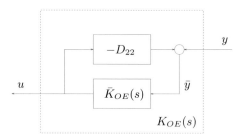

Figure 5.37: The controller structure when $D_{22} \neq 0$

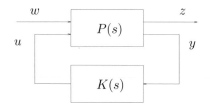

Figure 5.38: The partial information problem

is stable. In the above equations

$$A_c := A - B_2 D'_{12} C_1, \quad C_{1c} := (I - D_{12} D'_{12}) C_1 \tag{5.126}$$

a2) Existence of the symmetric, positive semidefinite and stabilizing solution Π_∞ of the Riccati equation (in the unknown Π)

$$0 = \Pi A'_f + A_f \Pi - \Pi (C'_2 C_2 - \gamma^{-2} C'_1 C_1) \Pi + B_{1f} B'_{1f} \tag{5.127}$$

that is, such that

$$A_{ff} := A_f - \Pi_\infty C'_2 C_2 + \gamma^{-2} \Pi_\infty C'_1 C_1 \tag{5.128}$$

is stable. In the above equations

$$A_f := A - B_1 D'_{21} C_2, \quad B_{1f} := B_1 (I - D'_{21} D_{21}) \tag{5.129}$$

a3)

$$r_s(P_\infty \Pi_\infty) < \gamma^2 \tag{5.130}$$

b) The set $\mathcal{F}_{\infty \gamma r}$ of the controllers $K_r(s)$ is defined by the diagram of fig. 5.39, where

$$S_\infty(s) := \left[\begin{array}{c|cc} A_{cc} + Z_\infty L_\infty (C_2 + \gamma^{-2} D_{21} B'_1 P_\infty) & -Z_\infty L_\infty & +Z_\infty B_{2\infty} \\ \hline F_\infty & 0 & I \\ -C_2 - \gamma^{-2} D_{21} B'_1 P_\infty & I & 0 \end{array} \right]$$

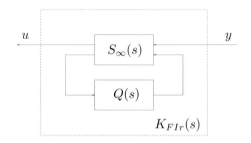

Figure 5.39: The set $\mathcal{F}_{\infty\gamma r}$ of the controllers $K_{FIr}(s)$

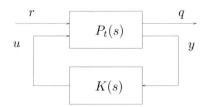

Figure 5.40: The equivalent output estimation problem

where

$$F_\infty := -B_2'P_\infty - D_{12}'C_1 \tag{5.131}$$

$$L_\infty := -\Pi_\infty C_2' - B_1 D_{21}' \tag{5.132}$$

$$Z_\infty := (I - \gamma^{-2}\Pi_\infty P_\infty)^{-1} \tag{5.133}$$

$$B_{2\infty} := B_2 + \gamma^{-2}\Pi_\infty C_1'D_{12}$$

and $Q(s) := \Sigma(A_q, B_q, C_q, D_q)$ *with* A_q *stable and* $\|Q(s)\|_\infty < \gamma$.

Proof First observe that Assumption 5.7 coincides with the necessary condition of Remark 5.6.

Sufficiency of parts a1) - a3) and part b) Assume that points (a1)-(a3) hold true and define the variables v and q as

$$w := r + \gamma^{-2}B_1'P_\infty x \tag{5.134}$$

$$q := u - F_\infty x \tag{5.135}$$

By utilizing these equations into eqs. (5.121) and (5.123) one obtains the system (see fig. 5.40)

$$P_t(s) := \left[\begin{array}{c|cc} A_t & B_1 & B_2 \\ \hline -F_\infty & 0 & I \\ C_{2t} & D_{21} & 0 \end{array}\right] \tag{5.136}$$

where

$$A_t := A + \gamma^{-2}B_1B_1'P_\infty \tag{5.137}$$

$$C_{2t} := C_2 + \gamma^{-2}D_{21}B_1'P_\infty \tag{5.138}$$

It is easy to ascertain that Lemma E.4 can be applied to the couple of systems shown in fig. 5.38 and 5.40, so that this part of the proof can be carried over by making reference to the system in fig. 5.40 only.

Observe that system $P_t(s)$ has the same structure as system $P_{OE}(s)$ dealt with in Section 5.4, where the output estimation problem was discussed. When dealing with system $P_t(s)$ Assumptions 5.3-5.5 become: (α1) the pair $(A_t - B_1 D'_{21} C_{2t}, B_{1f})$ is stabilizable; (α2) the pair (A_t, C_{2t}) is detectable; (α3) $D_{21} D'_{21} = I$; (α4) the matrix $A_t + B_2 F_\infty$ is stable. The stability of this last matrix trivially follows from eq. (5.125),(5.126),(5.131),(5.137). Condition (α3) is Assumption 5.9.

Condition (α1) is immediately derived by acknowledging that $A_t - B_1 D'_{21} C_{2t} = A_f + \gamma^{-2} B_{1f} B'_1 P_\infty$, so that such a condition is equivalent to the stabilizability of the pair $(A_f + \gamma^{-2} B_{1f} B'_1 P_\infty, B_{1f})$, which, in turn, is implied by Assumption 5.6, thanks to the invariance of the stabilizability property with respect to state feedback. The discussion on condition (α2) is postponed.

Therefore, the sufficiency part of the theorem and point (b) are proved once the existence is ascertained of a symmetric and positive semidefinite solution $\Pi_{t\infty}$ of the Riccati equation (in the unknown Π_t) which is relevant to the output estimation problem for system $P_t(s)$, namely, the equation

$$0 = A_{tt} \Pi_t + \Pi_t A'_{tt} + \Pi_t(\gamma^{-2} F'_\infty F_\infty - C'_{2t} C_{2t})\Pi_t + B_{1f} B'_{1f} \tag{5.139}$$

where

$$A_{tt} := A_t - B_1 D'_{21} C_{2t} \tag{5.140}$$

Moreover, the solution $\Pi_{t\infty}$ must be stabilizing, i. e. such that the matrix $A_{tt} + \Pi_{t\infty}(\gamma^{-2} F'_\infty F_\infty - C'_{2t} C_{2t})$ is stable. To this aim, consider the Hamiltonian matrix associated with the Riccati equation (5.139)

$$J_{t\infty} := \begin{bmatrix} A'_{tt} & \gamma^{-2} F'_\infty F_\infty - C'_{2t} C_{2t} \\ -B_{1f} B'_{1f} & -A_{tt} \end{bmatrix}$$

and the one associated with the Riccati equation (5.127)

$$J_\infty := \begin{bmatrix} A'_f & \gamma^{-2} C'_1 C_1 - C'_2 C_2 \\ -B_{1f} B'_{1f} & -A_f \end{bmatrix}$$

By exploiting the fact that P_∞ solves eq. (5.124), it is not difficult to verify that the two Hamiltonian matrices $J_{t\infty}$ and J_∞ are similar. Indeed, it results

$$J_{t\infty} = T J_\infty T^{-1}$$

with

$$T := \begin{bmatrix} I & -\gamma^{-2} P_\infty \\ 0 & I \end{bmatrix}$$

Being Π_∞ a solution of eq. (5.127), one has

$$J_\infty \text{Im} \left(\begin{bmatrix} I \\ \Pi_\infty \end{bmatrix} \right) \subseteq \text{Im} \left(\begin{bmatrix} I \\ \Pi_\infty \end{bmatrix} \right)$$

so that

$$J_{t\infty} T \text{Im} \left(\begin{bmatrix} I \\ \Pi_\infty \end{bmatrix} \right) = T J_\infty \text{Im} \left(\begin{bmatrix} I \\ \Pi_\infty \end{bmatrix} \right) \subseteq T \text{Im} \left(\begin{bmatrix} I \\ \Pi_\infty \end{bmatrix} \right)$$

Therefore, the subspace

$$\mathcal{S} := T\mathrm{Im}\left[\begin{bmatrix} I \\ \Pi_\infty \end{bmatrix}\right] = \mathrm{Im}\left[\begin{bmatrix} I - \gamma^{-2}P_\infty\Pi_\infty \\ \Pi_\infty \end{bmatrix}\right]$$

is $J_{t\infty}$-invariant. It is also complementary to $\mathrm{Im}[[0\ \ I]']$ because condition $(a3)$ guarantees that the matrix $I - \gamma^{-2}P_\infty\Pi_\infty$ is nonsingular (see the proof of Lemma B.11). Finally, the subspace $\mathcal{S}$ is generated by the (generalized) eigenvectors of $J_{t\infty}$ relative to the eigenvalues with negative real parts. Indeed, being Π_∞ a stabilizing solution,

$$J_\infty\mathrm{Im}\left[\begin{bmatrix} I \\ \Pi_\infty \end{bmatrix}\right] = \mathrm{Im}\left[\begin{bmatrix} I \\ \Pi_\infty \end{bmatrix}\right]\Lambda$$

with Λ stable, so that $J_{t\infty}\mathcal{S} = \mathcal{S}\Lambda$. Therefore, in view of Lemma C.2, it can be concluded that there exists a stabilizing solution $\Pi_{t\infty}$ of eq. (5.139) which is given by

$$\Pi_{t\infty} = \Pi_\infty(I - \gamma^{-2}P_\infty\Pi_\infty)^{-1} \tag{5.141}$$

Moreover, such a solution turns out to be positive semidefinite thanks to Lemma B.11.

It is left to prove condition $(a2)$, namely, the detectability of the pair (A_t, C_{2t}). Remark 5.17 applied to eq. (5.139) leads to the conclusion that matrix $A_{tt} - \Pi_tC'_{2t}C_{2t}$ is stable, which, in turn, implies that the pair (A_{tt}, C_{2t}) is detectable. By recalling the definition of the matrix A_{tt} (see eq. (5.140)), the detectability of the pair (A_t, C_{2t}), i. e. condition $(a2)$, follows too, since the stabilizability property of the pair (A'_t, C'_{2t}) is invariant with respect to the state feedback.

Finally, the form of the set $\mathcal{F}_{\infty\gamma r}$ given in the statement follows by applying Theorem 5.3 to system $P_t(s)$.

Necessity of part a) With reference to fig. 5.38, assume that there exists a controller $K(s)$ admissible in RH_∞ for system $P(s)$ such that $\|T(z, w; s)\|_\infty < \gamma$. Then, the controller

$$K_{FI}(s) := K(s)\,[C_2\ \ D_{21}]$$

is admissible in RH_∞ for the system $\bar{P}_{FI}(s)$ which is the system defined by eqs. (5.121) and (5.122) with the output vector constituted by $[x'\ w']'$. Moreover, $\|T(z, w; s)\|_\infty < \gamma$ also for system $\bar{P}_{FI}(s)$. This system coincides with the system $P_{FI}(s)$ considered in Section 5.3. The Assumptions 5.1 and 5.2 which are required for the result relevant to the full information problem (Theorem 5.2), are verified since they constitute a subset of Assumptions 5.6-5.9. Therefore, Theorem 5.2 guarantees the existence of the solution of eq. (5.124) endowed with all the properties specified in the statement of the present theorem. The request of the existence of the solution of eq. (5.127) can be proved in a similar manner by making reference to the system in fig. 5.41, where $\hat{P}(s) := P'(s)$ and $\hat{K}(s) := K'(s)$. Thus, system $\hat{P}(s)$ is described by

$$\dot{\xi} = A'\xi + C'_1\hat{z} + C'_2\hat{y} \tag{5.142}$$
$$\hat{w} = B'_1\xi + D'_{21}\hat{y} \tag{5.143}$$
$$\hat{u} = B'_2\xi + D'_{12}\hat{z} \tag{5.144}$$

From Lemma E.1, $\hat{K}(s)$ is admissible in RH_∞ for $\hat{P}(s)$ and $\|T(\hat{w}, \hat{z}; s)\|_\infty < \gamma$ because $K(s)$ is admissible in RH_∞ for $P(s)$ and $\|T(z, w; s)\|_\infty < \gamma$. The above exploited argument can be applied in the very same way leading to the conclusion that the controller

$$\hat{K}_{FI}(s) := \hat{K}(s)\,[B'_2\ \ D'_{12}]$$

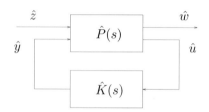

Figure 5.41: The transposed partial information problem

is admissible in RH_∞ for system $\hat{P}_{FI}(s)$ (described by eqs. (5.142) and (5.143) and having the vector $[\xi' \; \hat{z}']'$ as measurable output) and is such that $\|T(\hat{w}, \hat{z}; s)\|_\infty < \gamma$. System $\hat{P}_{FI}(s)$ has the same structure as system $P_{FI}(s)$ which has been defined for the full information problem (see Section 5.3). Since the assumptions required by Theorem 5.2 to be applied to system $\hat{P}_{FI}(s)$ are a subset of Assumptions 5.6-5.9, it is possible to claim that there exists a solution of eq. (5.127) which possesses the properties requested by the statement of the present theorem.

Further, it is easy to check that also the conditions for Lemma E.4 to be applied are verified, so that, with reference to the block-scheme of fig. 5.40, it is possible to conclude that the controller $K(s)$ is admissible in RH_∞ for system $P_t(s)$ and $\|T(q, r; s)\|_\infty < \gamma$ (recall the definition of $P_t(s)$ given in eqs. (5.136)-(5.138)). As already said, system $P_t(s)$ has the same structure as system $P_{OE}(s)$ which was introduced in Section 5.4 when dealing with the output estimation problem. Assumptions 5.3 - 5.5 which are relevant to Theorem 5.3, are now shown to be verified. These assumptions, if expressed in terms of system $P_t(s)$, are precisely those which have been denoted as $(\alpha 1)$, $(\alpha 2)$, $(\alpha 3)$ and $(\alpha 4)$ in the sufficiency part of the proof.

First observe that condition $(\alpha 2)$ (detectability of the pair (A_t, C_{2t})) is obviously satisfied. In fact, it is apparently necessary in view of the already established stability of the system in fig. 5.40 (again, recall the structure of system $P_t(s)$ given in eq. (5.136)). Second, in the sufficiency part of the proof it has already been shown that condition $(\alpha 1)$ is equivalent to Assumption 5.6. Condition $(\alpha 3)$ is straightforward. Finally, condition $(\alpha 4)$ precisely amounts to requiring the stability of matrix A_{cc} (see eq. (5.125)).

Thus, the conditions for Theorem 5.3 to be applied to system $P_t(s)$ are satisfied. Therefore, this theorem guarantees the existence of a positive semidefinite and stabilizing solution $\Pi_{t\infty}$ of the Riccati equation (5.139). By applying in reverse order what has been shown in passing from Π_∞ (a solution of eq. (5.127)) to $\Pi_{t\infty}$ (a solution of eq. (5.139)), it follows

$$\Pi_\infty := \Pi_{t\infty}(I + \gamma^{-2} P_\infty \Pi_{t\infty})^{-1} \tag{5.145}$$

Note that the existence of the inverse in eq. (5.145) is guaranteed by the fact that matrix $P_\infty \Pi_{t\infty}$ has nonnegative eigenvalues, thanks to Lemma B.10. From eq. (5.145) it follows

$$\Pi_\infty = (I - \gamma^{-2} \Pi_\infty P_\infty) \Pi_{t\infty} \tag{5.146}$$

so that

$$\Pi_{t\infty} = (I - \gamma^{-2} \Pi_\infty P_\infty)^{-1} \Pi_\infty$$

since the existence of the inverse in this last equation can easily be proved. In fact, let, by contradiction,

$$x'(I - \gamma^{-2}\Pi_\infty P_\infty) = 0 , \quad x \neq 0 \tag{5.147}$$

so that, from eq. (5.146), also $x'\Pi_\infty = 0$. But this would imply, in view of eq. (5.147), $x = 0$. By recalling that $\Pi_{t\infty} \geq 0$, one can conclude that condition $(a3)$ in the statement is verified, thanks to eq. (5.141) and Lemma B.11 .   □

The following result, which constitutes an interesting side-product of Theorem 5.4, can be stated in view of what has been presented in connection with the full information problem (Corollary 5.1) and the output estimation problem (Corollary 5.2).

Corollary 5.3 *Assume that Assumptions 5.8 and 5.9 hold. If, for a given positive γ, there exist the symmetric, positive semidefinite and stabilizing solutions of the Riccati equations (5.124) and (5.127) and such solutions satisfy eq. (5.130), then there exists a RH_∞ admissible controller $K(s)$ for $P(s)$, corresponding to which $\|T(z, w; s)\|_\infty < \gamma$. Moreover, the set of controllers described in point (b) of Theorem 5.4 constitutes the set $\mathcal{F}_{\infty\gamma r}$.*

Example 5.7 Consider system (5.121) - (5.123) with

$$A = \begin{bmatrix} 0 & 1 \\ 2 & 1 \end{bmatrix} , \quad B_1 = \begin{bmatrix} 0 \\ 0 \end{bmatrix} , \quad B_2 = \begin{bmatrix} 0 \\ 1 \end{bmatrix}$$

$$C_1 = \begin{bmatrix} 0 & 0 \end{bmatrix} , \quad C_2 = \begin{bmatrix} -1 & 1 \end{bmatrix} , \quad D_{12} = D_{21} = 1$$

Because of the particular form of matrices B_1 and C_1, the Riccati equations (5.124) and (5.127) do not depend on γ. The stabilizing solutions of these equations are

$$P_\infty = \begin{bmatrix} 4 & 4 \\ 4 & 4 \end{bmatrix} , \quad \Pi_\infty = \begin{bmatrix} 4 & 8 \\ 8 & 16 \end{bmatrix}$$

respectively. Moreover, $r_s(P_\infty\Pi_\infty) = 144$ so that, according to Corollary 5.3, an admissible controller in RH_∞ exists for the system if $\gamma > 12$ (recall eq. (5.130)).

Now consider the nominal plant $G_n(s) := (s - 1)/(s^2 - s - 2)$ subject to perturbations of the form (5.10),(5.11). Then, what is the maximum value of α corresponding to which a stabilizing controller exists? This question can be answered by applying Theorem 5.1 to an enlarged plant which coincides with the controlled system considered in the present example. Thus, a stabilizing controller exists only provided that $\alpha \leq 1/12$.   □

Remark 5.24 (Loop shifting) Theorem 5.4 can be applied to systems exhibiting a more general structure than that of system (5.121) - (5.123). Indeed, consider system $P^{(1)}(s)$ given by

$$P^{(1)}(s) := \left[\begin{array}{c|cc} A^{(1)} & B_1^{(1)} & B_2^{(1)} \\ \hline C_1^{(1)} & D_{11}^{(1)} & D_{12}^{(1)} \\ C_2^{(1)} & D_{21}^{(1)} & D_{22}^{(1)} \end{array} \right]$$

and depicted in fig. 5.42. Assume that the variables $w^{(1)}$, $u^{(1)}$, $z^{(1)}$, $y^{(1)}$ of such a system have dimension q, m, r, p, respectively, and suppose that $\text{rank}[D_{12}^{(1)}] = m$, $\text{rank}[D_{21}^{(1)}] = p$.

Then, if one performs the 6 *operations* described below (the effects of which can be easily understood by making reference to fig. 5.42), it is *generically* possible to successively transform system $P^{(1)}(s)$ into systems

$$P^{(i)}(s) := \left[\begin{array}{c|cc} A^{(i)} & B_1^{(i)} & B_2^{(i)} \\ \hline C_1^{(i)} & D_{11}^{(i)} & D_{12}^{(i)} \\ C_2^{(i)} & D_{21}^{(i)} & D_{22}^{(i)} \end{array} \right]$$

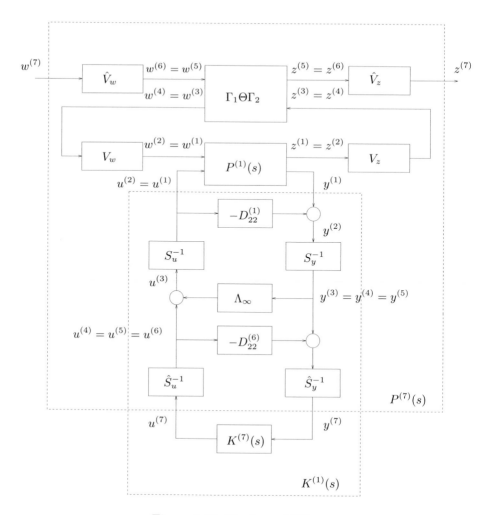

Figure 5.42: The loop shifting

which are characterized by the inputs $w^{(i)}$ and $u^{(i)}$ and the outputs $z^{(i)}$ and $y^{(i)}$, in such a way that

$$P^{(7)}(s) := \left[\begin{array}{c|cc} A^{(7)} & B_1^{(7)} & B_2^{(7)} \\ \hline C_1^{(7)} & 0 & D_{12}^{(7)} \\ C_2^{(7)} & D_{21}^{(7)} & 0 \end{array} \right]$$

with $D_{12}^{(7)\prime} D_{12}^{(7)} = I$, $D_{21}^{(7)} D_{21}^{(7)\prime} = I$.

Operation 1 : Let

$$y^{(2)} := y^{(1)} - D_{22}^{(1)} u^{(1)}$$
$$w^{(2)} := w^{(1)} , \quad u^{(2)} := u^{(1)} , \quad z^{(2)} := z^{(1)}$$

so that

$$A^{(2)} = A^{(1)} , \quad B_i^{(2)} = B_i^{(1)} , \quad i = 1, 2$$
$$C_i^{(2)} = C_i^{(1)} , \quad i = 1, 2$$
$$D_{11}^{(2)} = D_{11}^{(1)} , \quad D_{12}^{(2)} = D_{12}^{(1)} , \quad D_{21}^{(2)} = D_{21}^{(1)}$$

and

$$D_{22}^{(2)} = 0$$

Operation 2 : Find four matrices V, $V^\perp$, W, $W^\perp$ with orthonormal columns and such that

$$\text{Im}[V] = \text{Im}[D_{12}^{(2)}] , \quad \text{Im}[V^\perp] = \text{Im}[D_{12}^{(2)}]^\perp$$
$$\text{Im}[W] = \text{Im}[D_{21}^{(2)\prime}] , \quad \text{Im}[W^\perp] = \text{Im}[D_{21}^{(2)\prime}]^\perp$$

Such matrices do exist thanks to the assumptions $\text{rank}[D_{12}^{(1)}] = m$, $\text{rank}[D_{21}^{(1)}] = p$ and the fact that $D_{12}^{(2)} = D_{12}^{(1)}$ and $D_{21}^{(2)} = D_{21}^{(1)}$. Obviously, if $D_{12}^{(2)}$ is a square matrix (hence nonsingular), $V^\perp = \emptyset$ and $V = I$. A similar comment is in order if $D_{21}^{(2)}$ is a square matrix (hence nonsingular).
    Then it follows

$$V_z D_{12}^{(2)} := \begin{bmatrix} V^{\perp\prime} \\ V^\prime \end{bmatrix} D_{12}^{(2)} = \begin{bmatrix} 0 \\ S_u \end{bmatrix}$$
$$D_{21}^{(2)} V_w := D_{21}^{(2)} \begin{bmatrix} W^\perp & W \end{bmatrix} = \begin{bmatrix} 0 & S_y \end{bmatrix}$$

Observe that matrices V_w and V_z are orthogonal and matrices S_u and S_y are nonsingular. Then let

$$w^{(3)} := V_w^\prime w^{(2)} , \quad u^{(3)} := S_u u^{(2)}$$
$$z^{(3)} := V_z z^{(2)} , \quad y^{(3)} := S_y^{-1} y^{(2)}$$

so that

$$A^{(3)} = A^{(2)} , \quad B_1^{(3)} = B_1^{(2)} V_w , \quad B_2^{(3)} = B_2^{(2)} S_u^{-1}$$
$$C_1^{(3)} = V_z C_1^{(2)} , \quad C_2^{(3)} = S_y^{-1} C_2^{(2)} , \quad D_{11}^{(3)} = V_z D_{11}^{(2)} V_w$$
$$D_{22}^{(3)} = 0$$

and

$$D_{12}^{(3)} = V_z D_{12}^{(2)} S_u^{-1} = \begin{bmatrix} 0 \\ I \end{bmatrix} , \quad D_{21}^{(3)} = S_y^{-1} D_{21}^{(2)} V_w = \begin{bmatrix} 0 & I \end{bmatrix}$$

Note that, thanks to the orthogonality of matrices V_z and V_w, the transfer functions from $w^{(3)}$ to $z^{(3)}$ and from $w^{(2)}$ to $z^{(2)}$ have the same RH_∞ norm.
Operation 3 : Consider the following partition of the $r \times q$ matrix $D_{11}^{(3)}$

$$D_{11}^{(3)} := \begin{bmatrix} D_{1111} & D_{1112} \\ D_{1121} & D_{1122} \end{bmatrix}$$

where the submatrix D_{1122} is $m \times p$ (recall that the assumptions on the rank of matrices $D_{12}^{(1)}$ and $D_{21}^{(1)}$ imply $r \geq m$ and $q \geq p$). Let Λ_∞ be the matrix (whose existence is guaranteed by Theorem F.1 which supplies its expression, too) which minimizes $\|\Delta(\Lambda)\|$, where

$$\Delta(\Lambda) := \begin{bmatrix} D_{1111} & D_{1112} \\ D_{1121} & D_{1122} + \Lambda \end{bmatrix}$$

If

$$\alpha := \max \left(\left\| \begin{bmatrix} D_{1111} \\ D_{1121} \end{bmatrix} \right\| , \left\| \begin{bmatrix} D_{1111}^\prime \\ D_{1112}^\prime \end{bmatrix} \right\| \right)$$

then, thanks to the above quoted theorem, $\|\Delta(\Lambda_\infty)\| = \alpha$. By defining

$$w^{(4)} := w^{(3)} , \quad u^{(4)} := u^{(3)} - \Lambda_\infty y^{(3)}$$
$$z^{(4)} := z^{(3)} , \quad y^{(4)} := y^{(3)}$$

one obtains

$$A^{(4)} = A^{(3)} + B_2^{(3)} \Lambda_\infty C_2^{(3)}$$

$$B_1^{(4)} = B_1^{(3)} + B_2^{(3)} \Lambda_\infty D_{21}^{(3)} , \quad B_2^{(4)} = B_2^{(3)}$$

$$C_1^{(4)} = C_1^{(3)} + D_{12}^{(3)} \Lambda_\infty C_2^{(3)} , \quad C_2^{(4)} = C_2^{(3)}$$

$$D_{11}^{(4)} = \Delta(\Lambda_\infty)$$

$$D_{12}^{(4)} = D_{12}^{(3)} = \begin{bmatrix} 0 \\ I \end{bmatrix} , \quad D_{21}^{(4)} = D_{21}^{(3)} = \begin{bmatrix} 0 & I \end{bmatrix}$$

$$D_{22}^{(4)} = 0$$

Now consider the feedback connection of system $P^{(3)}(s)$ with the controller defined by $u_L^{(3)} = \Lambda(s)y_L^{(3)}$, with $\lim_{\omega \to \infty} \Lambda(j\omega) = \Lambda_\infty$. Then, the scalar α constitutes a lower bound to the RH_∞ norm of the transfer function from $w^{(3)}$ to $z^{(3)}$. Indeed, as $\omega \to \infty$, the limiting value of such a function is precisely

$$D_{11}^{(4)} = D_{11}^{(3)} + D_{12}^{(3)} \Lambda_\infty D_{21}^{(3)} = \Delta(\Lambda_\infty)$$

Operation 4 : Define the variables $z^{(5)}$ and $w^{(5)}$ by means of the equation

$$\begin{bmatrix} z^{(5)} \\ w^{(4)} \end{bmatrix} := \Gamma_1 \Theta \Gamma_2 \begin{bmatrix} w^{(5)} \\ z^{(4)} \end{bmatrix}$$

where

$$\Gamma_1 := \begin{bmatrix} I & 0 \\ 0 & \gamma^{-1}I \end{bmatrix} , \quad \Gamma_2 := \begin{bmatrix} \gamma I & 0 \\ 0 & I \end{bmatrix}$$

while

$$\Theta := \begin{bmatrix} \Theta_{11} & \Theta_{12} \\ \Theta_{21} & \Theta_{22} \end{bmatrix}$$

is any matrix such that: (i) $\Theta'\Theta = \Theta\Theta' = I$; (ii) Θ_{12} and Θ_{21} are square matrices of dimension r and q, respectively; (iii) $0 < \gamma^{-1} < \alpha^{-1}$. In particular, from these features it follows that $\|\Theta_{22}\| \leq \|\Theta\| = 1$ (recall Lemma 2.16), so that the matrices $\Psi_1 := (\gamma I - D_{11}^{(4)}\Theta_{22})^{-1}$ and $\Psi_2 := (\gamma I - \Theta_{22}D_{11}^{(4)})^{-1}$ are well defined because $\|D_{11}^{(4)}\| = \alpha$ (recall also Lemma 2.18, point (2) and Lemma 2.21). Further let

$$u^{(5)} := u^{(4)} , \quad y^{(5)} := y^{(4)}$$

In so doing, one obtains

$$A^{(5)} = A^{(4)} + B_1^{(4)} \Psi_2 \Theta_{22} C_1^{(4)}$$

$$B_1^{(5)} = \gamma B_1^{(4)} \Psi_2 \Theta_{21} , \quad B_2^{(5)} = B_2^{(4)} + B_1^{(4)} \Psi_2 \Theta_{22} D_{12}^{(4)}$$

$$C_1^{(5)} = \gamma \Theta_{12} \Psi_1 C_1^{(4)} , \quad C_2^{(5)} = C_2^{(4)} + D_{21}^{(4)} \Psi_2 \Theta_{22} C_1^{(4)}$$

$$D_{11}^{(5)} = \gamma(\Theta_{11} + \Theta_{12} D_{11}^{(4)} \Psi_2 \Theta_{21}) , \quad D_{22}^{(5)} = D_{21}^{(4)} \Psi_2 \Theta_{22} D_{12}^{(4)}$$

$$D_{12}^{(5)} = \gamma \Theta_{12} \Psi_1 D_{12}^{(4)} , \quad D_{21}^{(5)} = \gamma D_{21}^{(4)} \Psi_2 \Theta_{21}$$

Notice that if the feedback connection of system $P^{(4)}(s)$ with the controller described by $u_L^{(4)} = \Lambda(s)y_L^{(4)}$ results in a stable system and the RH_∞ norm of the transfer function $T(z^{(4)}, w^{(4)}; s)$ from the input $w^{(4)}$ to the output $z^{(4)}$ is less than γ, then also the feedback connection of system $P^{(5)}(s)$ with the controller described by by $u_L^{(5)} = \Lambda(s)y_L^{(5)}$ results in a stable system and the RH_∞ norm of the transfer function $T(z^{(5)}, w^{(5)}; s)$ from the input $w^{(5)}$ to the output $z^{(5)}$ is less than γ. In fact, notice that

$$\Gamma_1 \Theta \Gamma_2 = \begin{bmatrix} \gamma \Theta_{11} & \Theta_{12} \\ \Theta_{21} & \gamma^{-1}\Theta_{22} \end{bmatrix}$$

and $\|\gamma^{-1}\Theta_{22}\| \leq \gamma^{-1}$, so that it is possible to apply Theorem 5.1 and conclude about stability. As for the transfer functions norms, observe that, from the definition of the new variables $z^{(5)}$ and $w^{(5)}$, it follows

$$\begin{bmatrix} z^{(5)} \\ \gamma w^{(4)} \end{bmatrix} = \Theta \begin{bmatrix} \gamma w^{(5)} \\ z^{(4)} \end{bmatrix}$$

so that, thanks to the properties of Θ,

$$\|z^{(5)}\|_2^2 + \gamma^2\|w^{(4)}\|_2^2 = \|z^{(4)}\|_2^2 + \gamma^2\|w^{(5)}\|_2^2$$

that is

$$\|z^{(5)}\|_2^2 - \gamma^2\|w^{(5)}\|_2^2 = \|z^{(4)}\|_2^2 - \gamma^2\|w^{(4)}\|_2^2$$

The left (resp. right) hand side of this equation is negative if and only if $\|T(z^{(5)}, w^{(5)}; s)\|_\infty < \gamma$ (resp. $\|T(z^{(4)}, w^{(4)}; s)\|_\infty < \gamma$) (see Theorem 2.12), so that the RH_∞ norm of the first transfer function is less than γ if and only if the norm of the second one is such. Now let

$$\Theta_{11} := -\gamma^{-1}D_{11}^{(4)} , \quad \Theta_{12} := (I - \gamma^{-2}D_{11}^{(4)}D_{11}^{(4)\prime})^{1/2}$$
$$\Theta_{21} := (I - \gamma^{-2}D_{11}^{(4)\prime}D_{11}^{(4)})^{1/2} , \quad \Theta_{22} := \gamma^{-1}D_{11}^{(4)\prime}$$

By applying Lemma F.1 it is easy to verify that the matrix Θ satisfies the equation $\Theta'\Theta = \Theta\Theta' = I$.

Finally, one obtain

$$D_{11}^{(5)} = 0$$

Operation 5 : While performing Operation 4 it might be happened that $D_{22}^{(5)} \neq 0$. Should this be the case, one has to apply the procedure outlined in Operation 1 above by letting

$$y^{(6)} := y^{(5)} - D_{22}^{(5)}u^{(5)}$$
$$w^{(6)} := w^{(5)} , \quad u^{(6)} := u^{(5)} , \quad z^{(6)} := z^{(5)}$$

so that

$$A^{(6)} = A^{(5)} , \quad B_i^{(6)} = B_i^{(5)} , \quad i = 1, 2$$
$$C_i^{(6)} = C_i^{(5)}, \quad i = 1, 2$$
$$D_{11}^{(6)} = 0 , \quad D_{12}^{(6)} = D_{12}^{(5)} , \quad D_{21}^{(6)} = D_{21}^{(5)}$$

and

$$D_{22}^{(6)} = 0$$

Operation 6 : While performing Operation 4 it might be also happened that the matrices $D_{12}^{(5)\prime}D_{12}^{(5)}$ and/or $D_{21}^{(5)}D_{21}^{(5)\prime}$, though positive definite, are different from the identity matrices. Should this be the case, one has to apply the procedure outlined in Operation 2 above by letting

$$w^{(7)} := \hat{V}_w'w^{(6)} , \quad u^{(7)} := \hat{S}_u u^{(6)}$$
$$z^{(7)} := \hat{V}_z z^{(6)} , \quad y^{(7)} := \hat{S}_y^{-1}y^{(6)}$$

where the matrices $\hat{V}_w$, $\hat{S}_u$, $\hat{V}_z$ and $\hat{S}_y$ are computed in such a way as to comply with the same requirements expressed in Operation 2, so that

$$A^{(7)} = A^{(6)} , \quad B_1^{(7)} = B_1^{(6)}\hat{V}_w , \quad B_2^{(7)} = B_2^{(6)}\hat{S}_u^{-1}$$
$$C_1^{(7)} = \hat{V}_z C_1^{(6)} , \quad C_2^{(7)} = \hat{S}_y^{-1}C_2^{(6)} , \quad D_{11}^{(7)} = 0$$
$$D_{22}^{(7)} = 0$$

and

$$D_{12}^{(7)} = \hat{V}_z D_{12}^{(6)} \hat{S}_u^{-1} = \begin{bmatrix} 0 \\ I \end{bmatrix}, \quad D_{21}^{(7)} = \hat{S}_y^{-1} D_{21}^{(6)} \hat{V}_w = \begin{bmatrix} 0 & I \end{bmatrix}$$

Finally, given a controller $K^{(7)}(s)$ which is admissible in RH_∞ for system $P^{(7)}(s)$ and such that $\|T(z^{(7)}, w^{(7)}; s)\|_\infty < \gamma$, it is fairly apparent how to design a controller $K^{(1)}(s)$ which is admissible in RH_∞ for system $P^{(1)}(s)$ and such that $\|T(z^{(1)}, w^{(1)}; s)\|_\infty < \gamma$ (see also fig. 5.42). $\square$

Example 5.8 Consider system $P^{(1)}(s)$ with

$$A^{(1)} = \begin{bmatrix} 0 & 1 \\ 0 & 0 \end{bmatrix}, \quad B_1^{(1)} = \begin{bmatrix} 1 & 0 \\ 0 & 1 \end{bmatrix}, \quad B_2^{(1)} = \begin{bmatrix} 0 \\ 1 \end{bmatrix}$$

$$C_1^{(1)} = \begin{bmatrix} 1 & 0 \\ 1 & 1 \end{bmatrix}, \quad C_2^{(1)} = \begin{bmatrix} 1 & 0 \end{bmatrix}, \quad D_{22}^{(1)} = 1$$

$$D_{11}^{(1)} = \begin{bmatrix} 1 & 1 \\ 1 & 0 \end{bmatrix}, \quad D_{12}^{(1)} = \begin{bmatrix} 0 \\ 1 \end{bmatrix}, \quad D_{21}^{(1)} = \begin{bmatrix} 1 & 1 \end{bmatrix}$$

By performing the operations described in Remark 5.24 one obtains

$$V_z = I, \quad V_w = \frac{1}{\sqrt{2}} \begin{bmatrix} 1 & 1 \\ -1 & 1 \end{bmatrix}, \quad S_u = 1, \quad S_y = \sqrt{2}$$

$$\Lambda_\infty = -\frac{1}{\sqrt{2}}, \quad \gamma > \sqrt{2}, \quad \Theta_{11} = -\frac{1}{\gamma\sqrt{2}} \begin{bmatrix} 0 & 2 \\ 1 & 0 \end{bmatrix}$$

$$\Theta_{12} = \frac{1}{\gamma\sqrt{2}} \begin{bmatrix} \sqrt{2\gamma^2 - 4} & 0 \\ 0 & \sqrt{2\gamma^2 - 1} \end{bmatrix}$$

$$\Theta_{21} = \frac{1}{\gamma\sqrt{2}} \begin{bmatrix} \sqrt{2\gamma^2 - 1} & 0 \\ 0 & \sqrt{2\gamma^2 - 4} \end{bmatrix}$$

$$\Theta_{22} = \frac{1}{\gamma\sqrt{2}} \begin{bmatrix} 0 & 1 \\ 2 & 0 \end{bmatrix}$$

$$D_{22}^{(6)} = 0, \quad \hat{V}_z = \hat{V}_w = I$$

$$\hat{S}_u = \frac{\gamma\sqrt{2}}{\sqrt{2\gamma^2 - 1}}, \quad \hat{S}_y = \frac{\gamma}{\sqrt{\gamma^2 - 2}}$$

$\square$

Remark 5.25 It is worth noticing that the well known *separation* property of the control system eigenvalues (which has been discussed in Remark 4.17 within the framework of the RH_2 partial information problem) does not hold anymore in the RH_∞ setting. However, it is possible to read Theorem 5.4 in the light of a *weak* separation principle. In fact, the controller resulting from the choice $Q(s) = 0$ in fig. 5.39, may be written as

$$\dot{\xi} = A\xi + B_2 u + B_1 \hat{w}_w + Z_\infty L_\infty (C_2 \xi + D_{21} \hat{w}_w - y)$$
$$u = F_\infty \xi$$

where

$$\hat{w}_w := \gamma^{-2} B_1' P_\infty \xi$$

In view of eqs. (5.131) - (5.133), (5.138), (5.141), it is easy to verify that $Z_\infty L_\infty = -\Pi_{t\infty} C_{2t}' - B_1 D_{21}'$, so that it coincides with the gain of the filter for the linear combination $F_\infty x$ (recall the solution of the output estimation problem for system $P_t(s)$ defined by eqs. (5.136)-(5.138)). Further, $\hat{w}_w$ can be seen as the *worst* input for the control problem with full information (recall Remark 5.10 concerning differential games and see fig. 5.15, too). Therefore, it can be concluded that the controller for the partial information problem is the filter for the control law of the full information problem when the worst disturbance $w_p := \gamma^{-2}B_1'P_\infty x$ acts on the system. This claim is correct in the RH_2 setting too (recall Remark 4.19), if w_w is set equal to zero (hence $\hat{w}_w = 0$ as well). □

Remark 5.26 (Parametrization of the set $\mathcal{F}_{\infty\gamma}$) Observe that $\mathcal{F}_{\infty\gamma} = \mathcal{F}_{\infty\gamma r}$. In fact, the set of controllers which are admissible in RH_∞ for $P(s)$ has been shown to coincide with the set of controllers which are admissible in RH_∞ for $P_t(s)$ (see the proof of Theorem 5.4). Therefore, the claim is true thanks to Remark 5.15 which can be exploited because the structure of $P_t(s)$ is identical to that of $P_{OE}(s)$. □

Remark 5.27 Theorem 5.3 of Section 5.4 (which is relative to the output estimation problem in the RH_∞ setting) can be derived as a particular case of Theorem 5.4. In fact, it suffices to set $D_{12} = I$ in Assumptions 5.6 - 5.9 in order to conclude that Assumptions 5.3 - 5.5 are satisfied. In particular, it results $C_{1c} = 0$ so that the (unique) symmetric, positive semidefinite and stabilizing solution of the Riccati equation (5.124) is $P_\infty = 0$, which implies $Z_\infty = I$. In such a context, the conclusions of Theorem 5.4 are immediately redrawn to those of Theorem 5.3. In a similar way, the solution of the disturbance feedforward problem dealt with in Remark 5.12 can be viewed as the solution of a particular case of the partial information problem, namely the case in which $D_{21} = I$. Indeed, letting $D_{21} = I$ in Assumptions 5.6 - 5.9, it is easy to verify that the assumptions made in Remark 5.12 are satisfied. In particular, it results $B_{1f} = 0$ and hence $\Pi_\infty = 0$, $Z_\infty = I$, so that the conclusions of Theorem 5.4 coincide with those illustrated in Remark 5.12. □

Remark 5.28 (Partial information and parametric perturbations) Consider Problem 5.1 relative to the system $P_\Omega(s)$ described by

$$\dot{x} = (A + \Delta_A)x + \bar{B}_1 w_1 + B_2 u \tag{5.148}$$
$$z_1 = \bar{C}_1 x + \bar{D}_{12} u \tag{5.149}$$
$$y = C_2 x + \bar{D}_{21} w_1 \tag{5.150}$$

The structure of the perturbation Δ_A is taken to be

$$\Delta_A := L\Omega M , \quad \|\Omega\| < \beta^{-1} \tag{5.151}$$

The *robust* (with respect to the parametric perturbations) control problem in RH_∞ consists in designing a controller which, for all Ω, $\|\Omega\| < \beta^{-1}$, is such that: (*i*) it is admissible in RH_∞ relative to $P_\Omega(s)$; (*ii*) the RH_∞ norm of the transfer function from the input w_1 to the output z_1 (from now on denoted by $T_\Omega(z_1, w_1; s)$) is less than a certain positive scalar α. This problem can be tackled by considering the system $P(s)$

$$\dot{\xi} = A\xi + B_1 w + B_2 u$$
$$z = C_1 \xi + D_{12} u$$
$$y = C_2 \xi + D_{21} w$$

where

$$w := \begin{bmatrix} w_1 \\ w_2 \end{bmatrix} , \quad z := \begin{bmatrix} z_1 \\ z_2 \end{bmatrix}$$
$$B_1 := \begin{bmatrix} \bar{B}_1 & L \end{bmatrix} , \quad D_{21} := \begin{bmatrix} \bar{D}_{21} & 0 \end{bmatrix}$$
$$D_{12} := \begin{bmatrix} \bar{D}_{12} \\ 0 \end{bmatrix} , \quad C_1 := \begin{bmatrix} \bar{C}_1 \\ M \end{bmatrix}$$

Observe that $P(s) = P_\Omega(s)$ if $w_2 = \Omega z_2$. If Assumptions 5.6 - 5.9 are verified for system $P_0(s)$, that is for system $P_\Omega(s)$ in nominal conditions ($\Omega = 0$), then they are verified also for

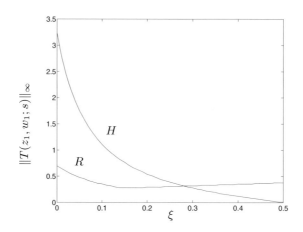

Figure 5.43: The performances of the controllers R and H

system $P(s)$. Therefore Theorem 5.4 can be applied to such a system with $\gamma := \min[\alpha, \beta]$. The solution of the relevant problem is then a solution also for the original problem. In fact, the controller which solves Problem 5.1 for system $P(s)$ is such that $\|T(z, w; s)\|_\infty < \gamma$ which implies $J := \|z\|_2^2 - \gamma^2 \|w\|_2^2 < 0, \forall w \in RH_2$. This controller, when applied to system $P_\Omega(s)$, is such that $\|T_\Omega(z_1, w_1; s)\|_\infty < \alpha$, which implies $J_1 := \|z_1\|_2^2 - \alpha^2 \|w_1\|_2^2 < 0, \forall w_1 \in RH_2$. Indeed, by recalling that $w_2 = \Omega z_2$, it follows

$$
\begin{aligned}
J_1 &= \|z_1\|_2^2 - \alpha^2 \|w_1\|_2^2 \\
&\leq \|z_1\|_2^2 - \gamma^2 \|w_1\|_2^2 \\
&\leq \|z\|_2^2 - \|z_2\|_2^2 - \gamma^2 \|w\|_2^2 + \gamma^2 \|w_2\|_2^2 \\
&\leq J - \|z_2\|_2^2 + \gamma^2 \|w_2\|_2^2 \\
&\leq J - \|z_2\|_2^2 + \gamma^2 \|\Omega z_2\|_2^2 \\
&\leq J - \|z_2\|_2^2 (1 - \gamma^2 \|\Omega\|^2) \\
&\leq J - \|z_2\|_2^2 (1 - \gamma^2 \beta^{-2}) \\
&\leq J < 0
\end{aligned}
$$

It is left to be proved that the controller which solves Problem 5.1 for $P(s)$ stabilizes $P_\Omega(s), \forall \Omega, \|\Omega\| < \beta^{-1}$ as well. This result is a direct consequence of Theorem 5.1. $\square$

Example 5.9 Consider the system (5.148)-(5.151) with

$$
A = \begin{bmatrix} 0 & 1 \\ -1 & -1 \end{bmatrix}, \quad \bar{B}_1 = \begin{bmatrix} \varphi^{-1} \\ 0 \end{bmatrix}, \quad B_2 = \begin{bmatrix} 0 \\ \varphi \end{bmatrix}, \quad L = \begin{bmatrix} 0 \\ 1 \end{bmatrix}
$$

$$
M = \begin{bmatrix} 0 & 1 \end{bmatrix}, \quad C_1 = \begin{bmatrix} \varphi^{-1} & \varphi^{-1} \end{bmatrix}, \quad C_2 = \begin{bmatrix} \varphi & 0 \end{bmatrix}
$$

$$
\bar{D}_{12} = \bar{D}_{21} = 1, \quad \beta = 1, \quad \varphi := \sqrt{2}
$$

A controller is sought which, for each $\Omega, \|\Omega\| < \beta^{-1}$, is admissible in RH_∞ for system $P_\Omega(s)$ and is such that $\|T(z_1, w_1; s)\|_\infty < \alpha$. Letting $\alpha = \beta$, a controller, labeled with R, is designed according to what has been shown in Remark 5.28. On the contrary, the label H denotes the controller resulting from Theorem 5.4 when applied to system $P_0(s)$ and for the choice $Q(s) = 0$. The performances of the two controllers are compared in fig. 5.43 where $\|T(z_1, w_1; s)\|_\infty$ is plotted against the system damping factor $\xi := (1 - \Omega)/2$. The controller R is weakly sensitive and apparently behaves in a much better way for low values of ξ. $\square$

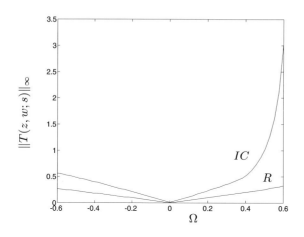

Figure 5.44: The performances of the controllers R and H $(M = MS)$

Example 5.10 Consider the system defined by the matrices

$$A = \begin{bmatrix} 0 & 1 \\ -1 & -1 \end{bmatrix} + L\Omega M , \quad B_1 = \begin{bmatrix} 0 \\ 1 \end{bmatrix} , \quad B_2 = \begin{bmatrix} 0 & 1 \\ 1 & 0 \end{bmatrix}$$

$$L = \begin{bmatrix} 0 \\ 1 \end{bmatrix} , \quad C_1 = \begin{bmatrix} 1 & 2 \end{bmatrix} , \quad C_2 = D_{12} = \begin{bmatrix} 1 & 0 \end{bmatrix} , \quad D_{21} = 1$$

while $M = MS := [2 \ 0]$ or $M = MU := [0 \ 2]$. The parameter Ω has nominal value equal to 0 and describes the uncertainty which is supposed to affect the system under consideration. A stabilizing controller is sought which makes small the effect of the disturbance w on the output $z = C_1 x + D_{12} u$. By utilizing the first component only of the control variable u it is possible to design a controller, labeled with IC, which makes zero the transfer function from w to z in nominal conditions. The transfer function of such a controller (which achieves the *perfect indirect compensation* of the disturbance) is

$$K_{IC}(s) = -\frac{1 + 2s}{s^2 + 3s + 3}$$

If, on the contrary, the uncertain knowledge of the parameter Ω has to be somehow taken into account, a controller, labeled with R, can be designed according to the discussion in Remark 5.28. Consistently, a new variable $z_a := Mx + u_2$ is added to the system. When $M = MS$ one finds, independently of the value of γ, $P_\infty = \Pi_\infty = 0$ because the matrices A_c and A_f are stable and $C_{1c} = 0$, $B_{1f} = 0$. Correspondingly,

$$F_\infty = \begin{bmatrix} -1 & -2 \\ -2 & 0 \end{bmatrix} , \quad L_\infty = \begin{bmatrix} 0 \\ -1 \end{bmatrix}$$

This is no more the case when $M = MU$, so that, for $\gamma = 0.275$ (notice that for $\gamma = 0.270$ a solution of eq. (5.25) with the required properties does not exist) one obtains

$$F_\infty = \begin{bmatrix} 7.67 & -4.43 \\ -30.87 & 6.67 \end{bmatrix} , \quad L_\infty = \begin{bmatrix} 0 \\ -1 \end{bmatrix}$$

The performances of the two controllers are compared in fig. 5.44 and 5.45 for the considered cases. The choice of the set of values of the parameter Ω reflects the need of stability for the control system resulting from the insertion of the controller IC. In both cases, the better behavior of the controller R is apparent, especially for small values of Ω. □

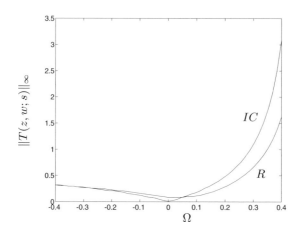

Figure 5.45: The performances of the controllers R and H $(M = MU)$

Remark 5.29 It is worth noticing that, also in the partial information case, when $\gamma \to \infty$ the RH_∞ controller of fig. 5.39 (with $Q(s) = 0$) tends to the RH_2 controller of fig. 4.19 (with $Q(s) = 0$). $\qquad\square$

Remark 5.30 The contents of Remark 5.13 (as for matrix D_{12}) and of Remark 5.21 (as for matrix D_{21}) apply with no changes to the problem dealt with in the present section. $\quad\square$

Remark 5.31 In view of Remarks 5.14 and 5.22 it can be said that, under Assumptions 5.8 and 5.9, Assumption 5.6 amounts to requiring that the two subsystems of $P(s)$ having transfer functions $P_{12}(s)$ (that is, system $\Sigma(A, B_2, C_1, D_{12})$ with input u and output z) and $P_{21}(s)$ (that is, system $\Sigma(A, B_1, C_2, D_{21})$ with input w and output y), respectively, do not have zeros in the closed right half plane. $\qquad\square$

5.6   The operatorial approach

This section is aimed at tackling the partial information problem in the RH_∞ setting from a point of view which is rather different from the one adopted in Section 5.5. In fact, reference will be made to the theory of linear *operators* applied to dynamical systems. The minimum norm problem will be presented by constraining the attention to the scalar case only, as the general multivariable situation is substantially more complex to be handled.

Recall that the problem at hand consists in designing a controller $K(s)$ which, with reference to fig. 5.46, is admissible in RH_∞ for $P(s)$ and such that the RH_∞ norm of the transfer function $T(z, w; s)$ from w to z is less than a given positive scalar γ. Indeed, it will be shown that, having restrained the attention to the case in which the variables w, u, z, y are scalar, it is fairly easy to design that controller $K^o(s)$ which *minimizes* $\|T(z, w; s)\|_\infty$. Preliminarily, partition the matrix $P(s)$ into the four (scalar) transfer functions $P_{ij}(s)$, so that

$$\begin{bmatrix} z_L \\ y_L \end{bmatrix} = \begin{bmatrix} P_{11}(s) & P_{12}(s) \\ P_{21}(s) & P_{22}(s) \end{bmatrix} \begin{bmatrix} w_L \\ u_L \end{bmatrix}$$

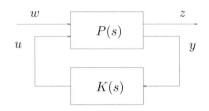

Figure 5.46: The partial information problem

and assume that system $P(s)$ can be made *internally stable* by implementing a feedback loop from y to u. It is obvious that a RH_∞ admissible controller for $P(s)$ exists if and only if such an assumption is verified. This is equivalent to the internal stabilizability (from y) of the subsystem $P_{22}(s)$. Further, if $K(s)$ is a RH_∞ admissible controller for $P_{22}(s)$, then it is RH_∞ admissible also for $P(s)$ and the set of controllers endowed with this property coincides with the set of controllers which stabilize $P_{22}(s)$.

Second, observe that, if $K(s)$ is any controller corresponding to which the control system in fig. 5.46 is well defined, that is if

$$\lim_{s \to \infty} \det[I - D_{22}K(s)] \neq 0$$

where $D_{22} := \lim_{s \to \infty} P_{22}(s)$, then

$$T(z, w; s) = P_{11}(s) + P_{12}(s)[I - K(s)P_{22}(s)]^{-1}K(s)P_{21}(s) \tag{5.152}$$

Hence, the problem of designing a RH_∞ admissible controller for $P(s)$ such that $\|T(z, w; s)\|_\infty < \gamma$ is equivalent to selecting a controller which verifies this last inequality among those which internally stabilize $P_{22}(s)$.

Theorem 3.7 supplies the parametrization of the set of all controllers which stabilize $P_{22}(s)$ on the basis of a double coprime factorization in RH_∞ of $P_{22}(s)$ (also recall Theorem 3.4). More precisely, if

$$P_{22}(s) = N(s)M^{-1}(s) = \hat{M}^{-1}(s)\hat{N}(s) \tag{5.153}$$

$$\begin{bmatrix} \hat{X}(s) & -\hat{Y}(s) \\ -\hat{N}(s) & \hat{M}(s) \end{bmatrix} \begin{bmatrix} M(s) & Y(s) \\ N(s) & X(s) \end{bmatrix} = I \tag{5.154}$$

where all the functions $N(s)$, $M(s)$, $\hat{M}(s)$, $\hat{N}(s)$, $\hat{X}(s)$, $\hat{Y}(s)$, $X(s)$, $Y(s)$ belong to RH_∞, then any controller $K(s)$ which stabilizes $P_{22}(s)$ can be given the form

$$K(s) = [\hat{X}(s) - Q(s)\hat{N}(s)]^{-1}[\hat{Y}(s) - Q(s)\hat{M}(s)] \tag{5.155}$$

where $Q(s) := \Sigma(A_q, B_q, C_q, D_q)$ is any element of RH_∞ such that $\det[I - D_{22}D_q] \neq 0$.

Thanks to the explicit expression of the stabilizing controllers given by eq. (5.155), it is easy to prove that the transfer function $T(z, w; s)$ is a *linear* function of the parameter $Q(s)$, as stated in the following theorem where reference is made to the functions

$$T_1(s) := P_{11}(s) + P_{12}(s)M(s)\hat{Y}(s)P_{21}(s) \tag{5.156}$$

$$T_2(s) := P_{12}(s)M(s) \tag{5.157}$$

$$T_3(s) := \hat{M}(s)P_{21}(s) \tag{5.158}$$

Theorem 5.5 *The functions $T_i(s)$, $i = 1, 2, 3$ defined in eqs. (5.156)-(5.158) belong to RH_∞. Moreover, if $K(s)$ is given by eq. (5.155), then*

$$T(z, w; s) = T_1(s) - T_2(s)Q(s)T_3(s)$$

Proof Consider a realization of $P(s)$

$$P(s) := \left[\begin{array}{c|cc} A & B_1 & B_2 \\ \hline C_1 & D_{11} & D_{12} \\ C_2 & D_{21} & D_{22} \end{array} \right]$$

with the pair (A, B_2) stabilizable and the pair (A, C_2) detectable. Further, let F and H be two matrices such that $A + B_2 F$ and $A + HC_2$ are stable. Consistently with the discussion in the proof of Theorem 3.4, choose

$$M(s) := \Sigma(A + B_2 F, B_2, F, I)$$
$$\hat{M}(s) := \Sigma(A + HC_2, H, C_2, I)$$
$$\hat{Y}(s) := \Sigma(A + HC_2, H, -F, 0)$$

Denote with φ_i (resp. η_i), $i = 1, \cdots, 5$ the state (resp. output) variables of the five systems $P_{21}(s)$, $\hat{Y}(s)$, $M(s)$, $P_{12}(s)$, $P_{11}(s)$, taken in their order of appearance. If ϑ_1 and ν_1 denote the output and input variables of system $T_1(s)$, respectively, then such a system is described by (recall eq. (5.156))

$$P_{21}(s) = \begin{cases} \dot{\varphi}_1 = A\varphi_1 + B_1\nu_1 \\ \eta_1 = C_2\varphi_1 + D_{21}\nu_1 \end{cases}$$

$$\hat{Y}(s) = \begin{cases} \dot{\varphi}_2 = (A + HC_2)\varphi_2 + H\eta_1 \\ \eta_2 = -F\varphi_2 \end{cases}$$

$$M(s) = \begin{cases} \dot{\varphi}_3 = (A + B_2 F)\varphi_3 + B_2\eta_2 \\ \eta_3 = F\varphi_3 + \eta_2 \end{cases}$$

$$P_{12}(s) = \begin{cases} \dot{\varphi}_4 = A\varphi_4 + B_2\eta_3 \\ \eta_4 = C_1\varphi_4 + D_{12}\eta_3 \end{cases}$$

$$P_{11}(s) = \begin{cases} \dot{\varphi}_5 = A\varphi_5 + B_1\nu_1 \\ \eta_5 = C_1\varphi_5 + D_{11}\nu_1 \end{cases}$$

$$\vartheta_1 = \eta_4 + \eta_5$$

Letting $\varepsilon_1 := \varphi_1 - \varphi_5$ and $\varepsilon_2 := \varphi_4 - \varphi_3$, it is easy to verify that

$$\dot{\varepsilon}_1 = A\varepsilon_1$$
$$\dot{\varepsilon}_2 = A\varepsilon_2$$

and that a realization of the transfer function $T_1(s)$ can be built up by exploiting the state variables φ_2, φ_3, φ_5 only. Further, by defining $\varepsilon_3 := \varphi_5 + \varphi_2$ and $\varepsilon_4 := \varphi_5 + \varphi_3$, one obtains

$$\dot{\varepsilon}_3 = (A + HC_2)\varepsilon_3 + (B_1 + HD_{21})\nu_1 \tag{5.159}$$
$$\dot{\varepsilon}_4 = -B_2 F\varepsilon_3 + (A + B_2 F)\varepsilon_4 + B_1\nu_1 \tag{5.160}$$
$$\vartheta_1 = -D_{12}F\varepsilon_3 + (C_1 + D_{12}F)\varepsilon_4 + D_{11}\nu_1 \tag{5.161}$$

These equations define a simpler realization of $T_1(s)$ which allows one to claim that $T_1(s) \in RH_\infty$, since the eigenvalues of the dynamic matrix of the system described by eqs. (5.159)-(5.161) are those of matrix $A + B_2 F$ together with those of matrix $A + HC_2$, both matrices being stable by construction.

Now, let φ_i (resp. η_i), $i = 6, 7$ be the state (resp. output) variables of the two systems $M(s)$ and $P_{12}(s)$, taken in their order of appearance. If ϑ_2 and ν_2 denote the output and input variables of system $T_2(s)$, respectively, then such a system is described by (recall eq. (5.157))

$$M(s) = \begin{cases} \dot{\varphi}_6 = (A + B_2 F)\varphi_6 + B_2 \nu_2 \\ \eta_6 = F\varphi_6 + \nu_2 \end{cases}$$

$$P_{12}(s) = \begin{cases} \dot{\varphi}_7 = A\varphi_7 + B_2 \eta_6 \\ \eta_7 = C_1 \varphi_7 + D_{12} \eta_6 \end{cases}$$

$$\vartheta_2 = \eta_7$$

Letting $\varepsilon_5 := \varphi_6 - \varphi_7$, it is easy to verify that

$$\dot{\varepsilon}_5 = A\varepsilon_5$$

and that a realization of the transfer function $T_2(s)$ can be built up by exploiting the state variables φ_6 only. The dynamic matrix of such a realization is $A + B_2 F$, so that $T_2(s) \in RH_\infty$ too.

Finally, let φ_i (resp. η_i), $i = 8, 9$ be the state (resp. output) variables of the two systems $P_{21}(s)$ and $\hat{M}(s)$, taken in their order of appearance. If ϑ_3 and ν_3 denote the output and input variables of system $T_3(s)$, respectively, then such a system is described by (recall eq. (5.158))

$$P_{21}(s) = \begin{cases} \dot{\varphi}_8 = A\varphi_8 + B_1 \nu_3 \\ \eta_8 = C_2 \varphi_8 + D_{21} \nu_3 \end{cases}$$

$$\hat{M}(s) = \begin{cases} \dot{\varphi}_9 = (A + HC_2)\varphi_9 + H\eta_8 \\ \eta_9 = C_2 \varphi_9 + \eta_8 \end{cases}$$

$$\vartheta_3 = \eta_9$$

Letting $\varepsilon_6 := \varphi_8 + \varphi_9$, it is easy to verify that

$$\dot{\varepsilon}_6 = (A + HC_2)\varepsilon_6 + (B_1 + HD_{21})\nu_3$$
$$\vartheta_3 = C_2 \varepsilon_6 + D_{21}\nu_3$$

so that a realization of the transfer function $T_3(s)$ is characterized the dynamic matrix $A + HC_2$. Hence, $T_3(s) \in RH_\infty$.

As for the expression of $T(z, w; s)$, first observe that, in view of eqs. (5.153)-(5.155), it is

$$[I - K(s)P_{22}(s)]^{-1} = \left[I - [\hat{X}(s) - Q(s)\hat{N}(s)]^{-1} \cdot \right.$$
$$\left. \cdot [\hat{Y}(s) - Q(s)\hat{M}(s)]N(s)M^{-1}(s)\right]^{-1}$$
$$= \left\{ [\hat{X}(s) - Q(s)\hat{N}(s)]^{-1} \left[\hat{X}(s) - Q(s)\hat{N}(s) - \right. \right.$$

$$-\hat{Y}(s)N(s)M^{-1}(s) + Q(s)\hat{M}(s)N(s)M^{-1}(s)\Big]\Big\}^{-1}$$

$$= \Big\{[\hat{X}(s) - Q(s)\hat{N}(s)]^{-1}\Big[\hat{X}(s)M(s) - \hat{Y}(s)N(s) -$$

$$- Q(s)[\hat{N}(s)M(s) - \hat{M}(s)N(s)]\Big]M^{-1}\Big\}^{-1}$$

$$= M(s)[\hat{X}(s) - Q(s)\hat{N}(s)]$$

and

$$[I - K(s)P_{22}(s)]^{-1} = \Big[I - [\hat{X}(s) - Q(s)\hat{N}(s)]^{-1}[\hat{Y}(s) -$$

$$- Q(s)\hat{M}(s)]N(s)M^{-1}(s)\Big]^{-1}$$

$$= \Big\{[\hat{X}(s) - Q(s)\hat{N}(s)]^{-1}\Big[\hat{X}(s) -$$

$$- Q(s)\hat{N}(s) - \hat{Y}(s)N(s)M^{-1}(s) +$$

$$+ Q(s)\hat{M}(s)N(s)M^{-1}(s)\Big]\Big\}^{-1}$$

$$= \Big\{[\hat{X}(s) - Q(s)\hat{N}(s)]^{-1}\Big[\hat{X}(s)M(s) -$$

$$- \hat{Y}(s)N(s) - Q(s)[\hat{N}(s)M(s) -$$

$$- \hat{M}(s)N(s)]\Big]M^{-1}\Big\}^{-1}$$

$$= M(s)[\hat{X}(s) - Q(s)\hat{N}(s)]$$

so that

$$[I - K(s)P_{22}(s)]^{-1}K(s) = M(s)[\hat{Y}(s) - Q(s)\hat{M}(s)]$$

Then, by recalling also eqs. (5.152),(5.156)-(5.158), it follows

$$\begin{aligned}
T(z, w; s) &= P_{11}(s) + P_{12}(s)[I - K(s)P_{22}(s)]^{-1}K(s)P_{21}(s) \\
&= P_{11}(s) + P_{12}(s)M(s)[\hat{Y}(s) - Q(s)\hat{M}(s)]P_{21}(s) \\
&= P_{11}(s) + P_{12}(s)M(s)\hat{Y}(s)P_{21}(s) - \\
&\quad - P_{12}(s)M(s)Q(s)\hat{M}(s)P_{21}(s) \\
&= T_1(s) - T_2(s)Q(s)T_3(s)
\end{aligned}$$

$\square$

In view of this theorem the problem of designing a RH_∞ admissible controller $K(s)$ for system $P(s)$ such that $\|T(z, w; s)\|_\infty < \gamma$ is apparently equivalent to that of finding a function $Q(s) \in RH_\infty$ such that $\|T_1(s) - T_2(s)Q(s)T_3(s)\|_\infty < \gamma$. Indeed, it will be shown that, under a suitable assumption, a function $Q^o(s) \in RH_\infty$ can be found which *minimizes* such a norm: the so called *model matching* problem is solved in this way and the original control problem has got an *optimal* answer (in the sense of making as small as possible $\|T(z, w; s)\|_\infty$) at the same time.

The following simple examples show that solving the model matching problem is not trivial in general, as there are cases where the solution exists and other ones where it does not.

Example 5.11 Let

$$T_3(s) = 1 , \quad T_2(s) = \frac{s-1}{s+1}$$

while $T_1(s)$ is any element of RH_∞. Then it follows, for any $Q(s) \in RH_\infty$ (recall Definition 2.24),

$$\|T_1(s) - T_2(s)Q(s)\|_\infty \geq |T_1(1) - T_2(1)Q(1)| = |T_1(1)|$$

so that

$$\inf_{Q(s)\in RH_\infty} \|T_1(s) - T_2(s)Q(s)\| := \alpha \geq |T_1(1)|$$

Chosen

$$Q^o(s) := \frac{T_1(s) - T_1(1)}{T_2(s)}$$

it is easy to verify that $Q^o(s) \in RH_\infty$ (observe that the zero of $T_2(s)$ is canceled out because the numerator of $Q^o(s)$ vanishes for $s = 1$). On the other side,

$$T_1(s) - T_2(s)Q^o(s) = T_1(1)$$

and therefore $\alpha \leq \|T_1(s) - T_2(s)Q^o(s)\|_\infty = |T_1(1)| \leq \alpha$, so that

$$\|T_1(s) - T_2(s)Q^o(s)\|_\infty = \alpha$$

□

Example 5.12 Let $T_3(s) = 1$ and

$$T_1(s) = \frac{1}{s+1} , \quad T_2(s) = \frac{1}{(s+1)^2}$$

Define

$$Q_\varepsilon(s) := \frac{s+1}{\varepsilon s + 1} , \quad 0 < \varepsilon < 1$$

so that

$$T_1(s) - T_2(s)Q_\varepsilon(s) = \frac{\varepsilon s}{(s+1)(\varepsilon s + 1)}$$

The diagram of $|T_1(j\omega) - T_2(j\omega)Q_\varepsilon(j\omega)|$ lies always below ε. Therefore, it is $\|T_1(s) - T_2(s)Q_\varepsilon(s)\|_\infty \leq \varepsilon$, so that

$$\inf_\varepsilon \|T_1(s) - T_2(s)Q_\varepsilon(s)\|_\infty = 0$$

This value of the norm is attained in correspondence of a $Q_\varepsilon^o(s)$ which is such that $T_1(s) - T_2(s)Q_\varepsilon^o(s) = 0$, namely $Q_\varepsilon^o(s) = s+1 \notin RH_\infty$.
□

The result to be presented in the next theorem makes reference to two functions $f(s)$ and $g(s)$ which are derived from a generic scalar function $F(s) \in RL_\infty$ in the following way. Let $F_a(s), F_s(s), F_\infty$ be such that

$$F(s) = F_a(s) + F_s(s) + F_\infty \tag{5.162}$$

where

$$F_a(s) \in RH_2^\perp , \quad F_s(s) \in RH_2 , \quad F_\infty := \lim_{s\to\infty} F(s) \tag{5.163}$$

Further, let

$$F_a(s) := \left[\begin{array}{c|c} A & B \\ \hline C & 0 \end{array} \right] \tag{5.164}$$

and suppose that the triple (A, B, C) constitutes a minimal realization of $F_a(s)$. Then, denote by $P_r > 0$ and $P_o > 0$ the unique solutions (recall Lemma C.1) of the two Lyapunov equations (in the unknown P)

$$0 = -PA' - AP + BB' \tag{5.165}$$
$$0 = -PA - A'P + C'C \tag{5.166}$$

respectively. Moreover, let λ_M^2 be the maximum eigenvalue of the matrix $P_o P_r$ (recall Lemma B.10 and the fact that all the eigenvalues of this matrix are different from 0, since it is nonsingular) and β a corresponding eigenvector, so that

$$P_o P_r \beta = \lambda_M^2 \beta , \quad \beta \neq 0 \tag{5.167}$$

and define

$$\chi := \lambda_M^{-1} P_r \beta \tag{5.168}$$

Now the following functions

$$f(s) := \left[\begin{array}{c|c} -A' & \beta \\ \hline B' & 0 \end{array} \right] \in RH_2 \tag{5.169}$$

$$g(s) := \left[\begin{array}{c|c} A & \chi \\ \hline C & 0 \end{array} \right] \in RH_2^\perp \tag{5.170}$$

can be associated with the scalar function $F(s) \in RL_\infty$. The two functions defined by eqs. (5.169),(5.170) are endowed with the property stated in the following lemma, where $\Gamma_{F\sim}$ and $\Gamma_{F\sim}^*$ are the Hankel operator with symbol $F^\sim$ and its adjoint, respectively (recall Definition 2.34 and Lemma 2.27).

Lemma 5.2 *Let $F(s) \in RL_\infty$ be a scalar function and $f(s) \in RH_2$ and $g(s) \in RH_2^\perp$ be two functions derived from $F(s)$ according to eqs. (5.162)-(5.170). Then*

$$\Gamma_{F\sim} g(s) = \lambda_M f(s)$$
$$\Gamma_{F\sim}^* f(s) = \lambda_M g(s)$$

Proof From eq. (5.165) it follows

$$C(sI - A)^{-1}(P_r A' + P_r s - s P_r + A P_r)(sI + A')^{-1}\beta =$$
$$= C(sI - A)^{-1} BB'(sI + A')^{-1}\beta$$

Therefore, by taking into account eqs. (5.164)-(5.169), one obtains

$$C(sI - A)^{-1}[P_r(sI + A') - (sI - A)P_r](sI + A')^{-1}\beta =$$
$$= F_a(s)f(s)$$

from which it follows

$$C(sI - A)^{-1} P_r \beta - C P_r (sI + A')^{-1}\beta = F_a(s)f(s) \tag{5.171}$$

On the other hand, eq. (5.168) is equivalent to

$$P_r \beta = \lambda_M \chi$$

so that, by recalling eqs. (5.170),(5.171), it results

$$\lambda_M g(s) - C P_r (sI + A')^{-1} \beta = F_a(s) f(s)$$

If the antistable orthogonal projection operator Π_a is applied to both sides of this equation (recall Definitions 2.32 - 2.34, Remark 2.25 and Lemmas 2.26, 2.27) one obtains

$$\lambda_M g(s) = \Gamma^*_{F_a^\sim} f(s) = \Gamma^*_{F^\sim} f(s)$$

In a similar way, from eq. (5.166) it follows

$$B'(sI + A')^{-1}(A'P_o + sP_o - P_o s + P_o A)(sI - A)^{-1}\chi =$$
$$= B'(sI + A')^{-1} C' C (sI - A)^{-1}\chi$$

so that, by taking into account eqs. (5.164), (5.167), (5.168), (5.170), one has

$$B'P_o(sI - A)^{-1}\chi - B'(sI + A')^{-1}\beta\lambda_M = -F_a^\sim(s) g(s)$$

In view of eq. (5.169), if the stable orthogonal projection operator Π_s is applied to both sides of this equation (recall Definitions 2.32 - 2.34 and Remark 2.25), one obtains

$$\lambda_M f(s) = \Gamma_{F_a^\sim} f(s) = \Gamma_{F^\sim} f(s)$$

$\square$

Theorem 5.6 *Let $F(s) \in RL_\infty$ be a given scalar function. Then, the function*

$$X^o(s) := F(s) - \frac{\|\Gamma_{F^\sim}\| g(s)}{f(s)} \tag{5.172}$$

is such that

$$\|F(s) - X^o(s)\|_\infty = \inf_{X(s) \in RH_\infty} \|F(s) - X(s)\|_\infty$$

In eq. (5.172) the two functions $f(s)$ and $g(s)$ are specified by eqs. (5.162)-(5.170).

Proof First observe that, thanks to Nehari's theorem (Theorem 2.19), there exists a function $X^o(s) \in RH_\infty$ such that $\|\Gamma_{F^\sim}\| = \|F(s) - X^o(s)\|_\infty$. Then, define $h(s) := [F(s) - X^o(s)]f(s)$ and notice that $h(s) \in RL_2$, because $F(s) - X^o(s) \in RL_\infty$ and $f(s) \in RH_2$. It follows

$$\|h(s) - \Gamma^*_{F^\sim} f(s)\|_2^2 = < h(s) - \Gamma^*_{F^\sim} f(s), h(s) - \Gamma^*_{F^\sim} f(s) >$$
$$= < h(s), h(s) > + < \Gamma^*_{F^\sim} f(s), \Gamma^*_{F^\sim} f(s) > -$$
$$-2 < h(s), \Gamma^*_{F^\sim} f(s) >$$

By taking into account that $\Gamma^*_{F^\sim} f(s) \in RH_2^\perp$, $X^o(s)f(s) \in RH_2$ and Lemma 2.27, one obtains

$$< h(s), \Gamma^*_{F^\sim} f(s) > = < \Pi_a h(s), \Gamma^*_{F^\sim} f(s) >$$
$$= < \Pi_a [F(s) - X^o(s)]f(s), \Gamma^*_{F^\sim} f(s) >$$
$$= < \Pi_a F(s) f(s), \Gamma^*_{F^\sim} f(s) >$$
$$= < \Gamma^*_{F^\sim} f(s), \Gamma^*_{F^\sim} f(s) >$$

Therefore,

$$
\begin{aligned}
\|h(s) - \Gamma_{F\sim}^* f(s)\|_2^2 &= <h(s), h(s)> - <\Gamma_{F\sim}^* f(s), \Gamma_{F\sim}^* f(s)> \\
&= \|[F(s) - X^o(s)]f(s)\|_2^2 - \\
&\quad - <f(s), \Gamma_{F\sim}\Gamma_{F\sim}^* f(s)> \\
&= \|[F(s) - X^o(s)]f(s)\|_2^2 - \lambda_M^2 \|f(s)\|_2^2
\end{aligned}
\tag{5.173}
$$

since, by Lemma 5.2, $\Gamma_{F\sim}\Gamma_{F\sim}^* f(s) = \lambda_M^2 f(s)$. In view of Theorem 2.12 it follows

$$
\|[F(s) - X^o(s)]f(s)\|_2^2 \le \|F(s) - X^o(s)\|_\infty^2 \|f(s)\|_2^2
$$

and therefore from eq. (5.173) one has

$$
\|h(s) - \Gamma_{F\sim}^* f(s)\|_2^2 \le \left[\|F(s) - X^o(s)\|_\infty^2 - \lambda_M^2 \right] \|f(s)\|_2^2 = 0
$$

because $\lambda_M = \|\Gamma_{F\sim}\| = \|F(s) - X^o(s)\|_\infty$ (recall Remark 2.27). Thus, $h(s) = \Gamma_{F\sim}^* f(s)$ and, thanks to Lemma 5.2, $[F(s) - X^o(s)]f(s) = h(s) = \lambda_M g(s)$, which implies, in turn,

$$
X^o(s) = F(s) - \frac{\lambda_M g(s)}{f(s)} = F(s) - \frac{\|\Gamma_{F\sim}\| g(s)}{f(s)}
$$

□

Example 5.13 Let

$$
F(s) = \frac{c}{s - a} \ , \quad a > 0 \ , \quad c \ne 0
$$

Then

$$
F_a(s) = \left[\begin{array}{c|c} a & 1 \\ \hline c & 0 \end{array} \right] \ , \quad F_s(s) = 0 \ , \quad F_\infty = 0
$$

and the solutions of eqs. (5.165), (5.166) are $P_r = (2a)^{-1}$ and $P_o = c^2(2a)^{-1}$, respectively. Therefore, $\lambda_M = |c|(2a)^{-1}$, $\beta = 1$, $\chi = |c|^{-1}$. From eqs. (5.169) and (5.170) one obtains

$$
f(s) = \frac{1}{s + a} \ , \quad g(s) = \frac{c}{s - a} |c|^{-1}
$$

so that

$$
X^o(s) = -\frac{c}{2a}
$$

□

Theorem 5.7 *Let $T_i(s) \in RH_\infty, i = 1, 2$, be two given scalar functions. If $T_2(s)$ does not have zeros on the extended imaginary axis, then there exists a function $Q^o(s) \in RH_\infty$ such that*

$$
\|T_1(s) - T_2(s)Q^o(s)\|_\infty = \inf_{Q(s) \in RH_\infty} \|T_1(s) - T_2(s)Q(s)\|_\infty
$$

Proof Let $T_{2i}(s) \in RH_\infty$ and $T_{2o}(s) \in RH_\infty$ be two functions, inner and outer, respectively, such that $T_2(s) = T_{2i}(s)T_{2o}(s)$ (recall Theorem 2.10) and notice that, being $T_2(j\omega) \ne 0$, $0 \le \omega \le \infty$, it is $T_{2o}^{-1}(s) \in RH_\infty$. Then, taking into account the identity $T_{2i}^\sim(s)T_{2i}(s) = I$, one obtains

$$
\begin{aligned}
\|T_1(s) - T_2(s)Q(s)\|_\infty &= \|T_{2i}(s)[T_{2i}^{-1}(s)T_1(s) - T_{2o}(s)Q(s)]\|_\infty \\
&= \|T_{2i}^{-1}(s)T_1(s) - T_{2o}(s)Q(s)\|_\infty
\end{aligned}
$$

Define $F(s) := T_{2i}^{-1}(s)T_1(s)$ and $X(s) := T_{2o}(s)Q(s)$. Then the problem of minimizing $\|F(s) - X(s)\|_\infty$ with respect to $X(s) \in RH_\infty$ is equivalent to that of minimizing $\|T_{2i}^{-1}(s)T_1(s) - T_{2o}(s)Q(s)\|_\infty$ with respect to $Q(s) \in RH_\infty$, since, once the optimal $X^o(s) \in RH_\infty$ has been found, it is

$$Q^o(s) = T_{2o}^{-1}(s)X^o(s) \in RH_\infty$$

Therefore, the theorem follows from Theorem 5.6. □

Remark 5.32 The request that the function $T_2(s)$ has no zeros on the extended imaginary axis is a sufficient condition only. Indeed, consider the case in which

$$T_2(s) = \frac{1}{s+1}$$

while $T_1(s)$ is an element of RH_∞. By defining $\hat{T}_1 := \lim_{s \to \infty} T_1(s)$, one has

$$\|T_1(s) - T_2(s)Q(s)\|_\infty \geq \lim_{s \to \infty} \|T_1(s) - T_2(s)Q(s)\|$$

$$\geq |\hat{T}_1|, \quad \forall Q(s) \in RH_\infty$$

so that

$$\inf_{Q(s) \in RH_\infty} \|T_1(s) - T_2(s)Q(s)\|_\infty := \alpha \geq |\hat{T}_1|$$

Chosen

$$\hat{Q}(s) := \frac{T_1(s) - \hat{T}_1}{T_2(s)}$$

it is easy to verify that such a function belongs to RH_∞ and

$$\alpha \leq \|T_1(s) - T_2(s)\hat{Q}(s)\|_\infty = |\hat{T}_1| \leq \alpha$$

which implies $\hat{Q}(s) = Q^o(s)$. □

Remark 5.33 In view of the above results a procedure for the computation of $Q^o(s)$ can be established as follows.

1) Find a double coprime factorization in RH_∞ of $P_{22}(s)$, thus getting $M(s)$, $\hat{M}(s)$, $\hat{Y}(s)$.

2) Compute $T_i(s)$, $i = 1, 2, 3$ from eqs.(5.156)-(5.158). If no zero of $T_4(s) := T_2(s)T_3(s)$ lies on the complete imaginary axis, go to point 3) below, otherwise stop.

3) Find an inner-outer factorization of $T_4(s)$, so that $T_4(s) = T_{4i}(s)T_{4o}(s)$.

4) Find the antistable and strictly proper part $F_a(s)$ of $F(s) := T_{4i}^{-1}(s)T_1(s)$.

5) Find a minimal realization (A, B, C) of $F_a(s)$.

6) Solve the Lyapunov equations (5.165) and (5.166).

7) Find the maximum eigenvalue λ_M^2 of P_oP_r and $f(s)$ and $g(s)$ from eqs. (5.167)-(5.170).

8) Compute $X^o(s)$ (eq. (5.172)).

9) Set $Q^o(s) = T_{4o}^{-1}(s)X^o(s)$.

□

Example 5.14 Consider the control system shown in fig. 5.47, where

$$G_n(s) = \frac{s-1}{(s-2)(s+1)}$$

A controller $K(s)$ is sought which stabilizes the system corresponding to the widest possible class of perturbations $\Delta(s)$ of the form

$$\Delta(s) \in \{\Delta(s) \mid \Delta(s) \in RH_\infty, \ \|\Delta(s)\|_\infty < \alpha\}$$

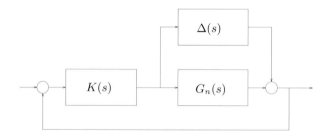

Figure 5.47: A control system with additive perturbations

that is, corresponding to the maximum possible value of the scalar α. Letting

$$P(s) := \begin{bmatrix} 0 & I \\ I & G_n(s) \end{bmatrix}$$

the problem can be solved, thanks to the here presented results, by following the procedure outlined in Remark 5.33. If the adopted (minimal) realization of $G_n(s)$ is

$$G_n(s) = \left[\begin{array}{cc|c} 0 & 1 & 0 \\ 2 & 1 & 1 \\ \hline -1 & 1 & 0 \end{array} \right] := \Sigma(A_c, B_c, C_c, 0)$$

and the chosen matrices F and H are

$$F = [-3 \ -3], \quad H = [-3 \ -6]'$$

(observe that $A_c + B_c F$ and $A_c + H C_c$ are stable), one obtains

$$M(s) = \hat{M}(s) = \frac{s-2}{s+1}, \quad \hat{Y}(s) = -\frac{27}{s+1}$$

Further, from eqs. (5.156)-(5.158) it results

$$T_1(s) = -27\frac{s-2}{(s+1)^2}, \quad T_2(s) = T_3(s) = \frac{s-2}{s+1}$$

so that

$$T_4(s) = \frac{(s-2)^2}{(s+1)^2}$$

The function $T_4(s)$ has no zeros on the complete imaginary axis. Therefore, starting from its inner-outer factorization given by

$$T_{4i}(s) = \frac{(s-2)^2}{(s+2)^2}, \quad T_{4o}(s) = \frac{(s+2)^2}{(s+1)^2}$$

it follows

$$F(s) = -27\frac{(s+2)^2}{(s+1)^2(s-2)}$$

A realization of the antistable part $F_a(s)$ of $F(s)$ is

$$F_a(s) = \left[\begin{array}{c|c} 2 & 8 \\ \hline -6 & 0 \end{array} \right].$$

so that, from eqs. (5.165)-(5.170), one obtains

$$P_r = 16 \; , \quad P_o = 9 \; , \quad \lambda_M = 12 \; , \quad \beta = 1 \; , \quad \chi = \frac{4}{3}$$

$$f(s) = \frac{8}{s+2} \; , \quad g(s) = -\frac{8}{s-2}$$

Then, from eq. (5.172) it follows

$$X^o(s) = 12 \frac{(s+2)(s+7/4)}{(s+1)^2}$$

which implies

$$Q^o(s) = 12 \frac{s+7/4}{s+2}$$

Finally, from eq. (5.155) and by taking into account what has been shown in the proof of Theorem 3.4, one obtains

$$K^o(s) = -12 \frac{s+1}{s-7}$$

Therefore, the widest class of perturbations corresponding to which stability is guaranteed is characterized by the value $\alpha = \lambda_M^{-1} = 1/12$. Compare this conclusion with the one given in Example 5.7. $\qquad\square$

5.7 Notes and references

Since the pioneering work of Zames [64], a great deal of attention has been devoted to robust control and related topics. The material collected in Sections 5.1 and 5.2 partly concerns classical issues of control theory suitably revisited so as to account for the basic instances of robustness. In particular, reference has been made to the books of Francis [19], Doyle et al. [16] and Maciejowski [43]. Further reading on related topics are Ackermann [1], Doyle [13] and Stein and Doyle [57]. No mention has been done to the somehow more realistic case of structured plant perturbations. The interested readers are referred to the papers by Doyle et al. [18], Doyle [14] and [15]. Sections 5.3, 5.4 are mainly based on the paper by Doyle et al. [17]. Again, the results concerning the parametrization of the admissible controllers follow the lines traced in the paper by Mita et al. [44]. More on the connections between differential games and H_∞ theory can be found in the book by Basar and Bernhard [2]. The robust stabilization problem of Remark 5.11 has been previously faced by Barmish [3], Khargonekar et al. [31] and Haddad and Bernstein [27]. The design of RH_2 filters in the presence of uncertainties (as in Remark 5.19) has been studied by Petersen and McFarlane [51] and Bolzern et al. [8], whereas the same problem in the RH_∞ context was tackled by De Souza et al. [12] and Fu et al. [20]. The paper by Safonov and Limebeer [56] has been exploited in writing down Remark 5.24 in Section 5.5. The rest of the section still relies on the paper by Doyle et al. [17]. A recent book on the H_∞ control problem is that by Stoorvogel [59] who has also explored the so called singular problem [58]. Finally, Section 5.6 deeply exploits the content of the book by Francis [19]. The here neglected approaches which rely on the gap metric and the polynomial framework are exploited in the paper by Georgiou and Smith [21] and by Kwakernaak [36].

Chapter 6

Nonclassical Problems in RH_2 and RH_∞

6.1 Introduction

This chapter introduces nonclassical problems in RH_2 and RH_∞ spaces which, roughly speaking, can not be solved in general by the machinery provided in the previous chapters. For a given stable transfer function the RH_2 norm is a measure of the level of a fixed input disturbance present in the output. The minimization of the RH_2 norm preserves the system against external disturbances. On the other hand, the RH_∞ norm of the same transfer function kept bounded above by a certain prescribed value $\gamma > 0$, imposes to the system a stable behavior against unmodeled dynamics with RH_∞ norm less than $1/\gamma$. The key observation is that RH_2 and RH_∞ norms compete one with other. Indeed, when $\gamma \to \infty$ the central RH_∞ controller approaches the optimal RH_2 controller, meaning that a level $\gamma < \infty$ can be imposed only at the expense of some performance level. Being so, it is in many cases important to determine a controller which, in some sense, expresses a desired tradeoff between both norms. This characterizes the so called mixed RH_2/RH_∞ optimal control problem. Its solution can not be addressed by means of classical methods based on Riccati equation solvers. Accordingly, our attention has to be moved to other numerical tools as for instance convex programming methods discussed in Appendix I.

6.2 Parameter Space Optimization

In this section, basic control synthesis problems involving stability, RH_2 and RH_∞ optimization are analyzed in the parameter space generated by the free elements of the feedback law. The main idea is to convert such nonconvex problems into convex ones in order to determine their global solutions by means of very powerful numeric procedures. First of all we need to introduce some concepts and definitions which are also discussed in Appendix H.

Definition 6.1 (Convex sets) *A set Ω in $R^{n \times m}$ is convex if $\forall\ X_1, X_2 \in \Omega$ the point $X = \alpha X_1 + (1 - \alpha)X_2 \in \Omega$ for every $\alpha \in [0,\ 1]$.* □

Definition 6.2 (Convex functions) *A function $f(\cdot) \in R$ defined in a convex set Ω is convex if $\forall\ X_1, X_2 \in \Omega$ and $X = \alpha X_1 + (1 - \alpha)X_2$ there holds $f(X) \leq \alpha f(X_1) + (1 - \alpha)f(X_2)$ for every $\alpha \in [0,\ 1]$.*   □

There is another characterization of convexity which is in many instances simpler to apply. A real valued function $f(\cdot)$ defined in a convex set $\Omega \subset R^{n \times m}$ is convex if and only if for each $X_0 \in \Omega$ there exists a matrix Λ_0 of appropriate dimension such that

$$f(X) \geq f(X_0) + < \Lambda_0, X - X_0 > \qquad (6.1)$$

for all $X \in \Omega$. In the above inequality (6.1) the inner product of matrices is defined by $< \Lambda, X >\ :=\ \text{trace}[\Lambda' X]$ which induces the Frobenius norm. The set $\partial f(X_0)$ of all matrices Λ_0 satisfying (6.1) is called the *subdifferential* of $f(\cdot)$ at $X = X_0$. If $f(\cdot)$ is differentiable at $X = X_0$ then $\Lambda_0 = \nabla f(X_0) \in \partial f(X_0)$ is unique.

Convexity is a key concept because any relative minimum of a convex programming problem formulated as to minimize $f(X)$ over a closed convex subset of Ω is a global minimum. Additionally, given a convex set Ω and a point $X_0 \notin \Omega$, it is always possible to determine an hyperplane which separates X_0 from Ω. The same result can be generalized to cope with X_0 in the boundary of Ω.

When dealing with linear system stability and optimization, one of the most important sets to be handled is $\mathcal{P}$, composed by all square, real, symmetric and positive semidefinite matrices with fixed and know dimension. Naturally, $\mathcal{P}$ is a subset of $\mathcal{S}$, the set of all real symmetric matrices with the same dimension and $\mathcal{P}$ is a convex set. To show this, let us recall that $X \in \mathcal{P}$ if and only if for any vector x of appropriate dimension $x'Xx \geq 0$. Hence, for every $X_1, X_2 \in \mathcal{P}$, the convex combination of them satisfies

$$x'Xx = \alpha x'X_1 x + (1 - \alpha)x'X_2 x$$

for all $\alpha \in [0,\ 1]$ implying that $X \in \mathcal{P}$. The convexity of $\mathcal{P}$ follows from Definition 6.1. This result is a particular case of a more general one given in the next lemma.

Lemma 6.1 *Suppose the matrix function $\mathcal{A}(X)$ is affine and its range is contained in $\mathcal{S}$. The set of all matrices X such that $\mathcal{A}(X)$ is positive semidefinite is convex.*

Proof Since $\mathcal{A}(X)$ is affine, for all $\alpha \in [0,\ 1]$ it follows that the convex combination X of arbitrary matrices X_1, X_2 satisfies

$$\mathcal{A}(X) = \alpha \mathcal{A}(X_1) + (1 - \alpha)\mathcal{A}(X_2) \qquad (6.2)$$

Using the fact that the range of $\mathcal{A}(\cdot)$ is a subset of $\mathcal{S}$, then $\mathcal{A}(X) \geq 0$ is equivalent to $x'\mathcal{A}(X)x \geq 0$ for all vectors x of compatible dimension. The above equality yields

$$x'\mathcal{A}(X)x = \alpha x'\mathcal{A}(X_1)x + (1 - \alpha)x'\mathcal{A}(X_2)x \qquad (6.3)$$

and so, $\mathcal{A}(X) \geq 0$ whenever $\mathcal{A}(X_1) \geq 0$ and $\mathcal{A}(X_2) \geq 0$, proving thus the lemma proposed.   □

Remark 6.1 Lemma 6.1 says that the set of all matrices X such that $\mathcal{A}(X) \in \mathcal{P}$ is convex. In particular, the choice $\mathcal{A}(X) = X$ shows once again that $\mathcal{P}$ is convex. Under the same assumptions of Lemma 6.1, the set of all matrices X such that $\mathcal{A}(X) \leq 0$ is also convex. The real valued function $f(X)$ defined as

$$f(X) = \max_i \lambda_i[\mathcal{A}(X)]$$

with $\mathcal{A}(X)$ affine and with range in $\mathcal{S}$ is convex as well. To show this, considering X as a convex combination of arbitrary X_1, X_2, the conclusion is that

$$f(X) = \max_{x \neq 0} \frac{x'\mathcal{A}(X)x}{x'x}$$

$$\leq \alpha \max_{x \neq 0} \frac{x'\mathcal{A}(X_1)x}{x'x} + (1 - \alpha) \max_{x \neq 0} \frac{x'\mathcal{A}(X_2)x}{x'x}$$

$$\leq \alpha f(X_1) + (1 - \alpha) f(X_2)$$

holds for all $\alpha \in [0, 1]$. □

Remark 6.2 It is important to keep clear that the range of $\mathcal{A}(X)$ being a subset of $\mathcal{S}$ is essential to get the above result. In fact, consider $\mathcal{A}(X)$ an affine matrix function but with its range not included in $\mathcal{S}$. This means that $\mathcal{A}(X)$ is not necessarily symmetric and the set of all $n \times n$ matrices X such that $\text{Re}[\lambda_i(\mathcal{A}(X))] < 0$, $i = 1, \cdots, n$ is not convex in general. This can be verified by means of a simple counterexample. For $\mathcal{A}(X) = X$ let us take

$$X_1 = \begin{bmatrix} -1 & 4 \\ 0 & -1 \end{bmatrix}, \quad X_2 = \begin{bmatrix} -1 & 0 \\ 4 & -1 \end{bmatrix}$$

and

$$X = 0.5X_1 + 0.5X_2 = \begin{bmatrix} -1 & 2 \\ 2 & -1 \end{bmatrix}$$

It is clear that matrices X_1 and X_2 belong to the previously defined set. However, matrix X which is a convex combination of X_1 and X_2 does not belong to it. The conclusion is that the set of all 2×2 matrices under consideration is not convex. □

Remark 6.3 Notice that the convex function $f(X)$ introduced in Remark 6.1 is not affine. It is not difficult to making use of inequality (6.1) for the determination of matrix Λ_0. As an example, consider the simplest case where $\mathcal{A}(X) = X \in \mathcal{S}$. Given $X_0 \in \mathcal{S}$ let x_0 be an unitary norm eigenvector associated to its largest eigenvalue. Obviously the equality $X_0 x_0 = f(X_0)x_0$ holds and for all $X \in \mathcal{S}$ we have

$$f(X) = \max_{\|x\|=1} x'Xx$$

$$\geq x_0'Xx_0$$

$$\geq f(X_0) + <x_0x_0', X - X_0>$$

Comparing this inequality with (6.1) it is apparent that $\Lambda_0 = x_0x_0'$. Function $f(\cdot)$ is not differentiable unless the multiplicity of the maximum eigenvalue of X_0 is one. In this case $\Lambda_0 = x_0x_0' \in \partial f(X_0)$ is the only matrix satisfying (6.1) at $X = X_0$.

Another important function is $f(X) = \text{trace}[B'X^{-1}B]$, defined for all nonsingular $X \in \mathcal{P}$. This function is differentiable and convex. Let us show this by means of inequality (6.1). Considering arbitrary nonsingular matrices $X, X_0 \in \mathcal{P}$ we have

$$f(X) = \text{trace}[B'X^{-1}B]$$

$$= f(X_0) - <X_0^{-1}BB'X_0^{-1}, X - X_0> +$$

$$+ \text{trace}[B'X_0^{-1}(X - X_0)X^{-1}(X - X_0)X_0^{-1}B]$$

$$\geq f(X_0) + <-X_0^{-1}BB'X_0^{-1}, X - X_0>$$

and again $\Lambda_0 = -X_0^{-1}BB'X_0^{-1} \in \partial f(X_0)$ is unique. These examples show that inequality (6.1) is an easy and useful way to characterize convexity of matrix functions.

Given matrices A and $Q = Q'$, the sets of all matrices $X \in \mathcal{S}$ such that

$$\mathcal{A}(X) = A'X + XA + Q \leq 0$$

or

$$\mathcal{A}(X) = AX + XA' + Q \leq 0$$

are both convex. This is an immediate consequence of Lemma 6.1. Furthermore, given matrices A, B and $Q = Q'$, the sets of all matrices $X = X' > 0$ and Y such that

$$\mathcal{B}(X, Y) = (A + BY)'X + X(A + BY) + Q \leq 0$$

or

$$\mathcal{B}(X, Y) = (A + BY)X + X(A + BY)' + Q \leq 0$$

are both nonconvex. The second one however can be converted into an equivalent convex set. Indeed, since X is symmetric and positive definite the change of variables $Y := ZX^{-1}$ provides

$$\mathcal{A}(X, Z) = AX + BZ + XA' + Z'B' + Q \leq 0$$

which is convex because $\mathcal{A}(X, Z)$ is jointly affine in the variables X, Z. Finally, the set of all (X, Y, Z) matrices such that $X = X' > 0$ and $Y'X^{-1}Y - Z \leq 0$ is convex as well. This follows immediately from the Schur complement formula (recall Lemma B.14) which states that this set is equivalent to the set of all (X, Y, Z) matrices such that

$$X > 0, \quad \begin{bmatrix} X & Y \\ Y' & Z \end{bmatrix} \geq 0$$

This is one of the most important result to be exhaustively used in the sequel. $\qquad \square$

6.2.1 Stabilizing controllers

A convex parametrization of all stabilizing matrix gains of a linear system is provided. Consider the following linear and time-invariant dynamic system

$$\dot{x} = Ax + B_2 u \tag{6.4}$$

with n states, m inputs and where the state variable is available for feedback. When dealing with optimal control problems in RH_2 and RH_∞ spaces, we have to restrict our attention to those $m \times n$ matrix gains F such that with $u = Fx$ the closed loop system is stable, that is

$$F \in \mathcal{K}_c := \{F \in R^{m \times n} : A + B_2 F \text{ stable}\} \tag{6.5}$$

The sentence "$A + B_2 F$ stable" means that the closed loop system is internally stable, that is $\text{Re}[\lambda_i(A + B_2 F)] < 0$, $i = 1, \cdots, n$. Since $\mathcal{A}(F) = A + B_2 F$ is affine with respect to F but its range is not in $\mathcal{S}$ the conclusion, as discussed before (recall Remark 6.2) is that the set $\mathcal{K}_c$ is not convex in general. Any optimal control problem formulated in the parameter space generated by the (free) elements of F and having $\mathcal{K}_c$ as the feasible set is not convex. At most only local optimal solutions can be numerically determined by the machinery available in the literature to date. The fact that the set $\mathcal{K}_c$ is not convex, introduces one of the major difficulties to be faced in this section.

To circumvent this difficulty let us proceed as follows. Define the extended $p \times p$ and $p \times m$ matrices ($p := n + m$)

$$M_c := \begin{bmatrix} A & B_2 \\ 0 & 0 \end{bmatrix}, \quad N_c := \begin{bmatrix} 0 \\ I \end{bmatrix} \tag{6.6}$$

where the null space of N_c' is spanned by all $v \in R^p$ such that $N_c'v = 0$ or equivalently by all vectors in R^p having the form $v = [x' \ 0]'$ with $x \in R^n$ arbitrary. Then, it is apparent that the set $\mathcal{N}_c$ of vectors v in the null space of N_c' with $\|v\| = 1$ equals the set of all $v = [x' \ 0]'$ with $x \in R^n$ and $\|x\| = 1$. Define also the $p \times p$ symmetric matrices W and Q_c partitioned as

$$W := \begin{bmatrix} W_1 & W_2 \\ W_2' & W_3 \end{bmatrix} , \quad Q_c := \begin{bmatrix} Q_{1c} & 0 \\ 0 & 0 \end{bmatrix} \tag{6.7}$$

where in both matrices, the $(1, 1)$ block has dimension $n \times n$.

Theorem 6.1 *Assume Q_{1c} is a positive definite matrix and consider the set*

$$\mathcal{C}_c := \{W \ : \ W \geq 0 , \quad v'\Theta_c(W)v \leq 0 , \ \forall v \in \mathcal{N}_c\} \tag{6.8}$$

where $\Theta_c(W) := M_c W + W M_c' + Q_c$. The following hold

a) $\mathcal{C}_c$ is a convex set.

b) Each $W \in \mathcal{C}_c$ is such that $W_1 > 0$.

c) $\mathcal{K}_c = \{W_2' W_1^{-1} \ : \ W \in \mathcal{C}_c\}$.

Proof Each part of the theorem is proved separately. Notice that W is symmetric and the matrix valued function $\Theta_c(\cdot)$ is affine.

Point a) Since the empty set is convex, let us proceed by considering X_1 and X_2 two arbitrary matrices belonging to $\mathcal{C}_c$. Taking $X = \alpha X_1 + (1 - \alpha)X_2$ with $\alpha \in [0, 1]$ we first notice that $X \geq 0$ since the set of all positive semidefinite matrices is convex. Furthermore, taking into account that the matrix valued function $\Theta_c(X)$ is affine we get

$$v'\Theta_c(X)v = \alpha v'\Theta_c(X_1)v + (1 - \alpha)v'\Theta_c(X_2)v \leq 0$$

for all $v \in \mathcal{N}_c$. Consequently $X \in \mathcal{C}_c$.

Point b) Let us prove this point by contradiction. Assume for some $W \in \mathcal{C}_c$ there exists $x \neq 0 \in R^n$ such that $W_1 x = 0$. Since $W \geq 0$, using the partitioning (6.7) we must have $W_2' x = 0$. On the other hand $W \in \mathcal{C}_c$ and $v = [x' \ 0]' \in \mathcal{N}_c$ imply that

$$0 \geq v'\Theta_c(W)v$$
$$\geq x' \left(AW_1 + W_1 A' + B_2 W_2' + W_2 B_2' + Q_{1c}\right) x$$
$$\geq x'Q_{1c}x$$

which is an impossibility because $Q_{1c} > 0$.

Point c) The proof of this part is done by construction. First assume $\mathcal{K}_c \neq \emptyset$, take $F \in \mathcal{K}_c$ and recall that Q_{1c} is positive definite. From the Extended Lyapunov lemma (see Appendix C), there exists a symmetric positive definite solution to the linear equation

$$(A + B_2 F)P + P(A + B_2 F)' + Q_{1c} = 0$$

Consequently, choosing

$$W = \begin{bmatrix} P & PF' \\ FP & FPF' \end{bmatrix}$$

it is readily seen that such a matrix W is feasible, that is $W \in \mathcal{C}_c$ and $W_2' W_1^{-1} = FPP^{-1} = F$. Conversely, assume $\mathcal{C}_c \neq \emptyset$ and take $W \in \mathcal{C}_c$. From point b) we already know that $W_1 > 0$ and $\forall v \in \mathcal{N}_c$ yields

$$0 \geq v'\Theta_c(W)v$$
$$\geq x' \left(AW_1 + W_1 A' + B_2 W_2' + W_2 B_2' + Q_{1c}\right) x$$
$$\geq x' \left[(A + B_2 W_2' W_1^{-1})W_1 + W_1(A + B_2 W_2' W_1^{-1})' + Q_{1c}\right] x \tag{6.9}$$

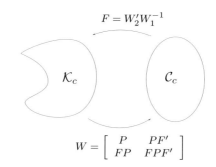

$$W = \left[\begin{array}{cc} P & PF' \\ FP & FPF' \end{array} \right]$$

Figure 6.1: Relationship between sets $\mathcal{K}_c$ and $\mathcal{C}_c$

The vector x in the third inequality of (6.9) being arbitrary implies that $F = W_2'W_1^{-1} \in \mathcal{K}_c$. In fact, assume by contradiction that there exist an eigenvalue λ and an eigenvector x of matrix $A + B_2W_2'W_1^{-1}$ such that $\mathrm{Re}(\lambda) \geq 0$ and $\|x\| = 1$. From inequality (6.9) it follows that $2\mathrm{Re}(\lambda)x^\sim W_1 x + x^\sim Q_{1c}x \leq 0$ which contradicts the assumption that Q_{1c} is positive definite. The proof is concluded because, from the above, if one of the sets $\mathcal{K}_c$ or $\mathcal{C}_c$ is empty then both are empty. $\qquad\square$

Figure 6.1 gives an interpretation of this result. Point c) of Theorem 6.1 provides a nonlinear mapping namely $W_2'W_1^{-1}$ which generates the set $\mathcal{K}_c$ from all matrices W in the convex domain $\mathcal{C}_c$. In some sense, the nonconvexity involved has been isolated in the nonlinearity $W_2'W_1^{-1}$. Furthermore, notice that for $\forall\, v \in \mathcal{N}_c$ the quantity $v'\Theta_c(W)v$ depends only upon the blocks W_1 and W_2 of matrix W. Consequently for given W_1 and W_2 such that $v'\Theta_c(W)v \leq 0$, $\forall\, v \in \mathcal{N}_c$, it is always possible to select $W_3 \geq W_2'W_1^{-1}W_2$ (recall Appendix B) in order to have $W \geq 0$ and $W \in \mathcal{C}_c$. This degree of freedom will be important for solving optimal control problems in RH_2 and RH_∞ spaces.

Remark 6.4 From Theorem 6.1 it is clear that the pair (A, B_2) is stabilizable if and only if $\mathcal{C}_c \neq \emptyset$. This follows from the fact that when W varies in $\mathcal{C}_c$ then all stabilizing state feedback gains are generated from $F = W_2'W_1^{-1}$. If such a matrix does not exist, it is clear that $\mathcal{C}_c = \emptyset$. $\qquad\square$

Let us now move our attention to the output feedback case. The linear system to be dealt with is

$$\dot{x} = Ax + B_2u \qquad (6.10)$$
$$y = C_2x \qquad (6.11)$$

with n states, m inputs and r outputs. The controller structure is given by

$$\dot{\xi} = (A + B_2F)\xi + L(C_2\xi - y) \qquad (6.12)$$
$$u = F\xi \qquad (6.13)$$

where both gains F and L have to be determined in order to assure the closed loop connection of the system and controller is internally stable. Simple algebraic manipulations (recall Section 4.1 and the separation property) bring to light that the dynamic matrix A_F of the resulting system satisfies

$$\det[sI - A_F] = \det[sI - (A + B_2F)]\det[sI - (A + LC_2)] \qquad (6.14)$$

Hence, as far as stability is concerned, the determination of both gains which defines completely the dynamic controller (6.12) - (6.13) have to be such that $F \in \mathcal{K}_c$ and

$$L \in \mathcal{K}_f := \{L \in R^{n \times r} : A + LC_2 \ \text{stable}\} \tag{6.15}$$

However, keeping in mind that the eigenvalues of any matrix and its transpose are the same, all elements of $\mathcal{K}_f$ can be generated by the dual version of Theorem 6.1. Indeed, following (6.6) let us define the $q \times q$ and $q \times r$ extended matrices ($q := n+r$) namely,

$$M_f := \begin{bmatrix} A & 0 \\ C_2 & 0 \end{bmatrix}, \quad N_f := \begin{bmatrix} 0 \\ I \end{bmatrix} \tag{6.16}$$

It is clear that the null space of N_f' spanned by all vectors $v \in R^q$ such that $N_f' v = 0$, exhibits again the structure pointed out before, that is $v = [x' \ 0]'$ with $x \in R^n$. The set $\mathcal{N}_f$ is composed by all vectors $v \in R^q$ in the null space of N_f' with unitary norm. Partitioned accordingly to the dimensions of the plant state and output vectors, the symmetric matrices

$$V := \begin{bmatrix} V_1 & V_2 \\ V_2' & V_3 \end{bmatrix}, \quad Q_f := \begin{bmatrix} Q_{1f} & 0 \\ 0 & 0 \end{bmatrix} \tag{6.17}$$

are on the basis for the next result, dual of the one provided in Theorem 6.1.

Theorem 6.2 *Assume Q_{1f} is a positive definite matrix and consider the set*

$$\mathcal{C}_f := \{V : V \geq 0, \ v' \Theta_f(V) v \leq 0, \ \forall v \in \mathcal{N}_f\} \tag{6.18}$$

where $\Theta_f(V) := M_f' V + V M_f + Q_f$. The following hold

a) $\mathcal{C}_f$ is a convex set.

b) Each $V \in \mathcal{C}_f$ is such that $V_1 > 0$.

c) $\mathcal{K}_f = \{V_1^{-1} V_2 : V \in \mathcal{C}_f\}$.

Remark 6.5 Form Theorem 6.2 it is clear that the pair (A, C_2) is detectable if and only if $\mathcal{C}_f \neq \emptyset$. This is the dual of the property discussed in Remark 6.4. □

Matrices F and L which define a stabilizing controller are thus determined from completely decoupled convex sets. This is possible because the dimension of the dynamic controller (6.12) - (6.13) equals the system dimension. The controller structure is based on the internal model of the plant and as expected, the error dynamics $\varepsilon := \xi - x$ is completely defined by the poles of matrix $A + LC_2$ and so independent upon the choice of $F \in \mathcal{K}_c$. For this reason, the controller (6.12) - (6.13) is called *observer-based controller* and, as will be seen in the next sections, it plays a central role in nonclassical control design.

6.2.2 RH_2 control design

Control design problems in RH_2 space have been studied in Chapter 4. The standard problem, divided in three related topics has been solved. The main purpose of this section is to analyze once again the same problems but in the context of convex analysis. Under the same assumptions introduced in Chapter 4 it is possible to show that

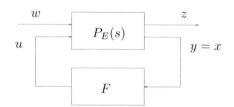

Figure 6.2: The state feedback control system

the standard problem in RH_2 can be converted to an equivalent convex programming problem.

Before giving an alternate solution to the Full information, Output estimation and Partial information problems, let us consider the following linear system

$$\dot{x} = Ax + B_1 w + B_2 u \tag{6.19}$$

$$z = C_1 x + D_{12} u \tag{6.20}$$

$$y = x \tag{6.21}$$

where, as indicated, the whole state vector is available for feedback. The structure of controller is simply a static state feedback gain so that

$$u = Fx \tag{6.22}$$

Calling $P_E(s)$ the transfer function of (6.19) - (6.21), the feedback system is depicted in fig. 6.2. Any state feedback gain matrix $F \in \mathcal{K}_c$ is called *feasible*. For any feasible gain, the transfer function $T(z, w; s)$ belongs to RH_2 and its norm can be evaluated as being

$$\|T(z, w; s)\|_2^2 = \|(C_1 + D_{12}F)[sI - (A + B_2 F)]^{-1} B_1\|_2^2$$
$$= \operatorname{trace}[B_1' P_o B_1]$$
$$= \operatorname{trace}[(C_1 + D_{12}F) P_r (C_1 + D_{12}F)']$$

where P_o and P_r are the unique solutions of the Lyapunov equations

$$0 = A_{cc}' P_o + P_o A_{cc} + C_{cc}' C_{cc} \tag{6.23}$$

$$0 = A_{cc} P_r + P_r A_{cc}' + B_1 B_1' \tag{6.24}$$

with $A_{cc} := A + B_2 F$ and $C_{cc} := C_1 + D_{12}F$. We are now in position to introduce the problem to be solved.

Problem 6.1 (State feedback problem in RH_2) *Find a feasible state feedback gain F which minimizes $\|T(z, w; s)\|_2$. In other terms, find the global optimal solution of the problem*

$$\min\ \{\|T(z, w; s)\|_2^2\ :\ F \in \mathcal{K}_c\}$$

Before solving this problem we observe that it is a particular case of the Full information problem treated in Chapter 4 and Assumptions 4.1 - 4.2 of the Full information problem are again made. That is, i) The pair (A, B_2) is stabilizable and no eigenvalue of the unobservable part of the pair $[(A - B_2 D_{12}' C_1), (I - D_{12} D_{12}') C_1]$

lies on the imaginary axis of the complex plane and ii) $D'_{12}D_{12} = I$. Under these assumptions, $\mathcal{K}_c \neq \emptyset$ and the Full information problem and the State feedback problem shares the same optimal solution because as we already know (recall Theorem 4.1) at the optimal solution of the former, the gain corresponding to w is zero. The solution of Problem 6.1 is thus

$$F^o = F_2 = -B'_2 P_2 - D'_{12}C_1 \tag{6.25}$$

where P_2 is the symmetric, positive semidefinite and stabilizing solution of the Riccati equation (in the unknown P)

$$0 = PA_c + A'_c P - PB_2 B'_2 P + C'_{1c}C_{1c} \tag{6.26}$$

with $A_c := A - B_2 D'_{12}C_1$ and $C_{1c} := (I - D_{12}D'_{12})C_1$. With $u = F_2 x$ the closed loop minimum performance is given by

$$\min \|T(z, w; s)\|_2^2 = \text{trace}[B'_1 P_2 B_1] \tag{6.27}$$

Remark 6.6 The solution of the State feedback problem (as well as the Full information problem) does not depend upon matrix B_1. Indeed, for any matrix B_1 the optimal feedback gain F_2 is completely characterized by means of (6.25) and (6.26) which do not depend of the aforementioned matrix. This aspect will be important in the sequel. In order to determine a convex problem equivalent to the State feedback problem it will be necessary to introduce an additional assumption, namely $B_1 B'_1 > 0$. To see that this can be done with no loss of generality, consider $B_1 B'_1 \geq 0$ and define $\bar{B}_1 := [B_1 \quad \sqrt{\epsilon}I]$ where ϵ is a positive constant. Obviously matrix $\bar{B}_1$ exhibits the desired property $\bar{B}_1 \bar{B}'_1 > 0$ for any $\epsilon > 0$ chosen. On the other hand, solving the State feedback problem we get

$$\min \|\bar{T}(z, w; s)\|_2^2 = \min \|T(z, w; s)\|_2^2 + \epsilon \text{trace}[P_2]$$

That is, the minimum value of the modified transfer function differs from the previous one by an amount of order ϵ which can be made arbitrarily small by a proper choice of this parameter. □

Remark 6.7 It is important to keep in mind that the set of admissible controllers for the State feedback and Full information problems in RH_2 are quite different. In the Full information problem we have considered as admissible, any dynamic controller $K(s)$ satisfying Definition 4.1 and we have proven (recall Theorem 4.1) that the optimal controller is given by

$$K(s) = K^o_{FI}(s) = \left[\begin{array}{c|cc} \emptyset & \emptyset & \emptyset \\ \hline \emptyset & F_2 & 0 \end{array}\right]$$

In the State feedback problem stated before, we have considered as admissible only the static and stabilizing state feedback gains, namely $u = Fx$ such that $F \in \mathcal{K}_c$. Since this structure is a particular case of the former and $K^o_{FI}(s)$ given above implies that $u = F_2 x$ then $F = F_2 \in \mathcal{K}_c$ is the optimal solution of the State feedback problem. The proof of this fact could be done in a different way which as expected is simpler than the proof of Theorem 4.1. Actually, making use of (6.23) we can restate the State feedback problem as to determine $F \in \mathcal{K}_c$ and $P = P' \geq 0$ such that

$$\min \left\{ \text{trace}[B'_1 PB_1] : 0 = A'_{cc}P + PA_{cc} + C'_{cc}C_{cc} \right\}$$

The Lagrangian function associated to this problem ($\Lambda = \Lambda'$ being the matrix of Lagrange multipliers associated to the equality constraint) is

$$\mathcal{L}(F, P, \Lambda) = \text{trace}[B'_1 PB_1] + \text{trace}[\Lambda(A'_{cc}P + PA_{cc} + C'_{cc}C_{cc})]$$

So, the necessary conditions for optimality are readily obtained by simple differentiation of $\mathcal{L}(\cdot)$ with respect to the unknown matrices F, P and Λ, providing

$$0 = A'_{cc}P + PA_{cc} + C'_{cc}C_{cc}$$
$$0 = A_{cc}\Lambda + \Lambda A'_{cc} + B_1 B'_1$$
$$0 = (F + D'_{12}C_1 + B'_2 P)\Lambda$$

A possible solution to this set of nonlinear equations is obtained by first setting $F = -B'_2 P - D'_{12}C_1$. This equality together with the first equation imply that P must solve the Riccati equation

$$0 = PA_c + A'_c P - PB_2 B'_2 P + C'_{1c}C_{1c}$$

Under the previous assumptions, this equation admits a stabilizing solution $P = P_2$ yielding $F = F_2$. Noticing that $F_2 \in \mathcal{K}_c$, the second equation has a solution with respect to Λ namely,

$$\Lambda_2 = \int_0^\infty e^{(A+B_2 F_2)t} B_1 B'_1 e^{(A+B_2 F_2)'t} dt$$

consequently the conclusion is that the triple (F_2, P_2, Λ_2) satisfies the necessary conditions for optimality. It remains to prove that this is in fact the global solution of the proposed problem. To this end, consider $F \in \mathcal{K}_c$ an arbitrary stabilizing state feedback gain. Simple algebraic manipulations show that

$$(A + B_2 F)'(P - P_2) + (P - P_2)(A + B_2 F) + (F - F_2)'(F - F_2) = 0$$

and (recall Appendix C) so

$$P = P_2 + \int_0^\infty e^{(A+B_2 F)'t}(F - F_2)'(F - F_2)e^{(A+B_2 F)t} dt \geq P_2$$

For $u = Fx$ the closed loop transfer function satisfies

$$\|T(z, w; s)\|_2^2 = \text{trace}[B'_1 P B_1]$$
$$\geq \text{trace}[B'_1 P_2 B_1] \,, \quad \forall F \in \mathcal{K}_c$$

proving thus that the static matrix gain F_2 generated by the stabilizing solution P_2 of the Riccati equation (6.26) is the global optimum of the State feedback problem indeed. Once again, it is noticed that its solution does not depend upon matrix B_1.

However, for B_1 given it is in principle possible to determine $F \neq F_2$ which equals the global minimum. In fact, P is equal to P_2 only if $F = F_2$ but $\text{trace}[B'_1 P B_1]$ may be equal to $\text{trace}[B'_1 P_2 B_1]$ even though $F \neq F_2$. Let $F \in \mathcal{K}_c$ satisfying the necessary conditions for optimality with $\Lambda \geq 0$ be such that $(F - F_2)\Lambda = 0$. Taking into account that

$$\Lambda = \int_0^\infty e^{(A+B_2 F)t} B_1 B'_1 e^{(A+B_2 F)'t} dt$$

together with the relationship between P and P_2 we get the desired result

$$0 = \text{trace}[(F - F_2)'(F - F_2)\Lambda]$$
$$= \text{trace}\left[\int_0^\infty B'_1 e^{(A+B_2 F)'t}(F - F_2)'(F - F_2)e^{(A+B_2 F)t} B_1 dt\right]$$
$$= \text{trace}[B'_1(P - P_2)B_1]$$

From this it is apparent that the way to prevent this pathological situation is to impose $\Lambda > 0, \forall F \in \mathcal{K}_c$. Thanks to the Extended Lyapunov lemma (recall Appendix C), this is always verified whenever $B_1 B'_1 > 0$, a condition just discussed in the previous remark. □

It is simple to verify that Problem 6.1 is not convex (recall Appendix H). This conclusion follows immediately from the fact that the set of admissible gains $\mathcal{K}_c$ is not convex. Furthermore, its objective function is not convex as well. Defining the symmetric matrices

$$R_c := \begin{bmatrix} C_1' \\ D_{12}' \end{bmatrix} \begin{bmatrix} C_1 & D_{12} \end{bmatrix}, \quad Q_c := \begin{bmatrix} B_1 B_1' & 0 \\ 0 & 0 \end{bmatrix} \tag{6.28}$$

the next theorem provides a convex problem equivalent to Problem 6.1.

Theorem 6.3 (State feedback) *Assume $B_1 B_1'$ is a positive definite matrix and consider the following convex programming problem*

$$\bar{W} := \operatorname{argmin} \{\operatorname{trace}[R_c W] \; : \; W \in \mathcal{C}_c\} \tag{6.29}$$

Then, $F = \bar{W}_2' \bar{W}_1^{-1}$ solves Problem 6.1.

Proof It is done in two main steps. First, consider matrix $\bar{W}$ given by

$$\bar{W} := \begin{bmatrix} \Lambda_2 & \Lambda_2 F_2' \\ F_2 \Lambda_2 & F_2 \Lambda_2 F_2' \end{bmatrix} \tag{6.30}$$

where (recall Remark 6.7) $\bar{W}_1 = \Lambda_2 > 0$ due to the assumption $B_1 B_1' > 0$. On the other hand, it is simple to verify that $\bar{W}_2' \bar{W}_1^{-1} = F_2$, $\bar{W} \geq 0$ and for all $v \in \mathcal{N}_c$

$$v'\Theta_c(\bar{W})v = x'[(A + B_2 F_2)\Lambda_2 + \Lambda_2(A + B_2 F_2)' + B_1 B_1']x = 0$$

implying that $\bar{W} \in \mathcal{C}_c$. The same matrix, together with (6.28) also provides

$$\begin{aligned} \operatorname{trace}[R_c \bar{W}] &= \operatorname{trace}\left[\begin{bmatrix} C_1 & D_{12} \end{bmatrix} \bar{W} \begin{bmatrix} C_1' \\ D_{12}' \end{bmatrix}\right] \\ &= \operatorname{trace}\left[(C_1 + D_{12} F_2)\Lambda_2(C_1 + D_{12} F_2)'\right] \\ &= \operatorname{trace}[B_1' P_2 B_1] \\ &= \min \|T(z, w; s)\|_2^2 \end{aligned}$$

which means that matrix $\bar{W}$ is feasible and generates the optimal solution of Problem 6.1. It remains to prove that matrix $\bar{W}$ is actually the optimal solution of the convex programming problem (6.29). To this end, let us keep in mind (recall Theorem 6.1) that all $W \in \mathcal{C}_c$ are such that $F = W_2' W_1^{-1} \in \mathcal{K}_c$ and

$$\|T(z, w; s)\|_2^2 = \operatorname{trace}\left[(C_1 + D_{12} W_2' W_1^{-1})P_r(C_1 + D_{12} W_2' W_1^{-1})'\right]$$

where $P_r > 0$ solves the linear matrix equation

$$0 = (A + B_2 W_2' W_1^{-1})P_r + P_r(A + B_2 W_2' W_1^{-1})' + B_1 B_1'$$

However, any $W \in \mathcal{C}_c$ is such that $W_1 > 0$ and satisfies the inequality

$$(A + B_2 W_2' W_1^{-1})W_1 + W_1(A + B_2 W_2' W_1^{-1})' + B_1 B_1' \leq 0$$

which enables us to conclude that $W_1 \geq P_r$. Furthermore, $W \geq 0$ and $W_1 > 0$ is equivalent to $W_3 \geq W_2' W_1^{-1} W_2$ and we finally get

$$\begin{aligned} \operatorname{trace}[R_c \bar{W}] &= \min \|T(z, w; s)\|_2^2 \\ &\leq \|T(z, w; s)\|_2^2 \\ &\leq \operatorname{trace}\left[(C_1 + D_{12} W_2' W_1^{-1})W_1(C_1 + D_{12} W_2' W_1^{-1})'\right] \\ &\leq \operatorname{trace}[R_c W] - \operatorname{trace}[D_{12}(W_3 - W_2' W_1^{-1} W_2)D_{12}'] \\ &\leq \operatorname{trace}[R_c W] \end{aligned}$$

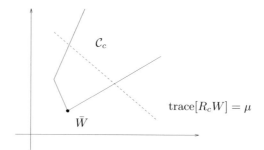

Figure 6.3: The state feedback control problem

which being true for all feasible $W \in \mathcal{C}_c$ completes the proof.   □

Remark 6.8 Theorem 6.3 opens a very attractive way to solve optimal control problems in RH_2 spaces by means of many powerful methods available to date. The convexity property is the key issue to be sure the global solution is always attained. However, in order to guarantee a selected numerical convex programming method can effectively solve Problem 6.1 it is necessary to prove that the feasible convex set $\mathcal{C}_c$ is bounded. Unfortunately, this is not true. In fact, the set $\mathcal{C}_c$ is a convex and unbounded cone. To show this take any $W \in \mathcal{C}_c$ and $\lambda \geq 1$. Then $\lambda W \geq 0$ and $\forall v \in \mathcal{N}_c$

$$v'\Theta_c(\lambda W)v = \lambda v'\Theta_c(W)v + (1 - \lambda)v'Q_c v$$
$$\leq \lambda v'\Theta_c(W)v \leq 0$$

implying that $\lambda W \in \mathcal{C}_c$.

     However, a weaker condition exists to guarantee Problem 6.1 is numerically solvable even though the feasible set is not bounded. The situation is illustrated in fig. 6.3 where the set $\mathcal{C}_c$ is an unbounded cone but the objective function $\mathrm{trace}[R_c W]$ has a global and finite optimum at $W = \bar{W}$. This occurs because the convex set

$$\mathcal{C}_{c\mu} := \mathcal{C}_c \cap \{W \; : \; \mathrm{trace}[R_c W] \leq \mu\}$$

is bounded for all μ such that $\mathrm{trace}[R_c \bar{W}] < \mu < \infty$. To show that Problem 6.1 has this property we proceed as follows. By contradiction assume that for some finite μ as specified before, the set $\mathcal{C}_{c\mu}$ is unbounded. In this case, thanks to its convexity, there exist $W \in \mathcal{C}_{c\mu}$ and $\tilde{W} \neq 0$ such that $W + \lambda \tilde{W} \in \mathcal{C}_{c\mu}$ for $\lambda > 0$ arbitrarily large. Then, $\tilde{W}$ must satisfy

$$\mathrm{trace}[R_c \tilde{W}] = 0$$
$$\tilde{W} \geq 0$$
$$v'(M_c \tilde{W} + \tilde{W} M_c')v \leq 0 \; , \quad \forall v \in \mathcal{N}_c$$

     Using the structure of the symmetric positive semidefinite matrix R_c, the first two equations are equivalent to $[C_1 \; D_{12}]\tilde{W}^{1/2} = 0$ and all its solutions are also solutions of

$$\begin{bmatrix} C_1 & D_{12} \end{bmatrix} \tilde{W} = \begin{bmatrix} C_1 & D_{12} \end{bmatrix} \begin{bmatrix} \tilde{W}_1 & \tilde{W}_2 \\ \tilde{W}_2' & \tilde{W}_3 \end{bmatrix} = 0$$

Making reference to Appendix B, this equation is solvable provided $C_{1c}\tilde{W}_1 = 0$ yielding

$$\tilde{W} = \begin{bmatrix} I \\ -D_{12}'C_1 \end{bmatrix} \tilde{W}_1 \begin{bmatrix} I & -C_1'D_{12} \end{bmatrix}$$

which replaced in the previous conditions allows us to express them in terms of $\tilde{W}_1$ only, that is we have to get $\tilde{W}_1 \neq 0$ such that

$$C_{1c}\tilde{W}_1 = 0$$
$$\tilde{W}_1 \geq 0$$
$$A_c\tilde{W}_1 + \tilde{W}_1 A'_c \leq 0$$

Once again making use of a result included in Appendix C, there is no matrix $\tilde{W}_1 \neq 0$ satisfying these conditions provided the pair $(-A_c, C_{1c})$ is detectable. Therefore, to be sure the set $\mathcal{C}_{c\mu}$ is bounded we need to change the assumption i) of the State feedback problem to i) the pair (A, B_2) is stabilizable and no eigenvalue of the unobservable part of the pair $(A_c, C_{1c}) = [(A - B_2 D'_{12}C_1), (I - D_{12}D'_{12})C_1]$ lies on left part (including the imaginary axis) of the complex plane. If the detectability assumption is violated then $\tilde{W}_1 \neq 0$ satisfying the above conditions may exist. In the example

$$A_c = \begin{bmatrix} 1 & 0 \\ 2 & -1 \end{bmatrix} , \quad C_{1c} = \begin{bmatrix} 1 & 0 \end{bmatrix} , \quad \tilde{W}_1 = \begin{bmatrix} 0 & 0 \\ 0 & 1 \end{bmatrix}$$

this occurs because the pair $(-A_c, C_{1c})$ fails to be detectable. □

Example 6.1 To illustrate the geometry of the convex problem introduced in Theorem 6.3, let us consider the system $P_E(s)$ given by

$$\dot{x} = x + \sqrt{2}w + u$$
$$z = \begin{bmatrix} 1 \\ 0 \end{bmatrix} x + \begin{bmatrix} 0 \\ 1 \end{bmatrix} u$$
$$y = x$$

The optimal solution of the State feedback problem is characterized by

$$P_2 = 1 + \sqrt{2} , \quad F_2 = -(1 + \sqrt{2}) , \quad \Lambda_2 = \frac{1}{\sqrt{2}}$$

and $\|T(z, w; s)\|_2^2 = 2(1 + \sqrt{2}) = 4.82$. For this simple example, Problem 6.29 is written as

$$\min \left\{ W_1 + W_3 \; : \; W_1 + W_2 + 1 \leq 0 , \quad \begin{bmatrix} W_1 & W_2 \\ W_2 & W_3 \end{bmatrix} \geq 0 \right\}$$

or in other terms, after using Schur complements

$$\min \left\{ W_1 + \frac{W_2^2}{W_1} \; : \; W_1 + W_2 + 1 \leq 0 , \quad W_1 > 0 \right\}$$

The optimal solution is given by (recall the proof of Theorem 6.3)

$$\bar{W} = \begin{bmatrix} \Lambda_2 & \Lambda_2 F'_2 \\ F_2\Lambda_2 & F_2\Lambda_2 F'_2 \end{bmatrix} = \begin{bmatrix} 0.70 & -1.70 \\ -1.70 & 4.12 \end{bmatrix}$$

Figure 6.4 illustrates the feasible set which lies below the line indicated and the set of points such that the objective function is equal to $\mu = 1, 2, 3$ and 4.82. For the last value of $\mu = 4.82$ the curve is tangent to the feasible region yielding the global minimum $W = \bar{W}$. This example illustrates also that it is possible to eliminate matrix W_3 from the set of variables by setting $W_3 = W'_2 W_1^{-1} W_2$. Of course this can only be done at the expense of changing a *linear* objective function to a *nonlinear* (although convex) objective function to be minimized. □

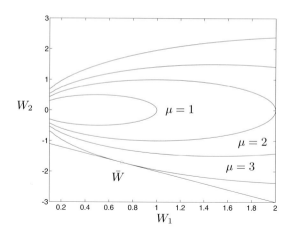

Figure 6.4: Feasible region and objective function

Remark 6.9 (Linear matrix inequalities - LMI) The State feedback problem can also be stated only in terms of Linear matrix inequalities, that is the feasible set is defined by affine constraints only. The importance of this approach is mainly the standard formulation of many optimal control problems.

As we already know, the solution of the State feedback problem depends upon the determination of matrix P_2, the positive semidefinite stabilizing solution of the Riccati equation (6.26). To this purpose, we assume that the pair (A_c, C_{1c}) is observable to guarantee that $P_2 > 0$ and notice that all $P = P'$ solving

$$0 \geq PA_c + A_c'P - PB_2B_2'P + C_{1c}'C_{1c}$$

are such that $0 < P_2 \leq P$. To show this notice that for any P satisfying this inequality, there exists $\bar{C}_{1c}$ such that $\bar{C}_{1c}'\bar{C}_{1c} \geq C_{1c}'C_{1c}$ for which the equality holds. Consequently $F = -B_2'P$ entails $A_c + B_2F$ stable and for all $x_0 \in R^n$ we get (recall that $C_{1c}'D_{12} = 0$)

$$x_0'(P - P_2)x_0 = x_0'Px_0 - \min_{F \in \mathcal{K}_c} \left\| (C_{1c} + D_{12}F)[sI - (A_c + B_2F)]^{-1}x_0 \right\|_2^2$$

$$\geq x_0'Px_0 - \left\| (C_{1c} - D_{12}B_2'P)[sI - (A_c - B_2B_2'P)]^{-1}x_0 \right\|_2^2$$

$$\geq x_0' \left[\int_0^\infty e^{(A_c - B_2B_2'P)'t}(\bar{C}_{1c}'\bar{C}_{1c} - C_{1c}'C_{1c})e^{(A_c - B_2B_2'P)t}dt \right] x_0$$

$$\geq 0$$

and $P = P_2$ is the minimum of trace$[B_1'PB_1]$ over all feasible P. Defining $X = P^{-1}$, the previous inequality can be rewritten as

$$A_cX + XA_c' - B_2B_2' + XC_{1c}'C_{1c}X \leq 0$$

which after use of the Schur complements is equivalent to

$$\mathcal{A}(X) = \begin{bmatrix} A_cX + XA_c' - B_2B_2' & XC_{1c}' \\ C_{1c}X & -I \end{bmatrix} \leq 0$$

Finally, from these calculations we have shown that matrix P_2 is the optimal global solution of the problem

$$\min \left\{ \text{trace}[B_1'X^{-1}B_1] \; : \; X > 0 \,, \; \mathcal{A}(X) \leq 0 \right\}$$

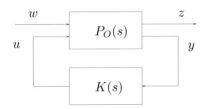

Figure 6.5: The output feedback control system

Clearly, the feasible set as well as the objective function are convex (recall Remark 6.3). Moreover, the feasible set is completely defined by affine functions only.   □

Remark 6.10 (Full information) As discussed before, the solution of the Full information problem is the same as the solution of the State feedback problem. For the sake of completeness the solution of the former is now restated. Consider the Problem 4.1 relative to system (4.10)-(4.14). Under Assumptions 4.1 and 4.2, it has the optimal solution :

a)
$$\min \|T(z, w; s)\|_2 = \sqrt{\mathrm{trace}[R_c \bar{W}]}$$

b)
$$K_{FI}^o(s) = \left[\begin{array}{c|cc} \varnothing & \varnothing & \varnothing \\ \hline \varnothing & \bar{W}_2' \bar{W}_1^{-1} & 0 \end{array}\right]$$

where matrix $\bar{W}$ partitioned as indicated in (6.7) solves the convex programming problem (6.29).   □

Now, our attention is moved to the following situation. Consider the linear system described by

$$\dot{x} = Ax + B_1 w + B_2 u \tag{6.31}$$
$$z = C_1 x + D_{12} u \tag{6.32}$$
$$y = C_2 x + D_{21} w \tag{6.33}$$

where only the partial information of the system state provided by the measured output y is available for feedback. Accordingly to the results of Chapter 4, the controller structure is fixed as being

$$K(s) := \left[\begin{array}{c|c} A + B_2 F + L C_2 & -L \\ \hline F & 0 \end{array}\right] \tag{6.34}$$

which as indicated is completely parametrized by matrices F and L. The rationale behind the choice of this structure is that the optimal solution of the Partial information problem can be obtained by a proper choice of the unknown matrices. With $P_O(s)$ being the transfer function of (6.31)-(6.33), the situation is illustrated by the block-scheme of fig. 6.5. The feedback connection has a state space realization of the form

$$\Sigma_F := \left[\begin{array}{c|c} \tilde{A} & \tilde{B} \\ \hline \tilde{C} & 0 \end{array}\right]$$

where the indicated matrices are

$$\tilde{A} = \begin{bmatrix} A + B_2 F & B_2 F \\ 0 & A + LC_2 \end{bmatrix}$$

and

$$\tilde{B} = \begin{bmatrix} B_1 \\ -B_1 - LD_{21} \end{bmatrix}, \quad \tilde{C} = \begin{bmatrix} C_1 + D_{12}F & D_{12}F \end{bmatrix}$$

being thus clear that to preserve internal stability we have to consider only matrices (F, L) such that $F \in \mathcal{K}_c$ and $L \in \mathcal{K}_f$. All pairs of matrices with this property are called *feasible* and for any feasible pair the transfer function $T(z, w; s)$ belongs to RH_2 and its norm is easily determined by

$$\begin{aligned} \|T(z, w; s)\|_2^2 &= \|\tilde{C}[sI - \tilde{A}]^{-1}\tilde{B}\|_2^2 \\ &= \text{trace}[\tilde{B}'\tilde{P}_o\tilde{B}] \\ &= \text{trace}[\tilde{C}\tilde{P}_r\tilde{C}'] \end{aligned}$$

where $\tilde{P}_o$ and $\tilde{P}_r$ are the unique solutions of the Lyapunov equations associated with the closed-loop system, that is

$$0 = \tilde{A}'\tilde{P}_o + \tilde{P}_o\tilde{A} + \tilde{C}'\tilde{C} \qquad (6.35)$$
$$0 = \tilde{A}\tilde{P}_r + \tilde{P}_r\tilde{A}' + \tilde{B}\tilde{B}' \qquad (6.36)$$

For all feasible pairs (F, L) the above equations are always solvable thanks to the stability of matrix $\tilde{A}$. As far as stability is concerned, the constraints on matrices F and L are completely decoupled. So, the control problem to be dealt with can be formulated as follows :

Problem 6.2 (Output feedback problem in RH_2) *Find a pair of feasible matrices (F, L) which minimizes $\|T(z, w; s)\|_2$. In other terms, find the global optimal solution of problem*

$$\min\ \left\{\|T(z, w; s)\|_2^2\ :\ F \in \mathcal{K}_c\ ,\quad L \in \mathcal{K}_f\right\}$$

The difficulty to solve this problem stems from the fact that the objective function depends in a very unusual way on the unknown matrices. Fortunately, taking into account that the constraints are decoupled the optimality conditions can be expressed in terms of two separable subproblems, yielding each one, the optimal matrices F and L. It should be clear that the Output feedback problem is a particular case of the Partial information problem introduced in Chapter 4 since the structure of the controller (6.34) has been fixed and completely described by means of only two unknown matrices. As a consequence, the Assumptions 4.6 - 4.9 are again considered, that is i) The pair (A, B_2) is stabilizable and the pair (A, C_2) is detectable, ii) $D_{12}'D_{12} = I$, iii) The eigenvalues of the unobservable and unreachable part of the pairs $[(A - B_2 D_{12}'C_1)(I - D_{12}D_{12}')C_1]$ and $[(A - B_1 D_{21}'C_2), B_1(I - D_{21}'D_{21})]$ respectively, do not lie on the imaginary axis and iv) $D_{21}D_{21}' = I$. Under these assumptions, $\mathcal{K}_c \neq \emptyset$, $\mathcal{K}_f \neq \emptyset$ and the optimal solution of the proposed problem is provided by Theorem 4.3, that is

$$F^o = F_2\ ,\quad L^o = L_2 \qquad (6.37)$$

which defines the optimal controller $K^o(s)$. With the optimal controller, the minimum cost is given by

$$\min \|T(z, w; s)\|_2^2 = \|P_c(s)L_2\|_2^2 + \|C_1 P_f(s)\|_2^2$$

where the transfer functions $P_c(s)$ and $P_f(s)$ are provided by Theorems 4.1 and 4.2 respectively. At this point, the key observation is that these two transfer functions defines completely the optimal solution of the problem under consideration and can be calculated by means of two decoupled problems. Indeed, simple calculations show that

$$P_c(s)L_2 = \left[\begin{array}{c|c} A_c - B_2 B_2' P_2 & L_2 \\ \hline C_{1c} - D_{12} B_2' P_2 & 0 \end{array}\right]$$

$$= \left[\begin{array}{c|c} A + B_2 F_2 & L_2 \\ \hline C_1 + D_{12} F_2 & 0 \end{array}\right]$$

and

$$C_1 P_f(s) = \left[\begin{array}{c|c} A_f - \Pi_2 C_2' C_2 & B_{1f} - \Pi_2 C_2' D_{21} \\ \hline C_1 & 0 \end{array}\right]$$

$$= \left[\begin{array}{c|c} A + L_2 C_2 & B_1 + L_2 D_{21} \\ \hline C_1 & 0 \end{array}\right]$$

which together with the solution of the State feedback problem, allows us to say that matrix $L^o = L_2$ solves the auxiliary state feedback problem

$$\|C_1 P_f(s)\|_2^2 = \min_{L \in \mathcal{K}_f} \|C_1[sI - (A + LC_2)]^{-1}(B_1 + LD_{21})\|_2^2 \qquad (6.38)$$

while having obtained matrix L_2, it is apparent that the remaining unknown matrix F_2 is the optimal solution of another auxiliary state feedback problem

$$\|P_c(s)L_2\|_2^2 = \min_{F \in \mathcal{K}_c} \|(C_1 + D_{12} F)[sI - (A + B_2 F)]^{-1} L_2\|_2^2 \qquad (6.39)$$

The above solution to the Output feedback problem admits a very important interpretation. First it can be decomposed in two decoupled problems. The first one depends only on the system data and can be readily solved. Its solution provides the gain matrix L_2 which is used to define the objective function of the other optimization problem yielding the matrix gain F_2. This decomposition of the Output feedback problem is on the basis of what is called the *Separation Principle* which is valid for many optimal control problems with the controller structure given by (6.34).

Remark 6.11 Following the same lines of Remark 6.7, it is worth noticing that the solution of the Output feedback problem can also be obtained by means of mathematical programming arguments. The key observation is that all feasible controllers are constrained to have the structure (6.34) and the goal is to determine matrices $F \in \mathcal{K}_c$, $L \in \mathcal{K}_f$ and $\tilde{P} \geq 0$, such that

$$\min \left\{ \text{trace}[\tilde{B}' \tilde{P} \tilde{B}] \; : \; 0 = \tilde{A}' \tilde{P} + \tilde{P} \tilde{A} + \tilde{C}' \tilde{C} \right\}$$

Using $\tilde{\Lambda} = \tilde{\Lambda}'$ as the matrix of Lagrange multipliers associated to the equality constraint, the partial derivatives of

$$\mathcal{L}(F, L, \tilde{P}, \tilde{\Lambda}) := \text{trace}[\tilde{B}' \tilde{P} \tilde{B}] + \text{trace}[\tilde{\Lambda}(\tilde{A}' \tilde{P} + \tilde{P} \tilde{A} + \tilde{C}' \tilde{C})]$$

give the necessary conditions for optimality

$$0 = \tilde{A}' \tilde{P} + \tilde{P} \tilde{A} + \tilde{C}' \tilde{C}$$
$$0 = \tilde{A} \tilde{\Lambda} + \tilde{\Lambda} \tilde{A}' + \tilde{B} \tilde{B}'$$
$$0 = \frac{\partial \mathcal{L}}{\partial F}$$
$$0 = \frac{\partial \mathcal{L}}{\partial L}$$

which are simple to be solved from the observation that for $F = F_2$ and $L = L_2$ the solution of the two first conditions are

$$\tilde{P} = \begin{bmatrix} P_2 & 0 \\ 0 & Y_2 \end{bmatrix}, \quad \tilde{\Lambda} = \begin{bmatrix} \Pi_2 + X_2 & -\Pi_2 \\ -\Pi_2 & \Pi_2 \end{bmatrix}$$

where matrices X_2 and Y_2 are solutions to the Lyapunov equations (in the unknown X and Y respectively)

$$0 = (A + B_2 F_2)X + X(A + B_2 F_2)' + L_2 L_2'$$
$$0 = (A + L_2 C_2)'Y + Y(A + L_2 C_2) + F_2 F_2'$$

At the same point under consideration, the last two partial derivatives of the Lagrangian with respect to F and L are given by

$$\frac{\partial \mathcal{L}}{\partial F} = (F - F_2)X_2 , \quad \frac{\partial \mathcal{L}}{\partial L} = Y_2(L - L_2)$$

showing that the pair (F_2, L_2) satisfies the necessary conditions for optimality indeed. The importance of the calculations introduced here is to make explicit the structure of the matrix variables $\tilde{P}$ and $\tilde{\Lambda}$ at the optimal solution of the Output feedback control problem. This result will be of great importance in the sequel.   □

We already know that the Output feedback problem can be converted in two decoupled State feedback problems, hence it is not convex. Defining the extended matrices

$$R_f := \begin{bmatrix} B_1 \\ D_{21} \end{bmatrix} \begin{bmatrix} B_1' & D_{21}' \end{bmatrix}, \quad Q_f := \begin{bmatrix} C_1'C_1 & 0 \\ 0 & 0 \end{bmatrix} \tag{6.40}$$

both problems (6.38) and (6.39) can be converted in two convex programming problems as is indicated in the next theorem.

Theorem 6.4 (Output feedback) *The global optimal solution of the Output feedback problem can be calculated by means of the following procedure involving convex programming problems only.*

a) *Solve the convex programming problem*

$$\bar{V} := \mathrm{argmin}\, \{\mathrm{trace}[R_f V]\ :\ V \in \mathcal{C}_f\}, \tag{6.41}$$

b) *Redefine the extended matrix*

$$Q_c := \begin{bmatrix} \bar{V}_1^{-1}\bar{V}_2\bar{V}_2'\bar{V}_1^{-1} & 0 \\ 0 & 0 \end{bmatrix} \tag{6.42}$$

and solve the convex programming problem

$$\bar{W} := \mathrm{argmin}\, \{\mathrm{trace}[R_c W]\ :\ W \in \mathcal{C}_c\} \tag{6.43}$$

Then, matrices $F = \bar{W}_2'\bar{W}_1^{-1}$ and $L = \bar{V}_1^{-1}\bar{V}_2$ solve Problem 6.2.

Proof It follows from the decomposition of Problem 6.2 in terms of (6.38) and (6.39). The result of Theorem 6.3 is used to convert them to equivalent convex programming problems.   □

Remark 6.12 Remark 6.6 applies to both optimization problems defined in Theorem 6.4. If matrices $C_1'C_1$ and LL' are not strictly positive definite then C_1' and $L = \bar{V}_1^{-1}\bar{V}_2$ have to be replaced, with no loss of generality, by $[C_1' \ \sqrt{\epsilon}I]$ and $[L \ \sqrt{\epsilon}I]$ respectively where ϵ is an arbitrarily small positive parameter. □

Remark 6.13 It is important to discuss under which condition problem (6.41) can be effectively solved by a chosen convex programming procedure. As discussed in Remark 6.8, we have to impose conditions assuring that the set

$$\mathcal{C}_{f\mu} := \mathcal{C}_f \ \cap \ \{V \ : \ \text{trace}[R_f V] \le \mu\}$$

is bounded for all μ such that $\text{trace}[R_f \bar{V}] < \mu < \infty$. Following the same reasoning, this occurs provided the pair $(-A_f, B_{1f})$ is stabilizable. Therefore to have both sets $\mathcal{C}_{c\mu}$ and $\mathcal{C}_{f\mu}$ bounded, the assumption iii) of the Output feedback problem has to be changed to iii) The eigenvalues of the unobservable and unreachable part of the pairs $[(A - B_2 D_{12}' C_1)(I - D_{12} D_{12}')C_1]$ and $[(A - B_1 D_{21}' C_2), B_1(I - D_{21}' D_{21})]$ respectively, do not lie on the left part (including the imaginary axis) of the complex plane. □

Remark 6.14 It is simple to be verified that in the State feedback problem, the set of all $F \in \mathcal{K}_c$ such that $\|T(z, w; s)\|_2$ is bounded above by a given positive scalar γ can be generated by (recall the definition of the convex set $\mathcal{C}_{c\mu}$ in Remark 6.8)

$$F = W_2'W_1^{-1} \ , \quad W \in \mathcal{C}_c \ \cap \ \left\{W \ : \ \text{trace}[R_c W] \le \gamma^2\right\}$$

Our purpose now is to generalize this result to the Output feedback problem in the case that the matrix gain L is fixed and equals the optimal value $L = L_2$. For any $F \in \mathcal{K}_c$ the transfer function $T(z, w; s)$ belongs to RH_2 and

$$\|T(z, w; s)\|_2^2 = \text{trace}[\tilde{C}\tilde{P}_r \tilde{C}']$$

where $\tilde{P}_r$ is the solution of the linear equation (6.36). Simple algebraic manipulations show that

$$\tilde{P}_r = \begin{bmatrix} \Pi_2 + X & -\Pi_2 \\ -\Pi_2 & \Pi_2 \end{bmatrix}$$

where matrix X is the solution of

$$0 = (A + B_2 F)X + X(A + B_2 F)' + L_2 L_2'$$

It is very interesting to compare this result to the ones given in Remark 6.11. Matrix $\tilde{P}_r$ has the same structure as matrix $\tilde{\Lambda}$ and the only difference, due to have considered now $F \in \mathcal{K}_c$ arbitrary, is restricted to the definition of matrix X above which equals matrix X_2 provided $F = F_2$. Using this we also have

$$\|T(z, w; s)\|_2^2 = \text{trace}[C_1 \Pi_2 C_1'] + \text{trace}[(C_1 + D_{12}F)X(C_1 + D_{12}F)']$$
$$= \text{trace}[C_1 \Pi_2 C_1'] + \|(C_1 + D_{12}F)[sI - (A + B_2 F)]^{-1}L_2 D_{21}\|_2^2$$

Finally, taking into account Theorem 6.4 it is clear that

$$F = W_2'W_1^{-1} \ , \quad W \in \mathcal{C}_c \ \cap \ \left\{W \ : \ \text{trace}[R_c W] \le \gamma^2 - \text{trace}[C_1 \Pi_2 C_1']\right\}$$

generates all matrices F such that, in the Output feedback problem, the internal stability is assured while $\|T(z, w; s)\|_2$ is bounded above by a given positive scalar γ. □

Remark 6.15 (Output estimation) The Output estimation problem as defined in the Chapter 4 can also be solved by means of convex programming tools. The main observation is that Assumptions 4.3 - 4.5 of the Output estimation problem imply the assumptions of the Output feedback problem are all verified with $D_{12} = I$. Furthermore, the assumption $A_c = A - B_2 C_1$ stable yields $P_2 = 0$ and the associated stabilizing matrix $F_2 = -C_1$. Then, Problem 4.1 relative to system (4.58)-(4.60) has the optimal solution :

a)
$$\min \|T(z, w; s)\|_2 = \sqrt{\text{trace}[R_f \bar{V}]}$$

b)
$$K_{OE}^o(s) = \left[\begin{array}{c|c} A - B_2 C_1 + \bar{V}_1^{-1} \bar{V}_2 C_2 & -\bar{V}_1^{-1} \bar{V}_2 \\ \hline - C_1 & 0 \end{array} \right]$$

where matrix $\bar{V}$ partitioned as indicated in (6.17) solves the convex programming problem (6.41). $\square$

Remark 6.16 (Partial information) For completeness we give here the solution of the Partial information problem. The additional result is the value of the objective function written in terms of matrices provided in Theorem 6.4. Consider the system (4.92)-(4.94) and Assumptions 4.6 - 4.9 then Problem 4.1 has the following solution :

a)
$$\min \|T(z, w; s)\|_2^2 = \text{trace}[R_f \bar{V}] + \text{trace}[R_c \bar{W}]$$

b)
$$K^o(s) = \left[\begin{array}{c|c} A + B_2 \bar{W}_2' \bar{W}_1^{-1} + \bar{V}_1^{-1} \bar{V}_2 C_2 & -\bar{V}_1^{-1} \bar{V}_2 \\ \hline \bar{W}_2' \bar{W}_1^{-1} & 0 \end{array} \right]$$

where matrices $\bar{W}$ and $\bar{V}$ solve the convex optimization problems introduced in Theorem 6.4. Finally, it is important to notice again that these matrices can be determined separately. $\square$

6.2.3 RH_∞ control design

This section presents the convex analysis counterpart of Chapter 5. The State feedback and Output feedback control design problems are again addressed but in a slight different setting. Indeed, not only the set of all controllers which impose to the plant a certain RH_∞ norm level γ is obtained but we also address the problem of determining the controller such that γ is minimized. Particular attention must be payed to the fact that in many instances, on the contrary of what has been done in Chapter 5, the constraints involving RH_∞ norms are not taken strictly.

Let the system under consideration be defined as

$$\dot{x} = Ax + B_1 w + B_2 u \tag{6.44}$$
$$z = C_1 x + D_{12} u \tag{6.45}$$
$$y = x \tag{6.46}$$

and illustrated in fig. 6.2 for which the controller is simply given by $u = Fx$ where matrix $F \in \mathcal{K}_c$ is such that $T(z, w; s)$ belongs to RH_∞. For all $F \in \mathcal{K}_c$, those that additionally satisfies the constraint $\|T(z, w; s)\|_\infty < \gamma$ with γ being a positive scalar are characterized from the result of Theorem 2.14 which states that there exists a symmetric and positive semidefinite stabilizing solution of the algebraic Riccati equation (in the unknown S)

$$0 = SA_{c\infty} + A_{c\infty}' S + \gamma^{-2} SB_1 B_1' S + C_{c\infty}' C_{c\infty} \tag{6.47}$$

with $A_{c\infty} := A + B_2 F$ and $C_{c\infty} := C_1 + D_{12} F$. An equivalent condition is obtained from Theorem 2.15 which requires the existence of a symmetric and positive definite feasible solution to the Riccati inequality

$$SA_{c\infty} + A_{c\infty}' S + \gamma^{-2} SB_1 B_1' S + C_{c\infty}' C_{c\infty} < 0 \tag{6.48}$$

in this case no further stabilizing condition is needed. However, it is a simple matter to see (recall Remark 2.22) that $\|T'(z, w; s)\|_\infty = \|T(z, w; s)\|_\infty$ meaning that the same set of controllers can also be characterized by the existence of a symmetric and positive semidefinite stabilizing solution of the algebraic Riccati equation (in the unknown P)

$$0 = A_{c\infty}P + PA'_{c\infty} + \gamma^{-2}PC'_{c\infty}C_{c\infty}P + B_1 B'_1 \tag{6.49}$$

or equivalently from Theorem 2.15 the existence of a symmetric and positive definite feasible solution to the Riccati inequality

$$A_{c\infty}P + PA'_{c\infty} + \gamma^{-2}PC'_{c\infty}C_{c\infty}P + B_1 B'_1 < 0 \tag{6.50}$$

As well as, the set of all $F \in \mathcal{K}_c$ such that additionally $\|T(z, w; s)\|_\infty \leq \gamma$ is completely characterized from the results of Theorem 2.16 which requires the existence of symmetric and positive definite feasible solutions to the nonstrict versions of the Riccati inequalities (6.48) and (6.50). Moreover, in the above Riccati inequalities each feasible solution are related one to the other by $P = \gamma^2 S^{-1}$. Although equivalent, to our purpose, inequality (6.50) or its nonstrict version are more convenient to handle.

With γ being an arbitrary and positive scalar (not fixed *a priori*), the pair (F, γ) is called *feasible* whenever $(F, \gamma) \in \mathcal{K}_{\gamma c}$ where

$$\mathcal{K}_{\gamma c} := \{(F, \gamma) : F \in \mathcal{K}_c , \ \gamma > 0 , \ \|T(z, w; s)\|_\infty \leq \gamma\} \tag{6.51}$$

which makes clear that the set $\mathcal{K}_{\gamma c}$ is a subset of $\mathcal{K}_c$. Based on our previous discussion, it is now a fact that the above defined set is not convex. Fortunately it can be converted to an equivalent convex set as indicated in the next theorem.

Theorem 6.5 *Assume $B_1 B'_1$ is a positive definite matrix and consider the set*

$$\mathcal{C}_{\gamma c} := \{(W, \mu) \ : \ W \geq 0 , \ \mu > 0 , \ v'\Theta_{\gamma c}(W, \mu)v \leq 0 , \ \forall v \in \mathcal{N}_c\} \tag{6.52}$$

where $\Theta_{\gamma c}(W, \mu) := \Theta_c(W) + \mu^{-1}W R_c W$. The following hold

a) $\mathcal{C}_{\gamma c}$ is a convex set.

b) Each $(W, \mu) \in \mathcal{C}_{\gamma c}$ is such that $W_1 > 0$.

c) $\mathcal{K}_{\gamma c} = \{(W'_2 W_1^{-1}, \sqrt{\mu}) \ : \ (W, \mu) \in \mathcal{C}_{\gamma c}\}$.

Proof Notice that function $\Theta_{\gamma c}(W, \mu)$ is not affine but it can be converted to an affine one by using Schur complements.

Point a) It is clear that we only need to prove that the set of all (W, μ) such that $v'\Theta_{\gamma c}(W, \mu)v \leq 0 , \ \forall v \in \mathcal{N}_c$ is convex. Since $R_c \geq 0$ and all $v \in \mathcal{N}_c$ can be obtained from $v = U_c x$ where $U_c = [I \ \ 0]'$ and $\|x\| = 1$, then this set is equivalently described as

$$0 \geq U'_c \Theta_c(W)U_c + \mu^{-1}U'_c W R_c^{1/2} R_c^{1/2} W U_c$$

which for $\mu > 0$, after using of the Schur complement formula becomes

$$\mathcal{A}(W, \mu) = \begin{bmatrix} U'_c \Theta_c(W)U_c & U'_c W R_c^{1/2} \\ R_c^{1/2} W U_c & -\mu I \end{bmatrix} \leq 0$$

Since $\mathcal{A}(W, \mu)$ is affine, the convexity of $\mathcal{C}_{\gamma c}$ follows directly from Lemma 6.1.

Point b) Since for each $(W, \mu) \in C_{\gamma c}$ the matrix W is positive semidefinite, if there exists a nonzero vector $x \in R^n$ such that $W_1 x = 0$ then for the same vector $W_2' x = 0$, yielding $W v = 0$ and obviously

$$0 \geq v' \Theta_{\gamma c}(W, \mu) v = \|B_1' x\|^2$$

which is an impossibility due to the fact that, by assumption, $B_1 B_1' > 0$.

Point c) Suppose $\mathcal{K}_{\gamma c} \neq \emptyset$ and the pair (F, γ) is feasible for the set $\mathcal{K}_{\gamma c}$. In this case, there exists a symmetric and positive definite matrix P satisfying the Riccati inequality (recall Theorem 2.16)

$$A_{c\infty} P + P A_{c\infty}' + \gamma^{-2} P C_{c\infty}' C_{c\infty} P + B_1 B_1' \leq 0$$

Choosing $\mu = \gamma^2$ and

$$W = \begin{bmatrix} P & PF' \\ FP & FPF' \end{bmatrix}$$

it is a simple matter to verify that $(W, \mu) \in C_{\gamma c}$ and $W_2' W_1^{-1} = F$. Conversely, assume $C_{\gamma c} \neq \emptyset$ and consider any $(W, \mu) \in C_{\gamma c}$, for all $v \in \mathcal{N}_c$ we get

$$
\begin{aligned}
0 &\geq v' \Theta_{\gamma c}(W, \mu) v \\
&\geq x' \left[(A + B_2 W_2' W_1^{-1}) W_1 + W_1 (A + B_2 W_2' W_1^{-1})' + \right. \\
&\quad \left. + \mu^{-1} W_1 (C_1 + D_{12} W_2' W_1^{-1})'(C_1 + D_{12} W_2' W_1^{-1}) W_1 + B_1 B_1' \right] x
\end{aligned}
\tag{6.53}
$$

where the factorization is possible since from point b) it has been already proven that $W_1 > 0$. Choosing $P = W_1$, $F = W_2' W_1^{-1}$ and $\gamma = \sqrt{\mu}$, this inequality assures that $F \in \mathcal{K}_c$ and the existence of a symmetric and positive definite matrix $P = W_1 > 0$ satisfying the Riccati inequality

$$A_{c\infty} P + P A_{c\infty}' + \gamma^{-2} P C_{c\infty}' C_{c\infty} P + B_1 B_1' \leq 0$$

Hence, using Theorem 2.16 it is verified that $\|T(z, w; s)\|_\infty \leq \gamma$ which, from (6.51) implies that the pair $(F, \gamma) = (W_2' W_1^{-1}, \sqrt{\mu}) \in \mathcal{K}_{\gamma c}$. From the above, if one set $\mathcal{K}_{\gamma c}$ or $C_{\gamma c}$ is empty both are empty. The proof is then complete. □

The joint convexity of the set $C_{\gamma c}$ with respect to both variables (W, μ) is of great importance. To get some insight on this fact let us consider the problem of determining the feasible pair (F, γ) which solves

$$\min \, \{\gamma \, : \, (F, \gamma) \in \mathcal{K}_{\gamma c}\} \tag{6.54}$$

This nonconvex problem has not been directly addressed in Chapter 5 although it is possible, in principle, to get its solution iteratively with the results provided in Theorem 5.2 (recall the forthcoming Remark 6.20). It is equivalent to the convex programming problem

$$\min \, \{\mu \, : \, (W, \mu) \in C_{\gamma c}\} \tag{6.55}$$

Indeed, at the optimal solution the value of μ is the minimum which preserves feasibility implying that for $\gamma = \sqrt{\mu}$ there exists a matrix F such that $(F, \gamma) \in \mathcal{K}_{\gamma c}$.

Let us now introduce the problem to be dealt with in the sequel. It resembles the Full information problem treated in Chapter 5. To this end it is first needed to put in evidence only the feasible pairs (F, γ) such that $\|T(z, w; s)\|_\infty < \gamma$, that is all $(F, \gamma) \in \text{int } \mathcal{K}_{\gamma c}$, where

$$\text{int } \mathcal{K}_{\gamma c} := \{(F, \gamma) : F \in \mathcal{K}_c \, , \, \gamma > 0 \, , \, \|T(z, w; s)\|_\infty < \gamma\}$$

Problem 6.3 (State feedback problem in RH_∞) *Given a scalar $\gamma > 0$, determine the conditions for the existence of a state feedback matrix F such that the pair (F, γ) is strictly feasible, that is $(F, \gamma) \in \text{int } \mathcal{K}_{\gamma c}$.*

The Full information problem as stated and solved in Theorem 5.2 is much more general than its State feedback version. However, the last captures the essential features of the former in the sense that the existence of a solution to the Full information problem, under Assumptions 5.1 - 5.2, is assured provided there exists a symmetric, positive semidefinite and stabilizing solution P_∞ of the Riccati equation (in the unknown P)

$$0 = PA_c + A_c'P - P(B_2 B_2' - \gamma^{-2} B_1 B_1')P + C_{1c}'C_{1c} \qquad (6.56)$$

As in the proof of Theorem 5.2 let us proceed by assuming that i) The pair $[(A - B_2 D_{12}'C_1), (I - D_{12}D_{12}')C_1]$ is observable and the pair (A, B_2) is stabilizable and ii) $D_{12}'D_{12} = I$.

Theorem 6.6 (State feedback) *Assume matrix $B_1 B_1'$ is positive definite and let γ a positive scalar be given. There exists a strictly feasible pair $(F, \gamma) \in \text{int } \mathcal{K}_{\gamma c}$ if and only if there exists W such that $(W, \gamma^2) \in \text{int } \mathcal{C}_{\gamma c}$, where*

$$\text{int } \mathcal{C}_{\gamma c} := \{(W, \mu) : W \geq 0, \ \mu > 0, \ v'\Theta_{\gamma c}(W, \mu)v < 0, \ \forall v \in \mathcal{N}_c\}$$

Proof Under the assumptions made, the State feedback problem is solvable if and only if there exists $P = P_\infty > 0$ solution of the Riccati equation (6.56). However, the existence of a stabilizing state feedback control such that the transfer function norm $\|T(z, w; s)\|_\infty$ is strictly less than γ implies that there exists $0 < \bar{\gamma} < \gamma$ for which the State feedback problem is also solvable. Consequently (recall Theorem 5.2) we can say that there exists a strictly feasible pair (F, γ) if and only if there exists a positive definite and stabilizing solution $\bar{P}_\infty$ of the Riccati equation (in the unknown P)

$$0 = PA_c + A_c'P - P(B_2 B_2' - \bar{\gamma}^{-2} B_1 B_1')P + C_{1c}'C_{1c}$$

Hence, the necessity is proved if we are able to calculate a matrix W such that $(W, \gamma^2) \in \text{int } \mathcal{C}_{\gamma c}$. Such a matrix is

$$W_\infty := \gamma^2 \begin{bmatrix} \bar{P}_\infty^{-1} & \bar{P}_\infty^{-1}\bar{F}_\infty' \\ \bar{F}_\infty \bar{P}_\infty^{-1} & \bar{F}_\infty \bar{P}_\infty^{-1}\bar{F}_\infty' \end{bmatrix} \geq 0$$

where $\bar{F}_\infty = -B_2'\bar{P}_\infty - D_{12}'C_1$. Actually, it suffices to observe that, from the previous Riccati equation, the inequality

$$v'\Theta_{\gamma c}(W_\infty, \gamma^2)v = (\gamma\bar{P}_\infty^{-1}x)' \big[A_c'\bar{P}_\infty + \bar{P}_\infty A_c -$$
$$- \bar{P}_\infty(B_2 B_2' - \gamma^{-2}B_1 B_1')\bar{P}_\infty + C_{1c}'C_{1c}\big](\gamma\bar{P}_\infty^{-1}x)$$
$$= \big[1 - (\gamma/\bar{\gamma})^2\big]\|B_1'x\|^2 < 0$$

holds for all $v \in \mathcal{N}_c$. The converse follows readily from Theorem 2.15. □

It is interesting to analyze the optimal solution of problem (6.55) on the light of this result. Suppose the global optimum of (6.55) has been calculated yielding $\mu = \mu^o$. Then, $F = W_2'W_1^{-1}$ and $\gamma^o = \sqrt{\mu^o}$ are such that $(F, \gamma^o) \in \mathcal{K}_{\gamma c}$. On the other hand, Theorem 6.6 assures that the Riccati equation (6.56) is solvable for any $\gamma > \gamma^o$ providing thus $F_\infty = -B_2'P_\infty - D_{12}'C_1$ which is also strictly feasible, that is $(F_\infty, \gamma) \in \text{int } \mathcal{K}_{\gamma c}$.

Remark 6.17 The discussion made in Remark 6.6 concerning the assumption on matrix B_1, namely, $B_1 B_1' > 0$ is still valid in the present case. Indeed, assume $B_1 B_1' \geq 0$ and define $\bar{B}_1 := [B_1 \ \sqrt{\epsilon}I]$ with $\epsilon > 0$ arbitrarily small. Clearly we now have $\bar{B}_1 \bar{B}_1' > 0$ and for all $F \in \mathcal{K}_c$, it is true that the transfer function $\bar{T}(z, w; s) := C_{c\infty}(sI - A_{c\infty})^{-1}\bar{B}_1$ can be factorized as

$$\bar{T}(z, w; -j\omega)\bar{T}'(z, w; j\omega) = T(z, w; -j\omega)T'(z, w; j\omega) + \epsilon G(-j\omega)G'(j\omega) \ , \quad \forall \ \omega \in R$$

where $G(s) := C_{c\infty}(sI - A_{c\infty})^{-1}$. The consequence is that

$$\|T(z, w; s)\|_\infty \leq \|\bar{T}(z, w; s)\|_\infty$$

and also

$$\|\bar{T}(z, w; s)\|_\infty^2 = \sup_\omega \|\bar{T}'(z, w; j\omega)\|^2$$
$$\leq \sup_\omega \|T'(z, w; j\omega)\|^2 + \sup_\omega \|G(j\omega)\|^2$$
$$\leq \|T(z, w; s)\|_\infty^2 + \epsilon\|C_{c\infty}(sI - A_{c\infty})^{-1}\|_\infty^2$$

which means that $\|\bar{T}(z, w; s)\|_\infty \leq \gamma$ implies $\|T(z, w; s)\|_\infty \leq \gamma$ and both norms differs one from the other by an amount of order ϵ. $\qquad \square$

Remark 6.18 Based on the result of the previous theorem, it is readily seen that if we want to involve the set int $\mathcal{C}_{\gamma c}$ in an optimization problem them it is possible to work with the approximation

$$\text{int } \mathcal{C}_{\gamma c} = \left\{(W, \mu) \ : \ W \geq 0 \ , \ \ \mu > 0 \ , \ \ v'\Theta_{\gamma c}(W, \mu)v \leq -\epsilon \ , \ \forall v \in \mathcal{N}_c\right\}$$

where the scalar $\epsilon > 0$ must be taken sufficiently small. Notice that this is exactly equivalent to replace matrix B_1 by matrix $[B_1 \ \sqrt{\epsilon}I]$. The obvious advantage is that in doing this the above approximation is always a closed convex set. $\qquad \square$

Remark 6.19 Theorem 5.2 makes possible the existence of a state feedback controller $u = F_\infty x$ called *central controller* which solves the Full information problem under the assumptions i) The pair $[(A - B_2 D_{12}'C_1), (I - D_{12}D_{12}')C_1]$ is detectable and the pair (A, B_2) is stabilizable and ii) $D_{12}'D_{12} = I$. Using mathematical programming arguments only, it can be obtained as follows. Given $\gamma > 0$, determine $F \in \mathcal{K}_c$ and $P = P' \geq 0$ such that

$$\min\left\{\text{trace}[B_1'PB_1] \ : \ 0 = A_{c\infty}'P + PA_{c\infty} + \gamma^{-2}PB_1B_1'P + C_{c\infty}'C_{c\infty}\right\}$$

Writing the associated Lagrangian (recall Remark 6.7) with $\Lambda = \Lambda'$ being the matrix of Lagrange multipliers associated to the equality constraint, the necessary conditions for optimality are

$$0 = A_{c\infty}'P + PA_{c\infty} + \gamma^{-2}PB_1B_1'P + C_{c\infty}'C_{c\infty}$$
$$0 = (A_{c\infty} + \gamma^{-2}B_1B_1'P)\Lambda + \Lambda(A_{c\infty} + \gamma^{-2}B_1B_1'P)' + B_1B_1'$$
$$0 = (F + D_{12}'C_1 + B_2'P)\Lambda$$

Restricting ourselves to the solutions such that matrix $A_{c\infty} + \gamma^{-2}B_1B_1'P$ is stable then with $B_1B_1' > 0$ we necessarily have $\Lambda > 0$. Solving the last equation for F, the first one yields $F = F_\infty = -B_2'P_\infty - D_{12}'C_1$ where, under the previous assumptions, P_∞ is the symmetric, positive semidefinite and stabilizing solution of the Riccati equation (in the unknown P)

$$0 = A_c'P + PA_c - P(B_2B_2' - \gamma^{-2}B_1B_1')P + C_{1c}'C_{1c}$$

Moreover, since $A_{cc} = A + B_2 F + \gamma^{-2} B_1 B_1' P$ calculated for $F = F_\infty$ and $P = P_\infty$ is stable then $B_1 B_1' > 0$ assures that

$$\Lambda_\infty := \int_0^\infty e^{A_{cc}t} B_1 B_1' e^{A_{cc}'t} dt > 0$$

and consequently $(F_\infty, P_\infty, \Lambda_\infty)$ is the unique solution of the necessary conditions for optimality. It remains to verify that it is actually the global solution of the proposed problem . To show this, let $F \in \mathcal{K}_{\gamma c}$ be arbitrary which is the same to say that only the first necessary condition is satisfied for some positive semidefinite and stabilizing matrix P. Some simple algebraic manipulations gives

$$0 = A_{cc}'(P - P_\infty) + (P - P_\infty)A_{cc} -$$
$$-\gamma^{-2}(P - P_\infty)B_1 B_1'(P - P_\infty) + (F - F_\infty)'(F - F_\infty)$$

The stability of matrix A_{cc} together with the result of Lemma C.4 imply that there exists $P - P_\infty \geq 0$ solution to this Riccati equation so that the objective function attains the global minimum at $P = P_\infty$. The problem introduced here deserves an additional interpretation to be used in the analysis of mixed RH_2/RH_∞ optimal control problems. For any $F \in \mathcal{K}_{\gamma c}$ it is simple to verify that

$$\mathrm{trace}[B_1' P B_1] = \mathrm{trace}\left[\int_0^\infty B_1' e^{A_{c\infty}'t}(\gamma^{-2} P B_1 B_1' P + C_{c\infty}' C_{c\infty})e^{A_{c\infty}t} B_1 dt\right]$$

$$\geq \mathrm{trace}\left[\int_0^\infty B_1' e^{A_{c\infty}'t} C_{c\infty}' C_{c\infty} e^{A_{c\infty}t} B_1 dt\right]$$

$$\geq \|T(z, w; s)\|_2^2$$

which enables us to say that the central controller actually minimizes an upper bound (and so provides a sub-optimal solution) to the problem

$$\inf\left\{\|T(z, w; s)\|_2^2 \ : \ F \in \mathcal{K}_c \ , \ \|T(z, w; s)\|_\infty < \gamma\right\}$$

which is a mixed RH_2/RH_∞ optimal control design problem to be deeply studied in the forthcoming section. For the moment notice that when $\gamma \to \infty$ the mixed problem tends to the State feedback problem in RH_2. The necessary optimality conditions reduce to those of Remark 6.7. □

Remark 6.20 The central controller can also be used to determine an approximate solution to the problem

$$\gamma_{opt} = \inf \ \{\gamma \ : \ (F, \gamma) \in \mathrm{int} \ \mathcal{K}_{\gamma c}\}$$

The basic algorithm can be stated as follows. For a given $\gamma_k > 0$ suppose that $P_\infty \geq 0$ is stabilizing. To make clear the dependence on γ_k of the transfer function $T(z, w; s)$ when the central controller is used, it is denoted as $T_k(z, w; s)$.

1) Choose $\gamma_0 > \gamma_{opt}$ (possibly $\gamma_0 = \infty$).

2) Iterate until convergence $\gamma_{k+1} = \|T_k(z, w; s)\|_\infty$.

It is immediate to see that this procedure approaches to γ_{opt} as k goes to infinite since $\gamma_{k+1} = \|T_k(z, w; s)\|_\infty < \gamma_k$, however the speed of convergence may be very slow when γ_k becomes close to γ_{opt}. This feature is numerically illustrated with the system of Example 5.2 for $\Omega = 0$. The system is

$$\dot{x} = Ax + B_1 w_1 + B_2 u$$
$$z = C_1 x + u$$
$$y = x$$

where

$$A = \begin{bmatrix} 0 & 1 \\ -1 & -1 \end{bmatrix} , \quad B_1 = \begin{bmatrix} 1 \\ 0 \end{bmatrix} , \quad B_2 = \begin{bmatrix} 0 \\ 1 \end{bmatrix} , \quad C_1 = \begin{bmatrix} 0 & -2 \end{bmatrix}$$

It has been calculated that $1.61 < \gamma_{opt} < 1.62$ while the first six iterations of the previous algorithm furnish

$$\{\gamma_k\}_{k=0}^6 = \{\infty, 2.30, 2.00, 1.88, 1.82, 1.78, 1.76\}$$

The importance of problem (6.55) is now apparent. It is jointly convex on the variables (W, μ) and so can be solved more efficiently. $\hspace{1cm}\square$

Remark 6.21 It is numerically attractive to get *a priori* bounds such that $\gamma_{min} \leq \gamma_{opt} \leq \gamma_{max}$. Indeed, a possible value to γ_{max} is simply determined from any $(F, \gamma_{max}) \in \mathcal{K}_{\gamma c}$, as for instance $F = F_2$ yields

$$\gamma_{max} := \|T(z, w; s)\|_\infty$$

The determination of γ_{min} is much more involved. It follows from the factorization

$$M_c W + W M_c' + \mu^{-1} W R_{c\epsilon} W + Q_c = Q_c - \mu M_c R_{c\epsilon}^{-1} M_c' +$$
$$+ \mu^{-1} (\mu M_c R_{c\epsilon}^{-1} + W) R_{c\epsilon} (\mu M_c R_{c\epsilon}^{-1} + W)'$$
$$\geq Q_c - \mu M_c R_{c\epsilon}^{-1} M_c'$$

where $R_{c\epsilon} := R_c + \epsilon I$ and $\epsilon > 0$. The left hand side of this inequality goes to $\Theta_{\gamma c}(W, \mu)$ as ϵ goes to zero then if we set

$$\gamma_m^2 := \lim_{\epsilon \to 0} \min \{\mu : v'(Q_c - \mu M_c R_{c\epsilon}^{-1} M_c')v \leq 0 , \quad \forall v \in \mathcal{N}_c\}$$

for any $\mu < \gamma_m^2$ there exists a vector $\bar{v} \in \mathcal{N}_c$ such that $\bar{v}'(Q_c - \mu M_c R_{c\epsilon}^{-1} M_c')\bar{v} > 0$ and so $\bar{v}'\Theta_{\gamma c}(W, \mu)\bar{v} > 0$ enabling the final conclusion that $(W, \mu) \notin \mathcal{C}_{\gamma c}$ which from Theorem 6.5 is the same to say that there is no F matrix such that the pair $(F, \gamma) \in \mathcal{K}_{\gamma c}$ for all $\gamma < \gamma_m$. Unfortunately, the calculation involved to get γ_m is prohibitive. However, assuming B_1 of full column rank, restricting $v = [x' \ 0]' \in \mathcal{N}_c$ to $x = B_1(B_1'B_1)^{-1}\xi$ and defining the matrix $V := (B_1'B_1)^{-1}B_1'[A \ B_2]$ we get

$$\gamma_m^2 \geq \lim_{\epsilon \to 0} \min \{\mu : \xi'(I - \mu V R_{c\epsilon}^{-1} V')\xi \leq 0 , \quad \forall \|\xi\| = 1\}$$
$$\geq \lim_{\epsilon \to 0} \min \{\mu : \mu^{-1} \leq \xi' V R_{c\epsilon}^{-1} V'\xi , \quad \forall \|\xi\| = 1\}$$
$$\geq \lim_{\epsilon \to 0} \lambda_{min}^{-1}(V R_{c\epsilon}^{-1} V') := \gamma_{min}^2$$

This limit can be easily calculated numerically and is a valid lower bound to γ_{opt} since we have shown $\mu = \gamma_{min}^2 < \gamma_m^2$ is always infeasible. $\hspace{1cm}\square$

Remark 6.22 The numerical solution of problem (6.55) by means of convex programming algorithms depends on the boundedness of the feasible set $\mathcal{C}_{\gamma c}$. Following the same lines of Remark 6.8, we have to analyze the conditions under which, for some $\mu > 0$, there exists $(W, \mu) \in \mathcal{C}_{\gamma c}$ and $\tilde{W} \neq 0$ such that $W + \lambda \tilde{W} \in \mathcal{C}_{\gamma c}$ with $\lambda > 0$ arbitrarily large. Unfortunately, the only constraint involving the sub-block W_3 imposed by $W \in \mathcal{C}_{\gamma c}$ is $W_3 \geq W_2' W_1^{-1} W_2$ and so the above conditions are always satisfied for the matrix (other possibilities exist)

$$\tilde{W} = \begin{bmatrix} 0 & 0 \\ 0 & I \end{bmatrix}$$

consequently, the set $\mathcal{C}_{\gamma c}$ is not bounded even though $\mu < \infty$. In order to circumvent this difficulty we first notice that

$$\mathcal{C}_{\gamma c} \subseteq \{\mu : \mu > 0\} \times \mathcal{C}_c$$

where with no loss of generality (recall Remark 6.21) the unbounded set $\mu > 0$ can be replaced by the bounded one $\gamma_{min}^2 \leq \mu \leq \gamma_{max}^2$. Second, we already know that the set $\mathcal{C}_c$ is not bounded but the set

$$\mathcal{C}_{c\beta} := \mathcal{C}_c \cap \{W : \text{trace}[R_c W] \leq \beta\}$$

is bounded for all finite β (recall Remark 6.8) provided the pair $(-A_c, C_{1c})$ is detectable. Under this assumption, the conclusion is that we can always solve an approximate version of problem (6.55), namely

$$\min \{\mu : (W, \mu) \in \mathcal{C}_{\gamma c} , \quad \text{trace}[R_c W] \leq \beta\}$$

which has a bounded feasible domain and the optimal solution is arbitrarily close to the optimal solution of problem (6.55) provided the parameter β is chosen appropriately large. This problem has another interesting and important interpretation for robust control design to be discussed in the next section. □

Remark 6.23 (Linear matrix inequalities - LMI) As in the RH_2 design (recall Remark 6.9), the State feedback problem in RH_∞ can also be stated only in terms of Linear matrix inequalities. Assuming the pair (A_c, C_{1c}) is observable, then any $P > 0$ satisfying the Riccati inequality

$$0 \geq A_c'P + PA_c - P(B_2B_2' - \gamma^{-2}B_1B_1')P + C_{1c}'C_{1c}$$

assures that with $F = -B_2'P - D_{12}'C_1$, the closed-loop transfer function is such that $\|T(z, w; s)\|_\infty \leq \gamma$. Defining $X = P^{-1}$ and applying the Schur complement formula, the above inequality turns out to be equivalent to the linear matrix inequality

$$\mathcal{A}(X) = \begin{bmatrix} A_c X + X A_c' - B_2 B_2' + \gamma^{-2} B_1 B_1' & X C_{1c}' \\ C_{1c} X & -I \end{bmatrix} \leq 0$$

On the other hand, from Remark 6.19 it is clear that $X = P_\infty^{-1}$ is the global solution of

$$\min \{\text{trace}[B_1' X^{-1} B_1] : X > 0 , \quad \mathcal{A}(X) \leq 0\}$$

which is a convex programming problem. Furthermore, it reduces to that of Remark 6.9 as γ goes to infinity. □

We consider now the Partial information problem. Let the linear system be defined by the standard state space representation

$$\dot{x} = Ax + B_1 w + B_2 u \tag{6.57}$$

$$z = C_1 x + D_{12} u \tag{6.58}$$

$$y = C_2 x + D_{21} w \tag{6.59}$$

where the output variable y is available for feedback. The situation is illustrated in fig. 6.5 where $K(s)$ is the controller transfer function. We call the controller $K(s)$ *strictly feasible* if it assures that $T(z, w; s)$ belongs to RH_∞ and $\|T(z, w; s)\|_\infty < \gamma$. The so called central controller is then obtained from part b) of Theorem 5.4 with $Q(s) = 0$. Notice that the central controller does not exhibits the classical observer-based structure. Indeed, it is given by

$$K(s) := \left[\begin{array}{c|c} A + B_2 F_\infty + \gamma^{-2} B_1 B_1' P_\infty + Z_\infty L_\infty C_{2\infty} & -Z_\infty L_\infty \\ \hline F_\infty & 0 \end{array} \right] \tag{6.60}$$

where $C_{2\infty} := C_2 + \gamma^{-2}D_{21}B_1'P_\infty$. However, the existence of a solution to the Partial information problem depends only upon the existence of a symmetric, positive semidefinite and stabilizing matrix P_∞ together with the existence of a symmetric, positive semidefinite and stabilizing matrix Π_∞ of the Riccati equation (in the unknown Π)

$$0 = \Pi A_f' + A_f\Pi - \Pi(C_2'C_2 - \gamma^{-2}C_1'C_1)\Pi + B_{1f}B_{1f}' \tag{6.61}$$

satisfying the additional constraint $r_s(P_\infty\Pi_\infty) < \gamma^2$. Based on this, the following problem is stated and solved.

Problem 6.4 (Output feedback problem in RH_∞) *Given a scalar $\gamma > 0$, determine the conditions for the existence of a strictly feasible output feedback controller $K(s)$.*

Recall Theorem 5.4 where the complete solution of the Partial information problem is provided. Here, we consider the following assumptions i) The pair $[A - B_2D_{12}'C_1), (I-D_{12}D_{12}')C_1]$ is observable and the pair $[(A-B_1D_{21}'C_2), B_1(I-D_{21}'D_{21})]$ is reachable, ii) The pair (A, B_2) is stabilizable and the pair (A, C_2) is detectable and iii) $D_{12}'D_{12} = I$ and $D_{21}D_{21}' = I$. The first assumption assures that matrices P_∞ and Π_∞ are both symmetric, positive definite and stabilizing. Furthermore, it is also assumed that B_1B_1' and $C_1'C_1$ are positive definite matrices.

Before proceed we need to define the dual version of the convex set $\mathcal{C}_{\gamma c}$, it is

$$\mathcal{C}_{\gamma f} := \{(V, \mu) : V \geq 0, \ \mu > 0, \ v'\Theta_{\gamma f}(V, \mu)v \leq 0, \ \forall v \in \mathcal{N}_f\} \tag{6.62}$$

where $\Theta_{\gamma f}(V, \mu) := \Theta_f(V) + \mu^{-1}VR_fV$. Clearly this set is convex as well (recall the proof of Theorem 6.5).

Theorem 6.7 (Output feedback) *Consider the previous assumptions and let γ a positive scalar be given. There exists a strictly feasible controller with transfer function $K(s)$ if and only if there exist W such that $(W, \gamma^2) \in \text{int } \mathcal{C}_{\gamma c}$, V such that $(V, \gamma^2) \in \text{int } \mathcal{C}_{\gamma f}$ satisfying the convex coupling constraint $(W, V, \gamma) \in \text{int } \mathcal{Z}_{cf}$, where*

$$\mathcal{Z}_{cf} := \left\{(W, V, \gamma) : \begin{bmatrix} W_1 & \gamma I \\ \gamma I & V_1 \end{bmatrix} \geq 0\right\} \tag{6.63}$$

Proof Under the assumptions made, the Output feedback problem is solvable if and only if $\bar{P}_\infty > 0$ solves the Riccati equation (6.56), $\bar{\Pi}_\infty > 0$ solves the Riccati equation (6.61), both with γ replaced by a suitable $0 < \bar{\gamma} < \gamma$ and $r_s(\bar{P}_\infty\bar{\Pi}_\infty) < \bar{\gamma}^2$. We have already shown (recall Theorem 6.6) that $(W_\infty, \gamma^2) \in \mathcal{C}_{\gamma c}$. On the other hand, it can be verified that matrix

$$V_\infty := \gamma^2 \begin{bmatrix} \bar{\Pi}_\infty^{-1} & \bar{\Pi}_\infty^{-1}\bar{L}_\infty \\ \bar{L}_\infty'\bar{\Pi}_\infty^{-1} & \bar{L}_\infty'\bar{\Pi}_\infty^{-1}\bar{L}_\infty \end{bmatrix} \geq 0$$

with $\bar{L}_\infty = -\bar{\Pi}_\infty C_2' - B_1D_{21}'$ is such that

$$v'\Theta_{\gamma f}(V_\infty, \gamma^2)v = (\gamma\bar{\Pi}_\infty^{-1}x)'\left[A_f\bar{\Pi}_\infty + \bar{\Pi}_\infty A_f' - \right.$$
$$\left. -\bar{\Pi}_\infty(C_2'C_2 - \gamma^{-2}C_1'C_1)\bar{\Pi}_\infty + B_{1f}B_{1f}'\right](\gamma\bar{\Pi}_\infty^{-1}x)$$
$$= \left[1 - (\gamma/\bar{\gamma})^2\right]\|C_1x\|^2 < 0$$

holds for all $v \in \mathcal{N}_f$, proving that $(V_\infty, \gamma^2) \in \text{int } \mathcal{C}_{\gamma f}$. Finally, using the Schur complement formula it is also verified that the coupling constraint is satisfied for

$W_1 = \gamma^2 \bar{P}_\infty^{-1}$ and $V_1 = \gamma^2 \bar{\Pi}_\infty^{-1}$ since $r_s(\bar{P}_\infty \bar{\Pi}_\infty) < \bar{\gamma}^2 < \gamma^2$. The converse is immediate from the previous result of Theorem 6.6. □

We have interest to reduce γ to the minimum value keeping feasibility of all constraints introduced in Theorem 6.7. The main goal is to generate a convex problem, similar to (6.55), valid in the output feedback case. Unfortunately, the change of variable defined previously, namely $\mu = \gamma^2$ destroys the convexity of the set $\mathcal{Z}_{cf}$. The other possibility comes to light from the simple observation that $(\gamma^{-1}W, \gamma^2) \in \mathcal{C}_{\gamma c}$ defines a convex set with respect to the new pair of variables $(\hat{W}, \gamma)$ where $\hat{W} := \gamma^{-1}W$. The same occurs for the set $\mathcal{C}_{\gamma f}$, that is $(\gamma^{-1}V, \gamma^2) \in \mathcal{C}_{\gamma f}$ defines a convex set with respect to the new pair of variables $(\hat{V}, \gamma)$ where $\hat{V} := \gamma^{-1}V$. Furthermore, the set $\mathcal{Z}_{cf}$, rewritten as

$$\mathcal{Z}_{cf} := \left\{ (\hat{W}, \hat{V}) \ : \ \begin{bmatrix} \hat{W}_1 & I \\ I & \hat{V}_1 \end{bmatrix} \geq 0 \right\}$$

makes clear that convexity is once again preserved. Based on these facts we are able to conclude that Theorem 6.7 can equivalently be stated in terms of the existence of a triple $(\hat{W}, \hat{V}, \gamma) \in \text{int } \mathcal{O}_{cf}$ with $\mathcal{O}_{cf}$ being a convex set. Naturally, the output feedback counterpart of (6.55) is the convex programming problem

$$\min \left\{ \gamma \ : \ (\hat{W}, \hat{V}, \gamma) \in \mathcal{O}_{cf} \right\} \tag{6.64}$$

which provides the minimum value γ^o keeping the Output feedback problem solvable for any $\gamma > \gamma^o$. In other words, for $\gamma > \gamma^o$ the associated Riccati equations admit stabilizing solutions P_∞ and Π_∞ respectively. These solutions may be used to calculate from (6.60), the (central) feasible controller.

Remark 6.24 Remark 6.17 applies to the set $\mathcal{C}_{\gamma f}$ also. If matrix $C_1'C_1$ is not strictly positive definite then C_1' must be replaced by $[C_1' \ \sqrt{\epsilon}I]$ where $\epsilon > 0$ is arbitrarily small. This is done with no loss of generality and is important to have $V_1 > 0$ for all $V \in \mathcal{C}_{\gamma f}$. □

Remark 6.25 The set $\mathcal{C}_{\gamma f}$ is not bounded but fortunately, the same reasoning of Remarks 6.21 and 6.22 applies to it. First the set $\mu > 0$ can be replaced by the bounded one $\gamma_{min}^2 \leq \mu \leq \gamma_{max}^2$ and second, the convex set

$$\mathcal{C}_{\gamma f} \ \cap \ \{V \ : \ \text{trace}[R_f V] \leq \beta\}$$

has a bounded domain for all finite β provided the pair $(-A_f, B_{1f})$ is stabilizable. □

The results of this section put in evidence the potentialities of the convex programming approach to deal with optimal control problems formulated in RH_2 and RH_∞ spaces. The same manipulations will be used in the next sections to handle more involved problems. To ease the presentation, the technical assumptions involving the positive definiteness of, for instance, matrices $B_1 B_1'$ and $C_1'C_1$ are assumed throughout. Actually, as discussed before they can be enforced with no loss of generality.

6.3 Mixed RH_2 / RH_∞ control

The main goal of the so called mixed RH_2/RH_∞ control problem is to take into account the two major features of any control system design. First it is desirable to optimize performance. Second it is important to be aware that the model at hand

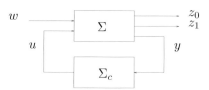

Figure 6.6: The mixed control design structure

always represents a nominal system while the true system is subject to uncertainties. In order to improve practical needs, these two aspects have to be accommodated in the same design problem. It consists on the optimization of the RH_2 norm of a transfer function while the RH_∞ norm of another transfer function is constrained to be less than a certain prescribed level.

The situation is represented in fig. 6.6 where Σ is a linear time-invariant system and Σ_c is the controller. The mixed RH_2/RH_∞ optimal control problem is to find (if one exists) a controller Σ_c such that, for $\gamma > 0$ given

$$\inf \left\{ \|T(z_0, w; s)\|_2^2 \ : \ \|T(z_1, w; s)\|_\infty < \gamma \right\} \tag{6.65}$$

A controller solving this problem imposes to the closed loop system an optimal performance against exogenous perturbation in the channel w to z_0 while robust stability is guaranteed to all model uncertainty expressed as $z_1 = \Delta w$ such that $\|\Delta\|_\infty \le 1/\gamma$. The solution of this problem can not be obtained from matrix Riccati manipulations. We show here the difficulties we have to face and the manipulations and approximations we have to introduce in order to solve it by means of convex programming methods. Before all, based on the results of the previous section, it is to be noticed that if instead of Problem (6.65) we solve

$$\min \left\{ \|T(z_0, w; s)\|_2^2 \ : \ \|T(z_1, w; s)\|_\infty \le \gamma \right\} \tag{6.66}$$

then its optimal solution (if any) provides a suboptimal solution to Problem (6.65) with γ replaced by $\bar{\gamma} > \gamma$ and the degree of suboptimality becomes arbitrarily small provided $\bar{\gamma}$ is chosen arbitrarily close to γ.

6.3.1 State feedback design

Consider $\gamma > 0$ be a fixed scalar and let the systems Σ be defined as follows

$$\dot{x} = Ax + B_1 w + B_2 u \tag{6.67}$$
$$z_0 = C_0 x + D_0 u \tag{6.68}$$
$$z_1 = C_1 x + D_{12} u \tag{6.69}$$
$$y = x \tag{6.70}$$

where the whole state vector is available for feedback. It is assumed that i) the pair (A, B_2) is stabilizable and ii) $D'_{12} D_{12} = I$. Furthermore, the controller Σ_c is assumed to be given by

$$u = Fx \tag{6.71}$$

where the gain matrix F is to be determined. To guarantee that the constraint $\|T(z_1, w; s)\|_\infty \le \gamma$ is satisfied, we have to consider $F \in \mathcal{K}_{\gamma c}$ which is the same to say that $F \in \mathcal{K}_c$ and there exists $P > 0$ satisfying the Riccati inequality (recall that Theorem 2.16 applies since $B_1 B_1' > 0$).

$$0 \geq A_{c\infty} P + P A_{c\infty}' + \gamma^{-2} P C_{c\infty}' C_{c\infty} P + B_1 B_1' \tag{6.72}$$

with $A_{c\infty} := A + B_2 F$ and $C_{c\infty} := C_1 + D_{12} F$. On the other hand, for the same control gain $F \in \mathcal{K}_c$, the transfer function from the input w to the output z_0 belongs to RH_2 and

$$\|T(z_0, w; s)\|_2^2 = \text{trace}\left[(C_0 + D_0 F) P_r (C_0 + D_0 F)'\right] \tag{6.73}$$

where $P_r > 0$ solves the Lyapunov equation

$$0 = A_{c\infty} P_r + P_r A_{c\infty}' + B_1 B_1' \tag{6.74}$$

Simple comparison of inequality (6.72) with equality (6.74) puts in evidence that $P \geq P_r$. Consequently, any feasible P for inequality (6.72) provides an upper bound for the norm of the transfer function under consideration, that is

$$\|T(z_0, w; s)\|_2^2 \leq \text{trace}\left[(C_0 + D_0 F) P (C_0 + D_0 F)'\right] \tag{6.75}$$

In the optimal control problem (6.66), the quantity $\|T(z_0, w; s)\|_2^2$ is then replaced by the quantity in right hand side of (6.75). It must be clear that this simplifies the problem to be dealt with at the expense of optimality. In fact, the minimum of the proposed upper bound may not coincide with the optimum of the true design problem (6.66). The next Theorem shows that this problem is convex which opens again the possibility of solving it by means of efficient numerical methods.

Theorem 6.8 *Let $\gamma > 0$ be given and define the symmetric and positive semidefinite extended matrix*

$$R_0 := \begin{bmatrix} C_0' \\ D_0' \end{bmatrix} \begin{bmatrix} C_0 & D_0 \end{bmatrix}$$

Let $\bar{W}$ be the optimal solution of the convex programming problem

$$J_{sub} := \min \left\{ \text{trace}[R_0 W] \; : \; (W, \gamma^2) \in \mathcal{C}_{\gamma c} \right\} \tag{6.76}$$

Then, $F = \bar{W}_2' \bar{W}_1^{-1} \in \mathcal{K}_{\gamma c}$ minimizes an upper bound of the objective function of problem (6.66) in the sense that $\|T(z_0, w; s)\|_2^2 \leq \text{trace}[R_0 W]$, for all $(W, \gamma^2) \in \mathcal{C}_{\gamma c}$.

Proof From Theorem 6.5 it follows that $(\bar{W}, \gamma^2) \in \mathcal{C}_{\gamma c}$ implies $F = \bar{W}_2' \bar{W}_1^{-1} \in \mathcal{K}_{\gamma c}$. On the other hand, for any $(W, \gamma^2) \in \mathcal{C}_{\gamma c}$ we have $W_1 \geq P_r$, consequently

$$\text{trace}[R_0 W] = \text{trace}[(C_0 + D_0 W_2' W_1^{-1}) W_1 (C_0 + D_0 W_2' W_1^{-1})'] +$$
$$+ \text{trace}[D_0 (W_3 - W_2' W_1^{-1} W_2) D_0']$$
$$\geq \text{trace}[(C_0 + D_0 W_2' W_1^{-1}) P_r (C_0 + D_0 W_2' W_1^{-1})']$$
$$\geq \|T(z_0, w; s)\|_2^2$$

and the proof is complete. $\qquad \square$

The joint convexity of the set $\mathcal{C}_{\gamma c}$ is crucial for the introduction of another mixed problem, called *the inverse* mixed RH_2/RH_∞ design problem, it is

$$\min \left\{ \mu \; : \; \text{trace}[R_0 W] \leq \beta \;, \; (W, \mu) \in \mathcal{C}_{\gamma c} \right\} \tag{6.77}$$

where β is a fixed positive scalar. For a given $\beta > 0$, the constraint $\|T(z_0, w; s)\|_2 \leq \sqrt{\beta}$ is satisfied while the upper bound on the admissible perturbation $\|\Delta\|_\infty < 1/\sqrt{\mu}$ is maximized. Needless to say that as β goes to infinity, its optimal solution goes to the solution of problem (6.55) already treated in the last section. Another problem of practical interest is

$$\min \{\mu + \theta \text{trace}[R_0 W] \, , \, (W, \mu) \in \mathcal{C}_{\gamma c}\} \tag{6.78}$$

where θ is a fixed positive scalar. In this case, the scalar $\theta > 0$ can be interpreted as the dual variable associated to the inequality constraint of problem (6.77), that is θ is the tradeoff, to be fixed by the designer, between the norms of the two transfer functions of system (6.67)-(6.70). Finally, it is important to stress that (recall Remark 6.18) if matrix B_1 is replaced by $[B_1 \, \sqrt{\epsilon}I]$ with $\epsilon > 0$ arbitrarily small them the optimal solutions of all previous mixed problems are strictly feasible, that is they belong to int $\mathcal{C}_{\gamma c}$.

Example 6.2 To get some feelings on the previous results, let us consider the following numerical example. The system is defined as

$$\dot{x} = Ax + B_1 w + B_2 u$$
$$z_0 = C_0 x + u$$
$$z_1 = C_1 x + u$$
$$y = x$$

where

$$A = \begin{bmatrix} 0 & 1 \\ -1 & -1 \end{bmatrix} \, , \quad B_1 = \begin{bmatrix} 1 \\ 0 \end{bmatrix} \, , \quad B_2 = \begin{bmatrix} 0 \\ 1 \end{bmatrix} \, , \quad C_1 = \begin{bmatrix} 0 & -2 \end{bmatrix}$$

Two different situations have been considerated :

1) With $C_0 = [1 \ 0]$ the mixed optimization problem (6.76) has been solved for $1.65 \leq \gamma \leq 3.00$. The optimal solution provided the controller M. For comparison purpose we have also determined the central controller C associated to the constraint $\|T(z_1, w; s)\|_\infty < \gamma$. Fig. 6.7 shows the optimal upper bound J_{sub} as well as the quantity $\|T(z_0, w; s)\|_2^2$ for each controller. It is useful to observe that, in this case, the controller M obtained by the mixed problem is always better than the central controller. This behavior is expected for large values of γ since in this situation, the gap between J_{sub} and $\|T(z_0, w; s)\|_2^2$ becomes arbitrarily small. On the contrary, for small values of γ the aforementioned gap is important and it is possible to build examples for which the central controller performs better than the one provided by the optimal solution of the mixed problem.

2) With $C_0 = C_1$ the mixed optimization problem (6.76) has been solved again for the γ in the same interval. As predicted in Remark 6.19, the optimal mixed controller M coincides with the central controller. For comparison, the performances of these controllers together with the upper bound J_{sub} are depicted in fig. 6.8.

Based on this, we can say that in many instances, the mixed problem (6.76) provides a controller that performs well when compared to the central controller. This suggestion becomes a fact for moderate and large values of γ. The main point to be retained however is that the controller M is only suboptimal for the true mixed RH_2/RH_∞ problem. Numerically speaking, one of the main attractive features of problem (6.76) is that it is convex and so easy to solve.   □

Remark 6.26 (Post-optimization procedure) The optimal solution of the mixed optimization problem (6.76) is, generally, a suboptimal solution to the true mixed problem

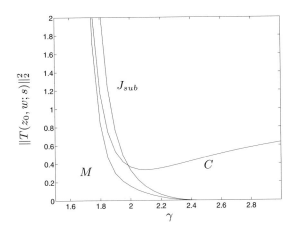

Figure 6.7: The performances of the controllers M and C

(6.66) since for $F = F_{sub}$ provided by (6.76) the constraint $\|T(z_1, w; s)\|_\infty \leq \gamma$ is not satisfied with equality. Hence we are now concerned to determine a new state feedback gain, still suboptimal, but with a smaller value of the objective function $\|T(z_0, w; s)\|_2$. To ease the algebraic manipulations involved we assume the pair (A, B_2) is stabilizable, the pair $(A - B_2 D_0' C_0, (I - D_0 D_0') C_0)$ is detectable and $D_0' D_0 = I$. For $F = F_{sub}$ we get $\|T(z_0, w; s)\|_2^2 = \text{trace}[B_1' P_{sub} B_1]$ where $P_{sub} \geq 0$ solves the Lyapunov equation

$$0 = (A + B_2 F_{sub})' P_{sub} + P_{sub}(A + B_2 F_{sub}) + (C_0 + D_0 F_{sub})'(C_0 + D_0 F_{sub}) \qquad (6.79)$$

Defining the state feedback gain

$$F_\alpha := (1 - \alpha)F_{sub} - \alpha(B_2' P_\alpha + D_0' C_0) \qquad (6.80)$$

where α is a scalar to be determined such that $A + B_2 F_\alpha$ is stable and P_α solves the Lyapunov equation

$$0 = (A + B_2 F_\alpha)' P_\alpha + P_\alpha(A + B_2 F_\alpha) + (C_0 + D_0 F_\alpha)'(C_0 + D_0 F_\alpha) \qquad (6.81)$$

the following conclusions can be drawn. First, it is clear that for $F = F_\alpha$ the associated cost function is written as $\|T(z_0, w; s)\|_2^2 = \text{trace}[B_1' P_\alpha B_1]$ and, the equality

$$0 = (A + B_2 F_{sub})'(P_{sub} - P_\alpha) + (P_{sub} - P_\alpha)(A + B_2 F_{sub}) + \alpha(2 - \alpha)\bar{F}_{sub}' \bar{F}_{sub}$$

valid for $\bar{F}_{sub} := F_{sub} + B_2 P_\alpha + D_0' C_0$ guarantees an improvement in the RH_2 norm provided $\alpha \in [0, 2]$ since under this condition $P_{sub} \geq P_\alpha$. Second, from (6.80) and (6.81) simple algebraic manipulations put in evidence that matrix P_α satisfies the Riccati equation (in the unknown P)

$$0 = A_\alpha' P + P A_\alpha - P B_{2\alpha} B_{2\alpha}' P + C_{0\alpha}' C_{0\alpha} \qquad (6.82)$$

where

$$A_\alpha := A + B_2 \left[(1 - \alpha)^2 F_{sub} + \alpha(\alpha - 2)D_0' C_0\right]$$
$$B_{2\alpha} := B_2 \sqrt{2\alpha - \alpha^2}$$
$$C_{0\alpha} := C_0 + D_0 \left[(1 - \alpha)F_{sub} - \alpha D_0' C_0\right]$$

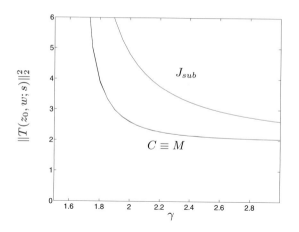

Figure 6.8: Performances of controllers M and C

The pairs $(A_\alpha, B_{2\alpha})$ and $(A_\alpha, C_{0\alpha})$ are stabilizable and detectable, respectively. In fact, stabilizability easily follows from stabilizability of (A, B_2), whereas detectability is implied by detectability of the pair $(A - B_2 D_0' C_0, (I - D_0 D_0')C_0)$ and by noticing that

$$A_\alpha = A - B_2 D_0' C_0 + B_2(1 - \alpha)[(1 - \alpha)(F_{sub} + D_0' C_0)]$$
$$C_{0\alpha} = (I - D_0 D_0')C_0 + D_0[(1 - \alpha)(F_{sub} + D_0' C_0)]$$
$$0 = D_0'(I - D_0 D_0')C_0$$

Such an equation (see Lemma C.4 of Appendix C) admits a stabilizing solution whenever $\alpha \in [0, 2]$. Actually, under the stated assumptions, this solution is also the unique positive semidefinite one. In order to check the stability of $A + B_2 F_\alpha$, with F_α given by (6.80), consider equation (6.81) and notice that

$$A + B_2 F_\alpha = A - B_2 D_0' C_0 + B_2[(1 - \alpha)(F_{sub} + D_0' C_0) - \alpha B_2' P_\alpha]$$
$$C_0 + D_0 F_\alpha = (I - D_0 D_0')C_0 + D_0[(1 - \alpha)(F_{sub} + D_0' C_0) - \alpha B_2' P_\alpha]$$
$$0 = D_0'(I - D_0 D_0')C_0$$

Thus, the detectability of the pair $(A - B_2 D_0' C_0, (I - D_0 D_0')C_0)$ together with the above equations imply that the pair $(A + B_2 F_\alpha, C_0 + D_0 F_\alpha)$ is detectable as well. This fact and the existence of a solution $P_\alpha \geq 0$, entails, (see Lemma C.1 of Appendix C) that $A + B_2 F_\alpha$ is a stable matrix.

The above results can be exploited in the following way. With the state feedback gain F_{sub} such that $A + B_2 F_{sub}$ is stable and $||T(z_1, w; s)||_\infty < \gamma$ an one dimensional search in the interval $[0, 2]$ for α with equation (6.82) taken into account allows to determine the value of α^o corresponding to which the control law $u = F_\alpha x$ minimizes the RH_2 norm while keeping the RH_∞ norm not greater than γ. Incidentally, notice that the choice $\alpha = 1$ corresponds to the optimal unconstrained RH_2 control law.

The solution P_α of equation (6.82) is a function of the parameter α which enjoys a symmetry property, namely $P_\alpha = P_{2-\alpha}$. Denoting with Γ_α the matrix which is the derivative of P_α with respect to α it is possible to verify that it satisfies the following Lyapunov equation

$$0 = \bar{A}_\alpha' \Gamma_\alpha + \Gamma_\alpha \bar{A}_\alpha - 2(1 - \alpha) \bar{F}_{sub}' \bar{F}_{sub} \tag{6.83}$$

where $\bar{A}_\alpha := A_\alpha - B_{2\alpha} B_{2\alpha}' P_\alpha$ is a stable matrix since P_α is a stabilizing solution of (6.82). Hence it follows from (6.83) that $\Gamma_\alpha \leq 0$ for $\alpha \in [0, 1]$ and $\Gamma_\alpha \geq 0$ for $\alpha \in [1, 2]$. Therefore

α^o is given by

$$\alpha^o := \min\{1 - \alpha_1, \alpha_2 - 1\}$$
$$\alpha_1 := \max_{\alpha \in [0,1]} \{\alpha \ : \ ||T(z_1, w; s)||_\infty \le \gamma\}$$
$$\alpha_2 := \min_{\alpha \in [1,2]} \{\alpha \ : \ ||T(z_1, w; s)||_\infty \le \gamma\}$$

Finally, it is important to stress that the above developments remain valid for any state feedback gain F_{sub} such that $A + B_2 F_{sub}$ is stable and $||T(z_1, w; s)||_\infty < \gamma$ and not just for the one provided by the solution of the mixed optimization problem (6.76). For instance, in view of the previous discussion (recall Example 6.2), we also have interest to adopt F_{sub} as the matrix gain defined by the central RH_∞ controller. Obviously, α^o and the relevant value for the RH_2 norm depend on the chosen gain F_{sub}.

To put in evidence the improvement obtained by this post-optimization procedure, let us consider the system defined in Example 6.2 with $C_0 = [1\ 0]$ and $\gamma = 2$. First take F_{sub} as the feedback gain which is the solution of the convex programming problem (6.76), that is $F_{sub} = [-1.1879\ \ -0.5965]$ and the associated cost $||T(z_0, w; s)||_2^2 = 0.1556$. Corresponding to the best choice α^o, we obtain

$$F_\alpha = [\ -1.1829 \quad -0.5829\ ], \ \ ||T(z_0, w; s)||_2^2 = 0.1504, \ \ ||T(z_1, w; s)||_\infty = 2$$

If, on the contrary, we adopt F_{sub} as the state feedback gain corresponding to the central controller, that is $F_{sub} = [-1.6\ \ -1.2]$, the best choice of α yields

$$F_\alpha = [\ -1.2786 \quad -0.6076\ ], \ \ ||T(z_0, w; s)||_2^2 = 0.1439, \ \ ||T(z_1, w; s)||_\infty = 2$$

In this case, it is apparent that the post-optimization procedure supplies a better result when starting from a worse gain. □

Remark 6.27 (Nash game approach) The mixed state feedback problem can also be approached by the following Nash game when the system (6.67) - (6.70) satisfies the additional assumptions $C_0 = C_1$, $D_0 = D_{12}$ and $D_{12}'C_1 = 0$. Defining the criteria

$$J_1(u, w) = \int_0^\infty [\gamma^2 w'(t)w(t) - z_1'(t)z_1(t)]dt$$

$$J_0(u, w) = \int_0^\infty z_0'(t)z_0(t)dt$$

the aim is to find the equilibrium strategies $(u^\star, w^\star)$ which satisfy the Nash equilibria conditions

$$J_1(u^\star, w^\star) \le J_1(u^\star, w), \ \ \forall\, w \in RH_2$$
$$J_0(u^\star, w^\star) \le J_0(u, w^\star), \ \ \forall\, u \in RH_2$$

It can be proven that the optimal strategies are given by

$$u^\star(t) = F_2 x(t), \ \ F_2 := -B_2'P_2$$
$$w^\star(t) = F_1 x(t), \ \ F_1 := -\gamma^{-2}B_1'P_1$$

provided there exist matrices $P_1 \le 0$ and $P_2 \ge 0$ solutions to the coupled Riccati equations

$$0 = A'P_1 + P_1 A - C_1'C_1 - \begin{bmatrix} P_1 & P_2 \end{bmatrix} \begin{bmatrix} \gamma^{-2}B_1 B_1' & B_2 B_2' \\ B_2 B_2' & B_2 B_2' \end{bmatrix} \begin{bmatrix} P_1 \\ P_2 \end{bmatrix}$$

$$0 = A'P_2 + P_2 A + C_1'C_1 - \begin{bmatrix} P_1 & P_2 \end{bmatrix} \begin{bmatrix} 0 & \gamma^{-2}B_1 B_1' \\ \gamma^{-2}B_1 B_1' & B_2 B_2' \end{bmatrix} \begin{bmatrix} P_1 \\ P_2 \end{bmatrix}$$

Based on these optimality conditions, the following interpretation can be drawn. The first Riccati equation rewritten as

$$0 = (A + B_2 F_2)' P_1 + P_1(A + B_2 F_2) - \gamma^{-2} P_1 B_1 B_1' P_1 - (C_1 + D_{12} F_2)'(C_1 + D_{12} F_2)$$

implies that $\|T(z_1, w; s)\|_\infty < \gamma$ since $P_1 \leq 0$ and stabilizing. On the other hand, the second Riccati equation, factorized as

$$0 = (A + B_1 F_1)' P_2 + P_2(A + B_1 F_1) - P_2 B_2 B_2' P_2 + C_1' C_1$$

shows that the criterion J_0 for $w = w^*$ attains its minimum value at $u = u^*$ as required by the Nash game. It is then clear that for $w \neq w^*$ the control $u = F_2 x$ is merely a suboptimal policy for the mixed design problem. Unfortunately this is frequently the case because for $\|T(z_1, w; s)\|_\infty < \gamma$ the worst input $w = w^*$ is not the one which produces the output z_1 such that $\|z_1\|_2 = \|T(z_1, w; s)\|_\infty \|w\|_2$. For comparison purpose let us consider the following numerical example

$$\dot{x} = 2x + w + 3u$$
$$z_1 = \begin{bmatrix} 3 \\ 0 \end{bmatrix} x + \begin{bmatrix} 0 \\ 1 \end{bmatrix} u$$
$$z_0 = z_1$$
$$y = x$$

and $\gamma = 0.4$. The pair $(P_1, P_2) = (-6.84, 10.05)$ is a solution of the coupled Riccati equations which provides the optimal Nash gain $F_2 = -30.16$. The closed-loop system exhibits the performances

$$\|T(z_0, w; s)\|_2^2 = 5.19 , \quad \|T(z_1, w; s)\|_\infty = 0.34 < \gamma$$

Then, problem (6.76) has also been solved. It provides the feedback gain $F = -8.05$ which imposes to the closed-loop system

$$\|T(z_0, w; s)\|_2^2 = 1.66 , \quad \|T(z_1, w; s)\|_\infty = 0.38 < \gamma$$

From these results it is clear that, in this case, the mixed design introduced in Theorem 6.8 is much better than the Nash game approach. □

Remark 6.28 (Structured robust stability and performance) Consider a linear system depending on uncertain parameters, more precisely

$$\dot{x} = (A + \Delta_A)x + B_0 w + (B_2 + \Delta_{B_2})u$$
$$z_0 = C_0 x + D_0 u$$
$$y = x$$

with

$$\Delta_A := B_{11} \Omega_1 C_{11} , \quad \Delta_{B_2} := B_{12} \Omega_2 C_{12}$$

where the only information available for matrices Ω_1 and Ω_2 is that

$$\|\Omega_1\| \leq 1 , \quad \|\Omega_2\| \leq 1$$

Defining the matrices

$$B_1 = \begin{bmatrix} B_{11} & B_{12} \end{bmatrix} , \quad C_1 = \begin{bmatrix} C_{11} \\ 0 \end{bmatrix} , \quad D_{12} = \begin{bmatrix} 0 \\ C_{12} \end{bmatrix} ,$$

the above system can be rewritten as

$$\dot{x} = Ax + B_0 w + B_1 w_1 + B_2 u$$
$$z_1 = C_1 x + D_{12} u$$
$$w_1 = \Omega z_1$$
$$z_0 = C_0 x + D_0 u$$
$$y = x$$

where matrix $\Omega := \text{diag}[\Omega_1, \Omega_2]$ satisfies $\|\Omega\| \leq 1$. Our first goal is to determine (if one exists) a matrix F such that with $u = Fx$ the closed-loop system stability is assured for all feasible parametric perturbations. A way we already know to solve this problem is to introduce the constraint $\|T(z_1, w_1; s)\|_\infty < 1$. Indeed, a controller which satisfies this constraint solve the stated problem. However, this approach produces frequently very conservative results since the block-diagonal structure of the uncertainty (represented by matrix Ω) is not taken into account. In a slightly more general setting, we consider matrix Ω composed by N square blocks, that is

$$\Omega := \text{diag}[\Omega_1, \cdots, \Omega_N] , \quad \|\Omega\| \leq 1$$

which together with the matrix

$$\Lambda := \text{diag}[\lambda_1 I, \cdots, \lambda_N I]$$

where each sub-block has the same dimension as the corresponding sub-block of matrix Ω and $\lambda_1 > 0, \cdots, \lambda_N > 0$, enable us to say that $\Omega \Lambda \Omega' \leq \Lambda$, $\forall \|\Omega\| \leq 1$.

The closed loop system stability depends on the stability of matrix

$$A_\Omega := A_{c\infty} + B_1 \Omega C_{c\infty}$$

where as before, $A_{c\infty} = A + B_2 F$ and $C_{c\infty} = C_1 + D_{12} F$. This property follows from the following inequality which holds for any symmetric matrix P

$$B_1 \Omega C_{c\infty} P + PC'_{c\infty} \Omega' B'_1 = B_1 \Omega \Lambda \Omega' B'_1 + PC'_{c\infty} \Lambda^{-1} C_{c\infty} P -$$
$$- (B_1 \Omega \Lambda^{1/2} - PC'_{c\infty} \Lambda^{-1/2})(B_1 \Omega \Lambda^{1/2} - PC'_{c\infty} \Lambda^{-1/2})'$$
$$\leq B_1 \Lambda B'_1 + PC'_{c\infty} \Lambda^{-1} C_{c\infty} P$$

Indeed, if there exist a symmetric and positive definite matrix P, a positive definite matrix Λ with the above structure and a matrix F such that the inequality

$$A_{c\infty} P + PA'_{c\infty} + PC'_{c\infty} \Lambda^{-1} C_{c\infty} P + B_1 \Lambda B'_1 < 0$$

holds then (recall Theorem 2.15) the transfer function

$$T_\Lambda(z_1, w_1; s) := \Lambda^{-1/2} T(z_1, w_1; s) \Lambda^{1/2}$$

belongs to RH_∞ and $\|T_\Lambda(z_1, w_1; s)\|_\infty < 1$. Since the *scaling* matrix Λ can also be used to redefine the parametric perturbation as $\Omega_\Lambda := \Lambda^{-1/2} \Omega \Lambda^{1/2}$ such that $\|\Omega_\Lambda\|_\infty \leq 1$, from Theorem 5.1 the stability of A_Ω follows. In addition, the above property is not lost if the previous Riccati inequality is replaced by (assuming again that $B_0 B'_0 > 0$)

$$A_{c\infty} P + PA'_{c\infty} + PC'_{c\infty} \Lambda^{-1} C_{c\infty} P + B_1 \Lambda B'_1 + B_0 B'_0 \leq 0$$

However, doing this we now have

$$A_\Omega P + PA'_\Omega + B_0 B'_0 \leq 0 , \quad \forall \|\Omega\| \leq 1$$

yielding an upper bound to the RH_2 norm of the transfer function from the input w to the output z_0, that is

$$\|T(z_0, w; s)\|_2^2 \leq \text{trace}[(C_0 + D_0 F)P(C_0 + D_0 F)']$$

which is valid for all $\|\Omega\| \leq 1$. The minimization of this upper bound keeping stability is a mixed RH_2/RH_∞ problem with an additional matrix variable, namely the scaling matrix Λ. It is somewhat surprising that this problem can be converted into a convex one. Actually, let us define the set

$$\mathcal{C}_{\Lambda c} := \left\{ (W,\Lambda) \ : \ W \geq 0 \ , \ \Lambda > 0 \ , \ v'\Theta_{\Lambda c}(W,\Lambda)v \leq 0 \ \forall v \in \mathcal{N}_c \right\}$$

where

$$\Theta_{\Lambda c}(W,\Lambda) := M_c W + W M_c' + W \begin{bmatrix} C_1' \\ D_{12}' \end{bmatrix} \Lambda^{-1} \begin{bmatrix} C_1 & D_{12} \end{bmatrix} W +$$

$$+ \begin{bmatrix} B_1 \\ 0 \end{bmatrix} \Lambda \begin{bmatrix} B_1' & 0 \end{bmatrix} + \begin{bmatrix} B_0 \\ 0 \end{bmatrix} \begin{bmatrix} B_0' & 0 \end{bmatrix}$$

Following the proof of Theorem 6.6, we see that it remains true in the present case. The set $\mathcal{C}_{\Lambda c}$ is convex and any feasible pair $(W,\Lambda) \in \mathcal{C}_{\Lambda c}$ provides $F = W_2' W_1^{-1}$, $P = W_1$ and Λ which satisfy the previous Riccati inequality. Finally, in the present context, the associated mixed RH_2/RH_∞ control design problem is (recall Theorem 6.8)

$$J_{sub} := \min \left\{ \text{trace}[R_0 W] \ : \ (W,\Lambda) \in \mathcal{C}_{\Lambda c} \right\}$$

which is jointly convex on both variables (W,Λ). Its optimal solution provides a robust control gain $F = W_2' W_1^{-1}$ imposing in addition $\|T(z_0,w;s)\|_2 \leq \sqrt{J_{sub}}$ for all structured parametric perturbations $\|\Omega\| \leq 1$. $\square$

Remark 6.29 Following the same lines of Remark 6.22, any convex programming method is effective to solve problem (6.76) provided the convex set

$$\mathcal{C}_{\gamma c} \cap \{ W \ : \ \text{trace}[R_0 W] \leq \beta \}$$

is bounded for any finite $\beta > 0$. Based on our previous discussion, this occurs whenever the pair $[-(A - B_2 D_0' C_0), (I - D_0 D_0')C_0]$ is detectable. $\square$

6.3.2   Output feedback design

In this case, with $\gamma > 0$ being a fixed scalar the system Σ is defined as follows

$$\dot{x} = Ax + B_1 w + B_2 u \tag{6.84}$$

$$z_0 = C_0 x + D_0 u \tag{6.85}$$

$$z_1 = C_1 x + D_{12} u \tag{6.86}$$

$$y = C_2 x + D_{21} w \tag{6.87}$$

and the controller Σ_c has to be determined from the solution of problem (6.65). As in the State feedback case, the complete solution to this problem is not known up to now. So, we search for a suboptimal and easy to calculate solution. The main idea to be pursued is to propose a structure to the controller Σ_c depending on only one unknown matrix in such way the Output feedback design problem reduces to the State feedback design problem already solved.

To this end, we make the following assumptions i) the pair (A, B_2) is stabilizable and the pair (A, C_2) is detectable and ii) $D_{12}' D_{12} = I$ and $D_{21} D_{21}' = I$. We also assume that there exists Π_∞ a positive semidefinite and stabilizing solution of the Riccati equation (in the unknown Π)

$$0 = \Pi A_f' + A_f \Pi - \Pi(C_2' C_2 - \gamma^{-2} C_1' C_1)\Pi + B_{1f} B_{1f}' \tag{6.88}$$

which, once it has been calculated, enables us to define matrices

$$A_\infty := A + \gamma^{-2}\Pi_\infty C_1' C_1$$
$$B_{2\infty} := B_2 + \gamma^{-2}\Pi_\infty C_1' D_{12}$$

and $L_\infty = -\Pi_\infty C_2' - B_1 D_{21}'$. The controller Σ_c with transfer function $K(s)$ given by

$$K(s) := \left[\begin{array}{c|c} A_\infty + B_{2\infty}F + L_\infty C_2 & -L_\infty \\ \hline F & 0 \end{array}\right] \tag{6.89}$$

has the important properties provided in the following lemma.

Lemma 6.2 *For all F such that there exists a symmetric, positive semidefinite and stabilizing solution to the Riccati equation*

$$0 = (A_\infty + B_{2\infty}F)X + X(A_\infty + B_{2\infty}F)' + $$
$$+ \gamma^{-2}X(C_1 + D_{12}F)'(C_1 + D_{12}F)X + L_\infty L_\infty' \tag{6.90}$$

the controller Σ_c with transfer function $K(s)$ given in (6.89) imposes to the closed loop system the following properties :

a) It is stable and $\|T(z_1, w; s)\|_\infty < \gamma$

b) $\|T(z_0, w; s)\|_2^2 \leq \mathrm{trace}[C_0\Pi_\infty C_0'] + \mathrm{trace}[(C_0 + D_0 F)X(C_0 + D_0 F)']$

Proof The feedback connection indicated in fig. 6.6 has the state space representation

$$\Sigma_F := \left[\begin{array}{c|c} \tilde{A} & \tilde{B} \\ \hline \tilde{C}_0 & 0 \\ \tilde{C}_1 & 0 \end{array}\right]$$

where the indicated matrices are

$$\tilde{A} = \left[\begin{array}{cc} A + B_2 F & B_2 F \\ \gamma^{-2}\Pi_\infty C_1'(C_1 + D_{12}F) & A + L_\infty C_2 + \gamma^{-2}\Pi_\infty C_1'(C_1 + D_{12}F) \end{array}\right]$$
$$\tilde{B} = \left[\begin{array}{c} B_1 \\ -B_1 - L_\infty D_{21} \end{array}\right]$$

and

$$\tilde{C}_0 = \left[\begin{array}{cc} C_0 + D_0 F & D_0 F \end{array}\right], \quad \tilde{C}_1 = \left[\begin{array}{cc} C_1 + D_{12}F & D_{12}F \end{array}\right]$$

Point a) Assuming there exists X_∞ satisfying (6.90), simple although tedious algebraic calculations show that the Riccati equation

$$0 = \tilde{A}\tilde{P} + \tilde{P}\tilde{A}' + \gamma^{-2}\tilde{P}\tilde{C}_1'\tilde{C}_1\tilde{P} + \tilde{B}\tilde{B}' \tag{6.91}$$

has a positive semidefinite stabilizing solution given by

$$\tilde{P} = \left[\begin{array}{cc} \Pi_\infty + X_\infty & -\Pi_\infty \\ -\Pi_\infty & \Pi_\infty \end{array}\right] \tag{6.92}$$

which yields the conclusion from Theorem 2.14 that the closed loop system is stable and $\|T(z_1, w; s)\|_\infty < \gamma$.

Point b) From the above, it is know that matrix $\tilde{A}$ is stable then, we proceed by making the calculation of the indicated norm, that is

$$\|T(z_0, w; s)\|_2^2 = \text{trace} \left[\tilde{C}_0 \int_0^\infty e^{\tilde{A}t} \tilde{B} \tilde{B}' e^{\tilde{A}'t} dt \tilde{C}_0' \right]$$

$$\leq \text{trace} \left[\tilde{C}_0 \tilde{P} \tilde{C}_0' \right]$$

$$\leq \text{trace} \left[C_0 \Pi_\infty C_0' \right] + \text{trace} \left[(C_0 + D_0 F) X (C_0 + D_0 F)' \right]$$

concluding thus the proof of the lemma proposed.   □

No major difficulty has to be faced to state and prove similar results using Riccati inequalities. Furthermore, it is to be noticed that the assumption on the existence of a solution for the Riccati equation (6.90) is always verified whenever the unknown matrix gain F is such that $A_\infty + B_{2\infty} F$ is stable and

$$\|(C_1 + D_{12} F)[sI - (A_\infty + B_{2\infty} F)]^{-1} L_\infty D_{21}\|_\infty < \gamma$$

Therefore, from the first part of Lemma 6.2 these values of F preserve admissibility of the controller Σ_c, that is $\|T(z_1, w; s)\|_\infty < \gamma$ and in the context of the mixed design we have to calculate one among them which minimizes the upper bound on $\|T(z_0, w; s)\|_2^2$ provided in the second part of the same lemma. This is accomplished if the State feedback design is applied to the *auxiliary* plant

$$\dot{x} = A_\infty x + L_\infty D_{21} w + B_{2\infty} u \tag{6.93}$$

$$z_0 = C_0 x + D_0 u \tag{6.94}$$

$$z_1 = C_1 x + D_{12} u \tag{6.95}$$

$$y = x \tag{6.96}$$

that is, problem (6.76) should be solved with the feasible set $\mathcal{C}_{\gamma c}$ being defined for matrices A, B_2 and B_1 replaced by matrices A_∞, $B_{2\infty}$ and $L_\infty D_{21}$ respectively.

Remark 6.30 The particular structure of matrix $\tilde{P}$ in (6.92) does not means that some conservativeness has been introduced in our calculations. To show this assume the Partial information problem is solvable and $P_\infty > 0$. It is possible to verify that for

$$F = F_c := F_\infty Z_\infty$$

the Riccati equation (6.90) is solvable in X providing

$$X_\infty = \gamma^2 Z_\infty^{-1} P_\infty^{-1} = \gamma^2 P_\infty^{-1} - \Pi_\infty$$

which is positive definite since $r_s(P_\infty \Pi_\infty) < \gamma^2$. Moreover, for $F = F_c$, the transfer function of the controller $K(s)$ turns out to be

$$K_c(s) = \left[\begin{array}{c|c} A_\infty + B_{2\infty} F_c + L_\infty C_2 & -L_\infty \\ \hline F_c & 0 \end{array} \right]$$

$$= \left[\begin{array}{c|c} Z_\infty (A_\infty + B_{2\infty} F_\infty Z_\infty + L_\infty C_2) Z_\infty^{-1} & -Z_\infty L_\infty \\ \hline F_\infty & 0 \end{array} \right]$$

$$= \left[\begin{array}{c|c} A_{cc} + Z_\infty L_\infty (C_2 + \gamma^{-2} D_{21} B_1' P_\infty) & -Z_\infty L_\infty \\ \hline F_\infty & 0 \end{array} \right]$$

showing that the transfer function $K_c(s)$ meets exactly the transfer function of the central RH_∞ controller.   □

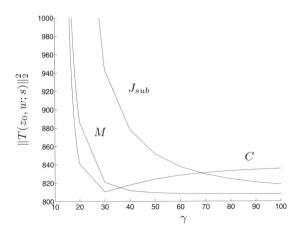

Figure 6.9: The performances of the controllers M and C

Example 6.3 We solve again the Example 5.7 but making use of the results of this section. Consider the system (6.84) - (6.87) with

$$A = \begin{bmatrix} 0 & 1 \\ 2 & 1 \end{bmatrix}, \quad B_1 = \begin{bmatrix} 0 \\ 0 \end{bmatrix}, \quad B_2 = \begin{bmatrix} 0 \\ 1 \end{bmatrix}$$

$$C_1 = \begin{bmatrix} 0 & 0 \end{bmatrix}, \quad C_2 = \begin{bmatrix} -1 & 1 \end{bmatrix}, \quad D_{12} = D_{21} = 1$$

$$C_0 = \begin{bmatrix} -1 & 0 \end{bmatrix}, \quad D_0 = 1$$

The Partial information problem is solvable for $\gamma > 12.0$ hence we consider the interval $12.5 \le \gamma \le 100.0$. In this particular case the auxiliary plant does not depend on γ, however the Riccati equation (6.90) admits a positive solution only if $\gamma > 12.0$. The State feedback design has been applied to the auxiliary plant and the mixed problem provided the controller M which minimizes the upper bound introduced in part b) of Lemma 6.2. Fig. 6.9 shows the actual value of $\|T(z_0, w; s)\|_2^2$ as well as the minimum upper bound J_{sub} as γ varies in the given interval. The same figure shows also the performance in terms of $\|T(z_0, w; s)\|_2^2$ when the RH_∞ central controller is used (it is indicated by C). It is interesting to observe that for moderate values of γ (approximately in the interval $12.5 \le \gamma \le 35.0$) the central controller is even better than the mixed controller as far as the RH_2 norm of the transfer function $T(z_0, w; s)$ is concerned. As expected, this behavior is reversed for large values of γ when so the central controller performs worse than the mixed controller. This example is important because it shows practically the existence of systems for which the mixed design does not furnish a good solution to the problem under consideration. However, it is important to stress that the mixed problem as introduced here, is convex and due to this fact it can be solved very efficiently. □

6.4 RH_2 control with regional pole placement

The performance of a system can be expressed in terms of RH_2 and RH_∞ of certain closed loop transfer functions. However, as it is simple to notice from the results of the last sections, the pole locations of the resulting controlled system are naturally defined by the optimality conditions of the associated optimal control problem. For

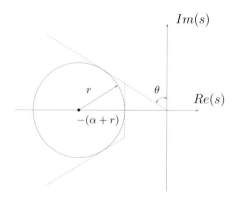

Figure 6.10: The pole placement region

instance, once a mixed RH_2/RH_∞ design problem is solved, the optimal controller imposes the closed-loop poles to be in certain places of the open left complex plane which may not be changed by the designer. The present section aims to combine the RH_2 design with regional pole placement.

In practice, a very popular design specification is expressed as

$$\xi \geq \xi_{min} \tag{6.97}$$

and

$$\xi \omega_n \geq \alpha \tag{6.98}$$

where the constraint (6.97) imposes a minimum damping ratio and the constraint (6.98) imposes a minimum decay of the time response of the closed-loop system. This is illustrated in Fig. 6.10 where it is also represented the circular region

$$\mathcal{R} := \{s \; : \; |s + (\alpha + r)| < r\} \tag{6.99}$$

with $\alpha > 0$ and radius $r > 0$. The radius r can be calculated such that this region is inside the sector defined by constraints (6.97) and (6.98) and so meets the closed loop poles location requirements. The circular region as indicated in Fig. 6.10, tangent to the sector boundary defined by $\alpha > 0$ and $\theta = sin^{-1}(\xi_{min})$, is given in (6.99) with

$$r = \frac{\alpha cos(\theta)}{1 - cos(\theta)}$$

The rationale behind the choice of the circular region $\mathcal{R}$ instead of the sector will be clear in the sequel. The main point is that the circular region imposes convex constraints on the optimal RH_2 control design problem. The next lemma characterizes by means of a modified Lyapunov inequality the matrices with all eigenvalues inside a given circular region $\mathcal{R}$.

Lemma 6.3 *Let the circular $\mathcal{R}$ be given. Matrix A with dimension $n \times n$ has all its eigenvalues inside $\mathcal{R}$ if and only if for any matrix $Q = Q' > 0$, there exists a matrix $P = P' > 0$ such that*

$$0 \geq A_\alpha P + P A_\alpha' + r^{-1} A_\alpha P A_\alpha' + Q \tag{6.100}$$

where $A_\alpha := A + \alpha I$.

Proof Let us first prove the necessity. Assume all eigenvalues of matrix A are in the circular region $\mathcal{R}$. It is a simple matter to verify that all eigenvalues $\tilde{\lambda}$ of the matrix $\tilde{A} := r^{-1}[A + (\alpha + r)I]$ are such that $|\tilde{\lambda}| < 1$ which implies that matrix

$$\bar{A} := (\tilde{A} - I)^{-1}(\tilde{A} + I)$$

is stable. From the Extended Lyapunov lemma we can say that for any $Q = Q' > 0$ chosen, there exits $P = P' > 0$ satisfying the linear inequality

$$0 \geq \bar{A}P + P\bar{A}' + 2r^{-1}(\tilde{A} - I)^{-1}Q(\tilde{A}' - I)^{-1}$$

or equivalently

$$0 \geq \tilde{A}P\tilde{A}' - P + r^{-1}Q$$

Using the definition of matrix A_α we then conclude that inequality (6.100) holds. For the sufficiency, take x an eigenvector of matrix A' associated to an arbitrary eigenvalue λ. Multiplying inequality (6.100) to left by $x^\sim$ and to the right by x it follows that

$$0 \geq \left[2Re(\lambda + \alpha) + r^{-1}|\lambda + \alpha|^2\right] x^\sim Px + x^\sim Qx$$

which together with the positive definiteness of both involved matrices provides

$$\left[2Re(\lambda + \alpha) + r^{-1}|\lambda + \alpha|^2\right] < 0$$

Finally, using this fact we get

$$\begin{aligned} r^{-2}|\lambda + \alpha + r|^2 &= r^{-2}\left[|\lambda + \alpha|^2 + 2Re(\lambda + \alpha)r + r^2\right] \\ &= r^{-1}\left[2Re(\lambda + \alpha) + r^{-1}|\lambda + \alpha|^2\right] + 1 \\ &< 1 \end{aligned}$$

which proves that $\lambda \in \mathcal{R}$. The sufficiency is proved because A and A' have the same eigenvalues. □

For a given circular region $\mathcal{R}$ and a given matrix A, the inequality (6.100) defines a convex constraint with respect to P. This is clearly true because its right hand side is an affine function of P. More surprisingly is that convexity still holds when A is not constant but depends upon a state feedback gain matrix.

6.4.1 State feedback design

Let a circular region $\mathcal{R}$ be given. The dynamic system under consideration has the following state space representation

$$\dot{x} = Ax + B_1w + B_2u \tag{6.101}$$
$$z = C_1x + D_{12}u \tag{6.102}$$
$$y = x \tag{6.103}$$

where the only assumption we a priori need is i) $D'_{12}D_{12} = I$. The goal is to determine a state feedback control law of the form

$$u = Fx \tag{6.104}$$

such that the gain matrix F solves the optimal control problem

$$\min\left\{\|T(z,w;s)\|_2^2 \ : \ F \in \mathcal{K}_R\right\}$$

where $\mathcal{K}_R$ denotes the set of matrices F such that all eigenvalues of matrix $A_{cc} = A + B_2 F$ are in the circular region $\mathcal{R}$. It is clear that since $\mathcal{R}$ is a circular region on the left part of the complex plane then $F \in \mathcal{K}_R$ always provides $T(z,w;s)$ in RH_2. Of course, we need the set $\mathcal{K}_R$ be nonempty which is the case if and only if the unreachable part of (A, B_2) has all eigenvalues belonging to $\mathcal{R}$. This problem is not exactly solved. Instead we propose here an overbounding objective function which has the main advantage to preserve convexity. Throughout this section we redefine the matrix

$$M_c := \left[\begin{array}{cc} A + \alpha I & B_2 \\ 0 & 0 \end{array} \right]$$

which has the same structure of matrix M_c defined in (6.6). All other matrices remain unchanged.

Theorem 6.9 *Consider $B_1 B_1' > 0$, let the circular region $\mathcal{R}$ be given and define the set*

$$\mathcal{C}_R := \{W \ : \ W \geq 0, \ v'\Theta_R(W)v \leq 0, \ \forall v \in \mathcal{N}_c\} \tag{6.105}$$

where $\Theta_R(W) := \Theta_c(W) + r^{-1} M_c W M_c'$. The following hold

a) *$\mathcal{C}_R$ is a convex set.*

b) *Each $W \in \mathcal{C}_R$ is such that $W_1 > 0$.*

c) *$\mathcal{K}_R = \{W_2' W_1^{-1} \ : \ W \in \mathcal{C}_R\}$.*

d) *The optimal solution $\bar{W}$ of the convex programming problem*

$$J_{sub} := \min\{\mathrm{trace}[R_c W] \ : \ W \in \mathcal{C}_R\} \tag{6.106}$$

provides $F = \bar{W}_2' \bar{W}_1^{-1} \in \mathcal{K}_R$ such that the upper bound of the RH_2 norm $\|T(z,w;s)\|_2 \leq \mathrm{trace}[R_c W]$, valid for all $W \in \mathcal{C}_R$, is minimized.

Proof The proof of *Points a) and b)* follow immediately from the fact that $\mathcal{C}_R \subset \mathcal{C}_c$ and the matrix function $\Theta_R(W)$ is affine together with Theorem 6.1.

Point c) Assume $\mathcal{K}_R \neq \emptyset$, for an arbitrary $F \in \mathcal{K}_R$, Lemma 6.3 applies for $Q = B_1 B_1'$ providing thus a symmetric and positive definite matrix P satisfying the inequality

$$0 \geq (A_{cc} + \alpha I)P + P(A_{cc} + \alpha I)' +$$
$$+r^{-1}(A_{cc} + \alpha I)P(A_{cc} + \alpha I)' + B_1 B_1'$$

Choosing

$$W = \left[\begin{array}{cc} P & PF' \\ FP & FPF' \end{array} \right]$$

it is seen that $W \geq 0$ and all $v \in \mathcal{N}_c$ yields

$$v'\Theta_R(W)v' = x'\left[(A_{cc} + \alpha I)P + P(A_{cc} + \alpha I)' +\right.$$
$$\left. +r^{-1}(A_{cc} + \alpha I)P(A_{cc} + \alpha I)' + B_1 B_1'\right] x$$
$$\leq 0$$

implying that $W \in \mathcal{C}_R$. Since, in addition $W_2'W_1^{-1} = FPP^{-1} = F$, the necessity is proved. Conversely, with $\mathcal{C}_R \neq \emptyset$, for any $W \in \mathcal{C}_R$ and any $v \in \mathcal{N}_c$ we get

$$
\begin{aligned}
0 &\geq v'\Theta_R(W)v \\
&\geq x'\left[(A_\alpha + B_2W_2'W_1^{-1})W_1 + W_1(A_\alpha + B_2W_2'W_1^{-1})' + \right.\\
&\quad +r^{-1}(A_\alpha + B_2W_2'W_1^{-1})W_1(A_\alpha + B_2W_2'W_1^{-1})' + B_1B_1' + \\
&\quad \left.+r^{-1}B_2(W_3 - W_2'W_1^{-1}W_2)B_2'\right]x \\
&\geq x'\left[(A_\alpha + B_2W_2'W_1^{-1})W_1 + W_1(A_\alpha + B_2W_2'W_1^{-1})' + \right.\\
&\quad \left.+r^{-1}(A_\alpha + B_2W_2'W_1^{-1})W_1(A_\alpha + B_2W_2'W_1^{-1})' + B_1B_1'\right]x
\end{aligned}
$$

which shows, from Lemma 6.3 that $F = W_2'W_1^{-1} \in \mathcal{K}_R$. The result follows from the fact that $\mathcal{K}_R = \emptyset$ implies $\mathcal{C}_R = \emptyset$ and vice versa.

Point d) This point follows from the fact that $\bar{W} \in \mathcal{C}_R$ generates $F = \bar{W}_2'\bar{W}_1^{-1} \in \mathcal{K}_R$. Furthermore, any $W \in \mathcal{C}_R$ satisfies

$$
0 \geq A_{cc}W_1 + W_1A_{cc}' + B_1B_1'
$$

which imposes that $W_1 \geq P_r$, where

$$
0 = A_{cc}P_r + P_rA_{cc}' + B_1B_1'
$$

Simple calculation of the RH_2 norm then shows that

$$
\begin{aligned}
\text{trace}[R_cW] &= \text{trace}\left[(C_1 + D_{12}W_2'W_1^{-1})W_1(C_1 + D_{12}W_2W_1^{-1})'\right] + \\
&\quad +\text{trace}\left[D_{12}(W_3 - W_2'W_1^{-1}W_2)D_{12}'\right] \\
&\geq \text{trace}\left[(C_1 + D_{12}W_2W_1^{-1})P_r(C_1 + D_{12}W_2W_1^{-1})'\right] \\
&\geq \|T(z,w;s)\|_2^2
\end{aligned}
$$

holds for all $W \in \mathcal{C}_R$ and the proof is complete. □

If the convex programming problem (6.106) admits a solution then the optimal feedback gain places the closed-loop poles in a desired circular region and minimizes an upper bound of $\|T(z,w;s)\|_2^2$. Clearly, we can not say that this norm has been minimized.

Remark 6.31 Since $\mathcal{C}_R \subset \mathcal{C}_c$, the same conclusion related to the boundedness of the feasible set of problem (6.106) applies. Its numeric solution by means of convex programming methods depends on the assumption that the pair $(-A_c, C_{1c})$ is detectable. □

Remark 6.32 As perhaps already occurred to the reader, there are many other regions in the complex plane that can be recasted in the same framework of the circular region.

For $\alpha > 0$ fixed and r arbitrarily large, the circular region degenerates to

$$
\mathcal{R} = \{s : Re(s) < -\alpha\}
$$

and Lemma 6.3 and Theorem 6.9 still hold.

Any region in the complex plane that can be written as a convex set of matrices W such that $W_1 \geq P_r$, is also handled with no additional theoretical difficulty. One of such regions is the vertical strip defined as

$$
\mathcal{R}_S := \{s : -\beta < Re(s) < -\alpha\}
$$

with $\beta > \alpha > 0$. Notice however, that we have to work with the intersection of two regions generating thus a feasible set which is the intersection of two convex constraints in the same variable W. Consequently, only the sufficient part of Theorem 6.9 still holds. □

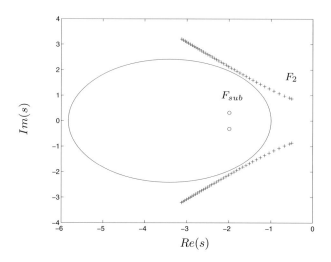

Figure 6.11: Closed-loop poles

Example 6.4 (Root locus) Let us consider the system (6.101) - (6.103) with

$$A = \begin{bmatrix} 0 & 1 \\ -1 & 1 \end{bmatrix}, \quad B_1 = \begin{bmatrix} 1 \\ 0 \end{bmatrix}, \quad B_2 = \begin{bmatrix} 0 \\ 1 \end{bmatrix}$$

$$C_1 = \begin{bmatrix} \rho & 0 \\ 0 & 0 \end{bmatrix}, \quad D_{12} = \begin{bmatrix} 0 \\ 1 \end{bmatrix}$$

where $0 \le \rho \le 20$.

First, the State feedback problem in RH_2 has been solved for all values of ρ in the given interval. The optimal feedback gain F_2 as a function of this parameter has been determinated. The root locus of the closed-loop system is then plotted in fig. 6.11. By inspection, it is possible to verify that the root locus never enters in the circular region defined by the parameters $\alpha = 1$ and $r = 1/(\sqrt{2} - 1)$. For $\rho = 0$ the optimal gain is $F_2 = [0 \ -2]$ which imposes to the system a performance such that $\|T(z,w;s)\|_2^2 = 2$.

Again, for $\rho = 0$ and imposing the above circular region for pole placement, the convex programming problem (6.106) has been solved, providing $F_{sub} = [-3.0678 \ -4.9841]$ and the associated upper bound on the minimum cost $J_{sub} = 51.9434$. The closed-loop poles are also shown in fig. 6.11. This example illustrates a very important fact. Indeed, there is no possibility to choose a penalty term in the (1,1) entry of matrix C_1 such that the poles of the closed-loop system are all inside the circular region which corresponds to a minimum damping factor $\xi_{min} = 1/\sqrt{2}$ or in other terms $\theta = \pi/4$ radians. Of course the design procedure introduced in Theorem 6.9 circumvents this drawback of the pure State feedback design in RH_2.   □

6.4.2   Output feedback design

The system to be dealt with is of the form

$$\dot{x} = Ax + B_1 w + B_2 u \tag{6.107}$$

$$z = C_1 x + D_{12} u \tag{6.108}$$

$$y = C_2 x + D_{21} w \tag{6.109}$$

and the controller with transfer function $K(s)$ has to be determined in order to place the closed-loop poles in a desired circular region while the quantity $\|T(z, w; s)\|_2$ is minimized. Let us elaborate more on this point. First all feasible controllers are parametrized as

$$K(s) := \left[\begin{array}{c|c} A + B_2 F + L C_2 & -L \\ \hline F & 0 \end{array} \right] \tag{6.110}$$

for all pairs of matrices (F, L). The rationale behind this choice has been put on evidence before. The closed-loop system has the state space representation (recall fig. 6.5)

$$\Sigma_F := \left[\begin{array}{c|c} \tilde{A} & \tilde{B} \\ \hline \tilde{C} & 0 \end{array} \right]$$

where the indicated matrices are

$$\tilde{A} = \left[\begin{array}{cc} A + B_2 F & B_2 F \\ 0 & A + L C_2 \end{array} \right]$$

and

$$\tilde{B} = \left[\begin{array}{c} B_1 \\ -B_1 - L D_{21} \end{array} \right], \quad \tilde{C} = \left[\begin{array}{cc} C_1 + D_{12} F & D_{12} F \end{array} \right]$$

From the very particular structure of matrix $\tilde{A}$ it is apparent that the closed loop poles are those of matrix $A + B_2 F$ and $A + L C_2$. As far as pole placement is under consideration, in principle it is possible to determine (F, L) such that all poles of the closed loop system lie in some region of the complex plane. However, since the state reconstruction from the output depends only on matrix L, in our present design procedure it is imposed as the optimal solution of the Output estimation problem, that is $L = L_2$. Thus, for a given circular region $\mathcal{R}$, the Output feedback design problem is formulated as

$$\min \left\{ \|T(z, w; s)\|_2^2 \ : \ F \in \mathcal{K}_R \right\}$$

where it is only necessary to make explicit the dependence of $\|T(z, w; s)\|_2^2$ with F. The optimal solution of this problem is not possible to be determined exactly. So, we proceed by overbounding its objective function. The assumptions i) The pair (A, C_2) is detectable and no eigenvalue of the unreachable part of the pair $[(A - B_1 D_{21}' C_2), B_1 (I - D_{21}' D_{21})]$ lies on the imaginary axis and ii) $D_{12}' D_{12} = I$ and $D_{21} D_{21}' = I$ are made. Under these assumptions, there exists Π_2 a positive semidefinite and stabilizing solution of the Riccati equation (in the unknown Π)

$$0 = \Pi A_f' + A_f \Pi - \Pi C_2' C_2 \Pi + B_{1f} B_{1f}' \tag{6.111}$$

which provides $L_2 = -\Pi_2 C_2' - B_1 D_{21}'$. The controller defined in this way has the following important design property which are obtained as a limit case of Lemma 6.2.

Lemma 6.4 *For all $F \in \mathcal{K}_c$, the symmetric and positive semidefinite solution to the Lyapunov equation*

$$0 = (A + B_2 F) X + X (A + B_2 F)' + L_2 L_2' \tag{6.112}$$

is such that the controller with transfer function $K(s)$ given in (6.110) imposes to the closed-loop system the performance

$$\|T(z, w; s)\|_2^2 = \text{trace}[C_1 \Pi_2 C_1'] + \text{trace}[(C_1 + D_{12} F) X (C_1 + D_{12} F)'] \tag{6.113}$$

Proof The state space representation of the closed-loop system being given by Σ_F, enables us to get immediately (recall Remark 6.14)

$$\|T(z,w;s)\|_2^2 = \text{trace}[\tilde{C}\tilde{P}_r\tilde{C}']$$

where $\tilde{P}_r$, solution of the linear equation (6.36) is given by

$$\tilde{P}_r = \begin{bmatrix} \Pi_2 + X & -\Pi_2 \\ -\Pi_2 & \Pi_2 \end{bmatrix}$$

Simple substitution shows that equality (6.113) holds, proving thus the lemma proposed.   □

This result opens the possibility to reduce the Output feedback design problem to the previous State feedback design problem. In fact, once the equality $L = L_2$ holds, the transfer function $K(s)$ of the output feedback controller is completely parametrized by matrix F only. The impact of this decision in the global cost is the first term in (6.113) which does not need to be considered further since it remains constant for all possible choices of F. Hence, if the state feedback design problem is applied to the *auxiliary* plant

$$\dot{x} = Ax + L_2 D_{21} w + B_2 u \tag{6.114}$$

$$z = C_1 x + D_{12} u \tag{6.115}$$

$$y = x \tag{6.116}$$

that is, if problem (6.106) is solved with B_1 replaced by $L_2 D_{21}$ then the optimal gain $F = W_2' W_1^{-1}$ is such that the controller

$$K(s) := \left[\begin{array}{c|c} A + B_2 W_2' W_1^{-1} + L_2 C_2 & -L_2 \\ \hline W_2' W_1^{-1} & 0 \end{array} \right] \tag{6.117}$$

imposes all the eigenvalues of matrix $A + B_2 W_2' W_1^{-1}$ in the circular region $\mathcal{R}$ while the upper bound

$$\|T(z,w;s)\|_2 \leq \sqrt{\text{trace}[C_1 \Pi_2 C_1'] + J_{sub}}$$

is globally minimized. Unfortunately, as illustrated before, the global minimization of this upper bound does not necessarily means that the global optimum of the true design problem is attained.

6.5   Time-domain specifications

One of the most important time-domain specifications of control systems design is the limitation, to some prespecified level, of the time-response overshoot. This section is completely devoted to generalize the previous results to this particular situation. Once again, the important feature is that convexity is preserved and similar manipulations for both the state and output feedback cases are allowed. Some few preliminary calculations are needed before we define and solve the associated optimal control problem.

Consider the system, specified by the following state space minimal realization

$$\dot{x} = Ax + Bw \ , \quad x(0) = 0 \tag{6.118}$$
$$z = Cx \tag{6.119}$$

where matrix A is assumed to be stable, the pair (A, B) is reachable and $w \in RL_2[0 \ \infty)$. The quantity we want to compute depends on the transfer function $T(z, w; s)$ of the above system and is defined as

$$\mathcal{G}(T(z, w; s)) := \sup_{\|w\|_2 \leq 1} \|z\|_\infty \tag{6.120}$$

where, for the norm of the output variable $\|z\|_\infty$, two different cases are considered, namely

$$\|z\|_\infty := \sup_{t \geq 0} \sqrt{z'(t) z(t)} \tag{6.121}$$

and

$$\|z\|_\infty := \sup_{t \geq 0} \max_i |z_i(t)| \tag{6.122}$$

where $z_i(t)$ denotes the i-th scalar component of $z(t)$. We are now in position to interpret the function $\mathcal{G}(\cdot)$ defined above. Suppose, for a certain positive scalar α,

$$\mathcal{G}(T(z, w; s)) \leq \alpha$$

then for each time $t \geq 0$ the worst case overshoot of $z(t)$ is limited by α. Hence, the possibility to take into account this time constraint in control system design is of great practical interest.

Lemma 6.5 *Let P be the symmetric and positive definite solution of the Lyapunov equation*

$$0 = AP + PA' + BB' \tag{6.123}$$

The following are true

a) *For the norm (6.121) then $\mathcal{G}(T(z, w; s)) = \lambda_{max}^{1/2}(CPC')$ where $\lambda_{max}(\cdot)$ denotes the maximum eigenvalue of $(\cdot)$.*

b) *For the norm (6.122) then $\mathcal{G}(T(z, w; s)) = d_{max}^{1/2}(CPC')$ where $d_{max}(\cdot)$ denotes the maximum diagonal element of $(\cdot)$.*

Proof Define $v(x) := x'P^{-1}x$ and consider the system (6.118) - (6.119) with an arbitrary input w satisfying $\|w\|_2 \leq 1$. The time derivative of $v(\cdot)$ along a trajectory of that system yields

$$\begin{aligned}
\dot{v}(x) &= x'(A'P^{-1} + P^{-1}A)x + 2w'B'P^{-1}x \\
&= -x'P^{-1}BB'P^{-1}x + 2w'B'P^{-1}x \\
&= w'w - (w - B'P^{-1}x)'(w - B'P^{-1}x) \\
&\leq \|w\|^2
\end{aligned}$$

which, after integration of both sides from 0 to $t \geq 0$ provides

$$v(x(t)) = x'(t)P^{-1}x(t) \leq \|w\|_2^2 \leq 1$$

This inequality means that, in the state space, the trajectories $x(t)$, for all $t \geq 0$ are confined in the set $x'P^{-1}x \leq 1$ whenever w remains bounded by $\|w\|_2 \leq 1$.

Point a) With $z = Cx$ and $\tilde{x} := P^{-1/2}x$, we have

$$\begin{aligned}
\mathcal{G}(T(z,w;s))^2 &= \sup_{\|w\|_2 \leq 1} \|z\|_\infty^2 \\
&\leq \max\left\{x'C'Cx \; ; \; x'P^{-1}x \leq 1\right\} \\
&\leq \max\left\{\tilde{x}'P^{1/2}C'CP^{1/2}\tilde{x} \; ; \; \tilde{x}'\tilde{x} \leq 1\right\} \\
&\leq \lambda_{max}(P^{1/2}C'CP^{1/2}) \\
&\leq \lambda_{max}(CPC')
\end{aligned}$$

and it remains to show that there exists a feasible trajectory $w(t)$ such that $\|z\|_\infty^2$ is arbitrarily close to $\lambda_{max}(CPC')$. To this end, consider $T > 0$ fixed but arbitrary

$$0 < S(T) := \int_0^T e^{At}BB'e^{A't}dt \leq P$$

and the input signal such that $w(t) = 0$ for all $t > T$ and

$$w(t) = B'e^{A'(T-t)}S(T)^{-1/2}\psi \; , \quad 0 \leq t \leq T$$

where ψ is a vector to be determined. Simple calculations show that

$$\|w\|_2^2 = \int_0^\infty w'(t)w(t)dt = \int_0^T w'(t)w(t)dt = \psi'\psi$$

and

$$z(T) = C\int_0^T e^{A(T-\tau)}Bw(\tau)d\tau = CS(T)^{1/2}\psi$$

Consequently, choosing ψ as being the unitary norm eigenvector associated to the maximum eigenvalue of matrix $S(T)^{1/2}C'CS(T)^{1/2}$, the feasibility of the input signal w is guaranteed and

$$\begin{aligned}
\|z\|_\infty^2 &= \sup_{t \geq 0} z'(t)z(t) \\
&\geq \psi'S(T)^{1/2}C'CS(T)^{1/2}\psi \\
&\geq \lambda_{max}(CS(T)C')
\end{aligned}$$

the proof is then concluded because $S(T)$ becomes arbitrarily close to P as T increases.

Point b) With $z_i = C_ix$ with C_i being the i-th row of matrix C and $\tilde{x} := P^{-1/2}x$, we have

$$\begin{aligned}
\mathcal{G}(T(z,w;s))^2 &= \sup_{\|w\|_2 \leq 1} \|z\|_\infty^2 \\
&\leq \max_{x,i}\left\{x'C_i'C_ix \; ; \; x'P^{-1}x \leq 1\right\} \\
&\leq \max_i \max_{\tilde{x}}\left\{\tilde{x}'P^{1/2}C_i'C_iP^{1/2}\tilde{x} \; ; \; \tilde{x}'\tilde{x} \leq 1\right\} \\
&\leq \max_i \lambda_{max}(P^{1/2}C_i'C_iP^{1/2}) \\
&\leq \max_i C_iPC_i' \\
&\leq d_{max}(CPC')
\end{aligned}$$

As before, it remains to determine a feasible input such that the equality holds. This is accomplished by the same function $w(t)$ already defined and a convenient choice of vector ψ. Indeed, take ψ_i as being the unitary norm eigenvector associated to the maximum eigenvalue of matrix $S(T)^{1/2}C_i'C_iS(T)^{1/2}$, and choose $\psi = \psi_l$ where $C_iS(T)C_i' \leq C_lS(T)C_l'$ for all index $i = 1, 2, \cdots$, then w is feasible and

$$\|z\|_\infty^2 = \sup_{t \geq 0} \max_i z_i'(t)z_i(t)$$
$$\geq \max_i \psi' S(T)^{1/2}C_i'C_iS(T)^{1/2}\psi$$
$$\geq C_lS(T)C_l'$$
$$\geq d_{max}(CS(T)C')$$

the proof is then concluded since as said before, $S(T)$ becomes arbitrarily close to P as T increases. $\qquad\square$

Remark 6.33 In the proof of Lemma 6.5, it is assumed that the solution of the linear matrix equation (6.123) is positive definite. This occurs whenever the pair (A, B) is reachable. If this assumption is not verified, the result still holds true. In this case, using Kalman's canonical controllability form, it is immediate to see that the output $z(t)$ in (6.119) depends only on the reachable part of the system. $\qquad\square$

Remark 6.34 The relationship between both $\mathcal{G}(\cdot)$ for the norms (6.121) and (6.122) are

$$d_{max}(CPC') \leq \lambda_{max}(CPC') \leq \text{trace}(CPC')$$

which also implies, in both cases, that $\mathcal{G}(T(z, w; s)) \leq \|T(z, w; s)\|_2$. $\qquad\square$

Remark 6.35 (Convexity) The real valued function $g(X) : \mathcal{P} \longrightarrow R$ defined as $g(X) := \lambda_{max}(X)$ is convex (recall Remark 6.3). The same is true for the function $g(X) := d_{max}(X)$. To prove this, take $X_0 \in \mathcal{P}$ and e_0 the column of the identity matrix such that $e_0'X_0e_0 = g(X_0)$. For all $X \in \mathcal{P}$ we get

$$g(X) = d_{max}(X)$$
$$\geq e_0'Xe_0$$
$$\geq g(X_0) + < e_0e_0', X - X_0 >$$

and inequality (6.1) is verified for $\Lambda_0 = e_0e_0'$, then convexity follows. Notice further that both functions are not differentiable in $\mathcal{P}$ and are *non decreasing* functions in the sense that for any $X_1, X_2 \in \mathcal{P}$ such that $X_1 \leq X_2$ then $g(X_1) \leq g(X_2)$. $\qquad\square$

Example 6.5 Consider the system (6.118) - (6.119) with

$$A = \begin{bmatrix} 0 & 1 & 0 \\ 0 & 0 & 1 \\ -10 & -9 & -4 \end{bmatrix} , \quad B = \begin{bmatrix} 2 \\ -2 \\ 4 \end{bmatrix} , \quad C = \begin{bmatrix} 0 & 1 & 0 \\ 0 & 0 & 1 \end{bmatrix}$$

Our purpose is to illustrate the result of Lemma 6.5. The definite positive solution P of the Lyapunov equation (6.123), provides

$$\mathcal{G}(T(z, w; s)) = \begin{cases} \lambda_{max}^{1/2}(CPC') &= 2.16 \\ d_{max}^{1/2}(CPC') &= 1.68 \end{cases}$$

On the other hand, taking $T = 5$ it can be verified that $S(T) \approx P$ and so, corresponding to the input

$$w(t) = \begin{cases} B'e^{A'(T-t)}P^{-1/2}\psi & , \quad 0 \leq t \leq T \\ 0 & , \quad\qquad t > T \end{cases}$$

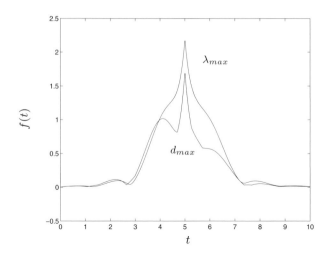

Figure 6.12: Norms of the output $z(t)$

the system produces the output $z(t)$ which enables us to calculate the time varying function

$$f(t) := \left\{ \begin{array}{l} \sqrt{z_1(t)^2 + z_2(t)^2} \\ \max\{|z_1(t)|, |z_2(t)|\} \end{array} \right.$$

for each norm used to define $\mathcal{G}(\cdot)$. These functions are shown in fig 6.12 where the labels λ_{max} and d_{max} identify the norms introduced in (6.121) and (6.122) respectively. In fig. 6.13 the corresponding inputs $w(t)$ are also shown. It is interesting to verify that in both cases

$$\max_{t \geq 0} f(t) \approx \mathcal{G}(T(z, w; s))$$

and $\|w\|_2 \approx 1$ as required in the proof of Lemma 6.5. $\square$

Throughout the remaining of this section, we define the convex function

$$g(CPC') := \mathcal{G}(T(z, w; s))^2 \tag{6.124}$$

to indicate both cases treated before. Since this function is convex in the domain $\mathcal{P}$, the constraint $g(CPC') \leq \alpha$, for $\alpha > 0$ fixed, can be handled with no additional difficulty because convexity is preserved. The same obviously occurs if $g(CPC')$ is used as a objective function to be minimized. This case, generalizes the control design problem in RH_2 in the sense that it is obtained from the above formulation for the choice $g(CPC') = \text{trace}(CPC')$.

6.5.1   State feedback design

The dynamic system is described by the equations

$$\dot{x} = Ax + B_1 w + B_2 u \tag{6.125}$$
$$z = C_1 x + D_{12} u \tag{6.126}$$
$$y = x \tag{6.127}$$

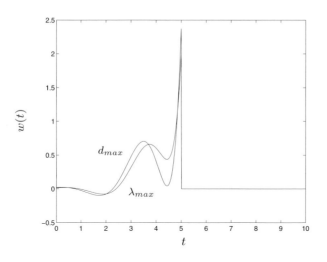

Figure 6.13: The inputs $w(t)$

where our purpose is to determine the optimal matrix F such that with $u = Fx$ the following control problem is solved

$$\min \left\{ \mathcal{G}(T(z, w; s)) \; : \; F \in \mathcal{K}_c \right\} \tag{6.128}$$

To this end we only consider the assumption i) $D'_{12} D_{12} = I$. For each $F \in \mathcal{K}_c$ then $T(z, w; s)$ is in RH_2 which means that if $\mathcal{K}_c \neq \emptyset$ then (6.128) is well-posed. The main feature of the result to be presented is that the optimal global solution of (6.128) is actually attained.

Theorem 6.10 *Consider $B_1 B'_1 > 0$ and define the matrix*

$$\bar{R}_c := \begin{bmatrix} C_1 & D_{12} \end{bmatrix} \tag{6.129}$$

The convex programming problem

$$J_{opt} := \min \left\{ g(\bar{R}_c W \bar{R}'_c) \; : \; W \in \mathcal{C}_c \right\} \tag{6.130}$$

is equivalent to problem (6.128) in the sense that both present the same global solution.

Proof Recall Theorem 6.1 where it is proved that all elements of the set $\mathcal{K}_c$ are generated from those of $\mathcal{C}_c$. Hence, as far as feasibility is concerned, both problems are equivalent. We proceed by assuming that problem (6.128) has an optimal solution. In this case, from Lemma 6.5 there exist matrices $\bar{P} > 0$ and $\bar{F} \in \mathcal{K}_c$ such that

$$g[(C_1 + D_{12}\bar{F})\bar{P}(C_1 + D_{12}\bar{F})'] = \min \mathcal{G}(T(z, w; s))$$

and

$$0 = (A + B_2\bar{F})\bar{P} + \bar{P}(A + B_2\bar{F})' + B_1 B'_1$$

Defining the matrix

$$\bar{W} := \begin{bmatrix} \bar{P} & \bar{P}\bar{F}' \\ \bar{F}\bar{P} & \bar{F}\bar{P}\bar{F}' \end{bmatrix}$$

simple algebraic manipulations yield the conclusion that $\bar{W} \in \mathcal{C}_c$ provides the minimum cost. It remains to prove that $\bar{W}$ is the global optimum of problem (6.130). To this end, taking any feasible W we have $F = W_2 W_1^{-1} \in \mathcal{K}_c$ and

$$\mathcal{G}(T(z,w;s)) = g[(C_1 + D_{12}W_2'W_1^{-1})P_r(C_1 + D_{12}W_2'W_1^{-1})]$$

where $W_1 \geq P_r$. Using again the fact that $W \geq 0$ together with the Schur complement formula we get

$$
\begin{aligned}
g(\bar{R}_c \bar{W} \bar{R}_c') = \min \mathcal{G}(T(z,w;s)) \\
\leq \mathcal{G}(T(z,w;s)) \\
\leq g[(C_1 + D_{12}W_2'W_1^{-1})W_1(C_1 + D_{12}W_2'W_1^{-1})] \\
\leq g[\bar{R}_c W \bar{R}_c' - D_{12}(W_3 - W_2'W_1^{-1}W_2)D_{12}'] \\
\leq g[\bar{R}_c W \bar{R}_c']
\end{aligned}
$$

which being true for all $W \in \mathcal{C}_c$ completes the proof. $\qquad\square$

The proof of this theorem is almost the same of that of Theorem 6.3 where the special case $g(\cdot) = \text{trace}(\cdot)$ has been considered. Notice further that the above proof depends basically on the convexity and on the non decreasing property of the function $g(\cdot)$ introduced in Remark 6.35. Consequently, the same result also applies to any other function with these properties.

6.5.2   Output feedback design

Once again consider the system

$$\dot{x} = Ax + B_1 w + B_2 u \tag{6.131}$$
$$z = C_1 x + D_{12} u \tag{6.132}$$
$$y = C_2 x + D_{21} w \tag{6.133}$$

where the transfer function $K(s)$ of the output controller is to be determined. It is adopted the same reasoning as before, that is all feasible controllers are completely parametrized by only one matrix F which is used to meet the design requirements. Its transfer function is given by

$$K(s) := \left[\begin{array}{c|c} A + B_2 F + L_2 C_2 & -L_2 \\ \hline F & 0 \end{array} \right] \tag{6.134}$$

being thus apparent that the choice $L = L_2$ is the best we can do as far as the reconstruction of the state from the output is concerned. Let us keep in mind that $L_2 = -\Pi_2 C_2' - B_1 D_{21}'$, where Π_2 is the positive semidefinite and stabilizing solution of the Riccati equation (in the unknown Π)

$$0 = \Pi A_f' + A_f \Pi - \Pi C_2' C_2 \Pi + B_{1f} B_{1f}' \tag{6.135}$$

The feedback connection drawn in fig. 6.5 puts in evidence that the internal stability of the closed-loop system is assured if $F \in \mathcal{K}_c$ and so our goal is to solve the associated optimal control problem

$$\min \{\mathcal{G}(T(z,w;s)) \ : \ F \in \mathcal{K}_c\}$$

The next lemma provides the generalization of the result introduced in Lemma 6.4 to deal with its objective function.

Lemma 6.6 *For all $F \in \mathcal{K}_c$ the symmetric and positive semidefinite solution to the Lyapunov equation*

$$0 = (A + B_2 F)X + X(A + B_2 F)' + L_2 L_2' \qquad (6.136)$$

is such that the controller with transfer function $K(s)$ given in (6.134) imposes to the closed-loop system the performance

$$\mathcal{G}(T(z, w; s)) = g[C_1 \Pi_2 C_1' + (C_1 + D_{12} F)X(C_1 + D_{12} F)'] \qquad (6.137)$$

Proof From Lemma 6.5 together with the state space realization of the closed-loop system, yields

$$\mathcal{G}(T(z, w; s)) = g[\tilde{C} \tilde{P}_r \tilde{C}']$$

where $\tilde{P}_r$, solution of the linear equation (6.36) is given by

$$\tilde{P}_r = \begin{bmatrix} \Pi_2 + X & -\Pi_2 \\ -\Pi_2 & \Pi_2 \end{bmatrix}$$

Furthermore, simple calculations enables us to write

$$\tilde{C} \tilde{P}_r \tilde{C}' = C_1 \Pi_2 C_1' + (C_1 + D_{12} F)X(C_1 + D_{12} F)' \qquad (6.138)$$

which proves the lemma proposed. $\qquad \square$

The equality (6.138) is of particular importance. With it, the specific properties of the function $g(\cdot)$ is not used in the proof of the above lemma. Consequently, the same result also holds for any other function $g(\cdot)$. For those functions under consideration in this section, the Output feedback problem is reduced to the State feedback problem applied to the *auxiliary* plant

$$\dot{x} = Ax + L_2 D_{21} w + B_2 u \qquad (6.139)$$
$$z = C_1 x + D_{12} u \qquad (6.140)$$
$$y = x \qquad (6.141)$$

Doing this, it is important to keep in mind that both, matrix B_1 should be replaced by $L_2 D_{21}$ and the objective function to be minimized over $\mathcal{C}_c$ should be replaced accordingly, leading to

$$J_{opt} = \min \left\{ g[C_1 \Pi_2 C_1' + \bar{R}_c W \bar{R}_c'] \ : \ W \in \mathcal{C}_c \right\} \qquad (6.142)$$

which again is a convex programming problem.

Remark 6.36 Due to the fact that

$$g[C_1 \Pi_2 C_1' + \bar{R}_c W \bar{R}_c'] \leq g[C_1 \Pi_2 C_1'] + g[\bar{R}_c W \bar{R}_c']$$

it is possible to simplify the objective function of Problem (6.142) by retaining in the optimization problem, only the second term in the above expression (since the first one is constant). Unfortunately, this may produces an important degree of sub-optimality because the overbound is, in general, very conservative. $\qquad \square$

Example 6.6 Consider the system (6.139) - (6.141) with the matrices as indicated below

$$A = \begin{bmatrix} 0 & 1 \\ 0 & 0 \end{bmatrix}, \quad B_1 = \begin{bmatrix} 1 & 0 \\ 1 & 0 \end{bmatrix}, \quad B_2 = \begin{bmatrix} 0 \\ 1 \end{bmatrix}, \quad D_{12} = \begin{bmatrix} 0 \\ 1 \end{bmatrix}$$

$$C_1 = \begin{bmatrix} 1 & 0 \\ 0 & 0 \end{bmatrix}, \quad C_2 = \begin{bmatrix} 1 & 0 \end{bmatrix}, \quad D_{21} = \begin{bmatrix} 0 & 1 \end{bmatrix}$$

Solving the Riccati equation (6.135) we get $L_2 = -[1.7321 \ 1]'$ and

$$C_1 \Pi_2 C_1' = \begin{bmatrix} 1.7321 & 0 \\ 0 & 0 \end{bmatrix}$$

Then we solved problem (6.142) considering $g(\cdot) = \text{trace}(\cdot)$, $g(\cdot) = \lambda_{max}(\cdot)$ and $g(\cdot) = d_{max}(\cdot)$. The optimal controllers and the minimum cost associated, for each case are respectively

$$K_{opt}(s) = \frac{-3.14s - 0.99}{s^2 + 3.14s + 4.44}, \quad J_{opt} = 10.85$$

$$K_{opt}(s) = \frac{-4.59s - 1.32}{s^2 + 4.03s + 6.31}, \quad J_{opt} = 8.59$$

$$K_{opt}(s) = \frac{-4.26s - 1.47}{s^2 + 3.44s + 5.44}, \quad J_{opt} = 5.73$$

Each controller and the associated cost are quite different which indicates that the optimum of each problem solved are distinct, even though they obey the inequality given in Remark 6.34. The optimal solution of the upper bound of all $g(\cdot)$ namely trace$(\cdot)$ may furnish a poor suboptimal solution to the other cases under consideration.

Following the discussion in Remark 6.36, we also calculated the following suboptimal cost

$$J_{sub} = g(C_1 \Pi_2 C_1') + \min \left\{ g[\bar{R}_c W \bar{R}_c'] \ : \ W \in \mathcal{C}_c \right\}$$

for each function $g(\cdot)$ as before. The result is

$$K_{sub}(s) = \frac{-3.14s - 0.99}{s^2 + 3.14s + 4.44}, \quad J_{sub} = 10.85$$

$$K_{sub}(s) = \frac{-4.16s - 1.16}{s^2 + 3.88s + 5.88}, \quad J_{sub} = 9.27$$

$$K_{sub}(s) = \frac{-3.60s - 1.19}{s^2 + 3.27s + 4.86}, \quad J_{sub} = 6.35$$

Comparing these controllers with the optimal ones, it can be verified an important loss on the performance index. This occurs in all cases but the first one on which the trace function is used.   □

6.6 Controllers with structural constraints

In practice one is frequently faced to control design problems where the controller must exhibit some desired structure. For instance, if the system to be controlled is composed by many coupled subsystems, robustness considerations requires the controller should use only local informations for feedback, that is, in the state feedback case, the matrix gain F must present a block-diagonal structure. This particular structure defines the important class of *Decentralized* control design problems. As well as, if only the system output is available for feedback we already know how to design a dynamic controller to meet certain performance criteria. However, it is also of great interest to determine (if any) a static feedback gain with the same paradigm. This is on the origin of the so called *Static output* control design problems. This section is entirely devoted to analyze these problems in the framework of convex analysis.

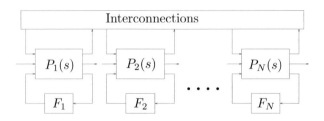

Figure 6.14: The interconnected System

6.6.1 Decentralized control design

The system to be dealt with is represented in Fig. 6.14. It is composed by a number N of subsystems each of them with transfer function $P_i(s)$, coupled together by means of an interconnection structure. The dynamic model of the i-th subsystem with transfer function $P_i(s)$ is given by

$$\dot{x}_i = A^{ii}x_i + B_1^i w + B_2^i u_i + \sum_{j \neq i}^{N} A^{ij}x_j \tag{6.143}$$

$$z_i = C_1^i x_i + D_{12}^i u_i \tag{6.144}$$

$$y_i = x_i \tag{6.145}$$

In order to impose to the overall system a desirable performance, it is asked to design N local controllers, each of then using only the information available in the local state variable x_i , $i = 1, 2, \cdots, N$. That is,

$$u_i = F_i x_i , \quad i = 1, 2, \cdots, N \tag{6.146}$$

Clearly, if the overall system is rewritten in the standard form

$$\dot{x} = Ax + B_1 w + B_2 u \tag{6.147}$$

$$z = C_1 x + D_{12} u \tag{6.148}$$

$$y = x \tag{6.149}$$

then the control is given by

$$u = Fx , \quad F = \text{blockdiag}[F_1, F_2, \cdots, F_N] \tag{6.150}$$

where each block of F is the local matrix gain with appropriate dimension. To ease the presentation, let us introduce the following notation. The subscript "D" in a matrix, for instance F_D means that this matrix is constraint to have a block-diagonal structure. In other words, F_D is obtained from any F by simply zeroing all off block-diagonal elements. The problem to be faced in this section is how to incorporate the above structural constraint in several design procedures of interest. It is important to keep in mind that the results of this section are related to the general linear system (6.147)-(6.149) and do not depend upon any particular system structure like that of system (6.143)-(6.145).

The first important feature to be analyzed is the internal stability of the closed-loop system. Clearly, internal stability is guaranteed whenever there exist F such that

$$F = F_D \in \mathcal{K}_c.$$

Furthermore, all decentralized matrices F_D with this property constitute the set of all stabilizing decentralized gains for the system under consideration. This set is not convex and in many cases may be constituted by disjoint subsets. So, we work here with a particular subset of it defined as follows

Definition 6.3 (Structural D - Stabilizability) *The pair (A, B_2) is said to be structurally D - stabilizable if there exist matrices P_D symmetric and positive definite and F_D such that*

$$0 \geq (A + B_2 F_D)P_D + P_D(A + B_2 F_D)' + Q \tag{6.151}$$

for some matrix $Q = Q' > 0$. The set of all such matrices F_D is denoted $\mathcal{K}_D$. ☐

We want to stress that $F_D \in \mathcal{K}_D$ implies that $F_D \in \mathcal{K}_c$ but the converse is not necessarily true. Moreover, it is possible to have $\mathcal{K}_D = \emptyset$ while there exists $F_D \in \mathcal{K}_c$. The additional constraint in Definition 6.3 is that the Lyapunov inequality (6.151) must present a block-diagonal solution $P = P_D$. At a first glance, it may appears that the existence of a pair (F_D, P_D) depends on a particular choice of matrix Q. Fortunately this is not true as can be simply demonstrated as follows. For a given matrix F_D, suppose $\bar{P}_D$ satisfies the Lyapunov inequality (6.151) with $Q = \bar{Q} > 0$. For any other matrix $Q > 0$, choosing the scalar $\beta > 0$ such that $\beta \bar{Q} \geq Q$, it is simple to verify that (6.151) is also satisfied for $P_D = \beta \bar{P}_D > 0$. Hence, matrix $Q > 0$ in Definition 6.3 can be chosen arbitrarily. Before presenting the next result we introduce the notation used for the partitioned matrix W. The subscript "D" is used as follows

$$W_D := \begin{bmatrix} W_{1D} & W_{2D} \\ W_{2D}' & W_3 \end{bmatrix}$$

which indicates that only the sub-blocks W_1 and W_2 have to present the decentralized structure, namely

$$W_{1D} = \text{blockdiag}[W_{11}, W_{12}, \cdots, W_{1N}]$$

$$W_{2D} = \text{blockdiag}[W_{21}, W_{22}, \cdots, W_{2N}]$$

with W_{1i} and W_{2i} being $n_i \times n_i$ and $n_i \times m_i$ matrices where n_i is the local state vector dimension and m_i is the local control vector dimension respectively for all $i = 1, 2, \cdots, N$.

Theorem 6.11 *Consider $B_1 B_1' > 0$ and define the set*

$$\mathcal{C}_D := \{W \ : \ W = W_D\} \cap \mathcal{C}_c \tag{6.152}$$

The following are true

 a) *$\mathcal{C}_D$ is a convex set.*

 b) *Each $W \in \mathcal{C}_D$ is such that $W_1 > 0$.*

 c) *$\mathcal{K}_D = \{W_2' W_1^{-1} \ : \ W \in \mathcal{C}_D\}$.*

Proof Since the constraint $W = W_D$ is linear and the set $\mathcal{C}_c$ is convex then point a) is proved. Moreover, the proof of point b) is a consequence of $B_1 B_1' > 0$ together with $\mathcal{C}_D \subset \mathcal{C}_c$.

Point c) For the necessity, assume $F_D \in \mathcal{K}_D \neq \emptyset$. Setting $Q = B_1 B_1'$ in (6.151), it is simple to verify that

$$W = W_D = \begin{bmatrix} P_D & P_D F_D' \\ F_D P_D & F_D P_D F_D' \end{bmatrix} \in \mathcal{C}_c$$

which is the same to say that $W \in \mathcal{C}_D$. Furthermore, $W_2' W_1^{-1} = F_D P_D P_D^{-1} = F_D$. Conversely, for any $W \in \mathcal{C}_D \neq \emptyset$ and any $v \in \mathcal{N}_c$ we get

$$0 \geq v' \Theta_c(W) v$$
$$\geq x' \left[(A + B_2 W_2' W_1^{-1}) W_1 + W_1 (A + B_2 W_2' W_1^{-1})' + B_1 B_1' \right] x$$

which shows that inequality (6.151) holds for $P_D = W_1 = W_{1D}$ and $F_D = W_2' W_1^{-1} = W_{2D}' W_{1D}^{-1}$, that is both matrices exhibit the desired block-diagonal structure. From the above, the equality in point c) also holds when $\mathcal{K}_D = \emptyset$ or $\mathcal{C}_D = \emptyset$. $\quad\square$

The linear constraint $W = W_D$ is essential to get this result. It provides all stabilizing decentralized matrices $F = F_D$ and the quadratic function

$$v(x) = x' W_{1D}^{-1} x = \sum_{i=1}^{N} x_i' W_{1i}^{-1} x_i$$

is a Lyapunov function associated to the closed-loop system with $w = 0$ since, its time-derivative along an arbitrary trajectory of the system is

$$\dot{v}(x) = -\|B_1' W_{1D}^{-1} x\|^2 < 0 , \quad \forall x \neq 0$$

The interpretation of Theorem 6.11 is now clear. It generates all stabilizing matrices F_D such that the closed-loop system stability is tested by an additively separable Lyapunov function.

In this framework, it is possible to solve an approximate version of the decentralized state feedback design problem in RH_2, written in the form

$$\min \left\{ \|T(z, w; s)\|_2^2 \ : \ F \in \mathcal{K}_D \right\} \tag{6.153}$$

which makes once again explicit that the feasible set is restricted to those gains satisfying Definition 6.3.

Theorem 6.12 *Assume $B_1 B_1' > 0$ and let $\bar{W}$ be the optimal solution of the convex programming problem*

$$J_{sub} := \min \left\{ \text{trace}[R_c W] \ : \ W \in \mathcal{C}_D \right\} \tag{6.154}$$

Then, $F = F_D = \bar{W}_2' \bar{W}_1^{-1} \in \mathcal{K}_D$ minimizes an upper bound of the objective function of Problem (6.153) in the sense that $\|T(z, w; s)\|_2^2 \leq \text{trace}[R_c W]$, for all $W \in \mathcal{C}_D$.

Proof The infeasibility of one problem implies the same is true to the other and vice versa. Hence, assuming they are feasible, that the optimal solution of Problem

(6.154) provides $\bar{W}_2'\bar{W}_1^{-1} \in \mathcal{K}_D$ is a consequence of Theorem 6.11. Additionally, for any $W \in \mathcal{C}_D$ we have $W_{1D} \geq P_r$ where

$$0 = (A + B_2 W_{2D}' W_{1D}^{-1})P_r + P_r(A + B_2 W_{2D}' W_{1D}^{-1})' + B_1 B_1'$$

yielding

$$\begin{aligned}
\text{trace}[R_c W] &= \text{trace}\left[(C_1 + D_{12}W_{2D}'W_{1D}^{-1})W_{1D}(C_1 + D_{12}W_{2D}'W_{1D}^{-1})'\right] + \\
&\quad + \text{trace}\left[D_{12}(W_3 - W_{2D}'W_{1D}^{-1}W_{2D})D_{12}'\right] \\
&\geq \text{trace}\left[(C_1 + D_{12}W_{2D}'W_{1D}^{-1})P_r(C_1 + D_{12}W_{2D}'W_{1D}^{-1})'\right] \\
&\geq \|T(z,w;s)\|_2^2
\end{aligned}$$

completing the proof.   □

At this point it is important to keep in mind why Problem (6.154) corresponds to minimize only an upper bound to the objective function of Problem (6.153). The reason is that even though $W \in \mathcal{C}_D$ generates all feasible gains $F = F_D$, the matrix P_r used to determine the corresponding value of $\|T(z,w;s)\|_2^2$ does not necessarily satisfies the decentralized constraint $P_r = (P_r)_D$. Generally the inequality $W_{1D} \geq P_r$ is only strictly satisfied for all $W \in \mathcal{C}_D$.

Remark 6.37 The exact Decentralized state feedback design in RH_2 is the optimal control problem

$$\min\left\{\|T(z,w;s)\|_2^2 \;:\; F = F_D \in \mathcal{K}_c\right\}$$

Using Theorem 6.1, this is equivalent to

$$\min\left\{\text{trace}[R_c W] \;:\; W \in \mathcal{C}_{cD}\right\}$$

where

$$\mathcal{C}_{cD} := \left\{W \;:\; W_2'W_1^{-1} = (W_2'W_1^{-1})_D\right\} \cap \mathcal{C}_c$$

Unfortunately, the nonlinear equality constraint present in the set $\mathcal{C}_{cD}$ makes it nonconvex. The way to circumvent this difficulty is to impose the decentralized structure on matrix W_2 and W_1 simultaneously, as required by Definition 6.3.   □

Remark 6.38 The convex constraint $W = W_D$ can be added to any other design problem in order to search a decentralized stabilizing control. For instance, any W belonging to the convex set,

$$\mathcal{C}_{\gamma D} := \{W \;:\; W = W_D\} \cap \mathcal{C}_{\gamma c}$$

provides $F = F_D \in \mathcal{K}_{\gamma c}$.   □

Example 6.7 Consider the interconnected system (6.143) - (6.145) composed by $N = 2$ local subsystems with matrices ($i = 1, 2$)

$$A^{ii} = \begin{bmatrix} 0 & 1 \\ 1 & 0 \end{bmatrix}, \quad B_1^i = \begin{bmatrix} 1 \\ 1 \end{bmatrix}, \quad B_2^i = \begin{bmatrix} 0 \\ 1 \end{bmatrix}$$

$$C_1^i = \begin{bmatrix} 0.5 & 0 \end{bmatrix}, \quad D_{12}^i = 1$$

coupled together by the interconnection matrices

$$A^{12} = A^{21} = \begin{bmatrix} -1 & 0 \\ 2 & 0 \end{bmatrix}$$

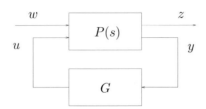

Figure 6.15: The static output feedback control system

Solving the optimal control problem in RH_2, the optimal feedback gain and the associated minimum cost are

$$F_2 = \begin{bmatrix} -2.84 & -2.15 & -0.34 & -0.15 \\ -0.34 & -0.15 & -2.84 & -2.15 \end{bmatrix}, \quad J_{opt} = \|T(z, w; s)\|_2^2 = 21.79$$

while the optimal solution of Problem (6.154) provides

$$F_D = \begin{bmatrix} -3.62 & -2.50 & 0 & 0 \\ 0 & 0 & -3.62 & -2.50 \end{bmatrix}, \quad J_{sub} = 30.27$$

However, for $F = F_D$ the exact value of the objective function can be calculated as being $\|T(z, w; s)\|_2^2 = 23.04$ which confirms numerically the fact that only an upper bound of the true cost has been minimized.   □

6.6.2 Static output control design

The system to be analyzed is given in fig. 6.15. The transfer function $P(s)$ has the state space realization

$$\dot{x} = Ax + B_1 w + B_2 u \tag{6.155}$$
$$z = C_1 x + D_{12} u \tag{6.156}$$
$$y = C_2 x \tag{6.157}$$

where in opposition to the Output feedback control problems solved before, the measured output variable $y(\cdot) \in R^r$ is not corrupted by the external disturbance $w(t)$. Moreover it is assumed that C_2 is full row rank. The goal is to design a static output feedback law of the form

$$u = Gy \tag{6.158}$$

where the internal stability as well as some previously defined performance behavior are assured.

The control law (6.158) can be rewritten as $u = F_O x$, where the subscript "O" in matrix F means that there exists G such that $F_O = GC_2$. In this case the internal stability of the closed-loop system is preserved by means of the static output control law (6.158) provided $F = F_O \in K_c$. Before proceed, let us made the assumption that matrix C_2 presents the following structure

$$C_2 = \begin{bmatrix} I & 0 \end{bmatrix}$$

If this is not the case, it is always possible to put the system in this form by a suitable choice of a similarity transformation. Under this assumption, matrix F_O must present the particular structure

$$F_O = \begin{bmatrix} F_1 & 0 \end{bmatrix}$$

which implies that $G = F_1$ as required before. For a symmetric matrix P the subscript
"O" has a slightly different meaning, that is

$$P = P_O = \left[\begin{array}{cc} P_{11} & 0 \\ 0 & P_{22} \end{array} \right]$$

where P_{11} and P_{22} are $r \times r$ and $n - r \times n - r$ symmetric sub-matrices respectively.
Unfortunately, the set of all stabilizing output feedback gains, being

$$F = F_O \in \mathcal{K}_c$$

is nonconvex and in many instances may be constituted of disjoint subsets. To circumvent this difficulty, we work here with a subset of it which as will be proven can
be converted to a convex set. This subset is characterized by the following definition

Definition 6.4 (Structural O - Stabilizability) *The triple (A, B_2, C_2) is said to be
structurally O - stabilizable if there exist matrices P_O symmetric and positive definite
and F_O such that*

$$0 \geq (A + B_2 F_O) P_O + P_O (A + B_2 F_O)' + Q \qquad (6.159)$$

for some matrix $Q = Q' > 0$. The set of all such matrices F_O is denoted $\mathcal{K}_O$. □

For any matrix such that $F = F_O$, it can be factorized as $F = G C_2$ for some G,
consequently structural O - stability implies that the eigenvalues of the closed-loop
matrix $A + B_2 F = A + B_2 G C_2$ are all in the open left hand side part of the complex
plane. However, as in the case of decentralized control, Definition 6.4 requires the
Lyapunov inequality solution, used to test stability presents a particular structure.
The consequence is that $F_O \in \mathcal{K}_O$ implies $F_O \in \mathcal{K}_c$ but the inverse is not true in
general. The price to be paid to handle convex sets only is to retain a subset of
the entire set of static output feedback stabilizing gains. With no loss of generality,
matrix $Q > 0$ in definition 6.4 can be *a priori* fixed. This fact is proved with no
major difficulty since for any $\bar{P} = \bar{P}_O > 0$ and any scalar $\beta > 0$ then $P = \beta \bar{P}_O$
satisfies $P = P_O$. Considering a matrix W partitioned in four blocks, the subscript
"O" stands for

$$W_O := \left[\begin{array}{cc} W_{1O} & W_{2O} \\ W'_{2O} & W_3 \end{array} \right]$$

which indicates that only the sub-blocks W_1 and W_2 have to present the output
structure, namely

$$W_{1O} = \left[\begin{array}{cc} W_{11} & 0 \\ 0 & W_{22} \end{array} \right] , \quad W_{2O} = \left[\begin{array}{c} W_{21} \\ 0 \end{array} \right]$$

with W_{11}, W_{22} and W_{21} being $r \times r$, $n - r \times n - r$ and $r \times m$ matrices where n is the
state vector dimension, m is the control vector dimension and r is the output vector
dimension respectively.

Theorem 6.13 *Consider $B_1 B'_1 > 0$ and define the set*

$$\mathcal{C}_O := \{W \; : \; W = W_O\} \cap \mathcal{C}_c \qquad (6.160)$$

The following are true

a) $\mathcal{C}_O$ is a convex set.

b) Each $W \in \mathcal{C}_O$ is such that $W_1 > 0$.

c) $\mathcal{K}_O = \{W_2' W_1^{-1} : W \in \mathcal{C}_O\}$.

Proof As in the proof of Theorem 6.11, the constraint $W = W_O$ is linear and hence convex. This fact together with $B_1 B_1' > 0$ and $\mathcal{C}_O \subset \mathcal{C}_c$ prove both points a) and b).

Point c) For the necessity, assume $F_O \in \mathcal{K}_O \neq \emptyset$. Setting $Q = B_1 B_1'$ in (6.159), it is simple to verify that

$$W = W_O = \begin{bmatrix} P_O & P_O F_O' \\ F_O P_O & F_O P_O F_O' \end{bmatrix} \in \mathcal{C}_O$$

and $W_2' W_1^{-1} = F_O P_O P_O^{-1} = F_O$. Conversely, for any $W \in \mathcal{C}_O \neq \emptyset$ and any $v \in \mathcal{N}_c$ we get

$$0 \geq v' \Theta_c(W) v$$
$$\geq x' \left[(A + B_2 W_2' W_1^{-1}) W_1 + W_1 (A + B_2 W_2' W_1^{-1})' + B_1 B_1' \right] x$$

which shows that inequality (6.159) holds for $P_O = W_1 = W_{1O}$ and $F_O = W_2' W_1^{-1} = W_{2O}' W_{1O}^{-1}$, that is both matrices exhibit the desired output feedback structure. In case one of the sets $\mathcal{K}_O$ or $\mathcal{C}_O$ is empty then the equality stated in point c) follows trivially. □

From this theorem, we can see that any feasible matrix $W \in \mathcal{C}_O$ generates an output feedback stabilizing gain which is very simple to be determined. Actually, the imposed structure constraint

$$W_{1O} = \begin{bmatrix} W_{11} & 0 \\ 0 & W_{22} \end{bmatrix}, \quad W_{2O} = \begin{bmatrix} W_{21} \\ 0 \end{bmatrix}$$

provides $F_O \in \mathcal{K}_O$ which can be factorized as $F_O = G C_2$ where

$$G = F_1 = W_{21}' W_{11}^{-1} \tag{6.161}$$

in addition, the complete parametrization of the set $\mathcal{K}_O$ by means of point c) opens the possibility, with no major difficulty, to involve it in an optimization procedure. For instance consider the Static output feedback control design problem

$$\min \left\{ \|T(z, w; s)\|_2^2 : F \in \mathcal{K}_O \right\} \tag{6.162}$$

which makes once again explicit that the feasible set is restrict to those gains satisfying Definition 6.4.

Theorem 6.14 *Assume $B_1 B_1' > 0$ and let $\bar{W}$ be the optimal solution of the convex programming problem*

$$J_{sub} := \min \left\{ \mathrm{trace}[R_c W] : W \in \mathcal{C}_O \right\} \tag{6.163}$$

Then, $F = F_O = \bar{W}_2' \bar{W}_1^{-1} \in \mathcal{K}_O$ minimizes an upper bound of the objective function of Problem (6.162) in the sense that $\|T(z, w; s)\|_2^2 \leq \mathrm{trace}[R_c W]$, for all $W \in \mathcal{C}_O$.

Proof From Theorem 6.13, it suffices to consider that both problems are feasible. In this case, the optimal solution of problem (6.163) provides $\bar{W}_2' \bar{W}_1^{-1} \in \mathcal{K}_O$. Additionally, for any $W \in \mathcal{C}_O$ we have $W_{1O} \geq P_r$ where

$$0 = (A + B_2 W_{2O}' W_{1O}^{-1}) P_r + P_r (A + B_2 W_{2O}' W_{1O}^{-1})' + B_1 B_1'$$

yielding immediately trace$[R_c W] \geq \|T(z, w; s)\|_2^2$. The proof is complete.   □

In general, the optimal solution of problem (6.162) is not generated by means of the global optimal solution of the convex problem (6.163). Even though all gains in $\mathcal{K}_O$ are generated by the proposed convex parametrization, matrix P_r needed to calculate $\|T(z, w; s)\|_2^2$ may not satisfy the constraint $P_r = (P_r)_O$, in this case there is no matrix $W \in \mathcal{C}_O$ for which the equality $W_{1O} = P_r$ holds.

Remark 6.39 Following the lines of Theorem 6.13 we notice that the set of all $F = F_O \in \mathcal{K}_c$ can be generated by $F = W_2' W_1^{-1}$ where

$$W \in \mathcal{C}_{cO} := \left\{ W \ : \ W_2' W_1^{-1} = (W_2' W_1^{-1})_O \right\} \cap \mathcal{C}_c$$

Unfortunately this set is not convex. The same theorem provides a subset of $\mathcal{C}_{cO}$, namely $\mathcal{C}_O$ which has the important property to be convex.   □

Remark 6.40 Consider the system (6.155) - (6.157) and the matrix

$$S := \begin{bmatrix} C_2 \\ E_2 \end{bmatrix}$$

where E_2 is any matrix of appropriate dimension such that S^{-1} exists. Defining the new state variable $\tilde{x} := Sx$, the new state space realization turns out to be such that $y = \tilde{C}_2 \tilde{x}$ where

$$\tilde{C}_2 = C_2 S^{-1} = \begin{bmatrix} I & 0 \end{bmatrix}$$

The concept of Structural O - stabilizability depends on the similarity matrix S by means of the arbitrary submatrix E_2. The next lemma puts in evidence this important point.

Lemma 6.7 *The following statements are equivalent:*

a) *The triple (A, B_2, C_2) is stabilizable by output feedback, that is there exists G such that the matrix $A + B_2 G C_2$ is stable.*

b) *There exists a matrix E_2 such that the triple $(\tilde{A}, \tilde{B}_2, \tilde{C}_2)$ in the new state space representation is structurally O - stabilizable.*

Proof To prove that a) implies b), let us suppose that there exists an output feedback gain G such that $A + B_2 G C_2$ is stable. From the Extended Lyapunov lemma, there exists $P = P' > 0$ solution to the linear equation

$$0 = (A + B_2 G C_2)P + P(A + B_2 G C_2)' + Q$$

for $Q = Q' > 0$ given. Chosen $E_2 = U_2' P^{-1}$ where U_2 defines an orthogonal basis to the null space of C_2, that is $C_2 U_2 = 0$ and $U_2' U_2 = I$ and multiplying the above equation to the right by S' and to the left by S we get

$$0 = (\tilde{A} + \tilde{B}_2 G \tilde{C}_2)\tilde{P} + \tilde{P}(\tilde{A} + \tilde{B}_2 G \tilde{C}_2)' + \tilde{Q}$$

where $\tilde{P} := SPS'$ and $\tilde{Q} := SQS'$. Simple calculations show that

$$\tilde{F} := G\tilde{C}_2 = \begin{bmatrix} G & 0 \end{bmatrix} = \tilde{F}_O$$

and

$$\tilde{P} = SPS' = \begin{bmatrix} C_2 P C_2' & 0 \\ 0 & U_2' P^{-1} U_2 \end{bmatrix} = \tilde{P}_O$$

meaning that point b) holds. The converse is immediate.   □

Remark 6.41 The numerical solution of problem (6.162) by means of convex programming methods is very efficient. However the main drawback of this approach is that only a subset of the true feasible set of stabilizing output feedback gains is considered. Let us verify the difficulties we have to face for the numerical solution of the true Output feedback design problem in RH_2

$$\min \left\{ \|T(z, w; s)\|_2^2 \ : \ GC_2 \in \mathcal{K}_c \right\}$$

which reduces to the determination of matrices G and $P = P' \geq 0$ such that

$$\min \left\{ \text{trace}[B_1' P B_1] \ : \ 0 = A_{cg}' P + P A_{cg} + C_{cg}' C_{cg} \right\}$$

where $A_{cg} := A + B_2 G C_2$ and $C_{cg} := C_1 + D_{12} G C_2$. Defining $\Lambda = \Lambda'$ the matrix of Lagrange multiplier associated to the equality constraint, the necessary optimality conditions are (recall Remark 6.7)

$$0 = A_{cg}' P + P A_{cg} + C_{cg}' C_{cg}$$
$$0 = A_{cg} \Lambda + \Lambda A_{cg}' + B_1 B_1'$$
$$0 = (G C_2 + D_{12}' C_1 + B_2' P) \Lambda C_2'$$

Assuming as before that $B_1 B_1' > 0$ then for any $G C_2 \in \mathcal{K}_c$ matrix Λ is positive definite, in which case the last equation yields

$$G = -(B_2' P + D_{12}' C_1) \Lambda C_2' (C_2 \Lambda C_2')^{-1}$$

This formula for the optimal gain couples the first two equations in a very nonlinear manner. The optimal gain can not be expressed in terms of a Riccati equation unless matrix C_2 is square and nonsingular. The following algorithm is useful for numerical purposes

1) Choose G_0 such that $F_0 := G_0 C_2 \in \mathcal{K}_c$ and iterate until convergence with $k = 0, 1, \cdots$.

2) Set $F_k := G_k C_2$ and let $P_k \geq 0$ be the solution of the Lyapunov equation (in the unknown P)

$$0 = (A + B_2 F_k)' P + P(A + B_2 F_k) + (C_1 + D_{12} F_k)'(C_1 + D_{12} F_k)'$$

   Determine $\tilde{F}_k := -B_2' P_k - D_{12}' C_1$.

3) Let $\Lambda_{k+1} > 0$ and G_{k+1} be the solution of the nonlinear equations (in the unknown Λ and G)

$$0 = (A + B_2 G C_2)\Lambda + \Lambda(A + B_2 G C_2)' + B_1 B_1'$$
$$G = \tilde{F}_k \Lambda C_2' (C_2 \Lambda C_2')^{-1}$$

Lemma 6.8 *With $B_1 B_1' > 0$, the above algorithm has the following properties*

a) *For all $k = 0, 1, \cdots$ the gain $F_{k+1} = G_{k+1} C_2 \in \mathcal{K}_c$.*

b) *For all $k = 0, 1, \cdots$ the criterion $J_{k+1} := \text{trace}[B_1' P_{k+1} B_1] \leq \text{trace}[B_1' P_k B_1] := J_k$*

Proof *Point a)* is a consequence of step 3). Actually, it implies that $\Lambda_{k+1} > 0$ satisfies

$$0 = (A + B_2 F_{k+1})\Lambda_{k+1} + \Lambda_{k+1}(A + B_2 F_{k+1})' + B_1 B_1'$$

thus, $F_{k+1} = G_{k+1} C_2$ is a stabilizing gain.

Point b) Simple but tedious algebraic manipulations show that for two subsequent iterations

$$0 = (A + B_2 F_{k+1})'(P_k - P_{k+1}) + (P_k - P_{k+1})(A + B_2 F_{k+1}) +$$
$$+ (F_k - \tilde{F}_k)'(F_k - \tilde{F}_k) - (F_{k+1} - \tilde{F}_k)'(F_{k+1} - \tilde{F}_k)$$

which allows us to determine the current value of the criterion as being

$$J_k - J_{k+1} = \text{trace} \left[\int_0^\infty B_1 e^{(A+B_2 F_{k+1})t} (F_k - \tilde{F}_k)'(F_k - \tilde{F}_k) e^{(A+B_2 F_{k+1})'t} B_1' dt \right] -$$

$$-\text{trace} \left[\int_0^\infty B_1 e^{(A+B_2 F_{k+1})t} (F_{k+1} - \tilde{F}_k)'(F_{k+1} - \tilde{F}_k) e^{(A+B_2 F_{k+1})'t} B_1' dt \right]$$

$$= \text{trace} \left[(F_k - \tilde{F}_k)\Lambda_{k+1}(F_k - \tilde{F}_k)' - (F_{k+1} - \tilde{F}_k)\Lambda_{k+1}(F_{k+1} - \tilde{F}_k)' \right]$$

$$= \text{trace} \left[(G_k C_2 - \tilde{F}_k)\Lambda_{k+1}(G_k C_2 - \tilde{F}_k)' \right] -$$

$$-\text{trace} \left[(G_{k+1} C_2 - \tilde{F}_k)\Lambda_{k+1}(G_{k+1} C_2 - \tilde{F}_k)' \right]$$

using now the value of G_{k+1} provided in step 3)

$$G_{k+1} = \tilde{F}_k \Lambda_{k+1} C_2' (C_2 \Lambda_{k+1} C_2')^{-1}$$

we get

$$J_k - J_{k+1} = \text{trace} \left[(G_k C_2 - \tilde{F}_k)\Lambda_{k+1}(G_k C_2 - \tilde{F}_k)' \right] -$$

$$-\text{trace} \left[\tilde{F}_k \left(\Lambda_{k+1} - \Lambda_{k+1} C_2'(C_2 \Lambda_{k+1} C_2')^{-1} C_2 \Lambda_{k+1} \right) \tilde{F}_k' \right]$$

$$= \text{trace} \left[G_k C_2 \Lambda_{k+1} C_2' G_k - G_{k+1} C_2 \Lambda_{k+1} C_2' G_k' - \right.$$

$$\left. -G_k C_2 \Lambda_{k+1} C_2' G_{k+1}' + G_{k+1} C_2 \Lambda_{k+1} C_2' G_{k+1}' \right]$$

$$= \text{trace} \left[(G_k - G_{k+1}) C_2 \Lambda_{k+1} C_2' (G_k - G_{k+1})' \right]$$

This equality finally enables us to conclude that

$$J_{k+1} = J_k - \text{trace} \left[(G_k - G_{k+1}) C_2 \Lambda_{k+1} C_2' (G_k - G_{k+1})' \right]$$
$$\leq J_k$$

which proves point b) of the lemma proposed.

It is important to recognize that the second part of Lemma 6.8 is also a consequence of the joint determination of matrices Λ_{k+1} and G_{k+1} in step 3) from the solution of two simultaneous equations, one of then nonlinear. It would be very attractive to change the previous algorithm in order to solve only linear equations. However, doing this it is no more possible to be sure that

$$F_k = G_k C_2 \in \mathcal{K}_c \Longrightarrow F_{k+1} = G_{k+1} C_2 \in \mathcal{K}_c$$

in which occurrence the properties introduced in Lemma 6.8 are both lost. Hence, we can say that the price to be paid to keep these convergence features is to solve in step 3) a nonlinear matrix equation.   □

Example 6.8 Consider the system (6.155) - (6.157) with the following data

$$A = \begin{bmatrix} 0 & 1 \\ -1 & 0 \end{bmatrix}, \quad B_1 = \begin{bmatrix} 1 & 0 \\ 0 & 1 \end{bmatrix}, \quad B_2 = \begin{bmatrix} 0 \\ 1 \end{bmatrix},$$

$$C_1 = \begin{bmatrix} 1 & 0 \\ 0 & 0 \end{bmatrix}, \quad D_{12} = \begin{bmatrix} 0 \\ 1 \end{bmatrix}, \quad C_2 = \begin{bmatrix} 0 & 1 \end{bmatrix}$$

Solving the State feedback control problem in RH_2 we get

$$F_2 = \begin{bmatrix} -0.41 & -0.91 \end{bmatrix}, \quad J_{opt} = \text{trace}[P_2] = 2.19$$

Since matrix C_2 is not in the standard form, from the discussion in Remark 6.39 we have first defined a similarity transformation with

$$S = \begin{bmatrix} 0 & 1 \\ 1 & 1 \end{bmatrix}$$

together with the optimal solution of problem (6.163) provides

$$G_{sub} = -2.0 \ , \quad J_{sub} = 3.50$$

In this case, the optimal solution of the exact design problem (recall Remark 6.41) is known to be $G_{opt} = -0.81$ with the associated cost $J_{opt} = 2.46$. This example shows that the quality of the solution provided by the solution of the convex problem (6.163) may be poor. Of course the quality, measured in terms of the deviation from the optimal solution depends on the similarity transformation matrix S which unfortunately is not easy to choose. Again, the main attractive feature of problem (6.163) is its convexity. □

6.7 Notes and references

The various aspects of Convex Analysis, from the basic facts to most important and deep results are included in the seminal book by Rockafellar [53]. The book [10] exhaustively treats the most important topics related to systems and control theory in the framework of Linear Matrix Inequalities so that the problems to be handled are convex. As well as, this book is also an important source of references to those interested in going deeper into optimal control design using convex programming techniques. Section 6.2 is mainly based on papers [6], [22], [47], [48] and on the results introduced in the previous chapters of this book. Section 6.3 follows the same lines of [32] and connections to the results of [5] are put in evidence. The approach proposed in [40] is summarized in Remark 6.28 where also a numeric example is included for comparison purpose. In Section 6.4 the pole placement problem in a circular region is analyzed in a convex programming view point. Other regions for pole placement as well as a different approach to solve the same problem are considered in [28]. The time-domain specifications treated in Section 6.5 is based on the result of Lemma 6.5 due to Wilson [62]. The proof of this result is different and new. The related optimal control design problems appeared first in [54]. Finally, the Decentralized control design in Section 6.6 parallels the results of [23]. In the same section the Static output control design problem considered, first appeared in [39]. From this reference comes both the Example 6.8 solved and the algorithm discussed in Remark 6.41. The proof of convergence is new.

Chapter 7

Uncertain Systems Control Design

7.1   Introduction

This chapter is devoted to uncertain systems control design. The main goal is to provide a simple and easy to follow introduction to this important subject of control theory. The necessity to analyze dynamic systems subject to uncertainties stems from the fact that the model to be used for design purpose is generally only an approximation of reality. In other words, for control design purpose we need to handle simple models. However, the controller obtained must work when connected to the real system. The way to take this feature into account is to consider a simplified *nominal* model which is corrupted by uncertainties belonging to a prespecified domain. To make clear this point it is interesting to remember the mixed RH_2/RH_∞ control design problem. The controller calculated from its solution imposes optimality to the nominal closed loop system while stability is preserved against the considered uncertainties. In this chapter, we go beyond this point. Indeed, not only robust stability is considered but also robust performance is taken into account. That is, the main purpose is to design controllers which preserve stability and minimize the performance loss due to existence of uncertainties. In this way, two important concepts are introduced, namely *quadratic stabilizability* and *guaranteed cost* which are both on the basis of the results that follow. Once again, the optimal control problems to be solved are all convex and so the same machinery provided in Chapter 6 is intensively used. It is important to make explicit that only linear control design is considered.

7.2   Robust stability and performance

In this section we generalize the concepts of stability and optimality to deal with uncertainties belonging to a precise although general domain. Control problems design involving norms of transfer functions in RH_2 and RH_∞ spaces are considered in the special case that the whole state is available for feedback.

The basic system structure is depicted in fig. 7.1 making explicit the linear state feedback control law to be used. The transfer function $P_{\mathcal{D}}(s)$ depends on the subscript $\mathcal{D}$ which will be made clear in the sequel. The linear system under consideration thus

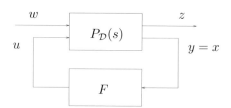

Figure 7.1: The control system structure

have the following state space representation

$$\dot{x} = Ax + B_1 w + B_2 u \tag{7.1}$$
$$z = C_1 x + D_{12} u \tag{7.2}$$
$$y = x \tag{7.3}$$

and the controller is completely defined by means of the matrix F which has to be determined (if any) such that with

$$u = Fx \tag{7.4}$$

some desired performance is assured. Equations (7.1) - (7.3) represent an uncertain linear system. That is, matrices A and B_2 are not exactly known. On the contrary we only know that they belong to some prespecified matrix set $\mathcal{D}$. In this sense, equations (7.1) - (7.3) represent in fact a family of linear systems any member of which has to be controlled by means of the same control law. Of course it is important, if possible, to determine F such that all members of the family are internally stable and, at the same time, some desired performance level is guaranteed.

To put this discussion in more precise terms, let us recall that for any pair $(A, B_2) \in \mathcal{D}$ the internal stability of the corresponding system is assured whenever $F \in \mathcal{K}_c$. Consequently, the matrix gain F stabilizes all members of the family if and only if

$$F \in \mathcal{K}_\mathcal{D} := \bigcap_{(A, B_2) \in \mathcal{D}} \mathcal{K}_c \tag{7.5}$$

This fact has a very clear interpretation. The set $\mathcal{K}_\mathcal{D}$ is composed by all matrices F which stabilize each feasible pair $(A, B_2) \in \mathcal{D}$. The set $\mathcal{K}_\mathcal{D}$ does not present any important property as for instance convexity. A crucial point is that although each element of the intersection above can be converted to a convex set (as has successfully been done in Chapter 6), the same is not true for the intersection itself, that is for the set $\mathcal{K}_\mathcal{D}$. A way to circumvent this difficulty is to work with a subset of it, characterized by means of the following definition.

Definition 7.1 (Quadratic stabilizability) *The pair (A, B_2) is said to be quadratic stabilizable if there exist matrices P, symmetric and positive definite, and F such that*

$$0 \geq (A + B_2 F)P + P(A + B_2 F)' + Q , \quad \forall\ (A, B_2) \in \mathcal{D} \tag{7.6}$$

for some matrix $Q = Q' > 0$. The set of all such matrices F is denoted $\mathcal{K}_Q$.   □

Once again it is important to observe that in this definition, the choice of matrix $Q > 0$ is immaterial as far as the existence of a matrix $P > 0$ satisfying (7.6) is concerned. If this inequality is satisfied for $P = \bar{P}$ when $Q = \bar{Q}$ then for any other $Q > 0$ the same inequality also holds true for $P = \beta\bar{P}$ provided $\beta > 0$ is such that $\beta\bar{Q} \geq Q$. Definition 7.1 has a very interesting interpretation. Pick from $\mathcal{D}$ a pair (A, B_2), consider the system (7.1) with $w(t) = 0$ and define the Lyapunov function

$$v(x) = x'P^{-1}x$$

If inequality (7.6) holds then the time derivative of $v(\cdot)$ along any trajectory of the closed loop system satisfies

$$\begin{aligned}
\dot{v}(x) &= x'\left[(A + B_2F)'P^{-1} + P^{-1}(A + B_2F)\right]x \\
&= x'P^{-1}\left[(A + B_2F)P + P(A + B_2F)'\right]P^{-1}x \\
&\leq -x'P^{-1}QP^{-1}x \\
&< 0 \,, \ \forall \, x \neq 0
\end{aligned}$$

showing that the closed-loop system is internally stable indeed. The point is that the above calculation holds for any pair $(A, B_2) \in \mathcal{D}$ and so quadratic stability means that only one Lyapunov function can be used to test stability of all systems generated by all pairs (A, B_2) in $\mathcal{D}$. Thus, it is clear that $\mathcal{K}_Q$ is only a subset of $\mathcal{K}_\mathcal{D}$ defined in (7.5) since for the latter many different Lyapunov functions can be used to test stability.

Remark 7.1 An uncertain linear system is said to be robustly stabilizable if $\mathcal{K}_\mathcal{D} \neq \emptyset$. This means that there exists a feedback gain matrix F such that for all $(A, B_2) \in \mathcal{D}$, the eigenvalues of $A + B_2F$ lie in the open left complex plane. □

At this point we have to move our attention to the uncertainty domain $\mathcal{D}$ since Definition 7.1 depends essentially on its mathematical description. There are several possibilities for that. Let us consider one of such description whose generality will be discussed in the sequel. Notice first that we are assuming only matrices A and B_2 of the open-loop system (7.1) - (7.3) to be uncertain. This is the same to say that matrix (recall Chapter 6)

$$M_c := \begin{bmatrix} A & B_2 \\ 0 & 0 \end{bmatrix} \tag{7.7}$$

is not exactly known but belongs to an uncertainty domain $\mathcal{D}_c$ such that

$$(A, B_2) \in \mathcal{D} \iff M_c \in \mathcal{D}_c$$

Then, we proceed by defining the uncertainty domain $\mathcal{D}_c$. It is characterized as a *polyhedral convex bounded domain* given by

$$\mathcal{D}_c := \mathrm{co}\,\{M_{ci} \,, \ i = 1, 2, \cdots, N\} \tag{7.8}$$

where co$\{\cdot\}$ denotes convex hull (recall Appendix H) generated by matrices $M_{ci}, i = 1, 2, \cdots, N$. From this, any matrix $M_c \in \mathcal{D}_c$ can be determined by a convex combination of the *extreme* matrices $M_{ci}, i = 1, 2, \cdots, N$. More precisely,

$$M_c = \sum_{i=1}^{N} \xi_i M_{ci} \,, \ \xi_i \geq 0 \,, \ \sum_{i=1}^{N} \xi_i = 1 \tag{7.9}$$

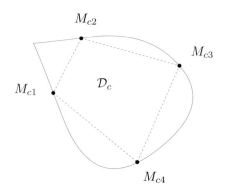

Figure 7.2: The uncertainty domain

This uncertainty description is very general. Any polyhedral and convex bounded domain can be exactly represented by a proper choice of the extreme matrices M_{ci}. On the other hand, a convex domain (not necessarily polyhedral) can be represented only approximately but the degree of approximation involved can be made arbitrarily small if the number of extreme matrices becomes sufficiently large. This fact is illustrated in fig. 7.2.

Remark 7.2 Another important uncertainty description is the so called *norm bounded domain* defined as

$$\mathcal{D}_n := \{M_c = M_{cn} + B_n \Omega C_n \; : \; \|\Omega\| \le 1\}$$

where matrix M_{cn} defines the open-loop *nominal* system and matrices B_n and C_n of appropriate dimension define the uncertainty structure. First of all, notice that this set is convex. Indeed, the generic element M_c is affine with respect to Ω which varies in a convex set defined by the norm constraint $\|\Omega\| \le 1$. Furthermore, this set is not polyhedral in general and so, as it has been commented before, only an approximation of it can be generated by means of the convex hull (7.8). However, in some particular although important cases, it degenerates to a polyhedral convex set. For instance, this occurs when the uncertain matrix Ω is constrained to be diagonal.   □

Example 7.1 Consider the following linear uncertain system

$$\dot{x} = Ax + B_2 u$$

where matrices A and B_2 are given by

$$A = \begin{bmatrix} 0 & \alpha - 1 \\ \beta & 0 \end{bmatrix} , \quad B_2 = \begin{bmatrix} \alpha \\ 1 - \beta \end{bmatrix}$$

and the parameters α and β, representing the uncertainties are such that

$$|\alpha - 0.5| \le 0.3 \; , \quad |\beta - 0.5| \le 0.3$$

Letting the nominal values for these parameters be $\alpha_0 = \beta_0 = 0.5$, the norm bounded domain $\mathcal{D}_n$ is completely defined by matrices

$$M_{cn} = \begin{bmatrix} 0 & -0.5 & 0.5 \\ 0.5 & 0 & 0.5 \\ 0 & 0 & 0 \end{bmatrix} , \quad B_n = \begin{bmatrix} 0.3 & 0 \\ 0 & 0.3 \\ 0 & 0 \end{bmatrix} , \quad C'_n = \begin{bmatrix} 0 & 1 \\ 1 & 0 \\ 1 & -1 \end{bmatrix}$$

since the uncertain parameters may be retained in matrix Ω, which for any $\|\Omega\| \leq 1$ exhibits the diagonal structure

$$\Omega = \operatorname{diag}\left\{\frac{\alpha - 0.5}{0.3},\ \frac{\beta - 0.5}{0.3}\right\}$$

The same system can be modeled by means of a convex bounded domain $\mathcal{D}_c$. This is done from the observation that $N = 4$ and the corresponding extreme matrices are generated from the four vertex of the rectangle $[0.2,\ 0.8] \times [0.2,\ 0.8]$, giving thus

$$M_{c1} = \begin{bmatrix} 0 & -0.8 & 0.2 \\ 0.2 & 0 & 0.8 \\ 0 & 0 & 0 \end{bmatrix}, \quad M_{c2} = \begin{bmatrix} 0 & -0.8 & 0.2 \\ 0.8 & 0 & 0.2 \\ 0 & 0 & 0 \end{bmatrix}$$

$$M_{c3} = \begin{bmatrix} 0 & -0.2 & 0.8 \\ 0.2 & 0 & 0.8 \\ 0 & 0 & 0 \end{bmatrix}, \quad M_{c4} = \begin{bmatrix} 0 & -0.2 & 0.8 \\ 0.8 & 0 & 0.2 \\ 0 & 0 & 0 \end{bmatrix}$$

From this example we conclude that any uncertain system represented by matrix M_c belongs to a polyhedral uncertain domain whenever each entry of M_c is an affine function of the uncertainty. The number of independent uncertain parameters is immaterial since the domain $\mathcal{D}_c$ depends only on the extreme matrices $M_{ci}, i = 1, 2, \cdots, N$ which are a priori determined from the data. □

We move now our attention to the characterization of uncertain systems optimality. Consider the closed-loop system of fig. 7.1 with the state space representation (7.1) - (7.3) and assume that the performance index can be expressed in terms of the transfer function from the input w to the output z, that is

$$J(M_c, F) := \mathcal{G}(T(z, w; s))$$

where it is made explicit the dependence of the performance index with respect to both the control gain F and the open-loop uncertain matrix M_c. The former has to be determined such that $J(\cdot)$ is optimized in a sense to be precisely defined. Thus, restricting our attention to the quadratic stabilizing gains $F \in \mathcal{K}_Q$, the following definition is of particular importance.

Definition 7.2 (Guaranteed cost) *The scalar ρ is said to be a guaranteed cost associated with the feedback gain matrix $F \in \mathcal{K}_Q$ if*

$$J(M_c, F) \leq \rho, \quad \forall\ M_c \in \mathcal{D}_c \tag{7.10}$$

The minimum value of ρ satisfying (7.10), denoted ρ_Q, is called the minimum guaranteed cost. □

From this definition, two points have to be kept in mind. The first one is that the guaranteed cost is a function of $F \in \mathcal{K}_Q$. Indeed, any $\rho \geq \rho(F)$ is also a guaranteed cost provided

$$\rho(F) := \max_{M_c \in \mathcal{D}_c} J(M_c, F)$$

and second, the best choice for the control gain matrix $F \in \mathcal{K}_Q$ is the one such that the minimum of $\rho(F)$ is achieved, that is

$$\rho_Q := \min_{F \in \mathcal{K}_Q} \rho(F)$$

Clearly, the scalar ρ_Q is the minimum guaranteed cost associated to the uncertain system under consideration, it does not depend on a particular value of $F \in \mathcal{K}_Q$ and it is the optimal solution of a min/max optimization problem. Actually, to show this, notice that

$$\rho_Q = \min_{F \in \mathcal{K}_Q} \rho(F)$$

$$= \min_{F \in \mathcal{K}_Q} \max_{M_c \in \mathcal{D}_c} \mathcal{J}(M_c, F)$$

$$= \min_{\rho, F \in \mathcal{K}_Q} \{ \rho \; : \; \mathcal{J}(M_c, F) \leq \rho , \;\; \forall \, M_c \in \mathcal{D}_c \} \tag{7.11}$$

which allows a very interesting interpretation of the minimum guaranteed cost, in terms of a game between *Man* and *Nature*. The Nature plays by fixing an open-loop model $M_c \in \mathcal{D}_c$ and the Man defines the best feedback gain matrix by choosing $F \in \mathcal{K}_Q$ such that $\mathcal{J}(M_c, F)$ is minimized. Once this information is available, the Nature plays again but trying to destroy the Man's action by choosing a new open-loop model $M_c \in \mathcal{D}_c$ such that $\mathcal{J}(M_c, F)$ is maximized. The guaranteed cost ρ_Q is the equilibrium of this game. The equilibrium value ρ_Q is generally very difficult to be calculated and so it must be thought as a paradigm for uncertain systems control design. A much more simple strategy, which is used in the sequel stems from the determination of a function $\bar{\rho}(F)$ such that

$$\mathcal{J}(M_c, F) \leq \bar{\rho}(F) , \;\; \forall \, M_c \in \mathcal{D}_c$$

which allows the determination of a guaranteed cost $\bar{\rho}_Q$ and the associated feedback gain by solving

$$\bar{\rho}_Q := \min_{F \in \mathcal{K}_Q} \bar{\rho}(F)$$

clearly implying that

$$\bar{\rho}_Q = \min_{F \in \mathcal{K}_Q} \bar{\rho}(F)$$

$$\geq \min_{F \in \mathcal{K}_Q} \max_{M_c \in \mathcal{D}_c} \mathcal{J}(M_c, F)$$

$$\geq \rho_Q$$

In the next sections, our goal will be to solve this problem for the class of linear uncertain systems with polyhedral convex bounded domains. Once again, it will be possible to convert the optimization problem to be dealt with into a convex programming problem.

7.2.1   Quadratic stabilizing controllers

The main purpose here is to analyze the geometry of the set of all quadratic stabilizing feedback gains $\mathcal{K}_Q$ defined before (recall Definition 7.1). Consider the uncertain linear system

$$\dot{x} = Ax + B_2 u \tag{7.12}$$

with n states, m inputs and where matrices A and B_2 of appropriate and known dimension are such that $(A, B_2) \in \mathcal{D}$ or equivalently $M_c \in \mathcal{D}_c$. The set $\mathcal{D}_c$ is a polyhedral convex bounded domain. Assuming the whole state variable is available for feedback, we want to characterize all state feedback gain matrices F such that with $u = Fx$, the uncertain closed-loop system is quadratic stable, that is

$$F \in \mathcal{K}_Q \tag{7.13}$$

To this end, we need to consider again the $p \times p$ matrices W and Q_c with $p := n + m$, partitioned as (recall Section 6.2.1)

$$W := \begin{bmatrix} W_1 & W_2 \\ W_2' & W_3 \end{bmatrix}, \quad Q_c := \begin{bmatrix} Q_{1c} & 0 \\ 0 & 0 \end{bmatrix} \tag{7.14}$$

where in both matrices, the $(1,1)$ block has dimension $n \times n$. We also consider, for $i = 1, 2, \cdots, N$ the convex sets

$$\mathcal{C}_{ci} := \{W \ : \ W \geq 0, \ v' \Theta_{ci}(W) v \leq 0, \ \forall v \in \mathcal{N}_c\} \tag{7.15}$$

where $\Theta_{ci}(W) := M_{ci} W + W M_{ci}' + Q_c$. In other terms, the convex set defined in (7.15) is exactly the same as $\mathcal{C}_c$ but with matrix M_c replaced by the extreme matrix M_{ci}. Of course, Theorem 6.1 remains valid for each extreme vertex of the polyhedral convex bounded domain $\mathcal{D}_c$. More interestingly, these convex sets can generate all quadratic stabilizing gains as is proved in the next theorem.

Theorem 7.1 *Assume Q_{1c} is a positive definite matrix and consider the set*

$$\mathcal{C}_Q := \bigcap_{i=1}^{N} \mathcal{C}_{ci} \tag{7.16}$$

The following hold

   a) $\mathcal{C}_Q$ is a convex set.

   b) Each $W \in \mathcal{C}_Q$ is such that $W_1 > 0$.

   c) $\mathcal{K}_Q = \{W_2' W_1^{-1} \ : \ W \in \mathcal{C}_Q\}$.

Proof The first two points follow immediately from Theorem 6.1 where it is proven that for each $i = 1, 2, \cdots, N$ the set $\mathcal{C}_{ci}$ is convex and each $W \in \mathcal{C}_{ci}$ is such that $W_1 > 0$. Consequently, point a) and point b) are both true from the definition of $\mathcal{C}_Q$ which is the intersection of the former convex sets. The proof of the last part of the theorem is much more involved.

   Point c) It is done by construction. First take $F \in \mathcal{K}_Q \neq \emptyset$ and remember that Q_{1c} is positive definite. From Definition 7.1, there exists a symmetric positive definite matrix P satisfying the matrix inequality

$$(A + B_2 F) P + P(A + B_2 F)' + Q_{1c} \leq 0$$

for all pairs (A, B_2) such that $M_c \in \mathcal{D}_c$. Consequently, the same inequality holds true for the N pairs (A_i, B_{2i}) corresponding to the extreme matrices $M_{ci}, i = 1, 2, \cdots, N$. Using this fact, it is readily seen that the matrix

$$W = \begin{bmatrix} P & PF' \\ FP & FPF' \end{bmatrix} \in \mathcal{C}_{ci}$$

for all $i = 1, 2, \cdots, N$ that is, $W \in \mathcal{C}_Q$ and $W_2' W_1^{-1} = FPP^{-1} = F$. Conversely, pick $W \in \mathcal{C}_Q \neq \emptyset$ and notice that from (7.16) $W \in \mathcal{C}_{ci}$ for all $i = 1, 2, \cdots, N$. On the other hand, we know that a generic matrix $M_c \in \mathcal{D}_c$ can be written as

$$M_c = \sum_{i=1}^{N} \xi_i M_{ci}, \quad \xi_i \geq 0, \quad \sum_{i=1}^{N} \xi_i = 1$$

which yields

$$\Theta_c(W) := M_c W + W M_c' + Q_{1c}$$

$$= \sum_{i=1}^{N} \xi_i \left(M_{ci} W + W M_{ci}' + Q_{1c} \right)$$

$$= \sum_{i=1}^{N} \xi_i \Theta_{ci}(W) \tag{7.17}$$

The consequence is that for all $v \in \mathcal{N}_c$ and $\forall \, (A, B_2) \in \mathcal{D}$

$$0 \geq \sum_{i=1}^{N} \xi_i v' \Theta_{ci}(W) v$$

$$\geq v' \Theta_c(W) v$$

$$\geq x' \left[(A + B_2 W_2' W_1^{-1}) W_1 + W_1 (A + B_2 W_2' W_1^{-1})' + Q_{1c} \right] x \tag{7.18}$$

This inequality, together with the fact that both matrices W_1 and Q_{1c} are positive definite implies, by Definition 7.1, that $F = W_2' W_1^{-1} \in \mathcal{K}_Q$ proving thus the theorem proposed since the case in which one of the sets is empty follows immediately. $\square$

This theorem is the uncertain systems stability counterpart of Theorem 6.1 proved in Chapter 6. The main point to be retained in mind is that it provides a convex description of the set of all quadratic stabilizing gains $\mathcal{K}_Q$ in terms of the same nonlinear mapping namely $W_2' W_1^{-1}$ but with W varying now in the convex set $\mathcal{C}_Q$. As discussed before, this theorem shows how to generate the set $\mathcal{K}_Q$ which is only a subset of $\mathcal{K}_D$. Thus, it is clear that using it we can not capture all robust stabilizing gains as defined in Remark 7.1 but only those inside $\mathcal{K}_Q$ which are easily obtained by a simple convex feasibility problem.

Remark 7.3 From part c) of Theorem 7.1 it is evident that the pair $(A, B_2) \in \mathcal{D}$ is quadratic stabilizable if and only if the convex set $\mathcal{C}_Q$ is not empty. Indeed, if there are no matrices $P > 0$ and F satisfying Definition 7.1 then $\mathcal{C}_Q = \emptyset$. $\square$

Remark 7.4 The matrix function $\Theta_c(W)$ defined in (7.17) is of great importance. For W fixed, it can be viewed as an affine function of $M_c \in \mathcal{D}_c$. Since the uncertain domain $\mathcal{D}_c$ is a polyhedral convex and bounded set, each matrix M_c can be written as a convex combination (although unknown) of the extreme matrices M_{ci}, $i = 1, 2, \cdots, N$. Consequently, it follows that

$$\Theta_c(W) = \sum_{i=1}^{N} \xi_i \Theta_{ci}(W)$$

which means that $\Theta_c(W) \leq 0$ if and only if $\Theta_{ci}(W) \leq 0$, $i = 1, 2, \cdots, N$. In other words, we can say that the uncertain system defined by the pair (A, B_2) is quadratic stabilizable if and only if the collection of N systems represented by the pairs (A_i, B_{2i}) is quadratic stabilizable. The quadratic stabilizability of any other pair $(A, B_2) \in \mathcal{D}_c$ is a mere consequence of the quadratic stabilizability of the extreme matrices which define the domain $\mathcal{D}_c$. Even though the set $\mathcal{D}_c$ is composed by an infinity number of matrices M_c, only N of them have to be used to check quadratic stabilizability.

To put in evidence the importance of Definition 7.1, let us proceed in the above discussion under the assumption that the collection of extreme matrices, namely $(A_i, B_{2i}), i = 1, 2, \cdots, N$ is robustly stabilizable (recall Remark 7.1). That is, there exists a matrix F

such that the eigenvalues of $A_i + B_{2i}F$ are all in the open left complex plane. From the Extended Lyapunov lemma, this is equivalent to the existence of matrices $P_i = P_i' > 0$ and $Q_i = Q_i' > 0$ such that

$$0 \geq (A_i + B_{2i}F)P_i + P_i(A_i + B_{2i}F)' + Q_i$$

for all $i = 1, 2, \cdots, N$. Since the right hand side of the inequality

$$0 \geq (A + B_2F)P + P(A + B_2F)' + Q$$

is not a convex function of the unknown matrices A, B_2, P and Q then the robust stabilizability of the collection of the extreme matrices does not provide enough information to guarantee the same for all matrices in the convex domain $\mathcal{D}_c$. □

Remark 7.5 (Comparison between $\mathcal{D}_n$ and $\mathcal{D}_c$) In many instances (recall Example 7.1) the uncertainty of a linear dynamic system can be modeled as $M_c \in \mathcal{D}_n$ only at expense to impose to matrix Ω a prespecified structure. In Example 7.1 this occurred since Ω was taken to be a diagonal matrix. Let us analyze, in this particular case, the degree of conservativeness introduced. From the definition of $\mathcal{D}_n$ in Remark 7.1, we notice that the *nominal* matrices must be partitioned as

$$M_{cn} = \begin{bmatrix} A_0 & B_{20} \\ 0 & 0 \end{bmatrix} , \quad B_n = \begin{bmatrix} B_1 \\ 0 \end{bmatrix} , \quad C_n = \begin{bmatrix} C_1 & D_{12} \end{bmatrix}$$

The consequence is that there exists a stabilizing gain F such that with $u = Fx$ the closed loop system is stable for all $M_c \in \mathcal{D}_n$ if $A_0 + B_{20}F$ is stable and

$$\|(C_1 + D_{12}F)[sI - (A_0 + B_{20}F)]^{-1}B_1\|_\infty < 1$$

Of course, if this is true then the closed-loop system is stable for all Ω such that $\|\Omega\| \leq 1$. In other words, this condition does not take into account the known fact that Ω is a diagonal matrix.

On the contrary, for this particular case (in which Ω is diagonal), the same uncertain domain can also be exactly described by $\mathcal{D}_c$ and consequently, applying Theorem 7.1 the whole set $\mathcal{K}_Q$ is generated which contributes to decrease the degree of conservativeness involved. This feature is numerically addressed in the example that follows. □

Example 7.2 Consider again the linear uncertain system (recall Example 7.1)

$$\dot{x} = Ax + B_2u$$

where matrices A and B_2 are given by

$$A = \begin{bmatrix} 0 & \alpha - 1 \\ \beta & 0 \end{bmatrix} , \quad B_2 = \begin{bmatrix} \alpha \\ 1 - \beta \end{bmatrix}$$

and the parameters α and β, representing the uncertainties are such that

$$|\alpha - 0.5| \leq \gamma , \quad |\beta - 0.5| \leq \gamma$$

Letting the nominal value for these parameters be $\alpha_0 = \beta_0 = 0.5$, the norm bounded domain $\mathcal{D}_n$ is completely defined by matrices (recall Remark 7.5)

$$A_0 = \begin{bmatrix} 0 & -0.5 \\ 0.5 & 0 \end{bmatrix} , \quad B_{20} = \begin{bmatrix} 0.5 \\ 0.5 \end{bmatrix}$$

$$B_1 = \sqrt{2} \begin{bmatrix} \gamma & 0 \\ 0 & \gamma \end{bmatrix} , \quad C_1 = \frac{1}{\sqrt{2}} \begin{bmatrix} 0 & 1 \\ 1 & 0 \end{bmatrix} , \quad D_{12} = \frac{1}{\sqrt{2}} \begin{bmatrix} 1 \\ -1 \end{bmatrix}$$

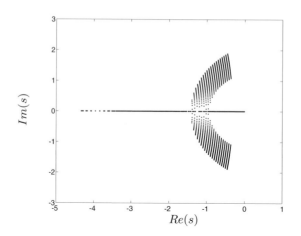

Figure 7.3: Uncertain system closed-loop poles

where matrix D_{12} is such that $D'_{12}D_{12} = 1$. In addition, matrix Ω such that $\|\Omega\| \leq 1$ exhibits the diagonal structure

$$\Omega = \text{diag}\left\{\frac{\alpha - 0.5}{\gamma}, \; \frac{\beta - 0.5}{\gamma}\right\}$$

The central controller which satisfies the RH_∞ constraint just discussed in the previous remark is provided by the solution (if any) of the Full information problem, that is $F = F_\infty = -B'_{20}P_\infty - D'_{12}C_1$ where P_∞ is the positive semidefinite and stabilizing solution of the Riccati equation (in the unknown P)

$$0 = PA_{c0} + A'_{c0}P - P(B_{20}B'_{20} - B_1B'_1)P + C'_{1c}C_{1c}$$

with $A_{c0} := A_0 - B_{20}D'_{12}C_1$ and $C_{1c} := (I - D_{12}D'_{12})C_1$. Numerically, it is possible to verify that the above Riccati equation admits a positive semidefinite and stabilizing solution for $\gamma \in [0, 0.27]$. Thus, setting $\gamma = 0.27$, we can conclude that the associated state feedback gain $F = F_\infty = [3.80 \; -21.49]$ is stabilizing for all $M_c \in \mathcal{D}_n$.

On the other hand, the same system can also be modeled by means of a convex bounded domain $\mathcal{D}_c$. This is done with $N = 4$ and the corresponding extreme matrices, whose entries, depending on the parameter γ, are easily determined. From Theorem 7.1, we observe that the convex set $\mathcal{C}_Q$ depends on γ. Numerically it is verified that with $Q_{1c} = I$, the set $\mathcal{C}_Q$ is not empty for all $\gamma \in [0, 0.36]$. The associated quadratic stabilizing gains, valid for all $M_c \in \mathcal{D}_c$ have been computed in two situations, namely

$$\gamma = 0.27 \Longrightarrow F = \begin{bmatrix} -0.13 & -1.91 \end{bmatrix}$$
$$\gamma = 0.36 \Longrightarrow F = \begin{bmatrix} -0.26 & -4.94 \end{bmatrix}$$

As we have said before, in this case, Theorem 7.1 provides a better result when compared to the former one. Figure 7.3 shows the root locus of the uncertain closed-loop system with the quadratic stabilizing gain above, corresponding to $\gamma = 0.36$ which is obtained with the parameters (α, β) varying in the rectangle $[0.14, \; 0.86] \times [0.14, \; 0.86]$. It is numerically confirmed that the closed-loop system is in the limit of stability since there exists a closed-loop pole with real part equal to -0.0058. Hence a small increase in γ will move some closed-loop poles to the right part of the complex plane, causing instability. $\square$

7.2.2 RH_2 guaranteed cost control

Let the uncertain linear system under consideration be

$$\dot{x} = Ax + B_1 w + B_2 u \tag{7.19}$$
$$z = C_1 x + D_{12} u \tag{7.20}$$
$$y = x \tag{7.21}$$

and represented in fig. 7.1. It is assumed that i) the pair (A, B_2) is quadratic stabilizable and ii) $D_{12}' D_{12} = I$. For this system, our interest is to determine a RH_2 guaranteed cost control which is defined as follows. Choosing

$$\mathcal{G}(T(z, w; s)) := \|T(z, w; s)\|_2^2 \tag{7.22}$$

we have to determine an upper bound, depending only on $F \in \mathcal{K}_Q$ such that

$$\|T(z, w; s)\|_2^2 = \mathcal{J}(M_c, F) \leq \bar{\rho}(F) , \quad \forall\, M_c \in \mathcal{D}_c \tag{7.23}$$

and solve the optimal state feedback control problem

$$\bar{\rho}_Q := \min_{F \in \mathcal{K}_Q} \bar{\rho}(F) \tag{7.24}$$

Before we proceed, it is important to keep in mind that our main objective is to convert the design problem (7.24) into an equivalent convex programming problem. This fact of course has to be kept in mind during the determination of the upper bound satisfying inequality (7.23). To this end we need to introduce the symmetric matrices

$$R_c := \begin{bmatrix} C_1' \\ D_{12}' \end{bmatrix} \begin{bmatrix} C_1 & D_{12} \end{bmatrix} , \quad Q_c := \begin{bmatrix} B_1 B_1' & 0 \\ 0 & 0 \end{bmatrix} \tag{7.25}$$

which together with the next lemma define the function $\bar{\rho}(\cdot)$ which exhibits the properties just discussed.

Lemma 7.1 *Assume $B_1 B_1'$ is a positive definite matrix, consider $F \in \mathcal{K}_Q$ arbitrary and define the convex set*

$$\bar{\mathcal{C}}_Q(F) := \mathcal{C}_Q \bigcap \{W \;:\; W_2' = F W_1\}$$

then

$$\bar{\rho}(F) := \min \left\{ \text{trace}[R_c W] \;:\; W \in \bar{\mathcal{C}}_Q(F) \right\} \tag{7.26}$$

is a valid upper bound to $\|T(z, w; s)\|_2^2$ for all feasible $M_c \in \mathcal{D}_c$.

Proof First notice that for any $F \in \mathcal{K}_Q$ given, the set $\bar{\mathcal{C}}_Q(F)$ is convex and nonempty. The convexity follows from the fact that it is the intersection of a convex set (recall Theorem 7.1) with another one defined by a linear and hence convex constraint. Again from Theorem 7.1, $W \in \mathcal{C}_Q$ generates all quadratic stabilizing state feedback gains and so for any $F \in \mathcal{K}_Q$ fixed the linear constraint $W_2' = F W_1$ must be satisfied for some $W \in \mathcal{C}_Q$. Let us now consider an arbitrary $M_c \in \mathcal{D}_c$ which can be written as a convex combination of the extreme matrices, that is

$$M_c = \sum_{i=1}^{N} \xi_i M_{ci} , \quad \xi_i \geq 0 , \quad \sum_{i=1}^{N} \xi_i = 1$$

and pick $W \in \bar{\mathcal{C}}_Q(F)$. For all $v \in \mathcal{N}_c$ we have

$$0 \geq \sum_{i=1}^{N} \xi_i v' \Theta_{ci}(W) v$$

$$\geq x' \left[(A + B_2 F) W_1 + W_1 (A + B_2 F)' + B_1 B_1' \right] x$$

which enables us to conclude that $W_1 \geq P_r$ where $P_r > 0$ solves the linear matrix equation

$$(A + B_2 F) P_r + P_r (A + B_2 F) + B_1 B_1' = 0$$

On the other hand, applying the Schur complement formula to $W \geq 0$ it is simple to see that $W_3 \geq F W_1 F'$. Then, for the same matrix M_c and using the fact that $W_2' = F W_1$, we also get

$$\mathrm{trace}[R_c W] = \mathrm{trace} \left[\begin{bmatrix} C_1 & D_{12} \end{bmatrix} W \begin{bmatrix} C_1' \\ D_{12}' \end{bmatrix} \right]$$

$$= \mathrm{trace} \left[(C_1 + D_{12} F) W_1 (C_1 + D_{12} F)' \right] +$$

$$+ \mathrm{trace} \left[D_{12} (W_3 - F W_1 F') D_{12}' \right]$$

$$\geq \mathrm{trace} \left[(C_1 + D_{12} F) P_r (C_1 + D_{12} F)' \right]$$

$$\geq \| T(z, w; s) \|_2^2$$

Since this inequality holds for an arbitrary $M_c \in \mathcal{D}_c$ and for all feasible $W \in \bar{\mathcal{C}}_Q(F)$, the final conclusion is that $\bar{\rho}(F)$ defined in (7.26) satisfies

$$\| T(z, w; s) \|_2^2 \leq \bar{\rho}(F) , \quad \forall \, M_c \in \mathcal{D}_c$$

which proves the lemma proposed. $\qquad \square$

Remark 7.6 There are two reasons to define $\bar{\rho}(F)$ as indicated in (7.26). First, any feasible W produces an upper bound to the RH_2 norm of the transfer function under consideration but the optimal solution of the convex problem indicated in the previous lemma provides the smaller upper bound as far as the choice of matrix W is concerned. Second, when $N = 1$, that is when $\mathcal{D}_c$ is composed by only on matrix M_c (there is no uncertainty at all) then from the proof above it is easily verified that the optimal solution of problem (7.26) is

$$W = \begin{bmatrix} P_r & P_r F' \\ F P_r & F P_r F' \end{bmatrix} \in \bar{\mathcal{C}}_Q(F)$$

providing thus

$$\bar{\rho}(F) = \| T(z, w; s) \|_2^2$$

In this special case, the function $\bar{\rho}(F)$ reduces to the RH_2 norm of the closed-loop transfer function. Thus, in the general case of uncertain systems with polyhedral convex bounded domain $\mathcal{D}_c$, this function can be interpreted as a *generalized RH_2 norm* which is finite for all quadratic stabilizing state feedback gains. $\qquad \square$

We are now in position to solve the basic design problem (7.24) by means of an equivalent convex programming problem. This result is summarized in the next theorem.

Theorem 7.2 *Assume $B_1 B_1'$ is a positive definite matrix. Then,*

$$\min_{F \in \mathcal{K}_Q} \bar{\rho}(F) = \min \{ \mathrm{trace}[R_c W] \; : \; W \in \mathcal{C}_Q \} \tag{7.27}$$

Furthermore, being $\bar{F}$ and $\bar{W}$ the optimal solution of each problem they are related one to each other by $\bar{F} = \bar{W}_2' \bar{W}_1^{-1}$.

Proof The proof is a mere consequence of Theorem 7.1 together with Lemma 7.1. Actually, the minimum of $\bar{\rho}(F)$ over $F \in \mathcal{K}_Q$ is obtained by the joint minimization of (7.26) with respect to both F and W. Due to the part c) of Theorem 7.1, this joint minimization can be calculated in two independent steps. In the first one, the optimal solution $\bar{W}$ of the convex programming problem in the right hand side of (7.27) is determined. This solution allows the determination of the optimal gain namely, $\bar{F} = \bar{W}_2'\bar{W}_1^{-1} \in \mathcal{K}_Q$ in the second and last step. □

Remark 7.7 Assuming $B_1 B_1' > 0$, the function $\bar{\rho}(F)$ introduced in Lemma 7.1 can also be calculated as follows

$$\bar{\rho}(F) = \min_{P>0} \left\{ \text{trace}[C_{cc} P C_{cc}'] \; : \; 0 \geq A_{ci}P + PA_{ci}' + B_1 B_1' \right\}$$

where $A_{ci} := A_i + B_{2i}F$ for $i = 1, 2, \cdots, N$ and $C_{cc} := C_1 + D_{12}F$. Since (recall Definition 7.1) the constraint set in the above problem is feasible if and only if $F \in \mathcal{K}_Q$ then it is simple to see that

$$\bar{\rho}_Q = \min_{F,P>0} \left\{ \text{trace}[C_{cc} P C_{cc}'] \; : \; 0 \geq A_{ci}P + PA_{ci}' + B_1 B_1' \right\}$$

Defining $\Lambda_i = \Lambda_i' \geq 0$ as being the Kuhn-Tucker multipliers associated to the $i-th$ inequality constraint, the Lagrangian function turns out to be

$$\mathcal{L} = \text{trace}[C_{cc} P C_{cc}'] + \sum_{i=1}^{N} \text{trace} \left[\Lambda_i (A_{ci}P + PA_{ci}' + B_1 B_1') \right]$$

from which the Kuhn-Tucker necessary conditions for optimality can be written in terms of the unknown matrices F, $P > 0$, $Q_i \geq 0$ and $\Lambda_i \geq 0$ satisfying

$$0 \geq A_{ci}P + PA_{ci}' + B_1 B_1' := -Q_i$$

$$0 = \sum_{i=1}^{N} \left(A_{ci}'\Lambda_i + \Lambda_i A_{ci} \right) + C_{cc}'C_{cc}$$

$$0 = \left(F + D_{12}'C_1 + \sum_{i=1}^{N} B_{2i}'\Lambda_i \right) P$$

$$0 = \text{trace}\left[\Lambda_i Q_i \right]$$

These conditions put in evidence several nonlinear relationship among the unknown matrices implying that they can be solved only in some special cases. Indeed, the third equation can be used to determine the optimal gain matrix F explicitly, however the second condition alone is not of great help in finding matrices $\Lambda_i, i = 1, 2, \cdots, N$ since it represents only one equation with N unknown. The consequence is that putting aside the third equation, the remaining ones have to be solved simultaneously. This is accomplished by solving the right hand side of (7.27) directly. It is important to stress that this discussion is no longer true in the particular case $N = 1$. The domain $\mathcal{D}_c$ degenerates to a single matrix and the above conditions reduce to those of Remark 6.7 with matrices P and Λ replaced one by the other since the problem now under consideration is exactly the dual of the one treated there. □

Remark 7.8 The use of any convex programming method to get the global solution of the optimal RH_2 guaranteed cost control problem

$$\bar{\rho}_Q = \min \left\{ \text{trace}[R_c W] \; : \; W \in \mathcal{C}_Q \right\}$$

depends on the boundedness of the following convex set (recall Remark 6.8)

$$\mathcal{C}_{Q\mu} := \mathcal{C}_Q \bigcap \{ W \; : \; \text{trace}[R_c W] \leq \mu \}$$

Since the set $\mathcal{C}_Q$ is itself an intersection of N convex sets, then the requirement that there exists an index $1 \leq i \leq N$ for which the pair $(-A_{ci}, C_{1c})$ is detectable is a sufficient condition for the boundedness of $\mathcal{C}_{Q\mu}$ for all μ such that $\bar{\rho}_Q < \mu < \infty$. $\square$

Remark 7.9 In this section we have defined the *generalized* norm $\bar{\rho}(F)$ satisfying

$$\|T(z, w; s)\|_2^2 \leq \bar{\rho}(F) , \quad \forall M_c \in \mathcal{D}_c$$

Let us now verify the difficulties we have to face if we desire to work, instead of $\bar{\rho}(F)$, with the *true* generalized RH_2 norm

$$\rho(F) = \max_{M_c \in \mathcal{D}_c} \|T(z, w; s)\|_2^2$$

Since $F \in \mathcal{K}_Q$ assures that all closed-loop transfer functions $T(z, w; s)$ are stable then with $A_{cc} := A + B_2 F$ and $C_{cc} := C_1 + D_{12} F$ we have

$$\rho(F) = \max_{M_c \in \mathcal{D}_c} \text{trace} \left[\int_0^\infty C_{cc} e^{A_{cc} t} B_1 B_1' e^{A_{cc}' t} C_{cc}' dt \right]$$

On the other hand, from Theorem 7.1 we already know that all $F \in \mathcal{K}_Q$ are generated by $F = W_2' W_1^{-1}$ with $W \in \mathcal{C}_Q$. Then we get

$$\rho_Q = \min_{F \in \mathcal{K}_Q} \rho(F)$$

$$= \min_{W \in \mathcal{C}_Q} \rho(W_2' W_1^{-1})$$

In the above problem the constraint $W \in \mathcal{C}_Q$ is convex. However, the difficulty is that it is not easy to prove if the objective function is convex or not. In the affirmative case, no major difficulty exists to approach the problem by means of convex programming methods. The convexity of $\rho(\cdot)$ remains until now an open question. $\square$

Remark 7.10 Let us now consider the RH_2 guaranteed cost control problem but with norm bounded uncertainty, that is $M_c \in \mathcal{D}_n$. The uncertain system (7.19) - (7.21) is now written with a slightly different notation

$$\dot{x} = Ax + B_0 w + B_2 u$$
$$z_0 = C_0 x + D_0 u$$
$$y = x$$

where $B_0 B_0' > 0$ and $D_0' D_0 = I$. Following Remarks 7.2 and 7.5, using $u = Fx$ the closed-loop system matrix can be expressed as

$$A + B_2 F = A_{cc} + B_1 \Omega C_{cc} , \quad \|\Omega\| \leq 1$$

where $A_{cc} := A_0 + B_{20} F$ and $C_{cc} := C_1 + D_{12} F$. Notice that in this setting, the external perturbation, and the parametric uncertainty enter in the system through two different matrices, namely B_0 and B_1 respectively. First of all, it is claimed that for the above uncertain system, the upper bound $\bar{\rho}(F)$ is given by

$$\bar{\rho}(F) = \min_{\beta \in \mathcal{B}} \text{trace}[B_0' P(\beta) B_0]$$

$$\geq \|T(z_0, w; s)\|_2^2 , \quad \forall M_c \in \mathcal{D}_n$$

where $\mathcal{B}$ is the set of all $\beta > 0$ such that there exists a symmetric, stabilizing and positive semidefinite solution $P(\beta)$ to the Riccati equation (in the unknown P)

$$0 = PA_{cc} + A_{cc}' P + \beta P B_1 B_1' P + \beta^{-1} C_{cc}' C_{cc} + C_{0c}' C_{0c}$$

where $C_{0c} := C_0 + D_0 F$. To show this, notice first that any $\beta > 0$ yields

$$C'_{cc}\Omega' B'_1 P + P B_1 \Omega C_{cc} = \beta^{-1} C'_{cc}\Omega'\Omega C_{cc} + \beta P B'_1 B_1 P - $$
$$-(\Omega C_{cc}\beta^{-1/2} - B'_1 P \beta^{1/2})'(\Omega C_{cc}\beta^{-1/2} - B'_1 P \beta^{1/2})$$
$$\leq \beta^{-1} C'_{cc} C_{cc} + \beta P B'_1 B_1 P$$

where to get the last inequality we have used the fact that $\Omega'\Omega \leq I$. Using this inequality together with the previously defined Riccati equation, for all $\beta \in \mathcal{B}$ we can draw two important conclusions. The first one is that (recall Remark 7.5)

$$\|C_{cc}[sI - (A_0 + B_{20} F)]^{-1} B_1\|_\infty < 1$$

meaning that the closed-loop system matrix $A + B_2 F$ is stable for all $M_c \in \mathcal{D}_n$. Second, from the above inequality we also have

$$(A + B_2 F)' P(\beta) + P(\beta)(A + B_2 F) + C'_{0c} C_{0c} \leq 0$$

for all $M_c \in \mathcal{D}_n$ implying that

$$\|T(z_0, w; s)\|_2^2 = \text{trace}\left[\int_0^\infty B'_0 e^{(A+B_2 F)'t} C'_{0c} C_{0c} e^{(A+B_2 F)t} B_0 dt\right]$$
$$\leq \text{trace}[B'_0 P(\beta) B_0]$$
$$\leq \bar{\rho}(F)$$

Based on this result, let us now determine the best guaranteed cost denoted $\bar{\rho}_N$, that is

$$\bar{\rho}_N = \min_{F, \beta \in \mathcal{B}} \text{trace}[B'_0 P(\beta) B_0]$$

The necessary conditions for optimality with respect to F is readily obtained (recall Remark 6.19) from the associated Lagrangian function. After tedious algebraic manipulations, a solution to these conditions can be written as

$$F(\beta) := \tilde{F} - (1 + \beta)^{-1} B'_{20} W(\beta)$$

where

$$\tilde{F} := -(1 + \beta)^{-1}(D'_{12} C_1 + \beta D'_0 C_0)$$

and $W(\beta) := \beta P(\beta)$ solves the Riccati equation (in the unknown W)

$$0 = W\tilde{A} + \tilde{A}' W + W\tilde{N} W + \tilde{M}$$

with

$$\tilde{A} := A_0 + B_{20}\tilde{F}$$
$$\tilde{N} := B_1 B'_1 - (1 + \beta)^{-1} B_{20} B'_{20}$$
$$\tilde{M} := (C_1 + D_{12}\tilde{F})'(C_1 + D_{12}\tilde{F}) + \beta(C_0 + D_0\tilde{F})'(C_0 + D_0\tilde{F})$$

Finally, using this result we get

$$\bar{\rho}_N = \min_{\beta \in \mathcal{B}} \beta^{-1} \text{trace}[B'_0 W(\beta) B_0]$$

which is numerically easy to solve since it involves the search of only one positive parameter. The main conclusion to be kept in mind is that the guaranteed cost associated to the uncertain domain $\mathcal{D}_n$ is numerically tractable by means of a unidimensional search procedure. It must be clear that in the present case, it may occur that $\mathcal{B} = \emptyset$, meaning that $\bar{\rho}_N$ is unbounded and so no guaranteed cost control is obtained. This aspect is illustrated in the next example. □

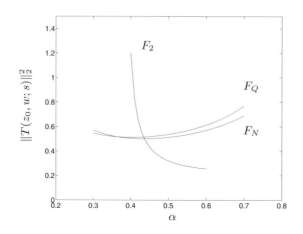

Figure 7.4: Actual value of the RH_2 norm

Example 7.3 Let the uncertain system be again the one of Example 7.2. The state space representation is

$$\dot{x} = Ax + B_0 w + B_2 u$$
$$z_0 = C_0 x + D_0 u$$
$$y = x$$

where the uncertainty is modeled as $M_c \in \mathcal{D}_c$ as well as $M_c \in \mathcal{D}_n$. The main goal is to determine the associated guaranteed RH_2 cost control. The data are those of Example 7.2 but with only one uncertain parameter obtained by imposing $\alpha = \beta$ such that

$$|\alpha - 0.5| \leq \gamma$$

where $\gamma = 0.20$. The controlled output is defined by matrices

$$B_0 = B_1 = \sqrt{2} \begin{bmatrix} \gamma & 0 \\ 0 & \gamma \end{bmatrix}, \quad C_0 = \begin{bmatrix} 1 & 1 \\ 0 & 0 \end{bmatrix}, \quad D_0 = \begin{bmatrix} 0 \\ 1 \end{bmatrix}$$

Three different situations have been considered :

1) First, assuming no uncertainty acts on the open-loop system, the Full information problem in RH_2 has been solved using the nominal data. The optimal solution found is

$$\min \|T(z_0, w; s)\|_2^2 = 0.32 \implies F_2 = \begin{bmatrix} -1.00 & -1.00 \end{bmatrix}$$

Using this state feedback gain, it is simple to verify that the closed-loop system remains stable for all α varying in the interval defined as above with $\gamma = (\sqrt{5} - 2)/2 = 0.1180$. This fact is illustrated in fig. 7.4 where it is clear that the corresponding value of $\|T(z_0, w; s)\|_2^2$, denoted by the symbol "F_2", becomes arbitrarily large as α goes to 0.3820. For all other values of α, the closed-loop system is unstable.

2) We considered $M_c \in \mathcal{D}_c$. In this case we have two extreme matrices only. The optimal solution of the convex programming problem in Theorem 7.2 provides

$$\bar{\rho}_Q = 1.18 \implies F_Q = \bar{F} = \begin{bmatrix} -0.02 & -2.39 \end{bmatrix}$$

As indicated in fig.7.4, the closed-loop system is now stable in the whole interval $0.3 \leq \alpha \leq 0.7$ implying that the associated cost is always finite. This is an important improvement of the guaranteed cost control design proposed when compared with the previous situation. It is interesting to notice that the closed-loop system transfer function satisfies

$$\max_{0.3 \leq \alpha \leq 0.7} \|T(z_0, w; s)\|_2^2 = 0.77 < \bar{\rho}_Q$$

which illustrates the fact discussed before (recall Lemma 7.1) that function $\bar{\rho}(F)$ is only an upper bound to the true value of the minimized cost.

3) In this situation we modeled the uncertainty as $M_c \in \mathcal{D}_n$. Following the results of Remark 7.10, it has been verified that $\mathcal{B} = (0, 0.7]$ in order to preserve $W(\beta) > 0$. Using this, an unidimensional search has been implemented to get the optimal $\bar{\beta} = 0.32$ which provides

$$\bar{\rho}_N = 2.45 \implies F_N = F(\beta = 0.32) = \begin{bmatrix} -0.13 & -2.52 \end{bmatrix}$$

Figure 7.4 shows that the closed-loop system is also stable for all values of α in the real interval $0.3 \leq \alpha \leq 0.7$. There is no big difference between the cost variation in the last two situations. However, comparing them, it is apparent that $\bar{\rho}_Q < \bar{\rho}_N$ which indicates that in this example the design based on $\mathcal{D}_c$ is preferable.

Figure 7.4 illustrates also the advantages of the guaranteed cost control design when compared with the nominal Full information control design. For the nominal system it is obvious that the latter is better. However, the performance imposed by the optimal Full information controller may become worse when a small change in the system parameters occurs. Even the closed-loop stability can be lost. On the contrary, the guaranteed cost controller assures closed-loop stability for the parameters varying in the prespecified range and impose a reduced level of performance deterioration. □

7.2.3 RH_∞ guaranteed cost control

We now move our attention to the RH_∞ counterpart of the results provided in the last section. The uncertain dynamic system under consideration is again the same depicted in fig. 7.1 where as indicated the full state vector is available for feedback. To ease the presentation the state space equations are given once again,

$$\dot{x} = Ax + B_1 w + B_2 u \tag{7.28}$$
$$z = C_1 x + D_{12} u \tag{7.29}$$
$$y = x \tag{7.30}$$

where A and B_2 are uncertain matrices such that is $M_c \in \mathcal{D}_c$ and as before we assume that i) the pair (A, B_2) is quadratic stabilizable and ii) $D_{12}' D_{12} = I$. The above equations define a family of linear dynamic systems for which our goal is to determine an associated RH_∞ guaranteed cost control. Following the results of Chapter 6, let γ denotes an arbitrary positive scalar and choose

$$\mathcal{G}(T(z, w; s)) := \|T(z, w; s)\|_\infty^2 \tag{7.31}$$

Then the minimum guaranteed RH_∞ cost can be calculated from (7.11), yielding

$$\rho_Q = \min_{\rho,\, F \in \mathcal{K}_Q} \{\rho \,:\, \mathcal{J}(M_c, F) \leq \rho \,,\; \forall\, M_c \in \mathcal{D}_c\} \tag{7.32}$$

Parallel to the case of RH_2 guaranteed cost control problem, the solution of (7.32) is extremely difficult to get, then the strategy to be used is to define a suboptimal easy to determine solution provided by

$$\bar{\rho}_Q := \min \{\gamma^2 \,:\, (F, \gamma) \in \mathcal{K}_{\gamma Q}\} \geq \rho_Q \tag{7.33}$$

where $\mathcal{K}_{\gamma Q}$ is a subset of the set of all pairs (F, γ) such that $F \in \mathcal{K}_Q$ and

$$\|T(z, w; s)\|_\infty^2 = \mathcal{J}(M_c, F) \leq \gamma^2 \ , \ \forall \ M_c \in \mathcal{D}_c \tag{7.34}$$

It is clear that the choice of the set $\mathcal{K}_{\gamma Q}$ is crucial to get good suboptimal solutions in terms of inequality (7.33). Besides it must be possible to convert it into an equivalent convex set. To this end, considering matrix W partitioned as indicated in (7.14) and matrices R_c and Q_c given in (7.25) let us define the following convex set (recall Theorem 6.5) associated with each extreme matrix of the polyhedral convex bounded uncertain domain $\mathcal{D}_c$,

$$\mathcal{C}_{\gamma i} := \{(W, \rho) \ : \ W \geq 0 \ , \ \rho > 0 \ , \ v' \Theta_{\gamma i}(W, \rho) v \leq 0 \ , \ \forall v \in \mathcal{N}_c\} \tag{7.35}$$

where $\Theta_{\gamma i}(W, \rho) := \Theta_{ci}(W) + \rho^{-1} W R_c W$ for each $i = 1, 2, \cdots, N$. Thus, it is apparent that the above set has been obtained from $\mathcal{C}_{\gamma c}$ by simply replacing matrix M_c by the extreme matrix M_{ci}. The next theorem provides the solution of the problem indicated in (7.33).

Theorem 7.3 *Assume $B_1 B_1'$ is a positive definite matrix and consider the set*

$$\mathcal{C}_{\gamma Q} := \bigcap_{i=1}^{N} \mathcal{C}_{\gamma i} \tag{7.36}$$

The following hold

   a) $\mathcal{C}_{\gamma Q}$ is a convex set.

   b) Each $(W, \rho) \in \mathcal{C}_{\gamma Q}$ is such that $W_1 > 0$.

   c) The subset $\mathcal{K}_{\gamma Q}$ defined above can be chosen as

$$\mathcal{K}_{\gamma Q} = \{(W_2' W_1^{-1}, \sqrt{\rho}) \ : \ (W, \rho) \in \mathcal{C}_{\gamma Q}\} \tag{7.37}$$

Proof The first two points are straightforward consequences of Theorem 6.5. Since, both are true for each set $\mathcal{C}_{\gamma i}$, $i = 1, 2, \cdots, N$ then the same is also true for their intersection.

   Point c) We have to prove that any $(W, \rho) \in \mathcal{C}_{\gamma Q}$ provides $F = W_2' W_1^{-1} \in \mathcal{K}_Q$ and $\gamma = \sqrt{\rho}$ such that (7.34) holds. This is proved by selecting an arbitrary pair $(W, \rho) \in \mathcal{C}_{\gamma Q}$ and noticing that $\Theta_{ci}(W) \leq \Theta_{\gamma i}(W, \rho)$ for all $i = 1, 2, \cdots, N$ implies

$$(W, \rho) \in \mathcal{C}_{\gamma Q} \Longrightarrow W \in \mathcal{C}_Q$$

which from part c) of Theorem 7.1 allows us to conclude that $F = W_2' W_1^{-1} \in \mathcal{K}_Q$. Let us now pick an arbitrary matrix $M_c \in \mathcal{D}_c$. Recalling Theorem 6.5, for this matrix the set $\mathcal{C}_{\gamma c}$ is well defined. So, expressing M_c as a convex combination of the extreme matrices

$$M_c = \sum_{i=1}^{N} \xi_i M_{ci} \ , \ \ \xi_i \geq 0 \ , \ \ \sum_{i=1}^{N} \xi_i = 1$$

yields (recall Theorem 6.5) for all $v \in \mathcal{N}_c$

$$\begin{aligned}
v' \Theta_{\gamma c}(W, \rho) v &:= v' \left[\Theta_c(W) + \rho^{-1} W R_c W \right] v \\
&= v' \left[\sum_{i=1}^{N} \xi_i \left(\Theta_{ci}(W) + \rho^{-1} W R_c W \right) \right] v \\
&= \sum_{i=1}^{N} \xi_i v' \Theta_{\gamma i}(W, \rho) v \leq 0
\end{aligned} \tag{7.38}$$

which means that $(W, \rho) \in \mathcal{C}_{\gamma c}$. Since the inequality (7.38) is always true for all $M_c \in \mathcal{D}_c$ we get from part c) of Theorem 6.5 that

$$(F, \gamma) = (W_2' W_1^{-1}, \sqrt{\rho}) \in \bigcap_{M_c \in \mathcal{D}_c} \mathcal{K}_{\gamma c}$$

which means that (7.34) actually holds. □

The importance of this theorem is twofold. First, Problem (7.33) which defines the guaranteed RH_∞ cost reduces to

$$\bar{\rho}_Q = \min \{\rho \; : \; (W, \rho) \in \mathcal{C}_{\gamma Q}\} \tag{7.39}$$

and this is a convex programming problem which can be solved efficiently. Second, the convex set $\mathcal{C}_{\gamma Q}$ allows the determination of a set of feedback matrix gains such that for a given $\gamma > 0$, the closed loop system is quadratically stable and $\|T(z, w; s)\|_\infty$ is bounded above by γ for all $M_c \in \mathcal{D}_c$.

Remark 7.11 The boundedness of the set $\mathcal{C}_{\gamma Q}$ can be analyzed following the same lines of Remark 6.22. The key property to assure the adequate use of convex programming methods in solving the approximate version of Problem (7.39), namely

$$\bar{\rho}_Q \approx \min \{\rho \; : \; (W, \rho) \in \mathcal{C}_{\gamma Q} \; , \; \text{trace}[R_c W] \leq \beta\}$$

where the scalar $\beta > 0$ is finite and appropriately large, is to impose that there exists an index $1 \leq i \leq N$ for which the corresponding pair $(-A_{ci}, C_{1c})$ is detectable. □

Remark 7.12 The convex set $\mathcal{C}_{\gamma Q}$ presents two additional properties. First, in the case of certain dynamic systems, the domain $\mathcal{D}_c$ is defined by only one matrix, then $\mathcal{C}_{\gamma Q}$ equals the set $\mathcal{C}_{\gamma c}$ introduced in Theorem 6.5. As a consequence the optimal value of $\bar{\rho}_Q$ reduces to the minimum possible RH_∞ norm level. Second, in the case of uncertain systems, with γ arbitrarily large $(W, \gamma) \in \mathcal{C}_{\gamma Q}$ implies $W \in \mathcal{C}_Q$. Once again, the set of all quadratic stabilizing feedback gains is obtained. □

7.2.4 Miscellaneous design problems

This section is devoted to treat in the framework of uncertain systems control design the other problems already solved in Chapter 6, namely Mixed RH_2/RH_∞ control, RH_2 control with regional pole placement, time domain specifications and controllers with structural constraints. The basic idea is to generalize the previous results on guaranteed cost control to cope with the new performance constraints. All design problems are solved for the special case of static linear state feedback controllers.

Let us first consider the Mixed RH_2/RH_∞ control problem associated to the linear dynamic system of fig. 7.1 and with the following state space representation

$$\dot{x} = Ax + B_1 w + B_2 u \tag{7.40}$$
$$z_0 = C_0 x + D_0 u \tag{7.41}$$
$$z_1 = C_1 x + D_{12} u \tag{7.42}$$
$$y = x \tag{7.43}$$

where it is assumed that the uncertainty is described by $M_c \in \mathcal{D}_c$. Furthermore, it is also assumed that i) the pair (A, B_2) is quadratic stabilizable and ii) $D_{12}' D_{12} = I$. The controller is given by $u = Fx$ where matrix F is to be determined. In the present context, the guaranteed mixed RH_2/RH_∞ cost control problem can be formulated as being to determine (if one exists) a state feedback gain matrix F such that for a given scalar $\gamma > 0$, the closed-loop system is quadratically stable and

$$\rho_Q := \min \left\{ \rho \ : \ \|T(z_0, w; s)\|_2^2 \le \rho \, , \ \ \|T(z_1, w; s)\|_\infty \le \gamma \, , \ \forall \, M_c \in \mathcal{D}_c \right\} \quad (7.44)$$

Based on the previous results, a suboptimal solution of this problem is given in the next theorem.

Theorem 7.4 *Let $\gamma > 0$ be given and let $\bar{W}$ be the optimal solution of the convex programming problem*

$$\bar{\rho}_Q = \min \left\{ \mathrm{trace}[R_0 W] \ : \ (W, \gamma^2) \in \mathcal{C}_{\gamma Q} \right\} \quad (7.45)$$

Then, $F = \bar{W}_2' \bar{W}_1^{-1} \in \mathcal{K}_Q$ and $\bar{\rho}_Q \ge \rho_Q$.

Proof It follows immediately from the result of Theorem 7.2 together with the ones of Theorem 7.3. Indeed, from the last theorem $(\bar{W}, \gamma^2) \in \mathcal{C}_{\gamma Q}$ implies that $F = \bar{W}_2' \bar{W}_1^{-1} \in \mathcal{K}_Q$ and

$$\|T(z_1, w; s)\|_\infty \le \gamma \, , \ \forall \, M_c \in \mathcal{D}_c$$

Furthermore, using the fact that

$$(\bar{W}, \gamma^2) \in \mathcal{C}_{\gamma Q} \implies \bar{W} \in \mathcal{C}_Q$$

from Theorem 7.2 we conclude that

$$\|T(z_0, w; s)\|_2^2 \le \mathrm{trace}[R_0 \bar{W}] = \bar{\rho}_Q \, , \ \forall \, M_c \in \mathcal{D}_c$$

which proves the theorem proposed since $(F, \bar{\rho}_Q)$ is feasible for Problem (7.44).   □

Example 7.4 Let us consider the following uncertain linear system (recall Example 6.2)

$$\dot{x} = Ax + B_1 w + B_2 u$$
$$z_0 = C_0 x + u$$
$$z_1 = C_1 x + u$$
$$y = x$$

where

$$A = \begin{bmatrix} 0 & 1 \\ -1 + \alpha & -1 + 2\alpha \end{bmatrix}, \ \ B_1 = \begin{bmatrix} 1 \\ 0 \end{bmatrix}$$

$$B_2 = \begin{bmatrix} 0 \\ 1 - \alpha/2 \end{bmatrix}, \ \ C_0 = C_1 = \begin{bmatrix} 0 & -2 \end{bmatrix}$$

The only available information concerning the uncertain parameter α is that $0 \le \alpha \le 1$. We have first verified numerically that the set $\mathcal{C}_{\gamma Q}$ is not empty for $\gamma \ge 4.10$. Then we choose $\gamma = 5.00$ for the calculations that follow. On the other hand, we observe that the dynamic system handled in Example 6.2 corresponds to the above one calculated for $\alpha = 0$. This system is called *nominal*. For the nominal system we solved the Mixed RH_2/RH_∞ control problem introduced in Theorem 6.8. The optimal solution provided the state feedback gain $F_{nominal}$ which has been used to calculate the closed-loop transfer function RH_2 norm indicated in fig. 7.5. Simple calculation shows that the closed-loop system is internally stable only for $0 \le \alpha < 0.5$ which implies that $F_{nominal} \notin \mathcal{K}_Q$.

Moreover, solving the convex programming problem (7.45) we got

$$\bar{\rho}_Q = 4.83 \implies F_{guaranteed} = \begin{bmatrix} -1.90 & -7.20 \end{bmatrix}$$

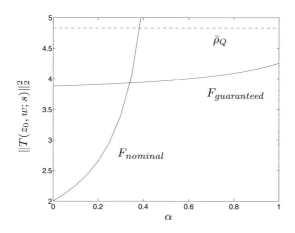

Figure 7.5: RH_2 norm and $\bar{\rho}_Q$

Now, the closed-loop is internally stable for all α in the prespecified interval. Figure 7.5 also shows the closed-loop system RH_2 norm compared with the minimum upper bound available. It is interesting to observe that the behavior of the cost as a function of the uncertain parameter is almost constant in the interval considered. Furthermore, fig. 7.6 shows the magnitude of the transfer function $T(z_1, w; j\omega)$ against frequency for several values of α in the given interval. From this figure we notice that the constraint $\|T(z_1, w; s)\|_\infty \leq \gamma = 5$ is binding and is always verified. $\qquad \square$

We now move our attention to the problem of uncertain system root clustering in a given circular region $\mathcal{R}$. It is stated as follows, given the uncertain linear system

$$\dot{x} = Ax + B_1 w + B_2 u \qquad (7.46)$$
$$z = C_1 x + D_{12} u \qquad (7.47)$$
$$y = x \qquad (7.48)$$

satisfying the assumption i) $D'_{12} D_{12} = I$, find (if one exists) an associated RH_2 guaranteed cost control such that in addition the closed-loop system poles with $u = Fx$ are all inside the circular region $\mathcal{R}$. More specifically, in the same framework considered before, we seek for the solution of the following optimization problem

$$\bar{\rho}_Q = \min \{\rho \ : \ (F, \rho) \in \mathcal{K}_{RQ}\} \qquad (7.49)$$

where $\mathcal{K}_{RQ}$ is a subset of the set of all pairs (F, ρ) such that $F \in \mathcal{K}_Q$,

$$\|T(z, w; s)\|_2^2 \leq \rho \ , \quad \forall \ M_c \in \mathcal{D}_c \qquad (7.50)$$

and the closed-loop system poles are all inside the circular region $\mathcal{R}$. As commented before, the key step towards the solution of problem (7.49) is the definition of the feasible set $\mathcal{K}_{RQ}$ (notice that now ρ is a variable to be determined). To this end let us redefine the matrix

$$M_c := \begin{bmatrix} A + \alpha I & B_2 \\ 0 & 0 \end{bmatrix}$$

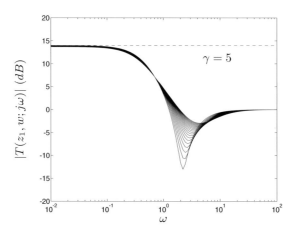

Figure 7.6: Bode diagram - magnitude

and accordingly the convex sets (recall Theorem 6.9) associated with each extreme matrix of the convex domain $\mathcal{D}_c$.

$$\mathcal{C}_{Ri} := \{W \ : \ W \geq 0 \ , \ v'\Theta_{Ri}(W)v \leq 0 \ , \ \forall v \in \mathcal{N}_c\} \tag{7.51}$$

where $\Theta_{Ri}(W) := \Theta_{ci}(W) + r^{-1}M_{ci}WM'_{ci}$ for each $i = 1, 2, \cdots, N$. In other words, each set above is the same as $\mathcal{C}_R$ with matrix M_c replaced by the extreme matrix M_{ci}.

Theorem 7.5 *Assume $B_1B'_1$ is a positive definite matrix, let the circular region $\mathcal{R}$ be given and consider the set*

$$\mathcal{C}_{RQ} := \bigcap_{i=1}^{N} \mathcal{C}_{Ri} \tag{7.52}$$

The following hold

a) *$\mathcal{C}_{RQ}$ is a convex set.*

b) *Each $W \in \mathcal{C}_{RQ}$ is such that $W_1 > 0$.*

c) *The subset $\mathcal{K}_{RQ}$ defined above can be chosen as*

$$\mathcal{K}_{RQ} = \{(W'_2W_1^{-1}, \text{trace}[R_cW]) \ : \ W \in \mathcal{C}_{RQ}\} \tag{7.53}$$

Proof The first two points are immediate consequences of Theorem 6.9. Both points are valid for the intersection defining $\mathcal{C}_{RQ}$.

Point c) It is apparent that $\mathcal{C}_{RQ} \subset \mathcal{C}_Q$ which from part c) of Theorem 7.1 means that $F = W'_2W_1^{-1} \in \mathcal{K}_Q$. Moreover, assume $\mathcal{C}_{RC} \neq \emptyset$ and take an arbitrary but fixed matrix $W \in \mathcal{C}_{RQ}$. For any $M_c \in \mathcal{D}_c$, from Theorem 6.9 the set $\mathcal{C}_R$ is well defined. Let us first prove that if $W \in \mathcal{C}_{QR}$ then $W \in \mathcal{C}_R$. Indeed, taking into account that $W \geq 0$, then $W \in \mathcal{C}_{QR}$ is equivalent to (by using Schur complement)

$$\mathcal{A}_i(W) = \left[\begin{array}{cc} U'_c\Theta_{ci}(W)U_c & U'_cM_{ci}W^{1/2} \\ W^{1/2}M'_{ci}U_c & -rI \end{array} \right] \leq 0$$

for all $i = 1, 2, \cdots, N$ where $U_c = [I \ \ 0]'$. Since $\mathcal{A}_i(W)$ is affine with respect to M_{ci}, writing $M_c \in \mathcal{D}_c$ as a convex combination of the extreme matrices we get

$$\mathcal{A}(W) := \sum_{i=1}^{N} \xi_i \mathcal{A}_i(W) \leq 0$$

which implies that $W \in \mathcal{C}_R$ indeed. From Theorem 6.9 the state feedback gain matrix $F = W_2' W_1^{-1}$ places the closed-loop system poles inside the given circular region $\mathcal{R}$. Furthermore, from part d) of the same theorem we conclude that the linear function trace$[R_c W]$ is a valid upper bound to the RH_2 norm of the closed-loop transfer function $T(z, w; s)$. Since M_c is an arbitrary matrix in $\mathcal{D}_c$, the result follows. From this, the case on which one of the sets is empty trivially holds and so the theorem is proved. $\square$

From this theorem, we can now solve the guaranteed cost optimal control problem stated before, namely (7.49)

$$\bar{\rho}_Q = \min \{ \rho \ : \ (F, \rho) \in \mathcal{K}_{RQ} \}$$
$$= \min \{ \text{trace}[R_c W] \ : \ W \in \mathcal{C}_{RQ} \} \tag{7.54}$$

which is once again a convex programming problem. It is interesting to observe that, in general lines, the proof of this result follows the same pattern of the previous ones. The main difference is that the matrix function $\Theta_R(\cdot)$ depends nonlinearly on matrix M_c but thanks to the Schur complement formula it can be converted to an affine and hence convex function from which the desired result is proved.

With this in mind it is possible to handle many other guaranteed cost optimal control problems with no big additional difficulty. An interesting and practically important case is to solve the guaranteed cost version of the Time-domain specification problem. Recalling Theorem 6.10, the optimal solution of the convex programming problem

$$\bar{\rho}_Q = \min \{ g(\bar{R}_c W \bar{R}_c') \ : \ W \in \mathcal{C}_Q \} \tag{7.55}$$

provides $F = W_2' W_1^{-1} \in \mathcal{K}_Q$ and the upper bound ρ satisfying

$$\mathcal{G}(T(z, w; s)) \leq \rho \ , \ \ \forall \ M_c \in \mathcal{D}_c$$

is minimized. Of course this result follows from the fact that the function $g(\cdot)$ defined in Chapter 6 is convex. The same reasoning can be adopted if the designer wants to include convex structural constraints in the state feedback gain matrix like decentralization or static output feedback. For that, it suffices to impose the linear constraint

$$W = W_D \ \ \text{or} \ \ W = W_O$$

in the problem to be solved. Since these constraints involve only the matrix variable, they do not change any property of the closed-loop system stability and the proposed upper bounds of the transfer function norms meaning that all results obtained before remain valid.

7.3 Actuators failure

The practical implementation of a given control is done by means of certain adequate devices or actuators. Of course, in the real world these devices are possible to fail

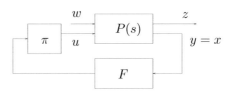

Figure 7.7: Practical control implementation

during the system operation. In this section we want to analyze the impact of actuators failure in control systems stability and performance. The basic tools to be used are the ones related to robust stability and performance already introduced in the previous section. It will be shown that the polyhedral convex bounded domain $\mathcal{D}_c$ plays a central role in this design problem.

Let us consider the linear dynamic system described by the state space equations

$$\dot{x} = Ax + B_1 w + B_2 u \tag{7.56}$$

$$z = C_1 x + D_{12} u \tag{7.57}$$

$$y = x \tag{7.58}$$

for which we seek a state feedback control of the form $u = Fx$ where F solves the Full information control problem in RH_2

$$\min\left\{\|T(z,w;s)\|_2^2 \ : \ F \in \mathcal{K}_c\right\} \tag{7.59}$$

Under the assumptions put in evidence in Chapter 4 this problem is feasible and admits an optimal solution given by

$$F = F_2 \tag{7.60}$$

With the optimal gain F_2 at hand, its practical implementation needs the use of m actuators (recall that m is the dimension of the control vector) as is indicated in fig. 7.7. Each actuator is ideally modeled as a a simple gain $\pi_i, i = 1, 2, \cdots, m$ such that

$$\pi_i := \{0, 1\} \tag{7.61}$$

meaning that it has only two states, namely *in operation* corresponding to $\pi_i = 1$ and *out of operation* corresponding to $\pi_i = 0$. From this, the control action in the closed-loop system instead of $u = F_2 x$ is

$$u(t) = \pi F_2 x(t) \ , \quad \pi := \text{diag}\left[\pi_1, \pi_2, \cdots, \pi_m\right] \tag{7.62}$$

Since the optimal gain F_2 has been calculated using only informations of the nominal open-loop system, during normal operation characterized by $\pi = I$, the closed-loop system evolves following an optimal trajectory. However, if a fail occurs in some control channel, say i, the actuator state changes to zero. In this situation the closed-loop system evolves with the control $u = \tilde{F}_2 x$ where $\tilde{F}_2$ is the same as F_2 but with the $i - th$ row equal to zero. Of course not only the optimality is lost but even instability may be observed. A way to cope with this problem is to apply the guaranteed control design procedure introduced before. To this end, consider instead of F_2 in (7.62) a matrix F to be determined. The closed-loop system matrix can be written as

$$A_{cc} = A + B_2 \pi F = A + B_2(\pi) F$$

where $B_2(\pi) := B_2 \pi$. This makes clear that the input matrix B_2 can be interpreted as an uncertain matrix depending upon the occurrence of actuators failure. Hence, in this framework the dynamic system (7.56) - (7.58) is an uncertain system where

$$B_2 \in \{B_2(\pi) : \pi \in \Pi\} \tag{7.63}$$

with Π being a set of diagonal matrices which defines all combinations of actuators failure. We assume that $I \in \Pi$ and $0 \notin \Pi$. The first condition is clearly necessary because it imposes feasibility to the nominal system. The latter one is also necessary because if it is violated then $B_2 = 0$ is feasible and the system is stabilizable only if it is stable. Suppose the set Π is composed by N matrices $\pi_i, i = 1, 2, \cdots, N$ defined before representing all actuators failure we want to take into account. With matrices

$$M_{ci} := \begin{bmatrix} A & B_2(\pi_i) \\ 0 & 0 \end{bmatrix} , \quad i = 1, 2, \cdots, N$$

the convex domain

$$\mathcal{D}_c := \mathrm{co}\,\{M_{ci} , \ i = 1, 2, \cdots, N\}$$

is the uncertain domain we have to consider since by construction it contains all input matrices $B_2(\pi)$ with $\pi \in \Pi$. Clearly, it contains many others, namely the ones generated by convex combination of the pairs $[A, B_2(\pi_i)]$. Fortunately, recalling Remark 7.4, the uncertain system defined by $M_c \in \mathcal{D}_c$ is quadratic stabilizable if and only if the collection of systems defined by all extreme matrices M_{ci} is quadratic stabilizable. Hence, as far as quadratic stability is concerned, modeling actuators failure as $M_c \in \mathcal{D}_c$ instead of by means of (7.63) does not introduce any kind of conservativeness in the results.

From the above discussion, let us proceed by replacing problem (7.59) by the associated guaranteed cost control problem. It is given by (recall Theorem 7.2)

$$\bar{\rho}_A := \min\,\{\mathrm{trace}[R_c W] \ : \ W \in \mathcal{C}_Q\} \tag{7.64}$$

and the optimal solution (if one exists) provides $F_A = W_2' W_1^{-1}$ such that with the state feedback $u = F_A x$ the closed-loop system is quadratically stable and $\|T(z, w; s)\|_2^2 \leq \bar{\rho}_A$ for all $M_c \in \mathcal{D}_c$. In other words, we can say that the closed-loop system with the control $u = \pi F_A x$ is such that all eigenvalues of $A_{cc} = A + B_2 \pi F_A$ lie in the left open part of the complex plane and

$$\|T(z, w; s)\|_2^2 \leq \bar{\rho}_A , \quad \forall\, \pi \in \Pi.$$

that is all control requirements are met. The next example illustrates this design policy for a simple example.

Example 7.5 Let the dynamic system (7.56)-(7.58) be given by

$$A = \begin{bmatrix} 0 & 1 & 1 \\ 1 & 1 & 2 \\ 1 & 1 & 1 \end{bmatrix} , \quad B_1 = \begin{bmatrix} 1 & 0 & 0 \\ 0 & 1 & 0 \\ 0 & 0 & 1 \end{bmatrix} , \quad B_2 = \begin{bmatrix} 1 & 0 \\ 0 & 0 \\ 0 & 1 \end{bmatrix}$$

$$C_1 = \begin{bmatrix} 2 & 2 & 0 \\ 1 & 0 & 1 \end{bmatrix} , \quad D_{12} = \begin{bmatrix} 1 & 0 \\ 0 & 1 \end{bmatrix}$$

Failure	Eigenvalues	$\|T(z,w;s)\|_2^2$
π_1	-1.91 , $-1.46 + j0.69$, $-1.46 - j0.69$	6.50
π_2	0.43 , -0.72 , -2.37	∞
π_3	1.94 , -1.99, -0.13	∞

Table 7.1: Closed-loop system poles and performance

Failure	Eigenvalues	$\|T(z,w;s)\|_2^2$
π_1	-10.56 , $-0.39 + j0.12$, $-0.39 - j0.12$	32.00
π_2	-3.11 , $-0.63 + j0.18$, $-0.63 - j0.18$	61.49
π_3	-2.88 , -1.62, -0.47	66.14

Table 7.2: Closed-loop system poles and performance

The optimal solution of Problem (7.59) provides

$$F_2 = \begin{bmatrix} -2.17 & -2.67 & -0.79 \\ -1.79 & -3.12 & -4.66 \end{bmatrix} \implies \min\|T(z,w;s)\|_2^2 \approx 6.50$$

For the practical implementation of this state feedback gain we need two actuators. It is supposed that they can fail during operation. Since the open-loop system is unstable, it is assumed that the actuators can not fail simultaneously. Hence the set Π is composed by matrices

$$\pi_1 = \begin{bmatrix} 1 & 0 \\ 0 & 1 \end{bmatrix} , \quad \pi_2 = \begin{bmatrix} 0 & 0 \\ 0 & 1 \end{bmatrix} , \quad \pi_3 = \begin{bmatrix} 1 & 0 \\ 0 & 0 \end{bmatrix}$$

corresponding to normal operation (π_1), first control channel actuator failure (π_2) and second control channel actuator failure (π_3) respectively.

Table 7.1 gives the eigenvalues of matrix $A_{cc} = A + B_2(\pi)F_2$ for the three fault matrices. Obviously, for $\pi = \pi_1 = I$ the closed-loop system is internally stable. However, if one of the two faults occurs then the closed-loop system becomes unstable. The optimal solution of the Full information problem is not robust for this kind of severe perturbation. On the contrary, the optimal solution of the guaranteed RH_2 cost control problem (7.64) gives

$$F_A = \begin{bmatrix} -6.97 & -10.17 & -13.56 \\ -3.70 & -5.00 & -6.37 \end{bmatrix} \implies \bar{\rho}_A \approx 73.18$$

which as indicated in Table 7.2 preserves internal stability in front of any prespecified failure occurrence. In this example, when compared with the nominal case, the price to be paid to keep internal stability under actuators failure is high. A possible interpretation of this fact is that the parametric perturbation caused by actuators failure is actually severe. $\square$

Remark 7.13 It is possible also to characterize actuators failure by means of norm bounded uncertainty. Actually, noticing that

$$B_2(\pi) = B_2\pi = B_2 + B_2(\pi - I) = B_2 + B_2\Omega$$

with $\Omega := \pi - I$ and that π is always a diagonal matrix, it is immediate to get

$$\|\Omega\| \leq \max_{\pi \in \Pi} \|\pi - I\| = 1$$

Hence, with matrices

$$M_{cn} = \begin{bmatrix} A & B_2 \\ 0 & 0 \end{bmatrix}, \quad B_n = \begin{bmatrix} B_2 \\ 0 \end{bmatrix}, \quad C_n = \begin{bmatrix} 0 & I \end{bmatrix}$$

any actuators failure such that $\pi \in \Pi$ may be alternatively described by (recall Remark 7.2)

$$M_c \in \mathcal{D}_n = \{M_c = M_{cn} + B_n \Omega C_n : \|\Omega\| \le 1\}$$

This characterization of the problem under consideration frequently leads, if any, to very conservative results and so worse than the use of the polyhedral convex bounded domain $\mathcal{D}_c$ as we have done before. Indeed, the particular choice $\Omega = -I$ which obviously is a diagonal matrix such that $\|\Omega\| = 1$ produces

$$\begin{bmatrix} A & 0 \\ 0 & 0 \end{bmatrix} \in \mathcal{D}_n$$

which is stabilizable if and only if the open-loop system is stable. In other words, the use of the uncertain domain $\mathcal{D}_n$ includes by construction the spurious matrix $\Omega = -I$ being thus equivalent to have $\pi = 0 \in \Pi$. This situation is avoided in the description of the set $\mathcal{D}_c$ as we have just shown in the solution of the previous example.   □

7.4 Nonlinear perturbations

In this section nonlinear robust stability and performance are considered. This topic is a generalization of the ones introduced before in the sense that advantages are taken into account from the a priori knowledgement of the class of nonlinear perturbations acting in the open-loop model. Two main classes are of importance, namely multiplicative and additive nonlinear perturbations, leading to what we call Persidiskii and Lur'e robust design procedures. In both cases, the control structure is assumed to be linear and the whole state vector is available for feedback.

7.4.1 Persidiskii design

Let us first consider the robust control design of a class of nonlinear systems subject to state dependent nonlinear perturbations called *multiplicative* perturbations. Once again the proposed design procedure will be expressed in terms of convex programming problems only. The block-scheme of the dynamic system to be dealt with is shown in fig. 7.8. The perturbed system Σ_p is subject to the nonlinear perturbation $f(x)$ to be precisely defined in the sequel. For the moment, it is important to keep clear that the open-loop system is subject to a class of perturbations such that, when they occur, the whole state vector x changes to $f(x)$. Then, the perturbation occurrence changes also the measured output y accordingly.

   Assuming the state vector has dimension n and the nonlinear function $f(x)$ is not exactly known, the only available information is that it belongs to the uncertain domain $\mathcal{D}_f$ composed by all vector valued functions with the following properties :

1) Each component of $f(x)$ namely $f_j(x)$, $j = 1, 2, \cdots, n$ is a real valued function such that

$$f_j(x) = f_j(x_j) \tag{7.65}$$

where $x_j \in R$ denotes the $j - th$ component of the vector $x \in R^n$.

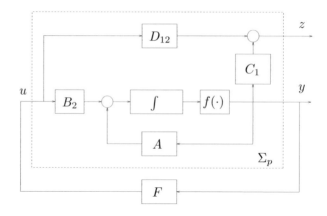

Figure 7.8: The bock-scheme of the perturbed system

2) Each component $f_j(x_j)$, $j = 1, 2, \cdots, n$ is such that

$$f_j(0) = 0 \tag{7.66}$$

$$f_j(\xi)\xi > 0 \ \ \forall \, \xi \neq 0 \in R \tag{7.67}$$

$$\int_0^\infty f_j(\xi)d\xi = \infty \tag{7.68}$$

The first condition defines the perturbation structure. Roughly speaking, the second condition says that the graph of $f(\cdot)$ must be contained in the first and third quadrants of the (f, ξ) plane. Since $f(x) = x \in D_f$ the corresponding linear system is called *nominal* system (Σ_n) and it has the following state space representation

$$\dot{x} = Ax + B_2 u \ , \ \ x(0) = x_0 \tag{7.69}$$

$$z = C_1 x + D_{12} u \tag{7.70}$$

$$y = x \tag{7.71}$$

Adapting the previous design goals to cope with nonlinear systems stability and performance, we proceed trying to determine (if one exists) a linear state feedback control law $u = Fx$ for Σ_n such that the origin $x = 0$ of Σ_p which has the state space representation

$$\dot{x} = (A + B_2 F)f(x) \ , \ \ x(0) = x_0 \tag{7.72}$$

$$z = (C_1 + D_{12} F)f(x) \tag{7.73}$$

is globally stable for all $f \in D_f$. Furthermore, among all state feedback gains with this property, find the one, namely F_f, which solves the associated guaranteed cost control problem

$$\bar{\rho}_f(x_0) := \min \bar{\rho}(F, x_0) \tag{7.74}$$

where

$$\int_0^\infty z(t)'z(t)dt \leq \bar{\rho}(F, x_0) \ , \ \ \forall \, f \in D_f$$

Similarly the minimum value of $\bar{\rho}(F, x_0)$ with respect to all F preserving stability is called the minimum guaranteed cost associated to the optimal feedback gain $F =$

F_f. To accomplish the first goal concerning the robust stability of Σ_n we need to introduce the following important result. To ease notation, for any square matrix P, the subscript "d" indicates that $P = P_d$ is constrained to be a diagonal matrix.

Theorem 7.6 (Persidiskii theorem) *For any given state feedback matrix F suppose there exists a positive definite matrix P_d such that*

$$0 \geq (A + B_2F)'P_d + P_d(A + B_2F) + Q \tag{7.75}$$

for some matrix $Q = Q' > 0$. Then, the origin $x = 0$ of the perturbed system Σ_p is globally asymptotically stable for all $f \in \mathcal{D}_f$.

Proof Using the given properties of $f \in \mathcal{D}_f$, let us take

$$v(x) := 2\sum_{j=1}^{n} P_{jj} \int_0^{x_j} f_j(\xi)d\xi$$

as a Lyapunov function candidate associated to an arbitrary trajectory of Σ_p such that $x(0) = x_0$. Its time derivative can be written as

$$\begin{aligned}
\dot{v}(x) &= 2\sum_{j=1}^{n} P_{jj}f_j(x_j)\dot{x}_j \\
&= 2f(x)'P_d\dot{x} \\
&= f(x)'\left[(A + B_2F)'P_d + P_d(A + B_2F)\right]f(x) \\
&\leq -f(x)'Qf(x) \\
&< 0 , \quad \forall\, x \neq 0
\end{aligned}$$

proving thus the theorem proposed. □

In the above calculations, it is clear that matrix Q must be positive definite but does not need to present any particular structure. Besides, the particular value of this matrix is immaterial to get $P = P_d$ satisfying inequality (7.75). In fact, if there exists $P = P_d$ satisfying (7.75) for a given $Q = \bar{Q} > 0$ then the same is true for any other choice $Q > 0$. This new degree of freedom is used in the next lemma to get the upper bound defined in (7.74).

Lemma 7.2 *Assume for all $f \in \mathcal{D}_f$ there exist n positive and finite parameters such that*

$$\int_0^{x_j(0)} f_j(\xi)d\xi \leq \frac{r_j}{2} , \quad j = 1, 2, \cdots, n \tag{7.76}$$

For any state feedback control gain F such that there exists $P = P_d$ satisfying the matrix inequality (7.75), it is possible to choose $Q = Q' > 0$ such that the upper bound $\bar{\rho}(F, x_0)$ is given by

$$\bar{\rho}(F, x_0) := \sum_{j=1}^{n} P_{jj}r_j \tag{7.77}$$

Proof Assume for F given, there exists $P = P_d$ such that the matrix inequality (7.75) holds. In this case, we can choose matrix $Q > 0$ as being

$$Q = (C_1 + D_{12}F)'(C_1 + D_{12}F) + \epsilon I$$

where $\epsilon > 0$ is an arbitrarily small parameter. From the previous theorem we already know that the origin of the perturbed system Σ_p is globally stable. Hence, using (7.72) and (7.73) we immediately have

$$\dot{v}(x(t)) \leq -f(x(t))' \left[(C_1 + D_{12}F)'(C_1 + D_{12}F) + \epsilon I\right] f(x(t))$$
$$\leq -z(t)'z(t) , \quad \forall\, t \geq 0$$

Integrating both sides of this inequality from $t = 0$ to $t = \infty$ we get

$$\int_0^\infty z(t)'z(t)dt \leq v(x(0))$$

Using this inequality together with the definition of the Lyapunov equation introduced in Theorem 7.6 and the upper bound on each element of function $f(x)$ yield

$$\int_0^\infty z(t)'z(t)dt \leq 2\sum_{j=1}^n P_{jj} \int_0^{x_j(0)} f_j(\xi)d\xi$$
$$\leq \sum_{j=1}^n P_{jj}r_j$$

which being true for all $f \in \mathcal{D}_f$ proves that the upper bound (7.77) is valid.   □

Up to now we have always worked with the dual version of inequality (7.75) where the system matrix transpose post-multiply the matrix variable. This fact was important to convert the associated optimal control problems to convex programming problems. Here, due to the nonlinearity of the perturbation $f(x)$ this is no longer possible. Even though, the guaranteed cost control problem (7.74) can be converted into a convex one by means of Schur complements. To this end, consider the affine matrix function which is defined by all pairs of matrices (X, Y) of appropriate dimension with the first one being symmetric

$$\mathcal{A}_f(X,Y) := \begin{bmatrix} AX + B_2Y + XA' + Y'B_2' & XC_1' + Y'D_{12}' \\ C_1X + D_{12}Y & -I \end{bmatrix} \tag{7.78}$$

The following preliminary result is of particular importance towards the complete solution of the guaranteed cost control problem stated before.

Theorem 7.7 *Define the convex set*

$$\mathcal{C}_f := \{(X,Y) \; : \; X = X_d > 0 , \;\; \mathcal{A}_f(X,Y) < 0\} \tag{7.79}$$

The set of all state feedback matrices F such that (7.75) holds for some $Q > 0$, denoted as $\mathcal{K}_f$ is alternatively given by

$$\mathcal{K}_f := \{YX^{-1} \; : \; (X,Y) \in \mathcal{C}_f\} \tag{7.80}$$

Proof It must be clear that the above defined set $\mathcal{C}_f$ is convex. Actually, it is defined by LMI's and the linear constraint $X = X_d$ which corresponds to impose that all off diagonal elements of matrix X are zero. For the necessity, let us take an arbitrary $F \in \mathcal{K}_f \neq \emptyset$ and observe that in this case, there exists $P = P_d$ positive definite such that

$$0 > (A + B_2F)'P_d + P_d(A + B_2F) + (C_1 + D_{12}F)'(C_1 + D_{12}F)$$

Multiplying this inequality to the left and to the right by P_d^{-1} and using the Schur complement formula, it is simple to verify that

$$(X, Y) = (P_d^{-1}, FP_d^{-1}) \in \mathcal{C}_f$$

and $YX^{-1} = FP_d^{-1}P_d = F$. Conversely, take any $(X, Y) \in \mathcal{C}_f \neq \emptyset$. Using again the Schur complement to the LMI $\mathcal{A}_f(X, Y) < 0$ and taking into account that $X = X_d$ is positive definite and diagonal, we conclude that with $P_d = X_d^{-1}$ and $F = YX^{-1}$ there exists $\epsilon > 0$ sufficiently small such that inequality (7.75) holds for

$$Q = (C_1 + D_{12}F)'(C_1 + D_{12}F) + \epsilon I$$

and so the proof of the theorem proposed is complete by noticing that if one of the sets in (7.80) is empty both are empty. □

From this result, we are able to generate by means of a feasibility convex problem all gains belonging to the nonconvex set $\mathcal{K}_f$. The elements of this set assure robust stability of the nominal closed-loop system against all nonlinear perturbations $f \in \mathcal{D}_f$. Besides, using Lemma 7.2 and defining the matrix

$$D := \text{diag}\left[\sqrt{r_1}, \sqrt{r_2}, \cdots, \sqrt{r_n}\right]$$

the elements of the set $\mathcal{C}_f$ allow the determination of the upper bound $\bar{\rho}(F, x_0)$ for all $F \in \mathcal{K}_f$ as being

$$\bar{\rho}(F, x_0) = \sum_{j=1}^{n} P_{jj} r_j$$
$$= \text{trace}\left[D'X^{-1}D\right] \tag{7.81}$$

valid for all $(X, Y) \in \mathcal{C}_f$ and $F = YX^{-1}$. From this fact the minimum guaranteed cost is readily calculated from

$$\bar{\rho}_f(x_0) = \min_{F \in \mathcal{K}_f} \bar{\rho}(F, x_0)$$
$$= \inf\left\{\text{trace}\left[D'X^{-1}D\right] : (X, Y) \in \mathcal{C}_f\right\} \tag{7.82}$$

which is a convex programming problem (recall Remark 6.3). Once the global solution of the right hand side of (7.82) is calculated, the corresponding state feedback gain, optimal solution of the left hand side of the same equation is provided simply by $F_f = YX^{-1}$.

Remark 7.14 It follows that $\mathcal{K}_f \subset \mathcal{K}_c$ and $\mathcal{K}_f \neq \emptyset$ if and only if $\mathcal{C}_f \neq \emptyset$. The stabilizability of the pair (A, B_2) is not sufficient to guarantee that $\mathcal{C}_f \neq \emptyset$ since besides internal stability, the linear constraint $X = X_d$ requires the inequality (7.75) to have a diagonal and positive definite feasible solution. □

Remark 7.15 The effectiveness to solve problem (7.82) by means of convex programming methods, under the assumption that $D_{12}'D_{12} = I$, is now addressed. The main point to be considered is the boundedness of the set $\mathcal{C}_f$. Notice that if there exists an unbounded feasible matrix $X > 0$ then it is one of the possible global solutions of Problem (7.82) because

$$0 \leq \text{trace}[D'X^{-1}D]$$

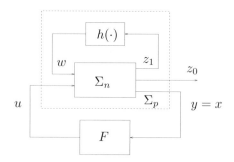

Figure 7.9: The bock-scheme of the perturbed system

and the right hand side approaches to zero as X increases. Using the same reasoning adopted in Remark 6.8 it is now determined under which conditions this occurrence is avoided. Let us assume that there exists $(X, Y) \in \mathcal{C}_f$ and $(\tilde{X}, \tilde{Y}) \neq 0$ such that $(X, Y) + \lambda(\tilde{X}, \tilde{Y}) \in \mathcal{C}_f$ for $\lambda > 0$ arbitrarily large. Then

$$\tilde{X} \geq 0$$

$$\begin{bmatrix} A\tilde{X} + B_2\tilde{Y} + \tilde{X}A' + \tilde{Y}'B_2' & \tilde{X}C_1' + \tilde{Y}'D_{12}' \\ C_1\tilde{X} + D_{12}\tilde{Y} & 0 \end{bmatrix} \leq 0$$

From Appendix C, these conditions are verified only if there exists a matrix $\tilde{X}$ such that

$$C_{1c}\tilde{X} = 0$$
$$\tilde{X} \geq 0$$
$$A_c\tilde{X} + \tilde{X}A_c' \leq 0$$

where $A_c = A - B_2D_{12}'C_1$ and $C_{1c} = (I - D_{12}D_{12}')C_1$. As in Remark 6.8, this situation is completely avoided if the open-loop system is such that the pair $(-A_c, C_{1c})$ is detectable. Under this mild assumption the optimal value of the objective function of Problem (7.82) is strictly positive and finite. □

7.4.2   Lur'e design

Let us consider now another important robust control design for the class of output dependent nonlinear *additive* perturbations. The basic block-scheme is given in fig. 7.9. It resembles the one used in the state feedback Mixed RH_2/RH_∞ control design. The *nominal* system is denoted by Σ_n. The perturbed dynamic system, namely Σ_p is subject to the nonlinear perturbation $h(\cdot)$ which is a vector valued function not a priori known. The available information is that it belongs to the uncertain domain $\mathcal{D}_h$ composed by all functions presenting the following properties :

1) The vector valued function $h(\cdot)$ is defined for all $\xi \in R^r$ and $h(\cdot) \in R^r$ where r is a positive integer less or equal the dimension of the state vector $x \in R^n$.

2) It is such that

$$h(0) = 0 \tag{7.83}$$
$$h(\xi)'\xi \leq 0 \quad \forall\, \xi \in R^r \tag{7.84}$$

The first condition imposes that, in fig. 7.9, the vectors w and z_1 have the same dimension. The second one implies that the nonlinear function $-h(\cdot)$ belongs to the sector $[0, \infty)$. In fact, in the one-dimensional case the graph of $-h(\xi)$ in the plane (h, ξ) is in the first and third quadrants. The state space equations of the nominal open-loop system Σ_n, corresponding to $h(\cdot) = 0 \in \mathcal{D}_h$ is the standard one

$$\dot{x} = Ax + B_1 w + B_2 u , \quad x(0) = x_0 \tag{7.85}$$
$$z_0 = C_0 x + D_0 u \tag{7.86}$$
$$z_1 = C_1 x + D_{12} u \tag{7.87}$$
$$y = x \tag{7.88}$$

As before, the goal is to design a state feedback control law, namely $u = Fx$ such that the closed-loop perturbed system Σ_p obtained from Σ_n together with $w = h(z_1)$ presents the following properties associated with its state space representation

$$\dot{x} = (A + B_2 F)x + B_1 h(z_1) , \quad x(0) = x_0 \tag{7.89}$$
$$z_0 = (C_0 + D_0 F)x \tag{7.90}$$
$$z_1 = (C_1 + D_{12} F)x \tag{7.91}$$

First, the origin $x = 0$ must be globally stable for all $h \in \mathcal{D}_h$. From all state feedback gains with this property select (if possible) one, namely F_h which solves the following guaranteed cost control problem

$$\bar{\rho}_h(x_0) := \min \bar{\rho}(F, x_0) \tag{7.92}$$

where

$$\int_0^\infty z_0(t)' z_0(t) dt \le \bar{\rho}(F, x_0) , \quad \forall\, h \in \mathcal{D}_h$$

The guaranteed cost control problem (7.92) is similar to the Mixed RH_2/RH_∞ control problem. The existence of the nonlinear function $h \in \mathcal{D}_h$ does not allow us to express it in the frequency domain. Instead the guaranteed cost is given in terms of an upper bound to the above integral of the controlled output. Accordingly, the exogenous signal is replaced by an arbitrary initial condition $x(0) \ne 0$.

Theorem 7.8 (Passivity theorem) *For any given state feedback matrix F suppose there exists a symmetric and positive definite matrix P such that*

$$0 \ge (A + B_2 F)' P + P(A + B_2 F) + Q \tag{7.93}$$
$$B_1' P = (C_1 + D_{12} F) \tag{7.94}$$

for some matrix $Q = Q' > 0$. Then, the origin $x = 0$ of the perturbed system Σ_p is globally asymptotically stable for all $h \in \mathcal{D}_h$.

Proof Consider the Lyapunov function candidate

$$v(x) := x' P x$$

Its time derivate along any trajectory of Σ_p is written as

$$\dot{v}(x) = 2x' P \dot{x}$$
$$= x' \left[(A + B_2 F)' P + P(A + B_2 F) \right] x + 2x' P B_1 h(z_1)$$

Making use of condition (7.94), for an arbitrary vector x we have

$$B_1' P x = (C_1 + D_{12} F) x = z_1$$

which together with the first condition (7.93) yields

$$\dot{v}(x) \leq -x'Qx + z_1'h(z_1)$$
$$< 0 , \quad \forall \, x \neq 0$$

where the last inequality holds due to (7.84). Hence, for any $h \in \mathcal{D}_h$ the origin $x = 0$ of the perturbed system Σ_p is asymptotically stable. $\qquad\square$

Remark 7.16 Theorem 7.8 also holds in a more general setting. Indeed the function $h(\cdot)$ can be considered time varying provided $h(\xi, t) \in \mathcal{D}_h$ for all $t \geq 0$. Furthermore, the assumption that $Q = Q' > 0$, yields

$$\dot{v}(x) \leq -\alpha \|x\|^2 , \quad \alpha := \min_i \lambda_i(Q) > 0$$

That is, the origin is in fact globally exponentially stable. $\qquad\square$

Remark 7.17 Conditions (7.93) and (7.94) can be rewritten as follows

$$0 \geq A_{cc}'P + P A_{cc} + Q$$
$$B_1'P = C_{cc}$$

where $A_{cc} := A + B_2 F$ and $C_{cc} := C_1 + D_{12} F$. Assume that for some $Q = Q' > 0$, there exist F and $P = P' > 0$ satisfying them. In this case, summing and subtracting $j\omega P$, with $\omega \in [0, \infty)$, in the right hand side of the first condition, after simple algebraic manipulation we have

$$0 \leq (-j\omega I - A_{cc}')^{-1}P + P(j\omega I - A_{cc})^{-1} - (-j\omega I - A')^{-1}Q(j\omega I - A)^{-1}$$

Finally, multiplying this inequality to the left by B_1' and to the right by B_1, making use of the second condition above and defining the closed-loop system transfer function

$$G(s) := C_{cc}(sI - A_{cc})^{-1}B_1$$

we get

$$G^\sim(j\omega) + G(j\omega) \geq B_1'(-j\omega - A')^{-1}Q(j\omega - A)^{-1}B_1$$
$$> 0 , \quad \forall \, \omega \in [0, \infty)$$

A transfer function with this property is called strictly positive real (SPR) and plays a central role in the stability analysis of nonlinear systems. In this remark we have shown that conditions (7.93) and (7.94) implies $G(s)$ is strictly positive real. Under certain mild additional assumptions the converse is also true and constitutes the well known Kalman-Yacubovitch lemma. Notice that the above frequency domain characterization of a SPR transfer functions does not include the limit point $\omega = \infty$ for which it is not satisfied because $G(j\infty) = 0$. $\qquad\square$

Remark 7.18 A transfer function $G_E(s)$ is called extended strictly positive real ($ESPR$) if it is SPR and

$$G_E^\sim(j\infty) + G_E(j\infty) > 0$$

From this definition, it is obvious that the transfer function $G(s)$ defined in the previous remark is SPR but not $ESPR$ due to the fact that $G(j\infty) = 0$. It is interesting to known that there is a very simple relationship between $ESPR$ and RH_∞ norm. Indeed, a transfer function $G_E(s)$ is $ESPR$ if and only if

$$\|[\alpha I - G_E(s)][\alpha I + G_E(s)]^{-1}\|_\infty < 1$$

where α is an arbitrary positive scalar. A proof of this property is not included here. We only mention that it is based on the fact that the above inequality holds if and only if

$$[\alpha I + G_E^\sim(j\omega)]^{-1}[\alpha I - G_E^\sim(j\omega)][\alpha I - G_E(j\omega)][\alpha I + G_E(j\omega)]^{-1} < 1 , \quad \forall \, \omega$$

or, equivalently

$$[\alpha I - G_E^\sim(j\omega)][\alpha I - G_E(j\omega)] < [\alpha I + G_E^\sim(j\omega)][\alpha I + G_E(j\omega)] , \quad \forall \, \omega$$

After obvious simplifications, it provides

$$2\alpha[G_E^\sim(j\omega) + G_E(j\omega)] > 0 , \quad \forall \, \omega$$

implying that $G_E(s)$ is $ESPR$. Indeed, for transfer functions which are only SPR, as for instance $G(s)$, it is violated at $\omega = \infty$. In this case, the consequence is

$$\|[\alpha I - G(s)][\alpha I + G(s)]^{-1}\|_\infty = \sup_\omega \bar{\sigma}\left[[\alpha I - G(j\omega)][\alpha I + G(j\omega)]^{-1}\right]$$

$$\geq \bar{\sigma}\left[[\alpha I - G(j\infty)][\alpha I + G(j\infty)]^{-1}\right]$$

$$\geq 1$$

which means that the previous constraint expressed in terms of a RH_∞ norm is also violated.

It is interesting to observe what happen if we do not take care of the above results and try to convert our control design problem to a state feedback problem in RH_∞ associated to the auxiliary transfer function

$$H(s) := [\alpha I - G(s)][\alpha I + G(s)]^{-1}$$

where $G(s)$ is the transfer function defined in the previous remark. With no loss of generality we adopt $\alpha = 1$. Routine calculations (recall in Chapter 2 the definition of inverse system) show that the following is true

$$[I + G(s)]^{-1} = \left[\begin{array}{c|c} A_{cc} - B_1 C_{cc} & B_1 \\ \hline -C_{cc} & I \end{array}\right]$$

$$G(s)[I + G(s)]^{-1} = \left[\begin{array}{c|c} A_{cc} - B_1 C_{cc} & B_1 \\ \hline C_{cc} & 0 \end{array}\right]$$

from which it follows that

$$H(s) = \left[\begin{array}{c|c} A_{cc} - B_1 C_{cc} & B_1 \\ \hline -2C_{cc} & I \end{array}\right]$$

Then, using the state space representation of the transfer function $H(s)$, the determination of a state feedback gain such that $\|H(s)\|_\infty < 1$ is reduced to the solution of the state feedback problem in RH_∞ for the auxiliary plant

$$\dot{x} = \tilde{A}x + \tilde{B}_1 w + \tilde{B}_2 u$$
$$z = \tilde{C}_1 x + \tilde{D}_{11} w + \tilde{D}_{12} u$$
$$u = Fx$$

where $\tilde{A} := A - B_1 C_1$, $\tilde{B}_1 := B_1$, $\tilde{B}_2 := B_2 - B_1 D_{12}$, $\tilde{C}_1 := -2C_1$, $\tilde{D}_{11} := I$ and $\tilde{D}_{12} := -2D_{12}$. However, this is impossible because with $\bar{\sigma}[\tilde{D}_{11}] = 1$, any state feedback gain such that $\tilde{A} + \tilde{B}_2 F$ is stable implies $\|T(z, w; s)\|_\infty \geq 1$. $\qquad\square$

In opposition to Theorem 7.6, the constraints (7.93) and (7.94) to be simultaneously satisfied, depend strongly on the particular choice of matrix $Q > 0$. Even
though, to be able to express the upper bound $\bar{\rho}(F, x_0)$ conveniently we need to impose

$$Q = (C_0 + D_0 F)'(C_0 + D_0 F) + \epsilon I \tag{7.95}$$

with $\epsilon > 0$ being an arbitrarily small parameter. Indeed, with this particular choice,
following the proof of Theorem 7.8 we have

$$\dot{v}(x(t)) \leq -x(t)'Qx(t)$$
$$\leq -z_0(t)'z_0(t) , \quad \forall \, t \geq 0$$

which after integration from $t = 0$ to $t = \infty$ provides

$$\int_0^\infty z_0(t)'z_0(t)dt \leq v(x(0)) = x_0'Px_0$$

Based on this, it is natural to define

$$\bar{\rho}(F, x_0) := x_0'Px_0 \tag{7.96}$$

as a valid upper bound for all $h \in \mathcal{D}_h$. Furthermore, let us denote as $\mathcal{K}_h$ the set of all
state feedback gains F such that with $Q > 0$ given in (7.95) both constraints (7.93)
and (7.94) are simultaneously satisfied for some $P > 0$ and introduce the affine matrix
functions defined for all pairs of matrices (X, Y) of appropriate dimension with the
first one being symmetric

$$\mathcal{A}_h(X, Y) := \begin{bmatrix} AX + B_2 Y + X A' + Y'B_2' & XC_0' + Y'D_0' \\ C_0 X + D_0 Y & -I \end{bmatrix}$$

and

$$\mathcal{B}_h(X, Y) := C_1 X + D_{12} Y - B_1'$$

The following theorem gives a complete parametrization of the set $\mathcal{K}_h$ in terms of a
convex set. It is the basis for the solution of the associated optimal guaranteed cost
control problem (7.92).

Theorem 7.9 *Define the convex set*

$$\mathcal{C}_h := \{(X, Y) \; : \; X > 0 , \quad \mathcal{A}_h(X, Y) < 0 , \quad \mathcal{B}_h(X, Y) = 0\} \tag{7.97}$$

The set $\mathcal{K}_h$ is alternatively given by

$$\mathcal{K}_h := \{Y X^{-1} \; : \; (X, Y) \in \mathcal{C}_h\} \tag{7.98}$$

Proof Since the set $\mathcal{C}_h$ is defined by means of affine functions only, it is convex. To
prove the necessity, assuming that $F \in \mathcal{K}_h \neq \emptyset$ then there exists $P > 0$ such that

$$0 > (A + B_2 F)'P + P(A + B_2 F) + (C_0 + D_0 F)'(C_1 + D_0 F)$$

Multiplying both sides of this inequality by P^{-1} and using the Schur complement it
is readily verified that

$$\mathcal{A}_h(P^{-1}, FP^{-1}) < 0$$

On the other hand, multiplying (7.94) to the right by P^{-1}, it can be rewritten as

$$\mathcal{B}_h(P^{-1}, FP^{-1}) = 0$$

hence, $(X, Y) = (P^{-1}, FP^{-1}) \in \mathcal{C}_h$ and $YX^{-1} = F$ from which the necessity follows. The sufficiency for $\mathcal{C}_h \neq \emptyset$ is immediate. From this, it is also immediate to see that (7.98) holds in case one of the indicated sets is empty. $\qquad\square$

We have now all elements to face the optimal guaranteed cost control problem (7.92). From Theorem 7.9 and (7.96) it reduces to the problem

$$\bar{\rho}_h(x_0) = \min_{F \in \mathcal{K}_h} \bar{\rho}(F, x_0)$$
$$= \inf \left\{ x_0' X^{-1} x_0 \ : \ (X, Y) \in \mathcal{C}_h \right\} \qquad (7.99)$$

which is a convex programming problem (recall Remark 6.3). Its global optimal solution provides both the minimum guaranteed cost $\bar{\rho}_h(x_0)$ and the associated optimal state feedback gain $F_h = YX^{-1}$.

Remark 7.19 The inclusion $\mathcal{K}_h \subset \mathcal{K}_c$ is obvious. It is also clear that the stabilizability of the pair (A, B_2) is necessary but not sufficient to $\mathcal{C}_h \neq \emptyset$. The necessary and sufficient condition is the existence of a state feedback gain $F \in \mathcal{K}_c$ such that the closed-loop system transfer function $G(s)$ is SPR (recall Remark 7.17). $\qquad\square$

Remark 7.20 Following the same reasoning of Remark 7.15, the boundedness of the set $\mathcal{C}_h$ is assured whenever the pair $[-(A - B_2 D_0' C_0), (I - D_0 D_0') C_0]$ is detectable. $\qquad\square$

Remark 7.21 Consider the convex programming problem

$$\inf \left\{ x_0' X^{-1} x_0 \ : \ X > 0 \ , \ \mathcal{A}_h(X, Y) < 0 \right\}$$

which is the same as problem (7.99) but without the linear constraint $\mathcal{B}_h(\cdot) = 0$. Let us search a solution of the above problem such that

$$Y = -B_2' - D_0' C_0 X$$

Using this formula, taking into account the standard assumption $D_0' D_0 = I$ the Schur complement implies that $\mathcal{A}_h(X, Y) < 0$ if and only if

$$\mathcal{A}(X) := \begin{bmatrix} A_c X + X A_c' - B_2 B_2' & X C_{0c}' \\ C_{0c} X & -I \end{bmatrix} < 0$$

where $A_c := A - B_2 D_0' C_0$ and $C_{0c} := (I - D_0 D_0') C_0$. Hence, the problem under consideration can be solved in two steps. The convex programming problem

$$\inf \left\{ x_0' X^{-1} x_0 \ : \ X > 0 \ , \ \mathcal{A}(X) < 0 \right\}$$

provides the optimal matrix $X > 0$ and the previous formula gives matrix Y. The corresponding state feedback gain turns out to be $F = YX^{-1}$. Comparing the last problem with respect to the determination of matrix X and Remark 6.9, it is seen that its optimal solution is exactly the associated Riccati equation and so, in this particular case, the guaranteed cost reduces exactly to the minimum RH_2 optimal cost. This gives a measure of the quality of the upper bound $\bar{\rho}(F, x_0)$ proposed. In the general case, this solution is no longer feasible and the minimization must be done taking into account the linear constraint $\mathcal{B}_h(\cdot) = 0$. $\qquad\square$

7.5   Notes and references

Uncertain systems design is nowadays a very wide topic in control theory. Concerning stability by means of linear state feedback control, the paper of Barmish [4] is important since, for the first time the author propose an effective and simple way to handle uncertainties acting on both A and B_2 system matrices. The notion of guaranteed cost has been introduced by Chang and Peng in [11] related to a simple LQ problem. Differently of what we have done in this chapter, the main idea was to get an (nonlinear) upper bound to the associated Riccati equation. In section 7.2 the problems already solved in Chapter 6 have been revisited. They have been solved in a manner to cope with parameter uncertainty. Two of the most important classes of uncertainty have been considered and compared, namely polyhedral convex bounded and norm bounded uncertainty. For the first type, results from [65] have been used in Remark 7.15 while the numerical example fully described in Example 7.65 and analyzed throughout the section is also given in [50]. With this last paper the reader can go deeper on the comparison of these two types of parameter uncertainty models just mentioned. The main part of Section 7.2 related to RH_2 and RH_∞ guaranteed cost control problems is based on [24] and [49] while the other results are the natural generalizations, to cope with parameter uncertainties, of problems introduced in Chapter 6. The stability and guaranteed cost control of dynamic linear systems subject to actuators failure has been analyzed in [23]. In Section 7.3 this problem is again solved but special attention is paid to the comparison and modeling this special kind of uncertainty by means of the domain $\mathcal{D}_n$. Once again, the convexity plays a central role and it is possible to verify that the uncertainty description by means of the convex domain $\mathcal{D}_c$ leads in many instances to better results. Section 7.4 is entirely devoted to control design problems involving nonlinear perturbations. The first one called Persidiskii design is based on papers [30], [25] and [26]. The former paper also provides many others and more general results and is an excellent reference on this topic. Finally the second control design procedure called Lur'e design is based on the classical results reported in the important book [61] where the notions of passivity and strictly positive real transfer functions are addressed in a general and complete setting.

Appendix A

Some Facts on Polynomials

The three results presented in this appendix can be found in many text of Algebra.

Lemma A.1 *Let $r_0(s)$ and $r_1(s)$ be two polynomials with $\deg[r_0(s)] \geq \deg[r_1(s)]$. Let $r(s)$ be a greatest common divisor of $r_0(s)$ and $r_1(s)$. Then, there exist two polynomials $\varphi_0(s)$ and $\varphi_1(s)$ such that*

$$\varphi_0(s)r_0(s) + \varphi_1(s)r_1(s) = r(s)$$

Proof The sequence of polynomials

$$r_0(s) = r_1(s)q_1(s) + r_2(s) , \quad \deg[r_2(s)] < \deg[r_1(s)]$$
$$r_1(s) = r_2(s)q_2(s) + r_3(s) , \quad \deg[r_3(s)] < \deg[r_2(s)]$$
$$r_2(s) = r_3(s)q_3(s) + r_4(s) , \quad \deg[r_4(s)] < \deg[r_3(s)]$$

$$\vdots \quad = \quad \vdots$$

$$r_{p-1}(s) = r_p(s)q_p(s) + r_{p+1}(s) , \quad \deg[r_{p+1}(s)] < \deg[r_p(s)]$$
$$r_p(s) = r_{p+1}(s)q_{p+1}(s)$$

is well defined for a certain $p \leq \deg[r_1(s)] - 1$.

From such a sequence, it is easily checked that the polynomial $r_{p+1}(s)$ divides $r_p(s)$ and then also $r_{p-1}(s)$ and so on and so forth till $r_1(s)$ and $r_0(s)$.

The single elements of the sequence can be written as

$$r_2(s) = r_0(s) - r_1(s)q_1(s)$$
$$r_3(s) = r_1(s) - r_2(s)q_2(s)$$
$$r_4(s) = r_2(s) - r_3(s)q_3(s)$$

$$\vdots \quad = \quad \vdots$$

Hence, $r_i(s) = r_1(s)\varphi_{p+2-i}(s) + \varphi_{p+1-i}(s)r_0(s)$, $i > 1$, where $\varphi_{p+1-i}(s)$ and $\varphi_{p-i+2}(s)$ are suitable polynomials, so that $r_{p+1}(s) = r_0(s)\varphi_0(s) + r_1(s)\varphi_1(s)$. Hence, any common factor of the two polynomials $r_0(s)$ and $r_1(s)$ is also a factor of $r_{p+1}(s)$ so that this last polynomial is a greatest common divisor of $r_0(s)$ and $r_1(s)$. Through a suitable choice of $q_{p+1}(s)$ such a polynomial can be made coincident with $r(s)$ (recall that two greatest common divisors of two assigned polynomials differ for a multiplicative factor). $\qquad \square$

Theorem A.1 *Two polynomials $r_0(s)$ and $r_1(s)$ are coprime if and only if there exist two polynomials $\varphi_0(s)$ and $\varphi_1(s)$ such that*

$$r_0(s)\varphi_0(s) + r_1(s)\varphi_1(s) = 1 \qquad (A.1)$$

Proof If the two polynomials are coprime, i.e. 1 is a greatest common divisor, then eq. (A.1) follows from Lemma A.1. Conversely, if eq. (A.1) holds, then $r_0(s)$ and $r_1(s)$ are obviously coprime, because any common factor should appear at the right hand side of this equation. $\qquad \square$

Lemma A.2 *Let $p(s)$, $q(s)$ and $r(s)$ be three polynomials with $p(s)$ and $q(s)$ coprime. Then there exist two polynomials $\varphi(s)$ and $\psi(s)$ with $\deg[\varphi(s)] < \deg[q(s)]$ such that*

$$\varphi(s)p(s) + \psi(s)q(s) = r(s)$$

Proof Being $p(s)$ and $q(s)$ coprime, there exist, in view of Theorem A.1, two polynomials $\bar{\varphi}(s)$ and $\bar{\psi}(s)$ such that $\bar{\varphi}(s)p(s) + \bar{\psi}(s)q(s) = 1$, so that, letting $\hat{\varphi}(s) := \bar{\varphi}(s)r(s)$, $\hat{\psi}(s) := \bar{\psi}(s)r(s)$, it follows

$$\hat{\varphi}(s)p(s) + \hat{\psi}(s)q(s) = r(s)$$

If $\deg[\hat{\varphi}(s)] < \deg[q(s)]$, the two polynomials $\hat{\varphi}(s)$ and $\hat{\psi}(s)$ satisfy the conclusion of the theorem, otherwise for each polynomial $\vartheta(s)$ it results

$$p(s)[\hat{\varphi}(s) - \vartheta(s)q(s)] + q(s)[\hat{\psi}(s) + \vartheta(s)p(s)] = r(s) \qquad (A.2)$$

If one performs the division of $\hat{\varphi}(s)$ by $q(s)$ it follows

$$\hat{\varphi}(s) = q(s)\beta(s) + \alpha(s)$$

with $\deg[\alpha(s)] < \deg[q(s)]$. It is immediate to ascertain that the polynomials $\varphi(s) := \hat{\varphi}(s) - \beta(s)q(s)$ and $\psi(s)) := \hat{\psi}(s) + \beta(s)p(s)$ (set $\vartheta(s) = \beta(s)$ in eq. (A.2)) verify the claim. $\qquad \square$

Appendix B

Singular Values of Matrices

In this section the proofs of the results presented in Section 2.6 are reported, along with a few useful matrix properties (Lemmas B.8-B.15) and hints on matrix manipulations. The book by Lawson and Hanson [38] is the main reference for the singular value decomposition. For the proof of Theorem 2.8 some preliminary results are needed (Lemmas B.1-B.7).

Lemma B.1 *Let* $v := [v_1 \ v_2 \ \cdots \ v_n]' \neq 0$. *Then there exists a unitary matrix* Q *(Householder matrix transformation) such that*

$$Qv = -\sigma \|v\| e_1$$

where $e_1 := [1 \ 0 \ \cdots \ 0]' \in R^n$, $\sigma := e^{j \arg(v_1)}$

Proof Take $u := v + \sigma \|v\| e_1$ and notice that $u := [u_1 \ u_2 \ \cdots \ u_n]' \neq 0$ since $u_1 = \sigma(|v_1| + \|v\|)$ and $v \neq 0$. Moreover, let

$$Q := I - 2 \frac{uu^\sim}{u^\sim u}$$

Then,

$$Q^\sim Q = (I - 2 \frac{uu^\sim}{u^\sim u})(I - 2 \frac{uu^\sim}{u^\sim u}) = I - 4 \frac{uu^\sim}{u^\sim u} + 4 \frac{uu^\sim uu^\sim}{u^\sim u u^\sim u} = I$$

since $uu^\sim uu^\sim = u^\sim u uu^\sim$ (actually $u^\sim u$ is scalar). Hence, it has been shown that Q is an unitary matrix. On the other hand,

$$
\begin{aligned}
u^\sim u &= (v^\sim + \sigma^\sim \|v\| e_1')(v + \sigma \|v\| e_1) \\
&= \|v\|^2 + |\sigma|^2 \|v\|^2 + \sigma^\sim \|v\| v_1 + v_1^\sim \sigma \|v\| \\
&= 2 \|v\|^2 + 2 |v_1| \|v\|
\end{aligned}
$$

since $|\sigma| = 1$ and $\sigma^\sim v_1 = \sigma v_1^\sim = |v_1|$. Therefore,

$$
\begin{aligned}
Qv &= Q(u - \sigma \|v\| e_1) \\
&= (I - \frac{2uu^\sim}{u^\sim u})(u - \sigma \|v\| e_1) \\
&= u - \sigma \|v\| e_1 - 2u + \frac{2u}{u^\sim u} u_1^\sim \sigma \|v\| \\
&= -u - \sigma \|v\| e_1 + \frac{2u(v_1^\sim + \sigma^\sim \|v\|) \sigma \|v\|}{2(\|v\|^2 + |v_1| \|v\|)} \\
&= -\sigma \|v\| e_1
\end{aligned}
$$

The last equality follows from $v_1^\sim + \sigma^\sim \|v\| = \sigma^\sim(|v_1| + \|v\|)$ and $\sigma^\sim \sigma = 1$.   □

Lemma B.2 *Let* $v := [v_1 \ \ v_2]' \neq 0$ *be an element of* C^2 *and let* $\vartheta_i := \arg(v_i)$. *Then there exists a unitary matrix* G *(Givens transformation matrix) such that*

$$Gv = \begin{bmatrix} \|v\|e^{j\vartheta_1} \\ 0 \end{bmatrix}$$

Proof Take

$$G := \begin{bmatrix} |v_1| & |v_2|e^{j(\vartheta_1 - \vartheta_2)} \\ |v_2|e^{j(\vartheta_2 - \vartheta_1)} & -|v_1| \end{bmatrix} \|v\|^{-1}$$

It is readily seen that $G^\sim G = I$ and Gv has the form claimed in the statement.   □

Lemma B.3 *Let* B *be a* $n \times m$ *matrix. Then there exists a unitary matrix* Q *such that* $R := QB$ *is upper triangular.*

Proof In view of Lemma B.1 there exists a unitary matrix Q_1 such that

$$Q_1 B = \begin{bmatrix} x & x & \cdots & x \\ 0 & & & \\ \vdots & & B_1 & \\ 0 & & & \end{bmatrix}$$

where the x's are generic scalar numbers. Again in view of Lemma B.1 there exists a unitary matrix P_2 such that $P_2 B_1$ has the same structure as $Q_1 B$. Then, defining the unitary matrix

$$Q_2 := \begin{bmatrix} I & 0 \\ 0 & P_2 \end{bmatrix}$$

it follows

$$Q_2 Q_1 B = \begin{bmatrix} x & x & x & \cdots & x \\ 0 & x & x & \cdots & x \\ 0 & 0 & & & \\ \vdots & \vdots & & B_2 & \\ 0 & 0 & & & \end{bmatrix}$$

Iterating this procedure at most $n - 1$ times, one can conclude that matrix Q, given by $Q := Q_{n-1}Q_{n-2}\cdots Q_2 Q_1$, is such that QB is upper triangular.   □

Lemma B.4 *Let* B *be a* $n \times m$ *matrix with* $\text{rank}[B] = k$. *Then, there exist two unitary matrices* Q *and* P *such that*

$$QBP = \begin{bmatrix} R & T \\ 0 & 0 \end{bmatrix} \} \, k \text{ rows}$$

Proof Let P be the permutation matrix (which is obviously unitary) such that the first k columns of BP are linearly independent. In view of Lemma B.3 there exists a unitary matrix Q such that $S := QBP$ is upper triangular. Obviously, the first k columns of S are linearly independent. The i-th row of S is zero for $i > k$, since, otherwise, the triangularity of S would imply that $\text{rank}[S]=\text{rank}[B] > k$, which is a contradiction.   □

Lemma B.5 *Let* $[R \quad T]$ *be a* $k \times m$ *matrix with* $\mathrm{rank}[R] = k$. *Then there exists a unitary matrix* W *such that* $[R \quad T]W = [\hat{R} \quad 0]$, *with* $\hat{R}$ *nonsingular and lower triangular.*

Proof Lemma B.4 implies that there exist two unitary matrices Q and P such that

$$Q \left[\begin{array}{c} R^{\sim} \\ T^{\sim} \end{array} \right] P = \left[\begin{array}{c} \hat{R}^{\sim} \\ 0 \end{array} \right]$$

with $\hat{R}^{\sim}$ upper triangular. Thanks to the assumption on the rank of R, the permutation matrix P is actually the identity matrix and $\hat{R}^{\sim}$ is nonsingular. The thesis now immediately follows by letting $W = Q^{\sim}$. □

Lemma B.6 *Let* B *be a* $n \times m$ *matrix with* $\mathrm{rank}[B] = k$. *Then there exist two unitary matrices* H *and* K *such that* $H^{\sim}BK = R$ *with*

$$R = \left[\begin{array}{cc} R_{11} & 0 \\ 0 & 0 \end{array} \right]$$

where R_{11} *is triangular and nonsingular.*

Proof The proof is straightforward in view of Lemmas B.4 and B.5. □

Lemma B.7 *Let* A *be a square nonsingular matrix. Then there exist two unitary matrices* U *and* V *such that* $U^{\sim}AV = S$, *where* S *is diagonal with real positive entries.*

Proof Let V be a unitary matrix such that $A^{\sim}A = VDV^{\sim}$ with D real diagonal and define S in such a way that its generic (i,j) element is the square root of the (i,j) element of D, so that $S^{\sim}S = S^2 = D$. Let now $U := AVS^{-1}$ (recall that, being $A^{\sim}A > 0$, matrices D and S are nonsingular). It follows that $U^{\sim}U = S^{-1}V^{\sim}A^{\sim}AVS^{-1} = S^{-1}V^{\sim}VDV^{\sim}VS^{-1} = S^{-1}DS^{-1} = I$, so that U is unitary. Finally, $USV^{\sim} = AVS^{-1}SV^{\sim} = AVV^{\sim} = A$. Notice that, by suitably choosing matrix V, it is possible to arrange the elements on the diagonal of D so that they are nonincreasing. □

Proof of Theorem 2.8 Thanks to Lemma B.6, there exist two unitary matrices H and K such that $A = HRK^{\sim}$ with

$$R = \left[\begin{array}{cc} R_{11} & 0 \\ 0 & 0 \end{array} \right]$$

where R_{11} is k-dimensional and nonsingular. In view of Lemma B.7 there exist two unitary matrices $\hat{U}$ and $\hat{V}$ such that $R_{11} = \hat{U}\hat{S}\hat{V}^{\sim}$ where matrix $\hat{S}$ is diagonal with positive nonincreasing elements. Letting

$$\bar{U} := \left[\begin{array}{cc} \hat{U} & 0 \\ 0 & I \end{array} \right] \} \, n - k \text{ rows} \quad , \quad \bar{V} := \left[\begin{array}{cc} \hat{V} & 0 \\ 0 & I \end{array} \right] \} \, m - k \text{ rows}$$

$$S := \left[\begin{array}{cc} \hat{S} & 0 \\ 0 & 0 \end{array} \right] \} \, n - k \text{ rows}$$

$$\underbrace{}_{m - k \text{ columns}}$$

it follows

$$\bar{U}S\bar{V}^{\sim} = \begin{bmatrix} R_{11} & 0 \\ 0 & 0 \end{bmatrix} = R$$

so that $A = HRK^{\sim} = H\bar{U}S\bar{V}^{\sim}K^{\sim} = USV^{\sim}$, with $U := H\bar{U}$ and $V := K\bar{V}$.   □

Proof of Lemma 2.16 *Point 1)* Let T be a unitary matrix such that

$$TA^{\sim}AT^{\sim} = D = \mathrm{diag}\{\lambda_i^2\}$$

Hence

$$\begin{aligned}
\max_{x \neq 0} \frac{\|Ax\|^2}{\|x\|^2} &= \max_{x \neq 0} \frac{x^{\sim}A^{\sim}Ax}{x^{\sim}T^{\sim}Tx} \\
&= \max_{x \neq 0} \frac{x^{\sim}T^{\sim}TA^{\sim}AT^{\sim}Tx}{x^{\sim}T^{\sim}Tx} \\
&= \max_{z=Tx \neq 0} \frac{z^{\sim}Dz}{z^{\sim}z} \\
&= \max_{z \neq 0} \frac{\sum_i z_i^{\sim}z_i\lambda_i^2}{\sum_i z_i^{\sim}z_i} \\
&\leq \max_{z \neq 0} \frac{\bar{\sigma}^2(A)\sum_i z_i^{\sim}z_i}{\sum_i z_i^{\sim}z_i} = \bar{\sigma}^2(A)
\end{aligned}$$

On the other hand, if $A^{\sim}A\xi = \bar{\sigma}^2(A)\xi$, $\xi \neq 0$, it results

$$\bar{\sigma}^2(A) = \frac{\xi^{\sim}A^{\sim}A\xi}{\xi^{\sim}\xi} \leq \max_{x \neq 0} \frac{x^{\sim}A^{\sim}Ax}{x^{\sim}x}$$

Point 2) Analogous considerations as in Point 1) lead to the conclusion.   □

Proof of Lemma 2.17 The conclusion is straightforward if rank$[A] = 0$. Then, let rank$[A] \neq 0$ so that $\bar{\sigma}(A) > 0$ and there exists $x \neq 0$ such that $A^{\sim}Ax = \bar{\sigma}^2(A)x$. From this relation it follows that $AA^{\sim}y = \bar{\sigma}^2(A)y$ with $y := Ax$. Vector y is nonzero otherwise $\bar{\sigma}^2(A)x = 0$. Hence, $\sigma^2(A)$ is an eigenvalue of $AA^{\sim}$ and then $\bar{\sigma}^2(A^{\sim}) \geq \bar{\sigma}^2(A)$. The same line of reasoning applied to $A^{\sim}$ instead of A leads to the conclusion that $\bar{\sigma}(A) \geq \bar{\sigma}(A^{\sim})$. Hence, the thesis follows.   □

Proof of Lemma 2.18 *Points 1) and 2)* Let x^i be an eigenvector associated with $\lambda_i(A)$. Then, recalling Lemma 2.16

$$\underline{\sigma}^2(A) = \min_{x \neq 0} \frac{x^{\sim}A^{\sim}Ax}{x^{\sim}x} \leq \frac{x^{i\sim}A^{\sim}Ax^i}{x^{i\sim}x^i} \leq \max_{x \neq 0} \frac{x^{\sim}A^{\sim}Ax}{x^{\sim}x} = \bar{\sigma}^2(A)$$

Finally

$$\frac{x^{i\sim}A^{\sim}Ax^i}{x^{i\sim}x^i} = \frac{x^{i\sim}\lambda_i^{\sim}(A)\lambda_i(A)x^i}{x^{i\sim}x^i} = |\lambda_i(A)|^2 \leq r_s^2(A)$$

leads to the conclusion.

Point 3) Recall that the eigenvalues of an inverse matrix are the inverses of the eigenvalues of the matrix. Then,

$$\underline{\sigma}^2(A) = \min_i \lambda_i(A^{\sim}A)$$

$$= \min_i \frac{1}{\lambda_i((A^\sim A)^{-1})}$$

$$= \frac{1}{\max\limits_i \ \lambda_i((A^\sim A)^{-1})} = \frac{1}{\bar{\sigma}^2(A^{-1})}$$

where the last equality follows from Lemma 2.17.

Point 4) The correctness of the claim directly follows by interchanging A and A^{-1} in Point 3).

Point 5) By exploiting the assumption that A is hermitian and recalling that $\lambda_i(A^2) = \lambda_i^2(A)$, it follows

$$\bar{\sigma}^2(A) = \max_i \lambda_i(A^\sim A) = \max_i \lambda_i(A^2) = \max_i \lambda_i^2(A) = r_s^2(A)$$

$\square$

Proof of Lemma 2.19 It turns out that

$$\sigma_i^2(\alpha A) = \lambda_i(\alpha^\sim A^\sim A\alpha) = |\alpha|^2 \lambda_i(A^\sim A) = |\alpha|^2 \sigma_i^2(A)$$

so that the thesis directly follows.

$\square$

Proof of Lemma 2.20 By recalling Lemma 2.16 and well known properties on the norm of a vector, it results

$$\bar{\sigma}(A + B) = \max_{x \neq 0} \frac{\|(A + B)x\|}{\|x\|}$$

$$\leq \max_{x \neq 0} \frac{\|Ax\| + \|Bx\|}{\|x\|}$$

$$\leq \max_{x \neq 0} \frac{\|Ax\|}{\|x\|} + \max_{x \neq 0} \frac{\|Bx\|}{\|x\|} = \bar{\sigma}(A) + \bar{\sigma}(B)$$

$\square$

Proof of Lemma 2.21 Preliminarily, by recalling Lemma 2.16 and the fact that in general $\text{Im}[B]$ may not coincide with C^m, it follows

$$\bar{\sigma}(A) = \max_{x \neq 0} \frac{\|Ax\|}{\|x\|} \geq \max_{z \neq 0} \frac{\|ABz\|}{\|Bz\|} \geq \frac{\|ABz\|}{\|Bz\|}, \ \forall z$$

Hence

$$\bar{\sigma}(AB) = \max_{z \neq 0} \frac{\|ABz\|}{\|z\|} \leq \max_{z \neq 0} \frac{\bar{\sigma}(A)\|Bz\|}{\|z\|} = \bar{\sigma}(A)\bar{\sigma}(B)$$

$\square$

Proof of Lemma 2.22 Preliminarily, observe that, $\forall x$

$$\frac{\|Ax\| + \|Bx\|}{\|x\|} \leq \frac{\|Ax\|}{\|x\|} + \max_{x \neq 0} \frac{\|Bx\|}{\|x\|} = \frac{\|Ax\|}{\|x\|} + \bar{\sigma}(B)$$

and that

$$\frac{\|Ax\| - \|Bx\|}{\|x\|} \geq \frac{\|Ax\|}{\|x\|} - \max_{x \neq 0} \frac{\|Bx\|}{\|x\|} = \frac{\|Ax\|}{\|x\|} - \bar{\sigma}(B)$$

From these expressions it follows that

$$\min_{x \neq 0} \frac{\|Ax\| + \|Bx\|}{\|x\|} \leq \underline{\sigma}(A) + \bar{\sigma}(B)$$

$$\min_{x \neq 0} \frac{\|Ax\| - \|Bx\|}{\|x\|} \geq \underline{\sigma}(A) - \bar{\sigma}(B)$$

As for the second inequality, it turns out

$$\underline{\sigma}(A + B) = \min_{x \neq 0} \frac{\|(A + B)x\|}{\|x\|}$$

$$= \min_{x \neq 0} \frac{\|Ax + Bx\|}{\|x\|}$$

$$\leq \min_{x \neq 0} \frac{\|Ax\| + \|Bx\|}{\|x\|} \leq \underline{\sigma}(A) + \bar{\sigma}(B)$$

Analogously, as for the first inequality it turns out

$$\underline{\sigma}(A + B) = \min_{x \neq 0} \frac{\|(A + B)x\|}{\|x\|}$$

$$= \min_{x \neq 0} \frac{\|Ax + Bx\|}{\|x\|}$$

$$\geq \min_{x \neq 0} \frac{\|Ax\| - \|Bx\|}{\|x\|} \geq \underline{\sigma}(A) - \bar{\sigma}(B)$$

$\square$

Proof of Lemma 2.23 As for the first inequality, Lemma 2.16 implies that

$$\bar{\sigma}([A\ B]) = \max_{\left[\begin{smallmatrix} x \\ y \end{smallmatrix}\right] \neq 0} \frac{\|Ax + By\|}{\left\| \left[\begin{smallmatrix} x \\ y \end{smallmatrix}\right] \right\|} \geq \max_{x \neq 0} \frac{\|Ax\|}{\|x\|} = \bar{\sigma}(A)$$

$$\bar{\sigma}([A\ B]) = \max_{\left[\begin{smallmatrix} x \\ y \end{smallmatrix}\right] \neq 0} \frac{\|Ax + By\|}{\left\| \left[\begin{smallmatrix} x \\ y \end{smallmatrix}\right] \right\|} \geq \max_{y \neq 0} \frac{\|By\|}{\|y\|} = \bar{\sigma}(B)$$

As for the second inequality, from Lemma 2.16 it follows that

$$\bar{\sigma}([A\ B]) = \max_{\left[\begin{smallmatrix} x \\ y \end{smallmatrix}\right] \neq 0} \frac{\|Ax + By\|}{\left\| \left[\begin{smallmatrix} x \\ y \end{smallmatrix}\right] \right\|}$$

$$\leq \max_{\left[\begin{smallmatrix} x \\ y \end{smallmatrix}\right] \neq 0} \frac{\|Ax\| + \|By\|}{\left\| \left[\begin{smallmatrix} x \\ y \end{smallmatrix}\right] \right\|}$$

$$\leq \max_{\left[\begin{smallmatrix} x \\ y \end{smallmatrix}\right] \neq 0} \frac{\bar{\sigma}(A)\|x\| + \bar{\sigma}(B)\|y\|}{\left\| \left[\begin{smallmatrix} x \\ y \end{smallmatrix}\right] \right\|}$$

$$\leq \max[\bar{\sigma}(A), \bar{\sigma}(B)] \max_{\left[\begin{smallmatrix} x \\ y \end{smallmatrix}\right] \neq 0} \frac{\|x\| + \|y\|}{\left\| \left[\begin{smallmatrix} x \\ y \end{smallmatrix}\right] \right\|}$$

The thesis then follows since, $\forall \, [x' \; y']'$ it results

$$\frac{\|x\| + \|y\|}{\left\| \begin{bmatrix} x \\ y \end{bmatrix} \right\|} \leq \sqrt{2}$$

Actually,

$$\begin{aligned}
0 \leq (\|x\| - \|y\|)^2 &= \|x\|^2 + \|y\|^2 - 2\|x\|\|y\| \\
&= 2(\|x\|^2 + \|y\|^2) - \|x\|^2 - \|y\|^2 - 2\|x\|\|y\| \\
&= 2 \left\| \begin{bmatrix} x \\ y \end{bmatrix} \right\|^2 - (\|x\| + \|y\|)^2
\end{aligned}$$

$\square$

Proof of Lemma 2.24 It turns out that

$$\sum_i \sigma_i^2(A) = \sum_i \lambda_i(A^\sim A) = \mathrm{trace}[A^\sim A]$$

$\square$

Proof of Lemma 2.25 As for the first inequality, from Lemma 2.16 it follows

$$\bar{\sigma}(A) = \max_{x \neq 0} \frac{\|Ax\|}{\|x\|} \geq \|Ae_i\| \,, \quad \forall i$$

where e_i is the i-th column of the identity matrix. But,

$$\|Ae_i\| = \sqrt{\sum_j |\{A\}_{j,i}|^2} \geq \max_j |\{A\}_{j,i}|$$

As for the second inequality, recalling Lemma 2.24 it results

$$\begin{aligned}
\bar{\sigma}^2(A) &\leq \sum_h \sigma_h^2(A) \\
&\leq \mathrm{trace}[A^\sim A] \\
&\leq \sum_i \{A^\sim A\}_{i,i} \\
&\leq m \, \max_i \{A^\sim A\}_{i,i} \\
&\leq m \, \max_i \sum_j |\{A\}_{i,j}|^2 \\
&\leq m \, \max_i \, m \, \max_j |\{A\}_{i,j}|^2 = m^2 \, \max_{i,j} |\{A\}_{i,j}|^2
\end{aligned}$$

$\square$

Lemma B.8 *Let A and B be two matrices with dimensions $n \times m$ and $m \times n$, respectively. If $\lambda \neq 0$ is an eigenvalue of AB, then it is also an eigenvalue of BA.*

Proof Let $x \neq 0$ such that $ABx = \lambda x \neq 0$. Hence $y := Bx \neq 0$, otherwise $\lambda x = 0$. Finally, $BAy = BABx = \lambda Bx = \lambda y$, that is λ is an eigenvalue of BA . $\square$

Lemma B.9 *Let A and B be two matrices with dimensions $n \times m$ and $m \times n$, respectively. Suppose also that 1 is not an eigenvalue of AB. Then*

i) $I + AB(I - AB)^{-1} = (I - AB)^{-1}$

ii) $AB(I - AB)^{-1} = (I - AB)^{-1}AB = A(I - BA)^{-1}B$

Proof Preliminarily observe that $(I - AB)$ and $(I - BA)$ are nonsingular thanks to the assumption on the eigenvalue of AB and Lemma B.8. Hence, the formulas are well defined.

Point i) It is

$$I + AB(I - AB)^{-1} = (I - AB)(I - AB)^{-1} + AB(I - AB)^{-1}$$
$$= (I - AB + AB)(I - AB)^{-1} = (I - AB)^{-1}$$

Point ii) As for the first equality, it is

$$AB(I - AB)^{-1} = -I + (I - AB)^{-1}$$
$$= -(I - AB)^{-1}(I - AB) + (I - AB)^{-1}$$
$$= (I - AB)^{-1}(-I + AB + I) = (I - AB)^{-1}AB$$

As for the second equality, it is $B - BAB = (I - BA)B = B(I - AB)$, so that $B(I - AB)^{-1} = (I - BA)^{-1}B$, which implies $AB(I - AB)^{-1} = A(I - BA)^{-1}B$. □

Lemma B.10 *Let $A = A' \geq 0$ and $B = B' \geq 0$ be two matrices with the same dimensions. Then the eigenvalues of $C := AB$ are real and nonnegative.*

Proof Let λ be an eigenvalue of C and $\xi \neq 0$ an associated eigenvector, i.e. $C\xi = AB\xi = \lambda\xi$, $\xi \neq 0$. From these expressions it follows that $BAB\xi = \lambda B\xi$ so that $\xi^\sim BAB\xi = \lambda\xi^\sim B\xi$. Being A and B positive semidefinite, the quantities $\alpha := \xi^\sim BAB\xi$ and $\beta := \xi^\sim B\xi$ are both real and nonnegative. If $\beta \neq 0$ then $\lambda = \frac{\alpha}{\beta}$ is real and nonnegative. If $\beta = 0$, then $B\xi = 0$ and $\lambda = 0$. □

Lemma B.11 *Let $A = A' \geq 0$ and $B = B' \geq 0$ be two matrices with the same dimensions and γ a positive scalar. If $r_s(\gamma^{-2}AB) < 1$, then*

i) $C := (I - \gamma^{-2}AB)^{-1}A \geq 0$

ii) If $A > 0$, then $C > 0$ and $A^{-1} > \gamma^{-2}B$

Conversely, if $C \geq 0$, then

iii) $r_s(\gamma^{-2}AB) < 1$

Proof Preliminarily observe that $I - \gamma^{-2}AB$ is singular if and only if $\gamma^{-2}AB$ has (at least) one eigenvalue equal to one. This is not possible thanks to the assumption. Hence, matrix C is well defined. It is also symmetric in view of what has been shown in the proof of Lemma B.9.

Point i) By exploiting Lemma B.9 and the definition of C, it is easy to check that

$$(-I + \gamma^{-2}AB)C + C(-I + \gamma^{-2}BA) + 2A = 0$$

Thanks to the assumption on the spectral radius of $\gamma^{-2}AB$, it turns out that matrix $-I + \gamma^{-2}AB$ has all its eigenvalue in the open left half plane, i.e. it is stable. The equation above is therefore a Lyapunov equation in the unknown C, with stable coefficient matrix and positive semidefinite known term $2A$. In view of Lemma C.1 one can conclude that C is positive semidefinite.

Point ii) If $A > 0$ then, recalling again Lemma C.1, it turns out that $C > 0$. Moreover,

$$C = (I - \gamma^{-2}AB)^{-1}A = [A(A^{-1} - \gamma^{-2}B)]^{-1}A = (A^{-1} - \gamma^{-2}B)^{-1}$$

so that $(A^{-1} - \gamma^{-2}B) > 0$, which is the thesis.

Point iii) Assume by contradiction that $\lambda \geq 1$ is an eigenvalue of $\gamma^{-2}AB$ (recall that, thanks to Lemma B.10 the eigenvalues of AB are real and nonnegative). Since $C = (I - \gamma^{-2}AB)^{-1}A$ exists, such an eigenvalue can not be equal to 1. In view of Lemma B.8, λ is also an eigenvalue of $\gamma^{-2}BA$ so that there exists a vector $\xi \neq 0$ such that $\gamma^{-2}BA\xi = \lambda\xi$, i.e. $(I - \gamma^{-2}BA)\xi = (1 - \lambda)\xi$ or, alternatively, $(I - \gamma^{-2}BA)^{-1}\xi = (1 - \lambda)^{-1}\xi$. Hence $\xi'(1 - \lambda)^{-1} = \xi'(I - \gamma^{-2}AB)^{-1}$. It follow that

$$\frac{\xi'A\xi}{(1 - \lambda)} = \xi'(I - \gamma^{-2}AB)^{-1}A\xi$$

The left hand side of this equation is a nonpositive number (actually, the numerator is nonnegative and the denominator is negative). The right hand side is nonnegative in view of the assumption on C. Hence, it must be $A\xi = 0$ and hence, from $(I - \gamma^{-2}BA)\xi = (1 - \lambda)\xi = \xi$ and $\lambda \neq 0$ it follows that $\xi = 0$, a contradiction. $\square$

Lemma B.12 *Let $A = A' > 0$ and $B = B' \geq 0$ be two matrices with the same dimensions and α a positive scalar. Then $r_s(BA) \geq \alpha$ if and only if there exists $x \neq 0$ such that $x'(B - \alpha A^{-1})x \geq 0$.*

Proof If $r_s(BA) \geq \alpha$, it must exist (recall Lemma B.10) a vector $\xi \neq 0$ such that $BA\xi = \lambda\xi, \lambda \geq \alpha$. Letting $x := A\xi$, it follows that $Bx = \lambda A^{-1}x$ and also $x'Bx = \lambda x'A^{-1}x \geq \alpha x'A^{-1}x$, which yields $x'(B - \alpha A^{-1})x \geq 0$.

Conversely, if $r_s(BA) < \alpha$, then, from Lemma B.11 it follows $A^{-1} > \alpha^{-1}B$ so that $x'(B - \alpha A^{-1})x < 0, \forall x \neq 0$ and hence it does not exist $x \neq 0$ such that $x'(B - \alpha A^{-1})x \geq 0$. $\square$

Lemma B.13 *Consider the block matrix*

$$\Phi = \begin{bmatrix} \Phi_1 & \Phi_2 \\ \Phi_3 & \Phi_4 \end{bmatrix}$$

where the submatrices Φ_1 and Φ_4 are square. Then

i) If Φ_1 is nonsingular, $\det[\Phi] = \det[\Phi_1]\det[\Phi_4 - \Phi_3\Phi_1^{-1}\Phi_2]$

ii) If Φ_4 is nonsingular, $\det[\Phi] = \det[\Phi_4]\det[\Phi_1 - \Phi_2\Phi_4^{-1}\Phi_3]$

Proof *Point i)* The identity

$$\begin{bmatrix} \Phi_1 & \Phi_2 \\ \Phi_3 & \Phi_4 \end{bmatrix} = \begin{bmatrix} I & 0 \\ \Phi_3\Phi_1^{-1} & I \end{bmatrix} \begin{bmatrix} \Phi_1 & \Phi_2 \\ 0 & \Phi_4 - \Phi_3\Phi_1^{-1}\Phi_2 \end{bmatrix}$$

is straightforward. By considering the determinant of the matrices on the left and right hand sides, point i) follows.

Point ii) The identity

$$\begin{bmatrix} \Phi_1 & \Phi_2 \\ \Phi_3 & \Phi_4 \end{bmatrix} = \begin{bmatrix} I & \Phi_2\Phi_4^{-1} \\ 0 & I \end{bmatrix} \begin{bmatrix} \Phi_1 - \Phi_2\Phi_4^{-1}\Phi_3 & 0 \\ \Phi_3 & \Phi_4 \end{bmatrix}$$

is straightforward. By considering the determinant of the matrices on the left and right hand sides, point *ii)* follows. □

Lemma B.14 (Schur complements) *Consider the block symmetric matrix*

$$\Phi = \begin{bmatrix} \Phi_1 & \Phi_2 \\ \Phi_2' & \Phi_3 \end{bmatrix}$$

where the submatrices Φ_1 and Φ_3 are square. Then

i) If Φ_1 is positive definite, $\Phi \geq 0 \Longleftrightarrow \Phi_3 \geq \Phi_2'\Phi_1^{-1}\Phi_2$

ii) If Φ_3 is positive definite, $\Phi \geq 0 \Longleftrightarrow \Phi_1 \geq \Phi_2\Phi_3^{-1}\Phi_2'$

Proof *Point i)* Considering the nonsingular matrix

$$T = \begin{bmatrix} I & 0 \\ \Phi_2'\Phi_1^{-1} & I \end{bmatrix}$$

it is straightforward to verify the identity

$$\Phi = T \begin{bmatrix} \Phi_1 & 0 \\ 0 & \Phi_3 - \Phi_2'\Phi_1^{-1}\Phi_2 \end{bmatrix} T'$$

from which point *i)* follows.

Point ii) Considering the nonsingular matrix

$$T = \begin{bmatrix} I & \Phi_2\Phi_3^{-1} \\ 0 & I \end{bmatrix}$$

it is straightforward to verify the identity

$$\Phi = T \begin{bmatrix} \Phi_1 - \Phi_2\Phi_3^{-1}\Phi_2' & 0 \\ 0 & \Phi_3 \end{bmatrix} T'$$

from which point *ii)* follows. □

Lemma B.15 *Consider A and D real matrices such that $D'D = I$. Then*

$$\min_Z \|A - DZ\| = \|(I - DD')A\|$$

and the optimal solution is $Z^o = D'A$.

Proof Define $\Delta := Z - Z^o = Z - D'A$. From the fact that matrices D and $(I - DD')$ are orthogonal we have

$$\|A - DZ\|^2 = \|(I - DD')A - D\Delta\|^2$$
$$= \max_{\|x\|=1} \left\{ \|(I - DD')Ax\|^2 + \|\Delta x\|^2 \right\}$$

However, due to

$$
\begin{aligned}
\min_{Z} \|A - DZ\|^2 &= \min_{\Delta} \max_{\|x\|=1} \left\{ \|(I - DD')Ax\|^2 + \|\Delta x\|^2 \right\} \\
&\geq \max_{\|x\|=1} \min_{\Delta} \left\{ \|(I - DD')Ax\|^2 + \|\Delta x\|^2 \right\} \\
&\geq \max_{\|x\|=1} \|(I - DD')Ax\|^2 \\
&\geq \|(I - DD')A\|^2
\end{aligned}
$$

the lemma is proved by simple verification that the equality holds for $Z = Z^o$. □

Remark B.1 From Lemma B.15 it follows that the linear equation $A - DZ = 0$, with $D'D = I$, admits a solution namely $Z = D'A$ if and only if $(I - DD')A = 0$. □

Appendix C

Riccati Equation

The Lyapunov and Riccati equations play an important role in the analysis and control of linear time-invariant systems. Here, for a given n-th dimensional system $\Sigma(A, B, C, D)$, a few basic results on these equations are reported. Standard references are the book by Bittanti et al. [7] and the paper by Doyle et al. [17].

Lemma C.1 (Extended Lyapunov lemma) *Consider the Lyapunov equation*

$$0 = PA + A'P + C'C$$

Then

i) *If A is stable, there exists a unique solution. Such a solution is symmetric and positive semidefinite.*

ii) *If the pair (A, C) is detectable and there exists a symmetric and positive semidefinite solution, A is stable.*

iii) *If the pair (A, C) is observable, the solution at point i) is actually positive definite.*

Proof *Points i) and iii)* If A is stable, then

$$P := \int_0^\infty e^{A't} C' C e^{At} dt$$

is well defined, symmetric and positive semidefinite. If (A, C) is observable, then P is positive definite. In fact, if, by contradiction, $Px_0 = 0$, $x_0 \neq 0$, then

$$x_0' P x_0 := \int_0^\infty x_0' e^{A't} C' C e^{At} x_0 dt = \int_0^\infty y' y \, dt = 0$$

where y is the free output of Σ with initial state $x(0) = x_0$. Hence, $y(t) = 0, \forall t \geq 0$ contradicts the observability assumption.

Based on the definition of P, it follows

$$A'P + PA = \int_0^\infty \frac{d}{dt}(e^{A't} C' C e^{At}) dt = -C'C$$

Matrix P is the unique solution of the Lyapunov equation. In fact, if $\hat{P}$ is any other solution, it results

$$0 = (P - \hat{P})A + A'(P - \hat{P})$$

Thanks to the stability of A and a well known result of linear algebra, this equation admits the unique solution $P - \hat{P} = 0$.

Point ii) Assume that $\bar{P} = \bar{P}' \geq 0$ is a solution of the Lyapunov equation and, by contradiction, that A is not stable, i.e. $Ax = \lambda x, x \neq 0$, $Re(\lambda) \geq 0$. From the equation it then follows

$$0 = x^\sim \bar{P} A x + x^\sim A' \bar{P} x + x^\sim C' C x = 2Re(\lambda)x^\sim \bar{P} x + x^\sim C' C x$$

Since the last term of this equation is the sum of two nonnegative elements, it turns out that $Cx = 0$. But $Ax = \lambda x, Cx = 0, x \neq 0, Re(\lambda) \geq 0$ contradicts the assumed detectability of (A, C) (recall Lemma D.2). □

Lemma C.2 (Stabilizing solution of the Riccati equation - 1) *Consider the Riccati equation*

$$0 = PA + A'P + PRP + Q$$

with R and Q real and assume that

  a) The matrix

$$Z := \begin{bmatrix} A & R \\ -Q & -A' \end{bmatrix}$$

   does not have eigenvalues lying on the imaginary axis.

  b) Matrices R and Q are symmetric and n-dimensional.

If the subspace

$$\mathcal{X} := \mathrm{Im}[L] , \quad L := \begin{bmatrix} X \\ Y \end{bmatrix} \tag{C.1}$$

generated by the (generalized) eigenvectors associated with the negative real part eigenvalues of Z is complementary to the n-dimensional subspace

$$\mathcal{I} := \mathrm{Im}[\bar{I}] , \quad \bar{I} := \begin{bmatrix} 0 \\ I \end{bmatrix} \tag{C.2}$$

then

  i) The matrix $P_s := YX^{-1}$ is a real, symmetric and stabilizing solution, namely is such that $(A + RP_s)$ is stable.

  ii) P_s is the unique stabilizing solution.

Proof Z is an *Hamiltonian* matrix, i.e. it satisfies

$$JZ + Z'J = 0 \tag{C.3}$$

where

$$J := \begin{bmatrix} 0 & I \\ -I & 0 \end{bmatrix}$$

is the so called *sympletic* matrix. Matrix Z has eigenvalues symmetric with respect to the imaginary axis, since $Z = -J^{-1}Z'J$. Assumption *a)* assures the existence of n

eigenvalues of Z in the open left half plane. It is well known that the subspace $\mathcal{X}$ is Z-invariant so that there exists a real matrix T with eigenvalues in the open left half plane (stable) such that

$$ZL = LT \tag{C.4}$$

where L satisfies eq. (C.1). The proof is divided into four main steps.

Step 1 It is here proved that $X^\sim Y = Y^\sim X$. Actually, letting $W := X^\sim Y - Y^\sim X$, it is immediate to check that $W = L^\sim JL$ so that, exploiting eqs. (C.3), (C.4), it follows $WT = L^\sim JZL = -L^\sim Z'JL = -T'W$, i.e.

$$WT + T'W = 0$$

Since T is stable, this Lyapunov equation admits the unique solution $W = 0$, which is the claim.

Step 2 Matrix X is invertible since the subspace $\mathcal{X}$ is complementary to the n-dimensional subspace $\mathcal{I}$ (see eq. (C.2)), so that $P_s = YX^{-1}$ is well defined.

Step 3 It is here proved that P_s is a real, symmetric and stabilizing solution of the Riccati equation.

As for reality, observe that the columns of $L = [X^\sim \ Y^\sim]^\sim$ can be chosen complex conjugate in pair, so that, if X_c and Y_c are the complex conjugate of X and Y, respectively, then

$$\begin{bmatrix} X_c^\sim & Y_c^\sim \end{bmatrix}^\sim \Gamma = \begin{bmatrix} X^\sim & Y^\sim \end{bmatrix}^\sim$$

where Γ is a permutation matrix. Hence $P_s = YX^{-1} = Y_c\Gamma\Gamma^{-1}X_c^{-1} = Y_cX_c^{-1}$ so that P_s coincides with its conjugate, i.e. P_s is real.

As for symmetry, observe that, thanks to the identity (proved in Step 1) $X^\sim Y = Y^\sim X$ it follows that $P_s' = P_s^\sim = (X^{-1})^\sim Y^\sim = (X^{-1})^\sim X^\sim YX^{-1} = YX^{-1} = P_s$.

As for the stabilizing property, from eq. (C.4) it follows that $AX + RY = AX + RYX^{-1}X = (A + RP_s)X = XT$ so that $A + RP_s = XTX^{-1}$. Therefore, $A + RP_s$ and T are similar. Since T is stable, matrix $A + RP_s$ is stable as well.

It is now shown that P_s is a solution of the equation. In fact, eq. (C.4) is equivalent to

$$AX + RY = XT$$
$$-QX - A'Y = YT$$

Premultiplying the first equation by $Y^\sim$ yields $Y^\sim AX + Y^\sim RY = Y^\sim XT$. Moreover, premultiplying the second equation by $X^\sim$ gives $-X^\sim QX - X^\sim A'Y = X^\sim YT$. Recalling that $X^\sim Y = Y^\sim X$ it then follows that

$$0 = X^\sim QX + X^\sim A'Y + Y^\sim AX + Y^\sim RY$$

Postmultiplying this equation by X^{-1} and premultiplying it by $(X^{-1})^\sim$ it follows

$$0 = Q + A'(YX^{-1}) + (YX^{-1})^\sim A + (YX^{-1})^\sim R(YX^{-1})$$
$$= Q + A'P_s + P_s^\sim A + P_s^\sim RP_s$$

Since it has been already shown that P_s is hermitian ($P_s = P_s^\sim$), the conclusion is straightforwardly derived.

Step 4 As for the uniqueness of the stabilizing solution, assume, by contradiction, that there exist two stabilizing solutions, namely P_1 and P_2. Then

$$0 = P_1A + A'P_1 + P_1RP_1 + Q$$
$$0 = P_2A + A'P_2 + P_2RP_2 + Q$$

Subtracting term by term, it is easy to see that

$$0 = (P_1 - P_2)(A + RP_1) + (A + RP_2)'(P_1 - P_2)$$

This equation can be considered as a linear equation in the unknown $P_1 - P_2$. Being both $A + RP_1$ and $A + RP_2$ stable, a well known result of linear algebra implies that there exists only one solution, precisely $P_1 - P_2 = 0$.   □

Lemma C.3 (Stabilizing solution of the Riccati equation - 2) *Consider the Riccati equation*

$$0 = PA + A'P + PRP + Q$$

and suppose that matrix R is square, n-dimensional and either equal to BB' or $-BB'$. Moreover, assume that

  a) The pair (A, B) is stabilizable

  b) The matrix

$$Z := \begin{bmatrix} A & R \\ -Q & -A' \end{bmatrix}$$

   does not have eigenvalues lying on the imaginary axis

  c) The matrix Q is real and symmetric

Then,

  i) The subspace

$$\mathcal{X} := \text{Im}[L] \ , \quad L := \begin{bmatrix} X \\ Y \end{bmatrix}$$

   generated by the (generalized) eigenvectors of Z associated with the negative real part is complementary to the n-dimensional subspace

$$\text{Im}[\bar{I}] \ , \quad \bar{I} := \begin{bmatrix} 0 \\ I \end{bmatrix}$$

  ii) $P_s := YX^{-1}$ is a solution of the Riccati equation. It is real, symmetric and stabilizing, i.e. such that $(A + RP_s)$ is stable.

 iii) P_s is the unique stabilizing solution of the Riccati equation.

Proof In view of Lemma C.2, the proof of points *ii)* and *iii)* is straightforward once point *i)* has been proved, which, in turn, derives once the invertibility of X is proved. To this aim, consider eq. (C.4) and assume, by contradiction, that there exists $\xi \neq 0$ such that $X\xi = 0$, i.e. $\xi \in \text{Ker}[X]$. It follows in particular that

$$AX + RY = XT \tag{C.5}$$

Premultiplying this equation by $Y^\sim$ and recalling that $Y^\sim X = X^\sim Y$, it follows $Y^\sim AX + Y^\sim RY = X^\sim YT$ so that $\xi^\sim(Y^\sim AX + Y^\sim RY - X^\sim YT)\xi = \xi^\sim Y^\sim RY\xi = 0$, namely (recall that $R = \pm BB'$)

$$RY\xi = 0 \ , \quad \forall \, \xi \in \text{Ker}[X] \tag{C.6}$$

This equation together with eq. (C.5) yields

$$XT\xi = 0 , \quad \forall \, \xi \in \text{Ker}[X] \tag{C.7}$$

On the other hand, eq. (C.4) entails that $-QX - A'Y = YT$ so that

$$- A'Y\xi = YT\xi , \quad \forall \, \xi \in \text{Ker}[X] \tag{C.8}$$

For each $\xi \in \text{Ker}[X]$ eqs. (C.6)-(C.8) hold. In particular, eq. (C.7) implies $T\xi \in \text{Ker}[X]$ so that $T^i\xi \in \text{Ker}[X], \forall i \geq 0$. Now it is proved, by induction, that, from eqs. (C.7),(C.8), it follows

$$R(A')^k Y\xi = 0 , \quad \forall \, k \geq 0 , \ \forall \, \xi \in \text{Ker}[X] \tag{C.9}$$

Actually, eq. (C.9) is true for $k = 0$ (eq. (C.6)) and, moreover, it results (see eq. (C.8))

$$R(A')^{k+1} Y\xi = R(A')^k A'Y\xi = -R(A')^k YT\xi = 0$$

since $T\xi \in \text{Ker}[X]$. Recalling that $T^i\xi \in \text{Ker}[X], \forall i \geq 0$, from eq. (C.8) it follows $YT^k\xi = YTT^{k-1}\xi = -A'YT^{k-1}\xi$. By repeatedly using this last equation, one obtains

$$YT^k\xi = (-A')^k Y\xi , \quad \forall \, k \geq 0 , \ \forall \, \xi \in \text{Ker}[X] \tag{C.10}$$

For each polynomial $\nu(s)$ it then follows $\forall \, \xi \in \text{Ker}[X]$

$$\nu(-A')Y\xi = Y\nu(T)\xi \tag{C.11}$$
$$R\nu(-A')Y\xi = 0 \tag{C.12}$$
$$\nu(T)\xi \in \text{Ker}[X] \tag{C.13}$$

Let now $\nu_m(s)$ be the monic polynomial of minimum degree such that $\nu_m(T)\xi = 0$. Notice that such a polynomial actually exists and its degree is not greater than the degree of the minimal polynomial $\varphi(s)$ of T (in fact $\varphi(T) = 0$). If λ is a root of $\nu_m(s)$ it follows that $(\lambda - s)\mu(s) = \nu_m(s)$, with $\deg[\mu(s)] < \deg[\nu_m(s)]$. Then,

$$\nu_m(T)\xi = (\lambda I - T)\mu(T)\xi = 0 \tag{C.14}$$

Observe that $\beta := \mu(T)\xi \neq 0$ due to the minimallity of $\nu_m(s)$, so that β is an eigenvector of T and λ, being an associated eigenvalue, must have negative real part (recall that T is stable). By exploiting eq. (C.14) and letting $\nu(s) = \nu_m(s)$ in eq. (C.11), it turns out that, for $\xi \in \text{Ker}[X]$

$$\nu_m(-A')Y\xi = (\lambda I + A')\mu(-A')Y\xi = Y\nu_m(T)\xi = 0$$

so that

$$A'\mu(-A')Y\xi = -\lambda\mu(-A')Y\xi$$

with $Re(-\lambda) > 0$, whereas eq. (C.12), letting $\nu(s) = \mu(s)$, becomes

$$R\mu(-A')Y\xi = \pm BB'\mu(-A')Y\xi = 0$$

which, in turn, implies $B'\mu(-A')Y\xi = 0$. Letting $\eta := \mu(-A')Y\xi$, it then follows

$$A'\eta = -\lambda\eta$$
$$B'\eta = 0$$

The assumption of stabilizability of the pair (A, B) and $Re(-\lambda) > 0$, implies, in view of the *PBH* test (Lemma D.4), that $\eta = 0$. From eq. (C.11) with $\nu(s) = \mu(s)$ it then follows that $Y\mu(T)\xi = 0$ so that $\mu(T)\xi \in \mathrm{Ker}[Y]$. On the other hand, $\mu(T)\xi \in \mathrm{Ker}[X]$, so that $L\beta = 0$ contradicts the fact that $\mathrm{Im}[L]$ is a n dimensional subspace. Hence X is invertible and point $i)$ is proved.   □

The preceding lemma allows proving the following result which settles in the context of the theory of optimal control with quadratic cost functionals.

Lemma C.4 *Consider the Riccati equation*

$$0 = PA + A'P + PRP + C'C, \ R := -BB'$$

If $\bar{P}$ is a solution, then

$$\mathrm{Ker}[\bar{P}] \subset \mathcal{X}_{no} := \mathrm{Ker}[C] \cap \mathrm{Ker}[CA] \cdots \cap \mathrm{Ker}[CA^{n-1}]$$

Moreover, there exists a solution P_s which is real, symmetric, stabilizing and positive semidefinite if and only if

a) *There do not exist eigenvalues of the unobservable part of the pair (A, C) lying on the imaginary axis.*

b) *The pair (A, B) is stabilizable*

Proof It is first proved that the kernel of $\bar{P}$ is contained in the unobservable subspace $\mathcal{X}_{no}$ of (A, C) . Let $\xi \in \mathrm{Ker}[\bar{P}]$, so that $\bar{P}\xi = 0$. From the Riccati equation it follows that $0 = \xi^{\sim}(\bar{P}A + A'\bar{P} + \bar{P}R\bar{P} + C'C)\xi = \xi^{\sim}C'C\xi$ so that $C\xi = 0$. Taking this into account, again from the Riccati equation one obtains $0 = (\bar{P}A + A'\bar{P} + \bar{P}R\bar{P} + C'C)\xi = \bar{P}A\xi$ so that $A\xi \in \mathrm{Ker}[\bar{P}]$. Hence it has been proved that $\xi \in \mathrm{Ker}[\bar{P}] \Rightarrow \xi \in \mathrm{Ker}[C]$ and $A\xi \in \mathrm{Ker}[\bar{P}]$. By repeating again this argument for $A\xi$ one obtains $A^2\xi \in \mathrm{Ker}[\bar{P}]$ and $\xi \in \mathrm{Ker}[CA]$ so that, in conclusion

$$\mathrm{Ker}[\bar{P}] \subset := \mathrm{Ker}[C] \cap \mathrm{Ker}[CA] \cdots \cap \mathrm{Ker}[CA^{n-1}] = \mathcal{X}_{no}$$

It is now proved that assumptions $a), b)$ imply condition $b)$ of Lemma C.3. Assume, by contradiction, that the Hamiltonian matrix Z associated with the Riccati equation has an eigenvalue $\lambda = j\omega$ (on the imaginary axis), i.e. $Zz = j\omega z$, with $z := [x^{\sim} \ y^{\sim}]^{\sim} \neq 0$. Hence,

$$Ax + Ry = j\omega x \tag{C.15}$$
$$-Qx - A'y = j\omega y \tag{C.16}$$

where $Q = C'C$. Now add to the first equation, premultiplied by $y^{\sim}$, the second, premultiplied by $x^{\sim}$. It follows

$$j\omega(x^{\sim}y + y^{\sim}x) - y^{\sim}Ax + x^{\sim}A'y = y^{\sim}Ry - x^{\sim}Qx \tag{C.17}$$

The left hand side is a purely imaginary number, whereas the right hand side is real. Hence, both sides must be equal to zero. In particular, being $Q \geq 0$ and $R \leq 0$, it follows that $Qx = 0$ and $Ry = 0$, which in turn imply that $Cx = 0$ and $B'y = 0$. These last equations, together with eqs. (C.15)-(C.17) entail

$$Ax = j\omega x \tag{C.18}$$
$$Cx = 0 \tag{C.19}$$
$$A'y = -j\omega y \tag{C.20}$$
$$B'y = 0 \tag{C.21}$$

In conclusion, since either x or y is different from zero, the above eqs. (C.18)-(C.19) violate either assumption *a)* (recall Lemma D.1) if $x \neq 0$ or the stabilizability assumption *b)* (recall Lemma D.4) if $y \neq 0$. Assumptions *a)-c)* of Lemma C.3 are so verified. Such lemma assures the existence of a real and symmetric solution P_s such that $A + RP_s$ is stable. Obviously, P_s is a solution of the Lyapunov equation (in the unknown P)

$$0 = PA + A'P + Q + P_sRP_s$$

which can be also rewritten as

$$0 = P(A + RP_s) + (A + RP_s)'P - P_sRP_s + Q$$

This is a Lyapunov equation whose coefficient matrix $A + RP_s$ is stable and the known term $Q - P_sRP_s$ is positive semidefinite. Thanks to Lemma C.1 this equation admits a unique solution, P_s, which is positive semidefinite.

Conversely, assume that there exists a real, symmetric and stabilizing solution P_s of the Riccati equation. Necessity of condition *b)* is then obvious. Now suppose by contradiction that condition *a)* does not hold, i.e. $Ax = j\omega x, Cx = 0$ and $x \neq 0$. Hence $Zz = j\omega z$ where $z := [x'\ 0] \neq 0$ and

$$Z := \begin{bmatrix} A & -BB' \\ -C'C & -A' \end{bmatrix}$$

Hence, the Hamiltonian matrix Z has an eigenvalue on the imaginary axis, so contradicting the existence of a stabilizing solution of the Riccati equation.   □

Lemma C.5 *Let A, B, C and D be four matrices with dimensions $n \times n$, $n \times m$, $p \times n$ and $p \times m$, respectively. Assume also that $C'D = 0$, $D'D = I$ and there exists the symmetric and stabilizing solution P_s of the Riccati equation*

$$0 = PA + A'P - PBB'P + C'C$$

Let $A_{cc} := A - BB'P_s$ and $C_{1c} := C - DB'P_s$ and define the three systems

$$G(s) := \Sigma(A_{cc}, I, C_{1c}, 0)$$
$$U(s) := \Sigma(A_{cc}, B, C_{1c}, D)$$
$$U^{\perp}(s) := \Sigma(A_{cc}, -P_s^{\dagger}C'D^{\perp}, C_{1c}, D^{\perp})$$

where $P_s^{\dagger}$ is the Moore Penrose pseudo-inverse of P_s, i.e. the matrix such that $P_s P_s^{\dagger} P_s = P_s$ and $P_s^{\dagger} P_s P_s^{\dagger} = P_s$, and $D^{\perp}$ is a matrix such that

$$\begin{bmatrix} D' \\ D^{\perp '} \end{bmatrix} \begin{bmatrix} D & D^{\perp} \end{bmatrix} = I$$

Then the system with transfer function

$$F(s) := \begin{bmatrix} U(s) & U^{\perp}(s) \end{bmatrix}$$

is square, inner and

$$H(s) := G^{\sim}(s)F(s) = \Sigma(A_{cc}, \begin{bmatrix} B & -P_s^{\dagger}C'D^{\perp} \end{bmatrix}, P_s, \begin{bmatrix} 0 & 0 \end{bmatrix})$$

Proof Preliminarily, observe that matrix $D^\perp$ with the requested properties actually exists since $D'D = I$. Systems $U(s)$ and $U^\perp(s)$ have the same dynamical matrix and output transformation. Hence, a realization of $F(s)$ is simply

$$F(s) = \left[\begin{array}{c|cc} A_{cc} & B & -P_s^\dagger C'D^\perp \\ \hline C_{1c} & D & D^\perp \end{array}\right]$$

The system with transfer function $F(s)$ is square since $\left[\begin{array}{cc} D & D^\perp \end{array}\right]$ is square. Now let η be the state of the system with transfer function $F(s)$ and ξ the state of the adjoint system, namely of the system with transfer function

$$F^\sim(s) = \left[\begin{array}{c|c} -A'_{cc} & -C'_{1c} \\ \hline B' & D' \\ -D^{\perp'}CP_s^\dagger & D^{\perp'} \end{array}\right]$$

Consider the series connection $F^\sim(s)F(s)$. By exploiting the fact that $C'_{1c}C_{1c} = C'C + P_sBB'P_s = -A'_{cc}P_s - P_sA_{cc}$ and the properties of matrices C, D and $D^\perp$, it follows

$$(\dot\xi - P_s\dot\eta) = -A'_{cc}(\xi - P_s\eta) - (I - P_sP_s^\dagger)C'D^\perp\psi_2 \qquad (C.22)$$
$$l_1 = B'(\xi - P_s\eta) + \psi_1 \qquad (C.23)$$
$$l_2 = -D^{\perp'}C(P_s^\dagger\xi - \eta) + \psi_2 \qquad (C.24)$$

where ψ_1 and ψ_2 are the inputs of $F(s)$ whereas l_1 and l_2 are the outputs of $F^\sim(s)$. Lemma C.4 entails that $\mathrm{Ker}[P_s] \subset \mathrm{Ker}[C]$ so that $C(I - P_s^\dagger P_s)\beta = 0$, if $\beta \in \mathrm{Ker}[P_s]$. On the other hand, $(I - P_s^\dagger P_s)\beta = 0$, if $\beta \in \mathrm{Ker}[P_s]^\perp$. In conclusion $C(I - P_s^\dagger P_s)\beta = 0$, $\forall\beta$. Hence, eqs. (C.22)-(C.24) can be rewritten as

$$(\dot\xi - P_s\dot\eta) = -A'_{cc}(\xi - P_s\eta)$$
$$l_1 = B'(\xi - P_s\eta) + \psi_1$$
$$l_2 = -D^{\perp'}CP_s^\dagger(\xi - P_s\eta) + \psi_2$$

These equations show that the transfer function from $[\psi'_1\ \psi'_2]'$ to $[l'_1\ l'_2]'$, i.e. $F^\sim(s)F(s)$, is the identity. Hence, $F(s)$ is inner.

Let now ϑ be the state of a realization of $G^\sim(s)$. A realization of $G^\sim(s)F(s)$ is given by

$$\dot\vartheta = -A'_{cc}\vartheta - C'_{1c}\nu$$
$$\dot\eta = A_{cc}\eta + B\psi_1 - P_s^\dagger C'D^\perp\psi_2$$
$$\nu = C_{1c}\eta + D\psi_1 + D^\perp\psi_2$$
$$\sigma = \vartheta$$

where ν and σ are the input and output of $G^\sim(s)$, respectively. It is easy to check that this realization is equivalent to

$$(\dot\vartheta - P_s\dot\eta) = -A'_{cc}(\vartheta - P_s\eta)$$
$$\dot\eta = A_{cc}\eta + B\psi_1 - P_s^\dagger C'D^\perp\psi_2$$
$$\sigma = (\vartheta - P_s\eta) + P_s\eta$$

so that a possible realization of $G^\sim(s)F(s)$ is

$$G^\sim(s)F(s) = \left[\begin{array}{c|cc} A_{cc} & B & -P_s^\dagger C'D^\perp \\ \hline P_s & 0 & 0 \end{array}\right]$$

as claimed in the statement.   □

Lemma C.6 *Let A and C be matrices with dimensions $n \times n$ and $p \times n$ respectively and assume the pair $(-A, C)$ is detectable. Then the unique symmetric solution of*

$$CP = 0$$
$$P \geq 0$$
$$AP + PA' \leq 0$$

is the trivial solution $P = 0$.

Proof By contradiction, suppose $P = P' \neq 0$ is a solution. With no loss of generality, one can assume that

$$P = \left[\begin{array}{cc} P_1 & 0 \\ 0 & 0 \end{array}\right], \quad A = \left[\begin{array}{cc} A_1 & A_2 \\ A_3 & A_4 \end{array}\right], \quad C = \left[\begin{array}{cc} C_1 & C_2 \end{array}\right]$$

with $P_1 > 0$. Actually, matrix P can be put in above form by means of a suitable orthogonal transformation. Then, $CP = 0$ implies $C_1 = 0$ and $AP + PA' \leq 0$ yields

$$\left[\begin{array}{cc} A_1 P_1 + P_1 A_1' & P_1 A_3' \\ A_3 P_1 & 0 \end{array}\right] \leq 0$$

so that $A_3 = 0$ and $A_1 P_1 + P_1 A_1' \leq 0$. Finally, the detectability of the pair

$$(-A, C) = \left(\left[\begin{array}{cc} -A_1 & -A_2 \\ 0 & -A_4 \end{array}\right], \left[\begin{array}{cc} 0 & C_2 \end{array}\right]\right)$$

entails that matrix $-A_1$ is stable. In this case, all possible solutions of the inequality

$$A_1 P_1 + P_1 A_1' \leq 0$$

are such that $-P_1 \geq 0$ proving thus that $P_1 > 0$ is an impossibility. The final conclusion is that the trivial solution $P = 0$ is unique.   □

Appendix D

Structural Properties

This section is devoted to the properties of reachability, observability, stabilizability and detectability of a linear time invariant continuous time n-dimensional system $\Sigma(A, B, C, D)$. All the material presented here can be found in any text of system theory, see, e.g., the book by Kailath [29].

Controllability and reconstructability will not be treated in the present context since they enjoy the same properties as reachability and observability, respectively.

In the forthcoming lemmas, whose proofs are easily available in specialized texts, the basic characterizations of the above properties will be provided.

Preliminarily, define the matrices $P_C(s)$, $P_B(s)$, K_o and K_r as

$$P_C(s) := \begin{bmatrix} sI - A \\ C \end{bmatrix} , \quad P_B(s) := \begin{bmatrix} sI - A & B \end{bmatrix}$$

$$K_o := \begin{bmatrix} C' & A'C' & \cdots & (A')^{n-1}C' \end{bmatrix}$$
$$K_r := \begin{bmatrix} B & AB & \cdots & A^{n-1}B \end{bmatrix}$$

Lemma D.1 (Observability) *System Σ or, equivalently, the pair (A, C), is observable if and only if the following equivalent conditions hold :*

a) *PBH test :*
$$\mathrm{rank}[P_C(s)] = n , \quad \forall s$$

The set of eigenvalues of the unobservable part of (A, C) coincides with the set of values of s in correspondence of which matrix $P_C(s)$ looses rank.

b) *Kalman test :*
$$\mathrm{rank}[K_o] = n$$

c) *Wonham test : Given an arbitrary symmetric set Λ of n complex number, there exists a matrix L such that the spectrum of $A + LC$ coincides with Λ*

Lemma D.2 (Detectability) *System Σ or, equivalently, the pair (A, C) is detectable if and only if the following equivalent conditions hold :*

a) *PBH test :*
$$\mathrm{rank}[P_C(s)] = n , \quad \mathrm{Re}(s) \geq 0$$

b) Kalman test : The unobservable part of the system is stable.

c) Wonham test : There exists a matrix L such that $A + LC$ is stable.

Lemma D.3 (Reachability) *System Σ or, equivalently, the pair (A, B) is reachable if and only if the following equivalent conditions hold :*

a) PBH test :
$$\text{rank}[P_B(s)] = n , \quad \forall s$$

The set of eigenvalues of the unreachable part of (A, B) coincides with the set of values of s in correspondence of which matrix $P_B(s)$ looses rank.

b) Kalman test :
$$\text{rank}[K_r] = n$$

c) Wonham test : Given an arbitrary symmetric set Λ of n complex number, there exists a matrix K such that the spectrum of $A + BK$ coincides with Λ

Lemma D.4 (Stabilizability) *System Σ or, equivalently, the pair (A, B) is stabilizable if and only if the following equivalent conditions hold :*

a) PBH test :
$$\text{rank}[P_B(s)] = n , \quad \text{Re}(s) \geq 0$$

b) Kalman test : The unreachable part of the system is stable.

c) Wonham test : There exists a matrix K such that $A + BK$ is stable.

At the light of what said in Section 2.5 for the zeros of Σ, one can now stress that Σ is reachable (resp. observable) if and only if it does not possess input (resp. output) decoupling zeros, and, analogously, system Σ is stabilizable (resp. detectable) if and only if it does not possess input (resp. output) decoupling zeros in the closed right half plane.

Appendix E

The Standard 2-Block Scheme

The material presented in this appendix is largely taken from the paper by Doyle et al. [17]. However, the proof of Lemma E.3 is partially original.

Lemma E.1 *Consider the systems $P(s)$ and $\hat{P}(s) := P'(s)$, where*

$$P(s) := \left[\begin{array}{c|cc} A & B_1 & B_2 \\ \hline C_1 & D_{11} & D_{12} \\ C_2 & D_{21} & D_{22} \end{array}\right]$$

These two systems are connected in a feedback configuration to $K(s)$ and $\hat{K}(s)$, respectively, according to the block schemes of fig. E.1. Correspondingly, let $T(a,b;s)$ denote the transfer function from the generic input b to the generic output a. Then, given $K(s)$ and letting $\hat{K}(s) := K'(s)$ it follows $T(\hat{z}, \hat{w}; s) = T'(z, w; s)$. Conversely, given $\hat{K}(s)$ and letting $K(s) := \hat{K}'(s)$, it follows $T(z, w; s) = T'(\hat{z}, \hat{w}; s)$. Moreover, the eigenvalues of the two systems in fig. E.1 coincide.

Proof Let $P_{ij}(s)$, $i = j = 1, 2$, be the four transfer functions of $P(s)$. It results $T(\hat{z}, \hat{w}; s) = P'_{11}(s) + P'_{21}(s)\hat{K}(s)[I - P'_{22}(s)\hat{K}(s)]^{-1}P'_{12}(s)$. If $\hat{K}(s) = K'(s)$ it follows that $T(\hat{z}, \hat{w}; s) = T'(z, w; s)$. Conversely, it results $T(z, w; s) = P_{11}(s) + P_{12}(s)K(s)[I - P_{22}(s)K(s)]^{-1}P_{21}(s)$. If $K(s) = \hat{K}'(s)$ it follows that $T(z, w; s) = T'(\hat{z}, \hat{w}; s)$.

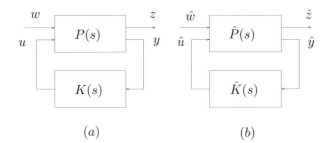

(a)

(b)

Figure E.1: A feedback system and its transpose

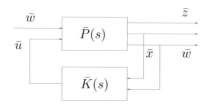

Figure E.2: A feedback system with full information

Now, let

$$K(s) = \left[\begin{array}{c|c} F & G \\ \hline H & E \end{array} \right]$$

be a controller such that the system of fig. E.1(a) is well defined, i.e. $\det[I - ED_{22}]$ is nonzero. It is easy to check that the dynamical matrix of the system of fig. E.1(a) is

$$A_a = \left[\begin{array}{cc} A + B_2(I - ED_{22})^{-1}EC_2 & B_2(I - ED_{22})^{-1}H \\ G[I + D_{22}(I - ED_{22})^{-1}E]C_2 & F + GD_{22}(I - ED_{22})^{-1}H \end{array} \right]$$

whereas that of system of fig. E.1(b) where $\hat{K}(s) = K'(s)$ is

$$A_b = \left[\begin{array}{cc} A' + C_2'(I - E'D_{22}')^{-1}E'B_2' & C_2'(I - E'D_{22}')^{-1}G' \\ H'[I + D_{22}'(I - E'D_{22}')^{-1}E']B_2' & F' + H'D_{22}'(I - E'D_{22}')^{-1}G' \end{array} \right]$$

Recalling Lemma B.9, it is easy to verify that $A_b' = A_a$ so that the two systems have the same eigenvalues.   □

Lemma E.2 *Consider the system*

$$\bar{P}(s) := \left[\begin{array}{c|cc} A & B_1 & B_2 \\ \hline C_1 & 0 & D_{12} \\ I & 0 & 0 \\ 0 & I & 0 \end{array} \right]$$

connected in feedback configuration with $\bar{K}(s)$ according to the block-scheme of fig. E.2. Moreover, consider system

$$\hat{P}(s) := \left[\begin{array}{c|cc} A & B_1 & B_2 \\ \hline C_1 & 0 & D_{12} \\ C_2 & I & 0 \end{array} \right]$$

connected in feedback configuration with $\hat{K}(s)$ as shown in fig. E.3. Finally suppose that matrix $A - B_1C_2$ is stable. Then,

  i) Given $\hat{K}(s)$ and letting $\bar{K}(s) := \hat{K}(s)[C_2\ I]$ it follows

   i1) $\bar{K}(s)$ stabilizes $\bar{P}(s)$ if and only if $\hat{K}(s)$ stabilizes $\hat{P}(s)$

   i2) $T(\hat{z}, \hat{w}; s) = T(\bar{z}, \bar{w}; s)$

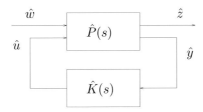

Figure E.3: A feedback system with partial information

ii) Given $\bar{K}(s)$ and defining $\hat{K}(s)$ through the block-scheme of fig. E.4 where

$$K_v(s) := \left[\begin{array}{c|cc} A - B_1C_2 & B_1 & B_2 \\ \hline 0 & 0 & I \\ I & 0 & 0 \\ -C_2 & I & 0 \end{array} \right]$$

it follows

ii1) $\hat{K}(s)$ stabilizes $\hat{P}(s)$ if and only if $\bar{K}(s)$ stabilizes $\bar{P}(s)$

ii2) $T(\hat{z}, \hat{w}; s) = T(z, w; s)$

Proof *Point i)* Let

$$\hat{K}(s) := \left[\begin{array}{c|c} \hat{A} & \hat{B} \\ \hline \hat{C} & \hat{D} \end{array} \right]$$

Then, by keeping in mind fig. E.3 and the definition of $\hat{P}(s)$ it follows

$$T(\hat{z}, \hat{w}; s) = \left[\begin{array}{cc|c} A + B_2\hat{D}C_2 & B_2\hat{C} & B_1 + B_2\hat{D} \\ \hat{B}C_2 & \hat{A} & \hat{B} \\ \hline C_1 + D_{12}\hat{D}C_2 & D_{12}\hat{C} & D_{12}\hat{D} \end{array} \right]$$

while $\bar{K}(s) = \hat{K}(s)[C_2 \ I]$ implies

$$\bar{K}(s) = \left[\begin{array}{c|cc} \hat{A} & \hat{B}C_2 & \hat{B} \\ \hline \hat{C} & \hat{D}C_2 & \hat{D} \end{array} \right]$$

It is easy to see (for instance by inspecting the state equations of the relevant systems with fig. E.2 and E.3 in mind) that the state representations of $T(\hat{z}, \hat{w}; s)$ and $T(\bar{z}, \bar{w}; s)$ are equal, so that point i) is proved.

Point ii) A state space representation of $T(\hat{z}, \hat{w}; s)$ in fig. E.3 is sought, when $\hat{K}(s)$ is built according to fig. E.4. Hence, let $\hat{x}$, $\bar{\mu}$ and x_v be the state variables of

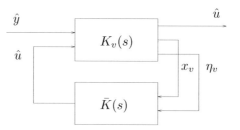

Figure E.4: A controller structure for the partial information case

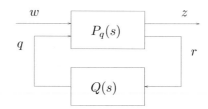

Figure E.5: A feedback system

$\hat{P}(s)$, $\bar{K}(s)$ and $K_v(s)$, respectively. A state description of $T(\hat{z}, \hat{w}; s)$ is then given by

$$\hat{P}(s) := \begin{cases} \dot{\hat{x}} &= A\hat{x} + B_1\hat{w} + B_2\hat{u} \\ \hat{z} &= C_1\hat{x} + D_{12}\hat{u} \\ \hat{y} &= C_2\hat{x} + \hat{w} \end{cases}$$

$$K_v(s) := \begin{cases} \dot{x}_v &= (A - B_1C_2)x_v + B_1\hat{y} + B_2\hat{u} \\ \eta_v &= -C_2x_v + \hat{y} \end{cases}$$

$$\bar{K}(s) := \begin{cases} \dot{\mu} &= \bar{A}\mu + \bar{B}_1 x_v + \bar{B}_2\eta_v \\ \hat{u} &= \bar{C}\mu + \bar{D}_1 x_v + \bar{D}_2\eta_v \end{cases}$$

Letting $\varepsilon := x_v - \hat{x}$ it results $\dot{\varepsilon} = (A - B_1C_2)\varepsilon$, which is by assumption a stable system. Then, in order to evaluate the stability of system of fig. E.3 and to compute the transfer function $T(\hat{z}, \hat{w}; s)$ one can, without loss of generality, put $\varepsilon = 0$, i.e. $x_v = \hat{x}$ in the relevant equations. It follows

$$T(\hat{z}, \hat{w}; s) = \left[\begin{array}{cc|c} A + B_2\bar{D}_1 & B_2\bar{C} & B_1 + B_2\bar{D}_2 \\ \bar{B}_1 & \bar{A} & \bar{B}_2 \\ \hline C_1 + D_{12}\bar{D}_1 & D_{12}\bar{C} & D_{12}\bar{D}_2 \end{array} \right]$$

On the other hand, it is simple to verify that the feedback system of fig. E.2 is described by the very same equations. Hence Point ii) immediately follows. □

Lemma E.3 *Let γ be a positive scalar and assume that there exists the symmetric, positive semidefinite solution P_∞ of the Riccati equation (in the unknown P)*

$$0 = PA_c + A_c'P + P(\gamma^{-2}B_1B_1' - B_2B_2')P + C_{1c}'C_{1c} \tag{E.1}$$

that is such that

$$A_{cc} := A_c + (\gamma^{-2}B_1B_1' - B_2B_2')P_\infty \tag{E.2}$$

is stable. Further, consider the block-scheme of fig. E.5 where

$$P_q(s) := \left[\begin{array}{c|cc} A_c - B_2 B_2' P_\infty & B_1 & B_2 \\ C_{1c} - D_{12} B_2' P_\infty & 0 & D_{12} \\ -\gamma^{-2} B_1' P_\infty & I & 0 \end{array} \right]$$

$$Q(s) := \left[\begin{array}{c|c} A_q & B_q \\ \hline C_q & D_q \end{array} \right]$$

with

$$D_{12}' C_{1c} = 0 \tag{E.3}$$
$$D_{12}' D_{12} = I \tag{E.4}$$

and $T(z, w; s) := \Sigma(A_z, B_z, C_z, D_z)$ is the transfer function from w to z. Then, A_z is stable and $\|T(z, w; s)\|_\infty < \gamma$ if and only if A_q is stable and $\|Q(s)\|_\infty < \gamma$.

Proof *Sufficiency* Let A_q be stable and $\|Q(s)\|_\infty < \gamma$. The system in fig. E.5 with transfer function $T(z, w; s)$ is described by

$$T(z, w; s) = \left[\begin{array}{c|c} A_z & B_z \\ \hline C_z & D_z \end{array} \right]$$

where

$$A_z := \left[\begin{array}{cc} A_c - B_2(B_2' + \gamma^{-2} D_q B_1') P_\infty & B_2 C_q \\ -\gamma^{-2} B_q B_1' P_\infty & A_q \end{array} \right] \tag{E.5}$$

$$B_z := \left[\begin{array}{c} B_1 + B_2 D_q \\ B_q \end{array} \right] \tag{E.6}$$

$$C_z := \left[\begin{array}{cc} C_{1c} - D_{12}(B_2' + \gamma^{-2} D_q B_1') P_\infty & D_{12} C_q \end{array} \right] \tag{E.7}$$

$$D_z := D_{12} D_q \tag{E.8}$$

Since $\|Q(s)\|_\infty < \gamma$, it results $\bar{\sigma}(D_q) < \gamma$. Then, by recalling that A_q is stable, Theorem 2.13, applied to system $\Sigma(A_q, B_q, C_q, D_q)$ entails that there exists the symmetric, positive semidefinite solution S_∞ of the Riccati equation (in the unknown S)

$$0 = S\hat{A}_q + \hat{A}_q' S + SB_q \Delta_{1q} B_q' S + C_q' \Delta_{2q} C_q \tag{E.9}$$

where

$$\Delta_{1q} := (\gamma^2 I - D_q' D_q)^{-1}, \quad \Delta_{2q} := (I - \gamma^{-2} D_q D_q')^{-1} \tag{E.10}$$

and

$$\hat{A}_q := A_q + B_q \Delta_{1q} D_q' C_q \tag{E.11}$$

Therefore, matrix

$$\tilde{A}_q := A_q + B_q \Delta_{1q}(D_q' C_q + B_q' S_\infty) \tag{E.12}$$

is stable. From the definition of D_z (eq. (E.8)) and eq. (E.4) it follows that

$$\bar{\sigma}(D_z) < \gamma \tag{E.13}$$

which is a necessary condition for $\|T(z, w; s)\|_\infty < \gamma$. It is now shown that

$$V_\infty := \begin{bmatrix} P_\infty & 0 \\ 0 & S_\infty \end{bmatrix}$$

is a symmetric and positive semidefinite solution of the Riccati equation (in the unknown V)

$$0 = V\hat{A}_z + \hat{A}'_z V + V B_z \Delta_{1z} B'_z V + C'_z \Delta_{2z} C_z \tag{E.14}$$

where

$$\Delta_{1z} := (\gamma^2 I - D'_z D_z)^{-1} , \quad \Delta_{2z} := (I - \gamma^{-2} D_z D'_z)^{-1} \tag{E.15}$$

and

$$\hat{A}_z := A_z + B_z \Delta_{1z} D'_z C_z \tag{E.16}$$

In fact, preliminarily observe that the right hand side of the Riccati equation (E.14) can be rewritten (if eqs.(E.15),(E.16) and Lemma B.9 are taken into account) as

$$V A_z + A'_z V + (V B_z + C'_z D_z)\Delta_{1z}(V B_z + C'_z D_z)' + C'_z C_z :=$$
$$:= f(V) := \begin{bmatrix} f_{11}(V) & f_{12}(V) \\ f_{21}(V) & f_{22}(V) \end{bmatrix}$$

In view of eqs. (E.5)-(E.8) it follows

$$V_\infty A_z = \begin{bmatrix} P_\infty A_c - P_\infty B_2 B'_2 P_\infty - \gamma^{-2} P_\infty B_2 D_q B'_1 P_\infty & P_\infty B_2 C_q \\ -\gamma^{-2} S_\infty B_q B'_1 P_\infty & S_\infty A_q \end{bmatrix}$$

and, by recalling eqs. (E.3),(E.4),

$$V_\infty B_z + C'_z D_z = \begin{bmatrix} \gamma^{-2} P_\infty B_1 \Delta_{1q}^{-1} \\ S_\infty B_q + C'_q D_q \end{bmatrix} \tag{E.17}$$

$$C'_z C_z = \begin{bmatrix} C_{z1} & C_{z2} \\ C'_{z2} & C_{z3} \end{bmatrix}$$

where

$$C_{z1} := C'_{1c} C_{1c} + P_\infty B_2 B'_2 P_\infty + \gamma^{-4} P_\infty B_1 D'_q D_q B'_1 P_\infty +$$
$$+\gamma^{-2} P_\infty B_2 D_q B'_1 P_\infty + \gamma^{-2} P_\infty B_1 D'_q B'_2 P_\infty$$
$$C_{z2} := -P_\infty B_2 C_q - \gamma^{-2} P_\infty B_1 D'_q C_q$$
$$C_{z3} := C'_q C_q$$

Thanks to eqs. (E.4),(E.8),(E.10) and (E.15) it is

$$\Delta_{1z} = \Delta_{1q} \tag{E.18}$$

so that

$$f_{11}(V_\infty) = P_\infty A_c + A'_c P_\infty - P_\infty(B_2 B'_2 - \gamma^{-2} B_1 B'_1)P_\infty +$$
$$+C'_{1c} C_{1c} = 0$$

since matrix P_∞ is a solution of eq. (E.1). By direct substitution one finds $f_{12}(V_\infty) = 0$. Finally,

$$f_{22}(V_\infty) = S_\infty A_q + A'_q S_\infty + (S_\infty B_q + C'_q D_q)\Delta_{1q} \cdot$$
$$\cdot (S_\infty B_q + C'_q D_q)' + C'_q C_q$$

By recalling Lemma B.9 and eqs. (E.10),(E.11) it is easy to check that $f_{22}(V_\infty)$ coincides with the right hand side of eq. (E.9) and is therefore zero.

Now, let

$$\tilde{A}_z := A_z + B_z \Delta_{1z}(D'_z C_z + B'_z V_\infty)$$
$$= \begin{bmatrix} A_c - B_2 B'_2 P_\infty + \gamma^{-2} B_1 B'_1 P_\infty & \Phi \\ 0 & A_q + B_q \Delta_{1z}(B'_q S_\infty + D'_q C_q) \end{bmatrix}$$

where Φ denotes a matrix of no interest in this context. In view of eqs. (E.2),(E.12) and (E.18), the conclusion is drawn that the eigenvalues of $\tilde{A}_z$ are those of A_{cc} and $\tilde{A}_q$. The solution V_∞ of eq. (E.14) is therefore the stabilizing one since matrices A_{cc} and $\tilde{A}_q$ are stable. Thus, by recalling eq. (E.13), it can be said that, thanks to Theorem 2.14, $\|T(z,w;s)\|_\infty < \gamma$ and A_z is stable.

Necessity Assume that $\|T(z,w;s)\|_\infty < \gamma$ and A_z is stable. It has to be proved that A_q is stable and $\|Q(s)\|_\infty < \gamma$. To this aim, consider the Hamiltonian matrix

$$Z_z := \begin{bmatrix} \hat{A}_z & B_z \Delta_{1z} B'_z \\ -C'_z \Delta_{2z} C_z & -\hat{A}'_z \end{bmatrix}$$

and the relevant Riccati equation (in the unknown V)

$$0 = V\hat{A}_z + \hat{A}'_z V + V B_z \Delta_{1z} B'_z V + C'_z \Delta_{2z} C_z \qquad (E.19)$$

In view of eqs. (E.15) and (E.16) matrix Z_z is associated with system $T(z,w;s) = \Sigma(A_z, B_z, C_z, D_z)$ defined by eqs. (E.5)-(E.8). By taking into account the equations (E.3),(E.4),(E.1),(E.10) and (E.11), it will be now verified that, chosen

$$\Psi := \begin{bmatrix} I & 0 \\ -\Omega & I \end{bmatrix}, \quad \Omega := \begin{bmatrix} P_\infty & 0 \\ 0 & 0 \end{bmatrix}$$

it is $\Psi Z_z \Psi^{-1} = \hat{Z}_z$, where

$$\hat{Z}_z = \begin{bmatrix} \Phi_1 & \Phi_2 & \Phi_3 & \Phi_4 \\ 0 & \hat{A}_q & \Phi_5 & B_q \Delta_{1q} B'_q \\ 0 & 0 & \Phi_6 & 0 \\ 0 & -C'_q \Delta_{2q} C_q & \Phi_7 & -\hat{A}'_q \end{bmatrix} \qquad (E.20)$$

The explicit expressions of the matrices $\Phi_i, i = 1, \cdots, 7$ appearing in $\hat{Z}_z$ are not given since of no interest in the subsequent discussion. In order to verify eq. (E.20) observe that by recalling eqs. (E.15),(E.16) and Lemma B.9, it is

$$\Psi Z_z \Psi^{-1} = \begin{bmatrix} \zeta_{11} & \zeta_{12} \\ \zeta_{21} & -\zeta'_{11} \end{bmatrix}$$

where

$$\zeta_{11} := A_z + B_z \Delta_{1z}(D'_z C_z + B'_z \Omega), \quad \zeta_{12} := B_z \Delta_{1z} B'_z$$
$$\zeta_{21} := -\Omega A_z - A'_z \Omega - (\Omega B_z + C'_z D_z)\Delta_{1z}(\Omega B_z + C'_z D_z)' - C'_z C_z$$

Further, by taking into account eqs. (E.3)-(E.8) and (E.10), one has

$$\Omega B_z + C_z' D_z = \left[\begin{array}{c} \gamma^{-2} P_\infty B_1 \Delta_{1q}^{-1} \\ C_q' D_q \end{array} \right]$$

so that, thanks to eq. (E.18),

$$\zeta_{11} = \left[\begin{array}{cc} \Phi_1 & \Phi_2 \\ 0 & A_q + B_q \Delta_{1q} D_q' C_q \end{array} \right]$$

In view of eqs. (E.11) and (E.20) it follows that ζ_{11} coincides with the submatrix of $\hat{Z}_z$ made up of the elements belonging to the first two blocks of rows and columns. In a similar way, thanks to eqs. (E.18) and (E.6),

$$\zeta_{12} = \left[\begin{array}{cc} \Phi_3 & \Phi_4 \\ \Phi_5 & B_q \Delta_{1q} B_q' \end{array} \right]$$

which coincides with the submatrix of $\hat{Z}_z$ made up of the elements belonging to the first two blocks of rows and the last two blocks of columns. Finally, one obtains,

$$\zeta_{21} = \left[\begin{array}{cc} g_1 & 0 \\ 0 & g_2 \end{array} \right]$$

with

$$g_1 := -[P_\infty A_c + A_c' P_\infty - P_\infty (B_2 B_2' - \gamma^{-2} B_1 B_1') P_\infty + C_{1c}' C_{1c}]$$
$$g_2 := -C_q' C_q - C_q' D_q \Delta_{1q} D_q' C_q$$

By recalling eq. (E.1) it results $g_1 = 0$, whereas from Lemma B.9 it follows $g_2 = -C_q' \Delta_{2q} C_q$. Therefore, ζ_{21} coincides with the submatrix of $\hat{Z}_z$ made up of the elements belonging to the last two blocks of rows and the first two blocks of columns.

In conclusion eq. (E.20) has been proved. Now notice that the eigenvalues of $\hat{Z}_z$ are those of the matrices Φ_1, Φ_6 and

$$Z_q := \left[\begin{array}{cc} \hat{A}_q & B_q \Delta_{1q} B_q' \\ -C_q' \Delta_{2q} C_q & -\hat{A}_q' \end{array} \right]$$

Condition (E.13) is satisfied since $\|T(z, w; s)\|_\infty < \gamma$ and matrix A_z is stable so that from Theorem 2.13 it follows that the eigenvalues of Z_z do not lie on the imaginary axis. Thus also the eigenvalues of $\hat{Z}_z$ do not have zero real part and the same can be said of the eigenvalues of Z_q, thanks to the above discussion.

Since the resulting system is stable (A_z is stable), the pair (A_q, B_q) is stabilizable and such is the pair $(\hat{A}_q, (B_q \Delta_{1q} B_q')^{1/2})$ too. In fact, first recall that in view of eq. (E.11) the stabilizability of the pair (A_q, B_q) implies that also the pair $(\hat{A}_q, B_q)$ is such (state feedback does not affect this property). Second, notice that if $(B_q \Delta_{1q} B_q')^{1/2} x = 0$, then also $B_q \Delta_{1q} B_q' x = 0$ and $B_q' x = 0$ (Δ_{1q} is positive definite), so that, in view of the PBH test (Lemma D.4), should the pair $(\hat{A}_q, (B_q \Delta_{1q} B_q')^{1/2})$ not be stabilizable then also the pair $(\hat{A}_q, B_q)$ would not be such.

Therefore Lemma C.3 can be applied to system $\Sigma(A_q, B_q, C_q, D_q)$ yielding the existence of the symmetric, positive semidefinite and stabilizing solution S_∞ of the Riccati equation (in the unknown S)

$$0 = S\hat{A}_q + \hat{A}_q' S + S B_q \Delta_{1q} B_q' S + C_q' \Delta_{2q} C_q \tag{E.21}$$

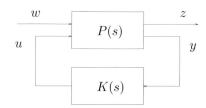

Figure E.6: A feedback system

The same lemma can be applied to eq. (E.19) since: (i) the pair $(\hat{A}_z, B_z)$ is stabilizable (indeed, matrix A_z is stable); (ii) the eigenvalues of matrix Z_z do not have zero real part; (iii) matrix $C'_z \Delta_{2z} C_z$ is symmetric; (iv) matrix $B_z \Delta_{1z} B'_z$ is positive semidefinite $(\Delta_{1z} > 0$ thanks to Lemma B.11 and $\bar{\sigma}(D_z) < \gamma$ $(\|T(z, w; s)\|_\infty < \gamma)$. Therefore, the stabilizing solution of eq. (E.19), which is unique in view of Lemma C.3, is given (see the sufficiency part of the proof of the lemma) by

$$V_\infty = \begin{bmatrix} P_\infty & 0 \\ 0 & S_\infty \end{bmatrix}$$

Matrix V_∞ is positive semidefinite (see Theorem 2.13), hence also S_∞ is such.

Finally, Theorem 2.14 applied to system $Q(s)$ (notice that eqs. (E.4) and (E.8) imply $\bar{\sigma}(D_q) = \bar{\sigma}(D_z)$) and eq. (E.21) lead to the desired conclusion, namely A_q stable and $\|Q(s)\|_\infty < \gamma$. □

Lemma E.4 *Consider the system*

$$P(s) := \left[\begin{array}{c|cc} A & B_1 & B_2 \\ \hline C_1 & 0 & D_{12} \\ C_2 & D_{21} & 0 \end{array} \right]$$

with $D'_{12} D_{12} = I$ and, for a given $\gamma > 0$, assume that there exists the symmetric, positive semidefinite and stabilizing solution P_∞ of the Riccati equation (in the unknown P)

$$0 = PA_c + A'_c P + P(\gamma^{-2} B_1 B'_1 - B_2 B'_2)P + C'_{1c} C_{1c} \tag{E.22}$$

that is such that

$$A_{cc} := A_c + (\gamma^{-2} B_1 B'_1 - B_2 B'_2)P_\infty \tag{E.23}$$

is stable, where

$$A_c := A - B_2 D'_{12} C_1, \quad C_{1c} := (I - D_{12} D'_{12})C_1 \tag{E.24}$$

Further, consider the block-scheme in fig. E.6 and E.7, where

$$P_t(s) := \left[\begin{array}{c|cc} A + \gamma^{-2} B_1 B'_1 P_\infty & B_1 & B_2 \\ \hline B'_2 P_\infty + D'_{12} C_1 & 0 & I \\ C_2 + \gamma^{-2} D_{21} B'_1 P_\infty & D_{21} & 0 \end{array} \right]$$

and let $T(z, w; s)$ and $T(q, r; s)$ be the transfer functions from w to z and from r to q, respectively. Then, a controller $K(s)$ is admissible in RH_∞ for $P(s)$ and such that $\|T(z, w; s)\|_\infty < \gamma$ if and only if it is admissible in RH_∞ for $P_t(s)$ and such that $\|T(q, r; s)\|_\infty < \gamma$.

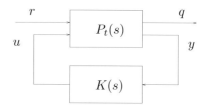

Figure E.7: An auxiliary feedback system

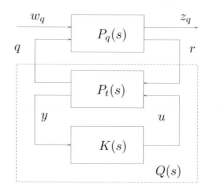

Figure E.8: A second auxiliary feedback system

Proof Consider the block-scheme in fig. E.8 where

$$P_q(s) := \left[\begin{array}{cc|cc} A_c - B_2 B_2' P_\infty & & B_1 & B_2 \\ C_{1c} - D_{12} B_2' P_\infty & & 0 & D_{12} \\ -\gamma^{-2} B_1' P_\infty & & I & 0 \end{array} \right]$$

and let x_q and x_t be the state variables of $P_q(s)$ and $P_t(s)$, respectively. It is easy to check that, by letting $\varepsilon := x_q - x_t$, it results

$$\dot{\varepsilon} = A_{cc}\varepsilon$$
$$\dot{x}_q = Ax_q + B_1 w_q + B_2 u - B_2(B_2' P_\infty + D_{12}' C_1)\varepsilon$$
$$z_q = C_1 x_q + D_{12} u - D_{12}(B_2' P_\infty + D_{12}' C_1)\varepsilon$$
$$y = C_2 x_q + D_{21} w_q - (C_2 + \gamma^{-2} D_{21} B_1' P_\infty)\varepsilon$$

Being matrix A_{cc} stable, the system in fig. E.8 is internally stable if and only if the system in fig. E.6 is such, since the equations for system $P(s)$ coincide with the last ones with $\varepsilon = 0$. For the same reason $T(z, w; s) = T(z_q, w_q; s)$, where $T(z_q, w_q; s)$ is the transfer function from w_q to z_q of the system in fig. E.8.

Thus the lemma is proved in view of Lemma E.3. Indeed, the system in fig. E.8 is equal to the one depicted in fig. E.5 if in the former figure the system with input r and output q is denoted by $Q(s)$. Further, the assumptions required by Lemma E.3 are satisfied.   □

Appendix F

Loop Shifting

This appendix presents some results which are (mainly) exploited in Remark 5.24. In particular, Lemmas F.1 and F.2 are almost standard in matrix algebra, while the remaining material has been taken from the paper by Parrott [46].

Lemma F.1 *Let A be a $m \times n$ dimensional matrix with $\|A\| \leq 1$. Then*

$$A^\sim (I - AA^\sim)^{1/2} = (I - A^\sim A)^{1/2} A^\sim$$

Proof Preliminarily observe that both the matrices $I - AA^\sim$ and $I - A^\sim A$ have real square roots since they are positive semidefinite thanks to the assumption on $\|A\|$ which implies the nonnegativity of their eigenvalues. In fact, if, for instance, $(I - AA^\sim)x = \lambda x$ with $\lambda < 0$ and $x \neq 0$, then $x^\sim AA^\sim x = x^\sim x(1 - \lambda) > x^\sim x$ which implies $\|A\| > 1$.

The proof of the lemma is carried out by assuming $m \geq n$, the reverse case being dealt with by modifying the subsequent discussion in an obvious way. By recalling Remark 2.11 it is possible to write $A = U_1 \Delta V^\sim$ with $U_1 = U[I\ 0]'$, $U^\sim U = UU^\sim = I$, $U_1^\sim U_1 = I$, $V^\sim V = VV^\sim = I$ and Δ diagonal and real. Therefore,

$$I - A^\sim A = I - V\Delta U_1^\sim U_1 \Delta V^\sim = I - V\Delta^2 V^\sim = V(I - \Delta^2)V^\sim = VD^2V^\sim$$

where $D^2 := I - \Delta^2$ is diagonal. Then

$$(I - A^\sim A)^{1/2} A^\sim = VDV^\sim V\Delta U_1^\sim = VD\Delta U_1^\sim = V\Delta DU_1^\sim$$

having exploited the fact that $VD^2V^\sim = VDV^\sim VDV^\sim$ (first equality sign) and the fact that the product of two diagonal matrices commutes (last equality sign). Being $U_1^\sim = [I\ 0]U^\sim$ it follows

$$(I - A^\sim A)^{1/2} A^\sim = V\Delta \begin{bmatrix} D & 0 \end{bmatrix} U^\sim \tag{F.1}$$

On the other hand, by again exploiting the relation between U and U_1, it is

$$I - AA^\sim = I - U_1 \Delta V^\sim V\Delta U_1^\sim$$
$$= I - U_1 \Delta^2 U_1^\sim$$
$$= I - U \begin{bmatrix} I \\ 0 \end{bmatrix} \Delta^2 \begin{bmatrix} I & 0 \end{bmatrix} U^\sim$$

$$= I - U \begin{bmatrix} \Delta^2 & 0 \\ 0 & 0 \end{bmatrix} U^{\sim}$$

$$= U \left[I - \begin{bmatrix} \Delta^2 & 0 \\ 0 & 0 \end{bmatrix} \right] U^{\sim}$$

$$= U \begin{bmatrix} D^2 & 0 \\ 0 & I \end{bmatrix} U^{\sim}$$

so that

$$A^{\sim}(I - AA^{\sim})^{1/2} = V\Delta U_1^{\sim}U \begin{bmatrix} D & 0 \\ 0 & I \end{bmatrix} U^{\sim}$$

$$= V\Delta \begin{bmatrix} I & 0 \end{bmatrix} U^{\sim}U \begin{bmatrix} D & 0 \\ 0 & I \end{bmatrix} U^{\sim}$$

$$= V\Delta \begin{bmatrix} I & 0 \end{bmatrix} \begin{bmatrix} D & 0 \\ 0 & I \end{bmatrix} U^{\sim}$$

$$= V\Delta \begin{bmatrix} D & 0 \end{bmatrix} U^{\sim} \tag{F.2}$$

From eqs. (F.1),(F.2) the lemma follows.   □

Lemma F.2 *Let A be a $m \times n$ dimensional matrix with $\mathrm{rank}[A] = r$. Further, let $A_R := (A^{\sim}A)^{1/2}$ and $A_L := (AA^{\sim})^{1/2}$. Then there exists a $m \times n$ dimensional matrix T such that*

a) $A = TA_R = A_L T$.

b) $A_R T^{\sim} T = A_R$.

c) $T^{\sim} T \leq I$.

Proof Consider the singular value decomposition (see Section 2.6) of A, namely $A = USV^{\sim}$ with

$$U := \begin{bmatrix} U_1 & U_2 \end{bmatrix} , \quad S := \begin{bmatrix} D & 0 \\ 0 & 0 \end{bmatrix}$$

where U_1 is $m \times r$ dimensional and D is $r \times r$ dimensional. Define

$$S_R := \begin{bmatrix} D & 0 \\ 0 & 0 \end{bmatrix} , \quad S_L := \begin{bmatrix} D & 0 \\ 0 & 0 \end{bmatrix}$$

where S_R and S_L are square and $n \times n$ and $m \times m$ dimensional, respectively, so that $S_R^2 = S'S$ and $S_L^2 = SS'$. Then it is now proved that

$$A_R = VS_R V^{\sim} , \quad A_L = US_L U^{\sim} , \quad T = \begin{bmatrix} U_1 & 0 \end{bmatrix} V^{\sim} \tag{F.3}$$

verify the theorem. First observe that

$$A_R^{\sim} A_R = VS_R V^{\sim} V S_R V^{\sim}$$
$$= VS_R^2 V^{\sim}$$
$$= VS'SV^{\sim}$$
$$= VS'U^{\sim}USV^{\sim} = A^{\sim}A$$

Thus A_R is a square root of $A^\sim A$. Second, notice that

$$A_L^\sim A_L = U S_L U^\sim U S_L U^\sim$$
$$= U S_L^2 U^\sim$$
$$= U S S' U^\sim$$
$$= U S V^\sim V S' U^\sim = A A^\sim$$

Thus, A_L is a square root of $A A^\sim$.

Point a) From eq. (F.3) it follows that

$$T A_R = \begin{bmatrix} U_1 & 0 \end{bmatrix} V^\sim V S_R V^\sim$$
$$= \begin{bmatrix} U_1 & 0 \end{bmatrix} S_R V^\sim$$
$$= \begin{bmatrix} U_1 & U_2 \end{bmatrix} S V^\sim = A$$

and

$$A_L T = U S_L U^\sim \begin{bmatrix} U_1 & 0 \end{bmatrix} V^\sim = U S_L \begin{bmatrix} I & 0 \\ 0 & 0 \end{bmatrix} V^\sim = A$$

Point b) It is

$$A_R T^\sim T = V S_R V^\sim V \begin{bmatrix} U_1^\sim \\ 0 \end{bmatrix} \begin{bmatrix} U_1 & 0 \end{bmatrix} V^\sim$$
$$= V S_R \begin{bmatrix} I & 0 \\ 0 & 0 \end{bmatrix} V^\sim$$
$$= V S_R V^\sim = A_R$$

Point c) It is

$$T^\sim T = V \begin{bmatrix} U_1^\sim \\ 0 \end{bmatrix} \begin{bmatrix} U_1 & 0 \end{bmatrix} V^\sim$$
$$= V \begin{bmatrix} I & 0 \\ 0 & 0 \end{bmatrix} V^\sim$$
$$\leq V \begin{bmatrix} I & 0 \\ 0 & I \end{bmatrix} V^\sim = I$$

$\square$

Lemma F.3 *Consider the real matrix*

$$M(Z) := \begin{bmatrix} P & Q \\ R & Z \end{bmatrix}$$

where the submatrices P, Q and R satisfy the equations

$$P'P + R'R = \alpha^2 I$$
$$PP' + QQ' = \alpha^2 I$$

with α a nonnegative scalar. Then

i)

$$\min_Z \|M(Z)\| = \alpha$$

ii) Letting $Z^o := -T_R P' T_Q$, where T_R and T_Q are matrices such that the equalities $R = T_R (R'R)^{1/2}$ and $Q = (QQ')^{1/2} T_Q$ hold, it is $\|M(Z^o)\| = \alpha$.

Proof Preliminarily observe that matrices T_R and T_Q exist thanks to Lemma F.2.
If $\alpha = 0$, then, necessarily, $P = 0$, $Q = 0$, $R = 0$, hence $Z^o = 0$ and $\|M(Z^o)\| = 0$.
If $\alpha > 0$, set $A := P/\alpha$, $B := Q/\alpha$, $C := R/\alpha$, so that

$$A'A + C'C = I \tag{F.4}$$
$$AA' + BB' = I \tag{F.5}$$

Moreover, for any real D set

$$N(D) := \begin{bmatrix} A & B \\ C & D \end{bmatrix}$$

By recalling the definition of the norm of a matrix and the relevant properties (see Section 2.7), it follows, for each D,

$$\|N(D)\|^2 = \max_{\begin{bmatrix} x \\ y \end{bmatrix} \neq 0} \frac{\begin{bmatrix} x' & y' \end{bmatrix} \begin{bmatrix} A'A + C'C & A'B + C'D \\ B'A + D'C & B'B + D'D \end{bmatrix} \begin{bmatrix} x \\ y \end{bmatrix}}{\left\| \begin{bmatrix} x \\ y \end{bmatrix} \right\|^2}$$

$$\geq \max_{x \neq 0} \frac{x'(A'A + C'C)x}{\|x\|^2} = \left\| \begin{bmatrix} A \\ C \end{bmatrix} \right\|^2$$

Analogously,

$$\|N(D)\|^2 = \|N'(D)\|^2 \geq \left\| \begin{bmatrix} A' \\ B' \end{bmatrix} \right\|^2$$

Hence the conclusion is drawn that

$$\|N(D)\| \geq \max \left[\left\| \begin{bmatrix} A' \\ B' \end{bmatrix} \right\|, \left\| \begin{bmatrix} A \\ C \end{bmatrix} \right\| \right] = 1 , \quad \forall D \tag{F.6}$$

It is now shown that, chosen

$$D^o := -T_C A' T_B \tag{F.7}$$

where T_C and T_B are matrices (whose existence is guaranteed by Lemma F.2) such that

$$C = T_C (C'C)^{1/2} \tag{F.8}$$
$$B = (BB')^{1/2} T_B \tag{F.9}$$

it is $\|N(D^o)\| \leq 1$, so that, in view of eq. (F.6), $\|N(D^o)\| = 1$ and D^o is optimal.
First notice that from eqs. (F.4),(F.5), (F.7)-(F.9), Lemmas F.1, F.2 (points *a*) and *b*)), it follows

$$\begin{aligned} C'D^o &= -(C'C)^{1/2} T_C' T_C A' T_B \\ &= -(C'C)^{1/2} A' T_B \\ &= -(I - A'A)^{1/2} A' T_B \\ &= -A'(I - AA')^{1/2} T_B \\ &= -A'(BB')^{1/2} T_B = -A'B \end{aligned}$$

Therefore, $C'D^o + A'B = 0$ and, by recalling eq. (F.4), one obtains

$$N'(D^o)N(D^o) = \begin{bmatrix} I & 0 \\ 0 & B'B + (D^o)'D^o \end{bmatrix}$$

However, by exploiting Lemma F.2 (points a) and c)), eqs. (F.5), (F.7) it is

$$\begin{aligned} B'B + (D^o)'D^o &= T'_B(BB')^{1/2}(BB')^{1/2}T_B + T'_B AT'_C T_C A'T_B \\ &\leq T'_B BB'T_B + T'_B AA'T_B \\ &\leq T'_B(BB' + AA')T_B \\ &\leq T'_B T_B \leq I \end{aligned}$$

This implies that the maximum eigenvalue of $B'B + (D^o)'D^o$ is not greater than 1: consequently the maximum eigenvalue of $N'(D^o)N(D^o)$ is equal to 1. Therefore, $\|N(D^o)\| = 1$.

Now observe that for all D, letting $Z := \alpha D$, it is $\alpha N(D) = M(\alpha D) = M(Z)$, so that

$$\min_Z \|M(Z)\| = \min_D \|M(\alpha D)\| = \alpha \min_D \|N(D)\|$$

and

$$Z^o = \alpha D^o \tag{F.10}$$

On the other side, $R = \alpha C = \alpha T_C(C'C)^{1/2} = T_C(R'R)^{1/2}$ and, analogously, $Q = \alpha B = \alpha(BB')^{1/2}T_B = (QQ')^{1/2}T_B$ which imply $T_C = T_R$ and $T_B = T_Q$. From eqs. (F.10) and (F.7) and the definition of A the expression of Z^o follows. □

Theorem F.1 (Parrott's theorem) *Let B, C, D, P, X be real matrices such that*
$$N(X) := \begin{bmatrix} P & C \\ B & D + X \end{bmatrix} \text{ and } \alpha := \max\left(\left\| \begin{bmatrix} P \\ B \end{bmatrix} \right\|, \left\| \begin{bmatrix} P' \\ C' \end{bmatrix} \right\|\right). \text{ Then}$$

i) There exists X^o such that $\alpha = \|N(X^o)\| = \min_X \|N(X)\|$.

ii) If $\|P\| < \alpha$, $X^o := -D - B(\alpha^2 I - P'P)^{-1}P'C$.

Proof Preliminarily, notice that, in view of the definition of α, the matrices

$$R := \begin{bmatrix} B \\ (\alpha^2 I - P'P - B'B)^{1/2} \end{bmatrix}$$
$$Q := \begin{bmatrix} C & (\alpha^2 I - PP' - CC')^{1/2} \end{bmatrix}$$

are real.

Point i) Letting

$$Z := \begin{bmatrix} Z_1 & Z_2 \\ Z_3 & Z_4 \end{bmatrix} \tag{F.11}$$

with

$$Z_1 := D + X \tag{F.12}$$

define the matrix

$$M(Z) := \begin{bmatrix} P & Q \\ R & Z \end{bmatrix}$$

Matrix $M(Z)$ satisfies the assumptions of Lemma F.3 since

$$P'P + R'R = \alpha^2 I \tag{F.13}$$
$$PP' + QQ' = \alpha^2 I \tag{F.14}$$

so that there exists a matrix

$$Z^o := \begin{bmatrix} Z_1^o & Z_2^o \\ Z_3^o & Z_4^o \end{bmatrix}$$

such that $\|M(Z^o)\| = \alpha$.

Note that for any matrix Z of the form (F.11),(F.12) it is $\|N(Z_1 - D)\| \leq \|M(Z)\|$. Hence, by recalling that $\|N(X)\| \geq \alpha$, one obtains

$$\alpha \leq \|N(Z_1^o - D)\| \leq \|M(Z^o)\| = \alpha$$

which proves point *i*).

Point ii) Preliminarily observe that, thanks to the assumption on the norm of P, matrix $\alpha^2 I - P'P$ is nonsingular, so that X^o is well defined.

By exploiting Lemma F.3 one has

$$Z_1^o = \begin{bmatrix} I & 0 \end{bmatrix} Z^o \begin{bmatrix} I \\ 0 \end{bmatrix}$$

$$= - \begin{bmatrix} I & 0 \end{bmatrix} T_R P' T_Q \begin{bmatrix} I \\ 0 \end{bmatrix}$$

$$= \begin{bmatrix} I & 0 \end{bmatrix} T_R P'(QQ')^{-1/2} Q \begin{bmatrix} I \\ 0 \end{bmatrix} \tag{F.15}$$

In writing down the last part of eq. (F.15) the nonsingularity of matrix $(QQ')^{1/2} = (\alpha^2 I - PP')^{1/2}$ ($\|P\| < \alpha$) and the equality $Q = (QQ')^{1/2} T_Q$ have been taken into account. In view of eqs. (F.13),(F.14) and Lemma F.1, from eq. (F.15) it follows

$$Z_1^o = - \begin{bmatrix} I & 0 \end{bmatrix} T_R P'(\alpha^2 I - PP')^{-1/2} C$$

$$= - \begin{bmatrix} I & 0 \end{bmatrix} T_R (\alpha^2 I - P'P)^{-1/2} P'C$$

$$= - \begin{bmatrix} I & 0 \end{bmatrix} T_R (R'R)^{-1/2} P'C$$

$$= - \begin{bmatrix} I & 0 \end{bmatrix} R(R'R)^{-1} P'C$$

$$= -B(\alpha^2 I - P'P)^{-1} P'C$$

By recalling that $\|N(Z_1^o)\| = \alpha$, point *ii*) follows.   □

Appendix G

Worst Case Analysis

This appendix makes basically reference to the paper by Doyle et al. [17].

Lemma G.1 *Consider the system*

$$\dot{x} = Ax + Bq$$
$$z = Cx$$

with A stable, $x(0) = x_0$ and $\|G(s)\|_\infty < \gamma$, where $G(s) := C(sI - A)^{-1}B$. Then

$$\sup_{q \in RH_2} [\|z\|_2^2 - \gamma^2 \|q\|_2^2] = x_0' Q_\infty x_0$$

where Q_∞ is the symmetric, positive semidefinite and stabilizing solution of the Riccati equation (in the unknown Q)

$$0 = QA + A'Q + \gamma^{-2}QBB'Q + C'C$$

that is such that the matrix $A + \gamma^{-2}BB'Q_\infty$ is stable.

Proof The existence of the solution of the Riccati equation is guaranteed by the assumption $\|G(s)\|_\infty < \gamma$ (Theorem 2.13). Further, it is

$$\frac{d}{dt} x'Q_\infty x = x'A'Q_\infty x + q'B'Q_\infty x + x'Q_\infty Ax + x'Q_\infty Bq$$
$$= x'(Q_\infty A + A'Q_\infty)x + q'B'Q_\infty x + x'Q_\infty Bq$$
$$= -x'(C'C + \gamma^{-2}Q_\infty BB'Q_\infty)x + q'B'Q_\infty x + x'Q_\infty Bq$$
$$= -x'C'Cx - (\gamma q - \gamma^{-1}B'Q_\infty x)'(\gamma q - \gamma^{-1}B'Q_\infty x) + \gamma^2 q'q$$
$$= -\|z\|^2 + \gamma^2 \|q\|^2 - \gamma^2 \|q - \gamma^{-2}B'Q_\infty x\|^2$$

Observe that $x \in RH_2$ and $z \in RH_2$ whenever $q \in RH_2$, since A is stable. Therefore, by integrating over the interval $[0, \infty)$ the first and last terms of the this series of equalities one obtains, for each $q \in RH_2$,

$$\|z\|_2^2 - \gamma^2 \|q\|_2^2 = x_0' Q_\infty x_0 - \gamma^2 \|q - \gamma^{-2}B'Q_\infty x\|_2^2 \le x_0' Q_\infty x_0$$

The choice $q = \gamma^{-2}B'Q_\infty x$ is consistent with the above derivation since $\dot{x} = (A + \gamma^{-2}BB'Q_\infty)x$ is a stable system by assumption. With this choice it is $\|z\|_2^2 - \gamma^2 \|q\|_2^2 = x_0' Q_\infty x_0$, the maximum possible value. $\square$

In the following lemma reference is made to the stable (Π_s) and antistable (Π_a) orthogonal projections which have been introduced in Definition 2.33.

Lemma G.2 *Consider the system*

$$\dot{x} = Ax + B_1 q_1 + B_2 q_2 \tag{G.1}$$

$$z = Cx \tag{G.2}$$

with A stable. Further, consider the subspace of RL_2

$$\mathcal{Q} := \{[q_1' \ q_2']' := q \ , \ q_1 \in RH_2^\perp \ , \ q_2 \in RL_2\}$$

and the operator $\Xi : \mathcal{Q} \to RH_2$ defined by

$$\Xi : q \mapsto \Xi q := \Pi_s[G_1(s)q_1 + G_2(s)q_2]$$

where, for $i = 1, 2$, $G_i(s) := C(sI - A)^{-1}B_i$. Then, denoting with γ a positive scalar,

$$\sup_{\substack{q \in \mathcal{Q} \\ \|q\|_2 = 1}} \|\Xi q\|_2 < \gamma \tag{G.3}$$

if and only if the following two conditions hold:

i) *There exists the symmetric, positive semidefinite and stabilizing solution Q_∞ of the Riccati equation (in the unknown Q)*

$$0 = QA + A'Q + \gamma^{-2}QB_2B_2'Q + C'C$$

   that is such that the matrix $A + \gamma^{-2}B_2B_2'Q_\infty$ is stable.

ii) *$r_s(Q_\infty L_r) < \gamma^2$, where L_r is the solution of the Lyapunov equation (in the unknown L)*

$$0 = LA' + AL + B_1B_1' + B_2B_2'$$

Proof Observe that

$$\sup_{\substack{q \in \mathcal{Q} \\ \|q\|_2 = 1}} \|\Xi q\|_2 \geq \sup_{\substack{q \in \mathcal{Q} \\ \|q\|_2 = 1 \ , \ q_1 = 0}} \|\Xi q\|_2$$

$$\geq \sup_{\substack{q_2 \in RL_2 \\ \|q_2\|_2 = 1}} \|\Pi_s G_2(s)q_2\|_2$$

$$\geq \sup_{\substack{q_2 \in RH_2 \\ \|q_2\|_2 = 1}} \|G_2(s)q_2\|_2 = \|G_2(s)\|_\infty$$

where the last equality sign follows from Theorem 2.12. If condition (G.3) holds, then Theorem 2.13 proves the necessity of condition (*i*). Therefore, the proof is carried out by showing that, under condition (*i*), condition (*ii*) is necessary and sufficient for condition (G.3) to hold.

Notice that, for any $q \in \mathcal{Q}$, it is

$$q = \Pi_s q + \Pi_a q = \begin{bmatrix} 0 \\ \Pi_s q_2 \end{bmatrix} + \Pi_a q$$

Having decomposed q into two orthogonal terms implies

$$\|\Xi q\|_2^2 - \gamma^2\|q\|_2^2 = \|\Xi q\|_2^2 - \gamma^2\|\Pi_s q_2\|_2^2 - \gamma^2\|\Pi_a q\|_2^2 \tag{G.4}$$

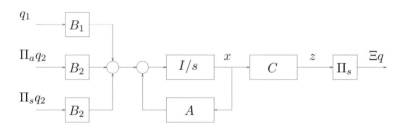

Figure G.1: The block-scheme representation of Ξ

Thanks to the linearity of Ξ, condition (G.3) is equivalent to (also recall Remark 2.13)

$$\sup_{0 \neq q \in \mathcal{Q}} \frac{\|\Xi q\|_2^2}{\|q\|_2^2} < \gamma^2$$

which implies that, for each $q \in \mathcal{Q}$, $q \neq 0$, $\|\Xi q\|_2^2 - \gamma^2 \|q\|_2^2 < 0$. Therefore, condition (G.3) is violated if and only if

$$\sup_{0 \neq q \in \mathcal{Q}} \left[\|\Xi q\|_2^2 - \gamma^2 \|q\|_2^2 \right] \geq 0$$

By taking into account eq. (G.4), condition (G.3) is violated if and only if

$$\sup_{0 \neq q \in \mathcal{Q}} \left[\|\Xi q\|_2^2 - \gamma^2 \|\Pi_s q\|_2^2 - \gamma^2 \|\Pi_a q\|_2^2 \right] \geq 0 \qquad (G.5)$$

With reference to fig. G.1 (which supplies a representation of the operator Ξ where the input q_2 has been decomposed into its stable and antistable component), recall that, in the time domain, a signal belonging to $RH_2^\perp$ is zero for $t > 0$, whereas a signal belonging to RH_2 is zero for $t < 0$. Therefore, the value of the state of system (G.1),(G.2) at time $t = 0$ only depends on the inputs which belong to $RH_2^\perp$ so that, given $x(0) := x_0$, the maximization of the first two terms in eq. (G.5) can be performed by ignoring the antistable part of the inputs to system (G.1),(G.2), so that

$$\sup_{0 \neq q \in \mathcal{Q}} \left[\|\Xi q\|_2^2 - \gamma^2 \|\Pi_s q\|_2^2 \right] \bigg|_{x(0)=x_0} = \sup_{0 \neq q_2 \in RH_2} \left[\|z\|_2^2 - \gamma^2 \|q_2\|_2^2 \right] \bigg|_{x(0)=x_0}$$

Therefore,

$$\sup_{0 \neq q \in \mathcal{Q}} \left[\|\Xi q\|_2^2 - \gamma^2 \|\Pi_s q\|_2^2 - \gamma^2 \|\Pi_a q\|_2^2 \right] \bigg|_{x(0)=x_0} =$$

$$= \sup_{0 \neq q_2 \in RH_2} \left[\|z\|_2^2 - \gamma^2 \|q_2\|_2^2 \right] \bigg|_{x(0)=x_0} - \inf_{q \in RH_2^\perp} \gamma^2 \|q\|_2^2 \bigg|_{x(0)=x_0} \qquad (G.6)$$

Observe that $\sup[M - N] = \sup[M] - \inf[N]$ whenever the choice of M is independent of the choice of N. This fact has been exploited in writing down eq. (G.6).

Assume, for the moment being, that the pair $(A, [B_1 \; B_2])$ is reachable. Thanks to Lemma G.1, the first term on the right hand side of eq. (G.6) equals $x_0' Q_\infty x_0$, while the second term equals, as it will be shown later on, $-\gamma^2 x_0' L_r^{-1} x_0$ (observe that

$L_r > 0$ since the pair $(A, [B_1 \ B_2])$ is reachable, see Lemma C.1). Thus, eq. (G.6) becomes

$$\sup_{0 \neq q \in Q} \left[\|\Xi q\|_2^2 - \gamma^2 \|\Pi_s q\|_2^2 - \gamma^2 \|\Pi_a q\|_2^2 \right] \Big|_{x(0)=x_0} = x_0'(Q_\infty - \gamma^2 L_r^{-1})x_0 \quad \text{(G.7)}$$

Therefore, condition (G.5) is verified if and only if there exists x_0 such that the right hand side of eq. (G.7) is nonnegative. Hence, condition (G.3) is violated if and only if there exists x_0 such that

$$x_0'(Q_\infty - \gamma^2 L_r^{-1})x_0 \geq 0$$

that is if and only if $r_s(Q_\infty L_r) \geq \gamma^2$ (recall Lemma B.12).

The equality

$$\inf_{q \in RH_2^\perp} \|q\|_2^2 \Big|_{x(0)=x_0} = x_0' L_r^{-1} x_0$$

can be proved by making reference to the (classical) optimal control problem relative to the system

$$\dot{x} = Ax + Bq$$
$$x(\tau) = 0, \quad x(0) = x_0, \quad \tau < 0$$

and the performance index

$$J_\tau = \int_\tau^0 q'q \, dt$$

where τ is a given parameter. The Hamilton-Jacobi theory leads to recognizing

$$q_\tau^o(t) = B'e^{-A't} \left[\int_\tau^0 e^{-Av} BB' e^{-A'v} dv \right]^{-1} x_0, \quad t \leq 0$$

as an optimal control, so that, as $\tau \to -\infty$ (recall eq. (2.21) and what has been said just after) it is

$$q^o(t) = \lim_{\tau \to -\infty} q_\tau^o(t)$$

$$= B'e^{-A't} \left[\int_{-\infty}^0 e^{-Av} BB' e^{-A'v} dv \right]^{-1} x_0$$

$$= B'e^{-A't} L_r^{-1} x_0, \quad t \leq 0$$

Notice that the so obtained control belongs to $RH_2^\perp$ since A is stable. Finally,

$$\inf_{q \in RH_2^\perp} \|q\|_2^2 \Big|_{x(0)=x_0} = \int_{-\infty}^0 q^o(t)' q^o(t) dt$$

$$= x_0' L_r^{-1} \left[\int_{-\infty}^0 e^{-At} BB' e^{-A't} dt \right] L_r^{-1} x_0$$

$$= x_0' L_r^{-1} x_0$$

The reachability assumption of the pair $(A, [B_1 \ B_2])$ is now relaxed. Assume to have already decomposed the system into the reachable and unreachable parts, so that $x := [x_r' \ x_u']'$ and

$$A := \begin{bmatrix} A_r & A_x \\ 0 & A_u \end{bmatrix}, \quad B := \begin{bmatrix} B_{1r} & B_{2r} \\ 0 & 0 \end{bmatrix}, \quad C := \begin{bmatrix} C_r & C_u \end{bmatrix}$$

where the pair $(A_r, [B_{1r} \ B_{2r}])$ is reachable. Obviously, the states x_0 to be considered in eq. (G.6) are only those belonging to the set $\mathcal{S}_0 := \{x \mid x = [x'_{r0} \ 0]'\}$, so that condition (G.5) is satisfied (hence condition (G.3) is violated) if and only if there exists $x_0 \in \mathcal{S}_0$ corresponding to which it is

$$\sup_{0 \neq q_2 \in RH_2} \left[\|z\|_2^2 - \gamma^2 \|q_2\|_2^2 \right] \bigg|_{x(0)=x_0} - \inf_{q \in RH_2^\perp} \gamma^2 \|q\|_2^2 \bigg|_{x(0)=x_0} \geq 0 \quad \text{(G.8)}$$

Observe that the stabilizing solution Q_∞ of the Riccati equation is given the form

$$Q_\infty = \begin{bmatrix} Q_r & Q_x \\ Q'_x & Q_u \end{bmatrix}$$

where Q_r is the symmetric, positive definite and stabilizing solution of the Riccati equation (in the unknown V)

$$0 = VA_r + A'_r V + \gamma^{-2} V B_{2r} B'_{2r} V + C'_r C_r$$

As already found, the first term on the left hand side of eq. (G.8) equals $x'_{r0} Q_r x_{r0}$, while, by again exploiting the previous discussion about the optimal control problem (relative to the reachable part only), it is easy to find that the second term on the left hand side of eq. (G.8) is given by $-\gamma^{-2} x'_{r0} L_{rr}^{-1} x_{r0}$, where L_{rr} is the solution of the Lyapunov equation (in the unknown W)

$$0 = WA'_r + A_r W + B_{1r} B'_{1r} + B_{2r} B'_{2r}$$

Therefore, condition (G.3) is violated if and only if $r_s(Q_r L_{rr}) \geq \gamma^2$. It is straightforward to check that, thanks to the peculiar structure of matrices A and $[B_1 \ B_2]$, the unique solution of the Lyapunov equation referred to in the statement of the theorem is

$$L_r = \begin{bmatrix} L_{rr} & 0 \\ 0 & 0 \end{bmatrix}$$

so that $r_s(Q_r L_{rr}) \geq \gamma^2$ is equivalent to the condition $r_s(Q_\infty L_r) \geq \gamma^2$. $\qquad \square$

Appendix H

Convex Functions and Sets

The material included in this appendix is standard in convex optimization. It is mainly based on books [37], [41] and [53]. However, to our present needs, it is important to keep in mind that here all variables and functions are defined in some subset of $R^{n \times m}$. In other terms, instead of vectors we work with $n \times m$ dimensional matrices. All sets to be handled are assumed closed.

First, let us introduce the algebraic structure to be dealt with. The inner product of two matrices X and Y belonging to $R^{n \times m}$ is defined as

$$< X, Y >:= \text{trace}[X'Y]$$

which induces the so called Frobenius norm, that is

$$\|X\|_F := \sqrt{< X, X >} = \sqrt{\text{trace}[X'X]}$$

The geometric interpretation is clear. Matrix $X \in R^{n \times m}$ may be thought as a point in $R^{n \times m}$ whose distance to the origin is $\|X\|_F$ just defined.

Definition H.1 (Convex sets) *A set Ω in $R^{n \times m}$ is convex if $\forall X_1, X_2 \in \Omega$ the point $X = \alpha X_1 + (1 - \alpha)X_2 \in \Omega$ for every $\alpha \in [0, 1]$.* □

In other words the line segment between any two points in a convex set is entirely contained in the set. The empty set is, by definition, a convex set. Furthermore, the above definition can be alternatively stated in terms of a number, say $N \geq 2$, of feasible points. Actually, for $X_i \in \Omega$, $i = 1, 2, \cdots, N$ the convexity of Ω assures that

$$X = \sum_{i=1}^{N} \xi_i X_i \in \Omega \tag{H.1}$$

for all scalars ξ_i, $i = 1, 2, \cdots, N$ such that

$$\xi_i \geq 0, \quad \sum_{i=1}^{N} \xi_i = 1 \tag{H.2}$$

This fact puts in evidence a very interesting property of convex sets. Given a bounded convex set Ω it is always possible to choose N (possibly infinite) points $X_i \in \Omega$, $i = 1, 2, \cdots, N$ such that all $X \in \Omega$ can be written as (H.1). The points X_i with this property are called *extreme points* of Ω. A precise definition of this concept is as follows.

Definition H.2 (Extreme point) *Let X be a point belonging to a convex set Ω. It is an extreme point of Ω if there are no points $X_i \neq X$ in Ω, $i = 1, 2, \cdots, N$ such that, (H.1) holds for some $\xi_1, \xi_2, \cdots, \xi_N$ satisfying (H.2).* □

Two convex sets of particular interest are the *convex polyhedron* and the *convex cone*. The first is defined as a convex set with a finite number of extreme points. The latter is a convex set such that if $X \in \Omega$ then $\lambda X \in \Omega$ for all $\lambda > 0$. With a slight abuse of notation, we also call a convex cone any set for which the above property holds for all $\lambda \geq \lambda_0$ with $\lambda_0 > 0$.

Definition H.3 (Convex hull) *Let Γ be a subset of $R^{n \times m}$. The convex hull of Γ denoted $\mathrm{co}(\Gamma)$ is the smallest convex set containing Γ.* □

It is clear, from the above definition that $\Omega = \mathrm{co}(\Omega)$ whenever Ω is a convex set. Furthermore, if Ω is a convex polyhedron then

$$\Omega = \mathrm{co}\left\{ X_i \, , \quad i = 1, 2, \cdots, N \right\}$$

where $X_1, X_2, \cdots, X_N$ are all extreme points of Ω. The set of all symmetric positive definite matrices

$$\Omega = \left\{ X \in R^{n \times n} \; : \; X > 0 \right\}$$

is a convex cone. Obviously, the same is true for the set of all symmetric negative definite matrices.

Let Λ be a nonzero matrix in $R^{n \times m}$ and let X_0 be an arbitrary matrix (point) in $R^{n \times m}$. An *hyperplane* is the set of all points $X \in R^{n \times m}$ such that $X - X_0$ is orthogonal to Λ, or in more precise terms

$$0 \; = \; <\Lambda, X - X_0> \; = \; \mathrm{trace}\left[\Lambda'(X - X_0) \right]$$

Defining the scalar $c := \mathrm{trace}[\Lambda' X_0]$ an hyperplane can be characterized by all matrices $X \in R^{n \times m}$ such that

$$\mathrm{trace}[\Lambda' X] \; = \; c$$

which puts in evidence that an hyperplane in $R^{n \times m}$ is nothing more than a linear variety of dimension $nm - 1$. This concept is useful to get the following results valid for closed convex sets, that is, for convex sets which contain all boundary points.

Lemma H.1 (Separating hyperplane) *Let Ω be a closed convex set and consider $X_0 \notin \Omega$. There exists a matrix Λ_0 and a scalar c_0 such that the hyperplane defined by $<\Lambda_0, X> = c_0$ separates X_0 from Ω. That is, the following two conditions hold simultaneously :*

i) $<\Lambda_0, X_0> < c_0$

ii) $<\Lambda_0, X> \geq c_0$, $\forall\, X \in \Omega$.

Proof The proof comes from the definition of the new matrix

$$Y_0 := \mathrm{argmin}\left\{ \|X - X_0\|_F \; : \; X \in \Omega \right\}$$

which enables us to get

$$\Lambda_0 := Y_0 - X_0 \, , \quad c_0 := <Y_0 - X_0, Y_0>$$

such that both conditions are simultaneously verified.

Point i) Using the above matrix we have

$$< \Lambda_0, X_0 > -c_0 = < Y_0 - X_0, X_0 > - < Y_0 - X_0, Y_0 >$$
$$= -\|Y_0 - X_0\|_F^2 < 0$$

where the last inequality follows from the fact that $Y_0 \neq X_0$.

Point ii) From the definition of matrix Y_0 we have

$$\|X - X_0\|_F^2 \geq \|Y_0 - X_0\|_F^2 , \quad \forall \ X \in \Omega$$

However, for any $X \in \Omega$ given, the point $\lambda X + (1 - \lambda)Y_0 \in \Omega$ for all $\lambda \in [0, 1]$ by a consequence of the convexity of Ω. In this case, the above inequality reveals that

$$< Y_0 - X_0, X - Y_0 >\geq -\left(\frac{\lambda}{2}\right) \|X - Y_0\|_F^2$$

holds. Using this inequality for $\lambda \to 0$ we finally obtain

$$< Y_0 - X_0, X - Y_0 >\geq 0 , \quad \forall \ X \in \Omega$$

This inequality, together with

$$< \Lambda_0, X > -c_0 = < Y_0 - X_0, X > - < Y_0 - X_0, Y_0 >$$
$$= < Y_0 - X_0, X - Y_0 >$$

conclude the proof of the theorem proposed.   □

Lemma H.2 (Supporting hyperplane) *Let Ω be a closed convex set and consider X_0 in the boundary of Ω. There exists a matrix Λ_0 and a scalar c_0 such that the hyperplane $< \Lambda_0, X >= c_0$ supports Ω at X_0. That is, the following two conditions hold simultaneously :*

i) $< \Lambda_0, X_0 >= c_0$

ii) $< \Lambda_0, X >\geq c_0 , \quad \forall \ X \in \Omega.$

Lemma H.1 says that given a convex set and an exterior point then it is always possible to determine an hyperplane which contains the point in one of its half spaces while the convex set lays entirely in the other half space. Lemma H.2 generalizes this result to cope with a point in the boundary of a closed convex set.

Let us now turns our attention to real valued functions defined in $R^{n \times m}$. A function $f(X) : R^{n \times m} \to R$ is *continuous* if for each $X_0 \in R^{n \times m}$

$$\lim_{X \to X_0} f(X) = f(X_0)$$

All functions handled here are continuous. The gradient of a continuous function $f(\cdot)$ at $X = X_0$ is a $n \times m$ matrix, denoted $\nabla f(X_0)$ and defined as

$$\nabla f(X_0) := \left\{ \frac{\partial f}{\partial X_{ij}}(X_0) , \quad i = 1, 2, \cdots, n ; \ j = 1, 2, \cdots m \right\}$$

For functions of matrices, a simple way to calculate gradients is from the concept of directional derivative. The *directional derivative* of a differentiable function $f(\cdot)$ at $X \in R^{n \times m}$ in the direction $Y \in R^{n \times m}$ is

$$Df(X,Y) := \lim_{\epsilon \to 0} \frac{f(X + \epsilon Y) - f(X)}{\epsilon}$$
$$= <\nabla f(X), Y>$$
$$= \text{trace}[\nabla f(X)'Y]$$

In other words, given a function $f(\cdot)$, if for any matrix $Y \in R^{n \times m}$, it is possible to write

$$f(X + \epsilon Y) = f(X) + \epsilon \text{trace}[G(X)'Y] + \mathcal{O}(\epsilon^2) \qquad \text{(H.3)}$$

such that $\lim_{\epsilon \to 0} \mathcal{O}(\epsilon^2)/\epsilon = 0$ then $f(\cdot)$ is differentiable and an adequate choice of directions leads to

$$\nabla f(X) = G(X) \qquad \text{(H.4)}$$

Example H.1 Using the above result, it is simple to calculate the gradient of the function $f(X) := \det[X]$ where $X \in R^{n \times n}$ and nonsingular. Indeed, elementary matrices properties yield, for $\epsilon \in R$ arbitrarily small

$$\det[X + \epsilon Y] = \det[X] \det\left[I + \epsilon X^{-1} Y\right]$$

$$= \det[X] \prod_{i=1}^{n} \left[1 + \epsilon \lambda_i \left(X^{-1}Y\right)\right]$$

$$= \det[X] \left(1 + \epsilon \text{trace}\left[X^{-1}Y\right]\right) + \mathcal{O}(\epsilon^2)$$

showing that $\nabla f(X) = \det[X](X')^{-1}$. The same steps can be adopted to evaluate the gradient of the function $f(X) := \log \det[X]$ defined for all $X \in R^{n \times n}$ such that $\det[X] > 0$. Indeed, from the above we get

$$f(X + \epsilon Y) = f(X) + \sum_{i=1}^{n} \log \left[1 + \epsilon \lambda_i \left(X^{-1}Y\right)\right]$$

$$= f(X) + \epsilon \text{trace}\left[X^{-1}Y\right] + \mathcal{O}(\epsilon^2)$$

implying that $\nabla f(X) = (X')^{-1}$.   □

Remark H.1 Equation (H.4) must be applied with care. In some important cases, it does not apply directly. To clarify its correct use, consider a function $f(\cdot) : \mathcal{S} \to R$, where $\mathcal{S}$ denotes the set of all $n \times n$ symmetric matrices. It is clear that this is a function of $n(n+1)/2$ independent variables. Assuming that (H.3) holds for $Y = Y'$ we observe that

$$\text{trace}[G(X)'Y] = \sum_{i=1}^{n} G(X)_{ii} Y_{ii} + 2 \sum_{i=1}^{n} \sum_{j>i}^{n} G(X)_{ij} Y_{ij}$$

which provides

$$\nabla f(X) = G(X) + G(X)' - \text{diag}[G(X)]$$

For instance, the gradient matrix of the linear function $f(X) := \text{trace}[R'X]$ defined for all square matrix $X \in R^{n \times n}$ is $\nabla f(X) = R$. On the other hand, the gradient of the same function defined in $\mathcal{S}$ is $\nabla f(X) = R + R' - \text{diag}[R]$.   □

Definition H.4 (Convex functions) *A function $f(\cdot) \in R$ defined in a convex set Ω is convex if $\forall\, X_1, X_2 \in \Omega$ and $X = \alpha X_1 + (1 - \alpha)X_2$ there holds $f(X) \leq \alpha f(X_1) + (1 - \alpha)f(X_2)$ for every $\alpha \in [0,\ 1]$.*   □

There are several different but equivalent ways to test a function for convexity. Of course each one depends on the previous informations we have for the function we are work with. The next lemma provides some equivalent tests for twice differentiable functions.

Lemma H.3 (Convexity) *Let $f(\cdot)$ be a twice differentiable function, defined in a convex set $\Omega \subset R^{n \times m}$. Assume that Ω contains an interior point. The following are equivalent :*

   i) $f(\cdot)$ is convex over Ω.

   ii) $f(Y) \geq f(X)+ < \nabla f(X), Y - X > , \quad \forall\ X, Y \in \Omega$.

   iii) The second order variation

$$V f(X, Y) := \left. \frac{d^2 f}{d\epsilon^2}(X + \epsilon Y) \right|_{\epsilon=0}$$

   is such that $V f(X, Y) \geq 0$ for all $X \in \Omega$ and all $Y \in R^{n \times m}$.

Dealing with optimal control problems, it is frequently necessary to handle non-differentiable functions. The next lemma characterizes convexity in this important case.

Lemma H.4 (Convexity) *Let $f(\cdot)$ be a function, defined in a convex set $\Omega \subset R^{n \times m}$. Assume that Ω contains an interior point. The following are equivalent :*

   i) $f(\cdot)$ is convex over Ω.

   ii) For each $X \in \Omega$, there exits a matrix $\Lambda(X)$ with finite norm such that

$$f(Y) \geq f(X)+ < \Lambda(X), Y - X > , \quad \forall\ Y \in \Omega.$$

   iii) The epigraph of function f, namely

$$\text{epi } f := \{(X, \gamma)\ :\ X \in \Omega ,\quad f(X) \leq \gamma\}$$

   is a convex set.

Proof The equivalence between points *i)* and *iii)* follows immediately from the observation that f is convex over Ω if and only if function $g(X, \gamma) := f(X) - \gamma$ is convex over $\Omega \times R$.

   ii) $\Longrightarrow$ *i)* Consider X_1 and X_2 two arbitrary matrices in Ω. Define $X = \alpha X_1 + (1 - \alpha)X_2$ for $\alpha \in [0,\ 1]$. Setting $Y = X_1$ and $Y = X_2$ we have

$$f(X_1) \geq f(X)+ < \Lambda(X), X_1 - X >$$
$$f(X_2) \geq f(X)+ < \Lambda(X), X_2 - X >$$

Multiplying the first inequality by α, the second by $1 - \alpha$ and adding terms we get

$$\alpha f(X_1) + (1 - \alpha)f(X_2) \geq f(X)$$

which is the desired result.

   i) $\Longrightarrow$ *ii)* Under the assumption that f is convex, from part *iii)* the set epi f is convex as well. Using Lemma H.2, there exists a supporting hyperplane to epi f in the boundary point (X_0, γ_0) where $\gamma_0 = f(X_0)$ and $X_0 \in \Omega$. That is, there exist a matrix Λ_0 and a scalar λ_0 such that all $(X, \gamma) \in \text{epi } f$ satisfy

$$< \Lambda_0, X - X_0 > + \lambda_0(\gamma - \gamma_0) \geq 0$$

We notice that the above inequality imposes $\lambda_0 > 0$. In fact, it is simple to check that $(X_0, \gamma) \notin \text{epi } f$ provided $\gamma < \gamma_0$ leading to $\lambda_0(\gamma - \gamma_0) < 0$. As a consequence we have

$$\gamma \geq f(X_0) + < -\lambda_0^{-1}\Lambda_0, X - X_0 >$$

which turns out to be true for all $(X, \gamma) \in \text{epi } f$. Then, it is verified for the particular point (X, γ) with $\gamma = f(X)$ of epi f. Defining $\Lambda(X_0) := -\Lambda_0/\lambda_0$, the proof is concluded. $\qquad\square$

The level set of a function $f(\cdot)$ defined in Ω is the set of all $X \in \Omega$ such that $f(X)$ is not greater than a fixed value. More precisely, let α be a fixed scalar, the level set of f is

$$L_\alpha f := \{X \ : \ X \in \Omega \ , \ f(X) \leq \alpha\}$$

It must be clear that the level set of a function is a subset of Ω for each value of α given. On the contrary, the epi f is a subset of $\Omega \times R$. Then, it is not surprising that while, convexity of f is equivalent to the convexity of epi f, the same is not true for $L_\alpha f$. Indeed, if f is convex in Ω the level set $L_\alpha f$ is convex for all $\alpha \in R$ but the latter statement is not generally sufficient to assure convexity of f.

Definition H.5 (Subgradient) *Let $f(\cdot)$ be a convex function defined in a convex set $\Omega \subset R^{n \times m}$. Matrix $\Lambda \in R^{n \times m}$ is said to be a subgradient of f at a point X if*

$$f(Y) \geq f(X) + < \Lambda, Y - X >$$

holds for all $Y \in \Omega$. The set of all subgradients of f at X is denoted $\partial f(X)$. $\qquad\square$

Lemma H.5 *The set $\partial f(X)$ is convex.*

Proof Consider $\Lambda_1, \Lambda_2 \in \partial f(X)$. From Definition H.5 we have, for all $Y \in \Omega$

$$f(Y) \geq f(X) + < \Lambda_1, Y - X >$$
$$f(Y) \geq f(X) + < \Lambda_2, Y - X >$$

Multiplying the first inequality by α, the second one by $1 - \alpha$ and adding terms, yields

$$f(Y) \geq f(X) + < \alpha\Lambda_1 + (1 - \alpha)\Lambda_2, Y - X >$$

implying that $\Lambda := \alpha\Lambda_1 + (1 - \alpha)\Lambda_2 \in \partial f(X)$, which is the stated property. $\qquad\square$

It is important to mention that the set of subgradients is in addition a closed set. An important concept to deal with nondifferentiable convex functions is the *one sided directional derivative*. The right directional derivative of function f at $X \in R^{n \times m}$ in the direction $Y \in R^{n \times m}$ is

$$D_+ f(X, Y) := \lim_{\epsilon \to 0^+} \frac{f(X + \epsilon Y) - f(X)}{\epsilon}$$

as well as, the left directional derivative in the same direction is

$$D_- f(X, Y) := \lim_{\epsilon \to 0^-} \frac{f(X + \epsilon Y) - f(X)}{\epsilon}$$

Simple calculations show that each one sided directional derivative is one related to the other by the equation

$$D_- f(X, Y) = -D_+ f(X, -Y) \tag{H.5}$$

indicating that, generally, they are different. Indeed, if they coincide then the function under consideration is differentiable at X. This important feature is now discussed from the result of the next lemma.

Lemma H.6 Let $f(\cdot)$ be a convex function defined in the convex set Ω. At any interior point of Ω, it follows that

$$D_+ f(X, Y) = \max \{ \text{trace}[\Lambda' Y] \ : \ \Lambda \in \partial f(X) \}$$

Proof For any $\Lambda \in \partial f(X)$, the definition of subgradient yields

$$D_+ f(X, Y) = \lim_{\epsilon \to 0^+} \frac{f(X + \epsilon Y) - f(X)}{\epsilon}$$
$$\geq \text{trace}[\Lambda' Y]$$

the proof is then concluded from the fact that at any interior point of Ω the set $\partial f(X)$ is closed and bounded and both sides of the above inequality must be equal for some $\Lambda \in \partial f(X)$. □

The above result used together (H.5) enables us to get

$$D_- f(X, Y) = \min \{ \text{trace}[\Lambda' Y] \ : \ \Lambda \in \partial f(X) \}$$

which implies that both, the right and the left directional derivatives are equal in any direction $Y \in R^{n \times m}$ whenever the set of subgradients contains only one element. In this case the function is differentiable, $\partial f(X) = \{ \nabla f(X) \}$ and the directional derivative is a *linear* function of the direction $Y \in R^{n \times m}$.

Example H.2 Let us calculate the set of subgradients $\partial g(X)$ of the function

$$g(X) := \max \{ f_i(X) \ : \ i = 1, 2, \cdots, N \}$$

where all functions $f_i(\cdot), i = 1, 2, \cdots, N$ are supposed to be convex and differentiable in all points of $R^{n \times m}$. It is a simple matter to verify that function g is convex as well. Defining the set

$$J(X) := \{ j \ : \ f_j(X) = g(X) \}$$

and taking $Y \in R^{n \times m}$ we get

$$g(Y) = \max \{ f_i(Y) \ : \ i = 1, 2, \cdots, N \}$$
$$\geq f_j(Y)$$
$$\geq f_j(X) + < \nabla f_j(X), Y - X >$$
$$\geq g(X) + < \nabla f_j(X), Y - X >$$

and so $\nabla f_j(X) \in \partial g(X)$ for all $j \in J(X)$. Even more, $\nabla f_j(X)$ is an extreme point of $\partial g(X)$. Taking into account the result of Lemma H.5 we then conclude that

$$\partial g(X) = \text{co } \{ \nabla f_j(X) \ : \ j \in J(X) \}$$

Accordingly, if there is only one index, say $k \in J(X)$ then g is differentiable and $\nabla g(X) = \nabla f_k(X)$. □

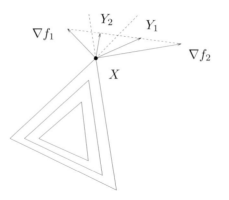

Figure H.1: Level sets of function $g(X)$

When dealing with nondifferentiable convex functions, it is important to keep in mind that many properties valid under differentiability may fail. For instance, if f is differentiable at $X \in R^{n \times m}$ then $-Y = \nabla f(X) \neq 0$ is a direction such that the directional derivative is negative, as a consequence, there exists a step length $\epsilon > 0$ such that $f(X + \epsilon Y) < f(X)$. On the other hand, if f is not differentiable at $X \in R^{n \times m}$ this property may be lost. In other words, it does not hold if we just consider an arbitrary direction $-Y \in \partial f(X)$. This common situation is illustrated in fig. H.1. The function $g(X)$ is the one treated in Example H.2 where $N = 3$ and $f_i(X), i = 1, 2, 3$ are linear. At point X the function is not differentiable because $f_1(X)$ and $f_2(X)$ coincide. Both indicated directions are in the set $\partial g(X)$. Even though, the right directional derivative in the direction $Y = -Y_2$ is clearly negative but the right directional derivative in the direction $Y = -Y_1$ is positive. Hence, for $-Y = Y_1 \in \partial g(X)$ the previously mentioned descent property does not hold. In the direction $Y = -Y_1$ the function increases.

For nondifferentiable functions, there is a way to determine the steepest descent direction. It is given by (notice the constraint on the maximum length of the direction)

$$Y_{st} := \operatorname{argmin} \left\{ D_+ f(X, Y) \; : \; \|Y\|_F^2 \leq 1 \right\}$$

which, from Lemma H.6 can be determined by solving

$$\min_{\|Y\|_F^2 \leq 1} \max_{\Lambda \in \partial f(X)} \operatorname{trace}[\Lambda' Y]$$

Recalling that all constraints are convex, closed and bounded, this problem is equivalent to

$$\max_{\Lambda \in \partial f(X)} \min_{\|Y\|_F^2 \leq 1} \operatorname{trace}[\Lambda' Y]$$

The minimization is readily solved by keeping in mind that the objective function is an inner product of two matrices whose minimum provides

$$Y_{st} := -\frac{\Lambda_{st}}{\|\Lambda_{st}\|_F}$$

where Λ_{st} is given by

$$\Lambda_{st} := \operatorname{argmin} \left\{ \|\Lambda\|_F \; : \; \Lambda \in \partial f(X) \right\}$$

The steepest descent direction is obtained from the selection among all subgradients, the one of minimum Frobenius norm. This task is, in general, very difficult to be accomplished numerically since the set $\partial f(X)$ may not be explicitly known. In Example H.2 (recall fig. H.1) $Y_{st} = -Y_2$ is the steepest direction at point X. Finally, it is also interesting to observe that under differentiability we get $Y_{st} = -\nabla f(X)/\|\nabla f(X)\|_F$ as expected. Furthermore, in the steepest descent direction the right directional derivative provides, for any other direction Y the lower bound

$$D_+ f(X,Y) \geq D_+ f(X, Y_{st}) = -\|\Lambda_{st}\|_F \tag{H.6}$$

It is obvious that if $0 \in \partial f(X)$, the calculation needed to determine the steepest descent direction Y_{st} can not be performed because $\Lambda_{st} = 0$. The possible occurrence of this fact is however of particular importance as indicated in the next lemma.

Lemma H.7 *Let $f(\cdot)$ be a convex function defined in $R^{n \times m}$. Matrix $X^\star$ minimizes f if and only if $0 \in \partial f(X^\star)$.*

Proof Suppose $0 \in \partial f(X^\star)$, from the convexity of f we have, for all $X \in R^{n \times m}$

$$f(X) \geq f(X^\star) + < 0, X - X^\star > \geq f(X^\star)$$

implying that $X^\star$ minimizes f indeed. Conversely, if $X^\star$ minimizes f then in any direction $Y \in R^{n \times m}$, the right directional derivative must be nonnegative. In view of (H.6) this occurs provided $\Lambda_{st} = 0$ implying that $0 \in \partial f(X^\star)$. □

We are now in position to solve the following optimization problem

$$\min \{ f(X) \; : \; X \in \mathcal{X} \} \tag{H.7}$$

which is called a *convex programming problem* provided i) the objective function $f(\cdot) : \Omega \to R$ is convex in Ω which by its turn is a convex subset of $R^{n \times m}$ and ii) the constraint set $\mathcal{X}$ is a closed convex subset of Ω. To avoid pathological cases, it is also assumed that $\mathcal{X}$ is bounded and contains an interior point.

Lemma H.8 *Consider the convex problem (H.7). The following hold :*

i) If $X^\star$ is a minimum then it is a global minimum.

ii) Matrix $X^\star$ minimizes f over $\mathcal{X}$ if and only if there exists $\Lambda^\star \in \partial f(X^\star)$ such that

$$< \Lambda^\star, X - X^\star > \geq 0 \; , \quad \forall \, X \in \mathcal{X}$$

Proof Only point ii) will be proved since the first point restates an important but very known property of convex problems.

Point ii) Suppose there exists $\Lambda^\star \in \partial f(X^\star)$ with the above property. From the convexity of f we have, for all $X \in \mathcal{X}$,

$$f(X) \geq f(X^\star) + < \Lambda^\star, X - X^\star > \geq f(X^\star)$$

implying that $X^\star$ minimizes f indeed. Conversely, if $X^\star$ minimizes f over $\mathcal{X}$ then, we first notice that due to the convexity of $\mathcal{X}$, for any $X \neq X^\star \in \mathcal{X}$, the direction $Y = X - X^\star$ is always a feasible direction in the sense that

$$X^\star + \epsilon Y \in \mathcal{X}$$

for all $\epsilon \in [0, \ 1]$. So, due to the fact that $X^\star$ is optimal we must have

$$D_+ f(X^\star, Y) \geq 0$$

Then, from Lemma H.6, it is true that for all $X \in \mathcal{X}$

$$\max_{\Lambda \in \partial f(X^\star)} < \Lambda, X - X^\star > \geq 0$$

that is

$$\min_{X \in \mathcal{X}} \max_{\Lambda \in \partial f(X^\star)} < \Lambda, X - X^\star > \geq 0$$

Using the fact that all constraints in the above problem are convex, bounded and closed, it can be written in the equivalent form

$$\max_{\Lambda \in \partial f(X^\star)} \min_{X \in \mathcal{X}} < \Lambda, X - X^\star > \geq 0$$

which allows us to conclude that there exists $\Lambda^\star \in \partial f(X^\star)$ such that

$$\min_{X \in \mathcal{X}} < \Lambda^\star, X - X^\star > \geq 0$$

which is the same to say that $< \Lambda^\star, X - X^\star > \geq 0$ for all $X \in \mathcal{X}$ and the proof is concluded. $\qquad \square$

Once again, it is to be noticed that under differentiability, Lemma H.7 reduces to the classical property $\nabla f(X^\star) = 0$ to characterize a minimum of a convex function. As well as, under the same assumption, the second part of Lemma H.8 imposes that $< \nabla f(X^\star), X - X^\star > \geq 0$, $\forall \ X \in \mathcal{X}$ for global optimality.

Problem (H.7) is frequently called *Primal*. With the purpose to solve it more efficiently, convexity allows us to determine another equivalent but in many instances easier problem to be solved called *Dual*. For convex problems the Primal and the Dual versions are equivalent in the sense that both provide the same optimal solution. Duality gap does not exist. The determination of the dual problem is based on the following result.

Lemma H.9 (Minimax) *Consider $\mathcal{X}$ and $\mathcal{Y}$ two closed and bounded convex sets. Let the continuous function $L(X, Y)$ be convex with respect to X and concave with respect to Y. The following equality holds*

$$\max_{Y \in \mathcal{Y}} \min_{X \in \mathcal{X}} L(X, Y) = \min_{X \in \mathcal{X}} \max_{Y \in \mathcal{Y}} L(X, Y) \tag{H.8}$$

Lemma H.9 states that under the given conditions, function $L(\cdot)$ admits a saddle point which is in fact the optimal solution of the min/max problem. Adding to (H.7) a new convex constraint, that is

$$\min_{X \in \mathcal{X}} \{ f(X) \ : \ g(X) \leq 0 \}$$

and defining the associated *Lagrangian* function

$$L(X, Y) := f(X) + < Y, g(X) >$$

in the domain

$$\mathcal{Y} := \left\{ Y \geq 0 \ : \ \min_{X \in \mathcal{X}} L(X, Y) > -\infty \right\}$$

the above lemma can be applied. Equality (H.8) provides a pair of equivalent problems. One of then is the primal while the other is the dual.

Appendix I

Convex Programming Numerical Tools

In this appendix some convex programming numerical tools are discussed. The material is mainly based on the natural generalization, to the nondifferentiable case, of two classical convex programming algorithms described in the classical books [37] and [41]. The method of centers is based on the book [10]. More efficient methods (but also much more involved from a theoretical point of view) can be found in reference [45].

The general form of a convex programming problem is

$$\min \{f(X) \; : \; X \in \mathcal{X}\}$$

where $f(\cdot)$ is a convex function and $\mathcal{X}$ is a convex set. The concept of epigraph allows us to rewrite it in the equivalent form

$$\min \{\gamma \; : \; (X, \gamma) \in \text{epi } f\}$$

which implies that, with no loss of generality, the objective function of a general convex programming problem can be considered to be a linear function. In addition, all problems we have manipulated in the previous chapters, are such that the convex set epi f can be alternatively written (possibly by adding new variables and constraints) as a LMI (Linear Matrix Inequality). Consequently, the general problem we have to solve numerically is

$$\min \{c'x \; : \; \mathcal{A}(x) \geq 0\} \tag{I.1}$$

where c and x are n dimensional real vectors and

$$\mathcal{A}(x) := A_0 + \sum_{i=1}^{n} x_i A_i$$

with $A_0, A_1, \cdots, A_n$ being $m \times m$ dimensional symmetric matrices and $x_i \in R$ denotes the i-th entry of the n dimensional vector x. It is assumed that the feasible set defined by all vectors $x \in R^n$ such that $\mathcal{A}(x) \geq 0$ is bounded. This assumption is verified in all problems of our interest already discussed in the previous chapters. Notice however that this set is convex but not polyhedral then, in general, it has an infinite number of extreme points.

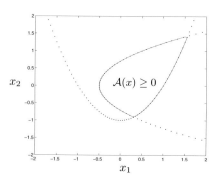

Figure I.1: The set $\mathcal{A}(x) \geq 0$

There is no difficulty to consider more general problems with the structure (I.1) and with an additional equality constraint $Bx = b$, where $B \in R^{r \times n}$ is a matrix of full row rank. If necessary, a preliminary change of variables, allows the partitioning $B = [B_1 \ B_2]$ where $B_1 \in R^{r \times r}$ is nonsingular. Doing the same with vector $x' = [x'_1 \ x'_2]$, the linear constraint provides

$$x_1 = B_1^{-1} b - B_1^{-1} B_2 x_2$$

and adopting again the above partitioning to $c' = [c'_1 \ c'_2]$, the feasibility with respect to the same equality constraint yields

$$c' x = c'_1 B_1^{-1} b + \bar{c}'_2 x_2$$

where $\bar{c}'_2 := c'_2 - c'_1 B_1^{-1} B_2$ and

$$\mathcal{A}(x) = \mathcal{A}\left(\begin{bmatrix} B_1^{-1} b - B_1^{-1} B_2 x_2 \\ x_2 \end{bmatrix} \right)$$
$$:= \bar{\mathcal{A}}_2(x_2)$$

The consequence is that the equality constraint can easily be handled by eliminating itself and formulating a new problem as (I.1) which depends exclusively on the variable $x_2 \in R^{n-r}$. From the numerical point of view the remaining problem to be solved is simpler since it presents a smaller number of free variables.

Example I.1 The geometry of the feasible set of problem (I.1) is now illustrated by means of a simple example. Consider the LMI $\mathcal{A}(x) \geq 0$ be given by $n = 2$ and matrices

$$A_0 = \begin{bmatrix} 1 & 0 & 0 & 0 \\ 0 & 1 & 0 & 0 \\ 0 & 0 & 1/2 & 0 \\ 0 & 0 & 0 & 1 \end{bmatrix}, \quad A_1 = \begin{bmatrix} 0 & 1 & 0 & 0 \\ 1 & 0 & 0 & 0 \\ 0 & 0 & 1 & 0 \\ 0 & 0 & 0 & 0 \end{bmatrix}, \quad A_2 = \begin{bmatrix} 1 & 0 & 0 & 0 \\ 0 & 0 & 0 & 0 \\ 0 & 0 & 0 & 1 \\ 0 & 0 & 1 & 0 \end{bmatrix}$$

Figure I.1 shows the convex set defined by this LMI. It is obtained by the intersection of the convex regions $x_2 \geq x_1^2 - 1$ and $x_1 \geq x_2^2 - 1/2$. □

Algorithm I.1 (Separating hyperplane algorithm) This is one of the simplest algorithms that can be applied to solve the stated convex problem (I.1). In the following it is discussed its convergence as well as its limitations to handle large scale problems.

1. Determine a convex polyhedral set $\mathcal{P}_0$ containing the overall feasible set of Problem I.1, that is

$$\{x \ : \ \mathcal{A}(x) \geq 0\} \subset \mathcal{P}_0$$

   and set the iteration index $k = 0$.

2. Solve the linear programming problem

$$\min \{c'x \ : \ x \in \mathcal{P}_k\}$$

   If it does not admit a feasible solution, the same is true for Problem I.1 - **stop**. Otherwise, let x_k be its optimal solution.

3. If $\mathcal{A}(x_k) \geq 0$ then x_k solves Problem I.1 - **stop**.

4. Determine a separating hyperplane $a_k'x = c_k$ which separates x_k from the feasible convex set $\{x : \mathcal{A}(x) \geq 0\}$. Define

$$\mathcal{P}_{k+1} := \mathcal{P}_k \cap \{x \ : \ a_k'x \geq c_k\}$$

   set the iteration index $k \leftarrow k + 1$ and go back to step 2.

When the algorithm stops the global optimal solution (if any) of Problem I.1 is provided or it is answered that it is infeasible. □

Several points have to be analyzed in details. The first one concerns the determination of the polyhedral set $\mathcal{P}_0$. It always exists from the boundedness assumption on the feasible set of Problem I.1 introduced before. For instance, it can be taken as

$$\mathcal{P}_0 := \{x \ : \ |x_i| \leq \rho \, , \ \ 1 = 1, 2, \cdots, n\}$$

where $\rho > 0$ is a scalar sufficiently large.

In step 2, it is more efficient to solve the stated linear programming problem by means of a dual method. Doing this, it is possible to take advantage from the fact that the sets $\mathcal{P}_{k+1}$ and $\mathcal{P}_k$ differ one to the other by only one new linear constraint added in step 4. Indeed, in iteration $k + 1$ a feasible dual solution is readily given by $\lambda_{k+1}' := [\lambda_k' \ 0]$, where λ_k is the optimal dual solution already obtained in the previous iteration.

In practice, the stopping condition in step 3 has to be changed to $\mathcal{A}(x_k) \geq -\epsilon$ where $\epsilon > 0$ is an arbitrarily small parameter. This is necessary because,

$$\{x \ : \ \mathcal{A}(x) \geq 0\} \subset \cdots \subset \mathcal{P}_{k+1} \subset \mathcal{P}_k \, , \ \ \forall \, k = 0, 1, \cdots \tag{I.2}$$

implies that the algorithm evolves from the outside of the feasible set. Of course, the number of iterations to reach a certain precision increases as ϵ decreases.

Finally, it must be clear how to calculate the separating hyperplane needed in step 4. We claim that the vector a_k and the scalar c_k can be taken as

$$a_k := \begin{bmatrix} z_k'A_1z_k & \cdots & z_k'A_nz_k \end{bmatrix}'$$
$$c_k := -z_k'A_0z_k$$

where $z_k \in R^m$ is any unitary norm eigenvector associated to the minimum eigenvalue of $\mathcal{A}(x_k)$. Actually, this important property follows immediately from the equality

$$a_k'x - c_k = z_k'\mathcal{A}(x)z_k$$

valid for all $x \in R^n$. Indeed, for all feasible vectors that is for those vectors such that $\mathcal{A}(x) \geq 0$ obviously $a_k' x - c_k \geq 0$. On the other hand, for $x = x_k$ we get

$$a_k' x_k - c_k = \lambda_{min}[\mathcal{A}(x_k)] < 0$$

since by construction the point x_k is not feasible as it has been previously tested in step 3. It is interesting to observe that if the feasible set of Problem I.1 is not empty, then $\|a_k\| \neq 0$. This is true because the function $g(x) := \lambda_{min}[\mathcal{A}(x)]$ is concave and so the inequality

$$
\begin{aligned}
g(x) &\leq z_k' \mathcal{A}(x) z_k \\
&\leq a_k' x - c_k \\
&\leq g(x_k) + a_k'(x - x_k) \ , \ \forall \ x \in R^n
\end{aligned}
$$

implies that $a_k \in \partial g(x_k)$ and $a_k = 0$ is possible if and only if x_k maximizes $g(\cdot)$. Since $g(x_k) < 0$ then it is impossible to have any other $x \in R^n$ such that $g(x) \geq 0$. So, $\|a_k\| = 0$ may occur if and only if the feasible set of Problem I.1 is empty.

It remains to prove the global convergence of the Separating hyperplane algorithm which is done in the next lemma.

Lemma I.1 *Suppose the feasible set of Problem (I.1) is not empty and the Separating hyperplane algorithm generates the sequence of points $\{x_k\}$. Any limit point of this sequence is a global optimal solution to Problem (I.1).*

Proof In the k-th iteration of the Separating hyperplane algorithm, the linear constraint (cut) $a_k' x \geq c_k$ has been added to the linear programming problem to be solved in step 2. Consequently in any subsequent iteration, say $l > k$, it must be verified for $x = x_l$, that is $a_k' x_l \geq c_k$ which yields

$$
\begin{aligned}
0 &\leq a_k' x_l - c_k \\
&\leq (a_k' x_k - c_k) + a_k'(x_l - x_k) \\
&\leq \lambda_{min}[\mathcal{A}(x_k)] + a_k'(x_l - x_k)
\end{aligned}
$$

or equivalently

$$\lambda_{min}[\mathcal{A}(x_k)] \geq a_k'(x_k - x_l) \geq -\|a_k\| \ \|x_k - x_l\|$$

Now, since $\|a_k\| < \infty$ in all iterations, the conclusion is that as k and l go to infinity, the algorithm provides a vector x such that $\mathcal{A}(x) \geq 0$, that is x is a feasible solution to Problem (I.1). On the other hand, due to (I.2), in any iteration we have

$$c' x_k \leq \min \{c'x \ : \ \mathcal{A}(x) \geq 0\}$$

which shows, by continuity, that the limit point x is optimal indeed.   □

The Separating hyperplane algorithm has some important features. First, unlike many nonlinear programming methods, it does not require any line search. The practical implementation is simple and depends basically on the development of a powerful dual-simplex routine to solve the linear programming problem on step 2. The use of a dual method also provides useful informations to drop non binding constraints, which of course, keeps reduced the number of constraints to be handled and so increases numerical efficiency. Unfortunately, the Separating hyperplane algorithm as presented before, may have a poor rate of convergence. Indeed, it can be estimated that it converges arithmetically or at most geometrically with a ratio that goes to unity as the dimension of the problem increases. More specifically

$$\|x_k - x_{opt}\| \le \alpha \beta^k$$

for some constants $\alpha > 0$ and $0 < \beta < 1$ where $\beta = \beta(n)$ goes to one as n increases. Based on this, it is predicted (and practically verified) that it does not perform well for solving large scale problems.

It is also possible to recognize that the determination of the separating hyperplane plays a central role in the numerical efficiency of the method. The deepness of the cut added in each iteration, appears to be of great importance. This leads to the introduction of the so called Supporting hyperplane algorithm which works with the deepest cut that can be calculated in each iteration. For that we assume the feasible set of Problem (I.1) is not empty and we have previously calculated an interior point $y \in R^n$ such that $\mathcal{A}(y) > 0$. At any infeasible point x_k, it is possible to calculate a step size $0 < \alpha_k < 1$ which defines a point in the line $d_k := y - x_k$ and, in the same time, is on the boundary of the constraint $\mathcal{A}(x) \ge 0$. To this end, we have to find the scalar α_k solution to the nonlinear equation (in the unknown α)

$$0 = \lambda_{min}[\mathcal{A}(y - \alpha d_k)]$$
$$= \lambda_{min}[\mathcal{A}(y) - \alpha(\mathcal{A}(d_k) - \mathcal{A}(0))]$$

which after simple algebraic manipulations provides

$$\alpha_k := \frac{1}{\mu_k}$$

where

$$\mu_k := \lambda_{max}\left[\mathcal{A}(y)^{-1/2}\left(\mathcal{A}(d_k) - \mathcal{A}(0)\right)\mathcal{A}(y)^{-1/2}\right]$$

All operations indicated above are well defined since $\mathcal{A}(y) > 0$ implies that $\mathcal{A}(y)$ is nonsingular. Furthermore, from the above, $\mu_k \le 0$ implies that

$$0 \ge \mathcal{A}(d_k) - \mathcal{A}(0)$$
$$\ge \mathcal{A}(y) - \mathcal{A}(x_k)$$

that is $\mathcal{A}(x_k) \ge \mathcal{A}(y) > 0$ which puts in evidence that for any infeasible vector x_k we necessarily have $\mu_k > 0$. The point

$$\bar{x}_k := y - \alpha_k d_k$$

is, by construction, on the boundary of the feasible set of Problem (I.1) and for its determination only an additional eigenvalue calculation is needed. The Supporting hyperplane algorithm is stated in the sequel. The main idea is to use the above information to define iteratively a supporting hyperplane at the boundary point $\bar{x}_k$.

Algorithm I.2 (Supporting hyperplane algorithm) It is completely based on the previous algorithm where the separating hyperplane is replaced by a supporting hyperplane. Although more efficient, the limitation to handle large scale problems remains.

1. Determine an interior feasible point $y \in R^n$, a convex polyhedral set $\mathcal{P}_0$ containing the overall feasible set of Problem I.1, that is

$$\{x \ : \ \mathcal{A}(x) \ge 0\} \subset \mathcal{P}_0$$

and set the iteration index $k = 0$.

2. Solve the linear programming problem

$$\min\{c'x \ : \ x \in \mathcal{P}_k\}$$

and let x_k be its optimal solution.

3. If $\mathcal{A}(x_k) \geq 0$ then x_k solves Problem I.1 - **stop**. Otherwise, calculate the point $\bar{x}_k$ on the boundary of the feasible set and go to the next step.

4. Determine a supporting hyperplane $a'_k x = c_k$ at the boundary point $\bar{x}_k$. Define

$$\mathcal{P}_{k+1} := \mathcal{P}_k \cap \{x \ : \ a'_k x \geq c_k\}$$

set the iteration index $k \leftarrow k + 1$ and go back to step 2.

Once again, when the algorithm stops the global optimal solution of Problem I.1 is provided.   □

Let us now discuss an important class of convex programming methods, called *Interior point methods*. These methods apply to the solution of an approximate version of Problem (I.1) given in the form

$$\inf\{c'x \ : \ \mathcal{A}(x) > 0\} \tag{I.3}$$

Clearly, this problem is equivalent to Problem (I.1) provided the LMI $\mathcal{A}(x) \geq 0$ admits an interior point, that is a vector $x \in R^n$ such that $\mathcal{A}(x) > 0$. In this case, the equivalence between problems (I.1) and (I.3) holds in the sense that their optimal solutions are arbitrarily close one to the other. In the developments that follow, we work with Problem (I.3).

The main idea comes from the definition of the analytic center of a LMI. The analytic center of the LMI

$$\mathcal{A}(x) = A_0 + \sum_{i=1}^{n} x_i A_i > 0$$

is the vector $x_{ac} \in R^n$ such that

$$x_{ac} := \operatorname{argmin}\{-\log \det[\mathcal{A}(x)] \ : \ \mathcal{A}(x) > 0\} \tag{I.4}$$

The objective function of the above problem can be interpreted as a barrier function for the LMI under consideration. Indeed, as x goes to the boundary of the feasible set, at least one eigenvalue of $\mathcal{A}(x)$ goes to zero and enforces the objective function to be arbitrarily large. Moreover, the following properties are of great importance in the calculations that follow.

Lemma I.2 *The function* $p(x) := -\log \det[\mathcal{A}(x)]$, *defined in the open convex set* $\mathcal{A}(x) > 0$ *is such that :*

i) Function $p(x)$ is convex.

ii) At any $x \mid \mathcal{A}(x) > 0$, the gradient of $p(x)$ is

$$\nabla p(x)_i := -\operatorname{trace}\left[\mathcal{A}(x)^{-1} A_i\right] \ , \ i = 1, 2, \cdots, n$$

iii) At any $x \mid \mathcal{A}(x) > 0$, the Hessian matrix of $p(x)$ is

$$H(x)_{ij} := \text{trace}\left[\mathcal{A}(x)^{-1}A_i\mathcal{A}(x)^{-1}A_j\right] \ , \ i,j = 1, 2, \cdots, n$$

Proof The proof is based on the concavity of the scalar function $\log(z)$ in the interval $z > 0$ which implies that $\log(z) \leq z - 1$ for all $z > 0$. Using this and any two vectors such that $\mathcal{A}(x) > 0$ and $\mathcal{A}(y) > 0$, we get

$$\begin{aligned}
p(y) - p(x) &= -\log \det\left[\mathcal{A}(x)^{-1}\mathcal{A}(y)\right] \\
&= -\sum_{l=1}^{m} \log \lambda_l\left[\mathcal{A}(x)^{-1}\mathcal{A}(y)\right] \\
&\geq -\sum_{l=1}^{m}\left\{\lambda_l\left[\mathcal{A}(x)^{-1}\mathcal{A}(y)\right] - 1\right\}
\end{aligned}$$

and consequently

$$\begin{aligned}
p(y) &\geq p(x) - \text{trace}\left[\mathcal{A}(x)^{-1}\mathcal{A}(y) - I\right] \\
&\geq p(x) - \text{trace}\left[\mathcal{A}(x)^{-1}\mathcal{A}(y) - \mathcal{A}(x)^{-1}\mathcal{A}(x)\right] \\
&\geq p(x) - \sum_{i=1}^{n} \text{trace}\left[\mathcal{A}(x)^{-1}A_i\right](y_i - x_i) \\
&\geq p(x) + \nabla p(x)'(y - x)
\end{aligned}$$

which proves the first two points of the lemma proposed. The last one is proved by simple partial differentiation of $\nabla p(x)_i$ with respect to the variable x_j. The proof is concluded. $\square$

Example I.2 For the same LMI of example I.1, figure I.2 shows the level set of $\det[\mathcal{A}(x)] = \alpha > 0$ which for $\mathcal{A}(x) > 0$ and $\beta = -\log(\alpha)$ coincides with that of $p(x) = \beta$. It is clearly seen that in the interior of the LMI the level set for each value of $\alpha > 0$ defines a convex set. Moreover, the closed region approaches to the boundary of the LMI as α goes to zero. Outside this region, there exist points for which the determinant of the affine function $\mathcal{A}(x)$ attains the same level but with an even number of negative eigenvalues. Finally, using part *ii)* of Lemma I.2 we solve $\nabla p(x) = 0$ to get the analytic center $x_{ca} = [0.5902\ 0.4236]$. $\square$

Further inspection reveals that function $p(x)$ is in fact strictly convex which means that the Hessian matrix $H(x)$ is positive definite whenever the vector $x \in R^n$ is such that $\mathcal{A}(x) > 0$. Hence, the analytic center of the LMI can very efficiently be calculated by the following well known Modified Newton's method.

Algorithm I.3 (Modified Newton's method) Assume an initial point x_0 such that $\mathcal{A}(x_0) > 0$ is given. Then, perform the following iterations until convergence.

1. Determine the gradient vector $\nabla p(x_k)$ and the Hessian matrix $H(x_k)$.

2. Determine the descent direction $d_k := H(x_k)^{-1}\nabla p(x_k)$. If within some prespecified precision $\|d_k\| = 0$ - **stop**. Otherwise, go to the next step.

3. Determine the optimal step length α_k given by

$$\alpha_k := \text{argmin } p(x_k - \alpha d_k)$$

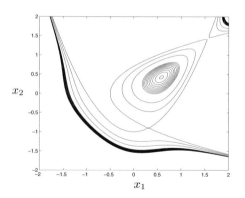

Figure I.2: The level set of $\det[\mathcal{A}(x)]$

3. Update $x_{k+1} = x_k - \alpha_k d_k$, set the iteration index $k \leftarrow k + 1$ and go back to step 1.

When the algorithm stops the analytic center is approximately given by $x_{ac} = x_k$. □

For the complete implementation of this algorithm, it remains to calculate the optimal step length α_k defined in step 3. This is accomplished with no great difficulty. Indeed, simple calculations put in evidence that

$$p(x_k - \alpha d_k) = p(x_k) - \sum_{l=1}^{m} \log[1 - \alpha e_{kl}] \qquad (I.5)$$

where

$$e_{kl} := \lambda_l \left[\mathcal{A}(x_k)^{-1/2} \left(\mathcal{A}(d_k) - \mathcal{A}(0) \right) \mathcal{A}(x_k)^{-1/2} \right]$$

is the l-th component of the m dimensional real vector $e_k := [\ e_{k1}\ e_{k2}\ \cdots\ e_{km}\ ]'$. Hence, the derivative of $p(x_k - \alpha d_k)$ with respect to α provides

$$\frac{dp}{d\alpha}(x_k - \alpha d_k) = \sum_{l=1}^{m} \frac{e_{kl}}{1 - \alpha e_{kl}}$$

$$= \sum_{l=1}^{m} \left[\frac{e_{kl}^2}{\beta - e_{kl}} + e_{kl} \right]$$

where $\beta := 1/\alpha$. Setting the right hand side of the above equation to zero and taking into account that

$$\delta_k^2 := -\sum_{l=1}^{m} e_{kl} = -\frac{dp}{d\alpha}(x_k - \alpha d_k)\bigg|_{\alpha=0} > 0$$

since, d_k is a nonzero descent direction, the optimal step size $\alpha = 1/\beta$ solves the nonlinear equation

$$\sum_{l=1}^{m} \left[\frac{e_{kl}^2}{\beta - e_{kl}} \right] - \delta_k^2 = 0$$

The solution of this nonlinear equation is not simple to be determined unless we realize it can be rewritten in terms of the following determinantal equation

$$\det\left[\delta_k^2 - e_k'(\beta I - E_k)^{-1}e_k\right] = 0$$

where matrix $E_k \in R^{m \times m}$ is defined as $E_k := \text{diag}[e_{k1}, e_{k2}, \cdots, e_{km}]$. Finally, the last equation together with some elementary determinant manipulations provides

$$\det\left[\beta I - (E_k + \delta_k^{-2}e_k e_k')\right] = 0$$

which makes clear that the optimal step size α_k is given by

$$\alpha_k = \lambda_{max}^{-1}\left[E_k + \left(\frac{e_k}{\delta_k}\right)\left(\frac{e_k}{\delta_k}\right)'\right] \tag{I.6}$$

This shows that to determine the optimal steep size, we have to calculate all eigenvalues of a symmetric matrix in order to define the vector e_k and finally to calculate the maximum eigenvalue of the symmetric matrix indicated in (I.6). To reduce this amount of calculations, in some cases, we have a great advantage to get a suboptimal step size as indicated in the sequel. It comes to light from the observation that any positive step size less than the optimal one may also be used to assure that function $p(x)$ is reduced in the direction $-d_k$. To get such a suboptimal step size, we proceed by establishing the following equality

$$\begin{aligned}
\sum_{l=1}^{m} e_{kl}^2 &= \frac{d^2 p}{d\alpha^2}(x_k - \alpha d_k)\Big|_{\alpha=0} \\
&= d_k' H(x_k) d_k \\
&= \nabla p(x_k)' d_k \\
&= -\frac{dp}{d\alpha}(x_k - \alpha d_k)\Big|_{\alpha=0} \\
&= \delta_k^2
\end{aligned}$$

which together with (I.6) implies that

$$\begin{aligned}
\alpha_k^{-1} &= \lambda_{max}\left[E_k + \left(\frac{e_k}{\delta_k}\right)\left(\frac{e_k}{\delta_k}\right)'\right] \\
&\leq \lambda_{max}[E_k] + \delta_k^{-2}\sum_{l=1}^{m} e_{kl}^2 \\
&\leq 1 + \lambda_{max}[E_k]
\end{aligned}$$

consequently, a suboptimal step size, denoted as α_k^+ is given by

$$\alpha_k^+ := \frac{1}{1 + \mu_k}$$

where

$$\mu_k := \lambda_{max}\left[\mathcal{A}(x_k)^{-1/2}\left(\mathcal{A}(d_k) - \mathcal{A}(0)\right)\mathcal{A}(x_k)^{-1/2}\right]$$

It is to be noticed that $\mu_k > 0$ since otherwise all $e_{kl} \leq 0$ which implies from (I.5) that the feasible set is unbounded, a situation avoided by our previous assumption.

It is also interesting to see that the above formula for the suboptimal step size is very similar to the one introduced before for the calculation of a point on the boundary of the feasible set of the LMI under consideration.

Let us now use the concept of analytic center to calculate the optimal solution of Problem (I.3). Obviously it can be equivalently stated in the form

$$\inf \{ \gamma \ : \ \mathcal{A}(x) > 0 \ , \ \ \gamma - c'x > 0 \}$$

or, in terms of only one *augmented* LMI

$$\inf \{ \gamma \ : \ \mathcal{B}(x, \gamma) > 0 \} \tag{I.7}$$

where

$$\mathcal{B}(x, \gamma) := \left[\begin{array}{cc} \mathcal{A}(x) & 0 \\ 0 & \gamma - c'x \end{array} \right]$$

This LMI depends on both variables namely the vector $x \in R^n$ and the scalar γ. However, for γ fixed, let us define as before the analytic center

$$x_{ac}(\gamma) := \operatorname{argmin} \{ -\log \det[\mathcal{B}(x, \gamma)] \ : \ \mathcal{B}(x, \gamma) > 0 \}$$

where it is indicated the dependence of the analytic center with respect to the scalar γ and that the minimization must be done with respect to $x \in R^n$ only. The curve $x_{ac}(\gamma)$ obtained for all possible values of γ is called the *Path of centers* and plays a central role to the numerical solution of Problem (I.3) as indicated in the next algorithm.

Algorithm I.4 (Method of centers) Assume an initial pair (x_0, γ_0) is given, such that simultaneously $\mathcal{A}(x_0) > 0$ and $\gamma_0 > c'x_0$. Choose $0 < \theta < 1$ and $\epsilon > 0$ sufficiently small and perform the following iterations until convergence.

1. $\gamma_{k+1} = (1 - \theta)c'x_k + \theta\gamma_k$

2. $x_{k+1} = x_{ac}(\gamma_{k+1})$

3. If $\gamma_{k+1} - c'x_{k+1} < \epsilon/m$ - **stop**. Otherwise set the iteration index $k \leftarrow k+1$ and go back to step 1.

When the algorithm stops the optimal solution to Problem (I.7) is found within ϵ. □

It is important to recognize that the rule in step 1, never produces infeasibility on the analytic center determination in step 2. Actually, assume that in a generic iteration $k \geq 0$ we have $\mathcal{B}(x_k, \gamma_k) > 0$. With the formula stated in step 1, we get

$$\gamma_{k+1} - c'x_k = \theta(\gamma_k - c'x_k) > 0$$

which implies that $\mathcal{B}(x_k, \gamma_{k+1}) > 0$ and consequently the vector x_k can be used to initialize the Modified Newton's method for the determination of the analytic center $x_{ac}(\gamma_{k+1})$. In practice, it is verified that this simple initialization procedure is very effective as far as numerical efficiency is concerned.

Lemma I.3 *The Method of centers converges geometrically to the optimal solution of Problem (I.3).*

Proof Denote (x_{opt}, γ_{opt}) the optimal solution of Problem (I.3). For $\gamma = \gamma_{k+1}$ fixed, Lemma I.2 enables us to write the optimality conditions to characterize the analytic center $x_{ac}(\gamma_{k+1})$. So, due to step 2 we must have

$$\frac{c_i}{\gamma_{k+1} - c'x_{k+1}} = \text{trace}\left[\mathcal{A}(x_{k+1})^{-1} A_i\right] , \quad i = 1, 2, \cdots, n$$

which gives (recall that $\gamma_{opt} = c'x_{opt}$)

$$\frac{c'x_{k+1} - \gamma_{opt}}{\gamma_{k+1} - c'x_{k+1}} = \text{trace}\left[\mathcal{A}(x_{k+1})^{-1}\left(\mathcal{A}(x_{k+1}) - \mathcal{A}(x_{opt})\right)\right]$$

Now, define the scalar ϕ as being

$$\phi := \sup_{\mathcal{A}(x)>0} \text{trace}\left[\mathcal{A}(x)^{-1}\left(\mathcal{A}(x) - \mathcal{A}(x_{opt})\right)\right]$$

and observe that $0 \le \phi \le m$. Actually, the lower bound is obtained from the simple observation that $x = x_{opt}$ is feasible and the upper bound is a consequence of the fact that $\mathcal{A}(x_{opt}) > 0$. Then, the inequality

$$\phi\gamma_{k+1} + \gamma_{opt} \ge (1 + \phi)c'x_{k+1}$$

holds in all iterations. Using the update of step 1, namely

$$c'x_{k+1} = \frac{\gamma_{k+2} - \theta\gamma_{k+1}}{1 - \theta}$$

simple algebraic manipulations put in evidence that

$$\gamma_{k+2} - \gamma_{opt} \le \frac{\phi + \theta}{\phi + 1}\left(\gamma_{k+1} - \gamma_{opt}\right)$$

which proves that the Method of centers converges geometrically. This concludes the proof of the Lemma proposed. □

This proof is of great practical importance for two main reasons. First, if the stopping criterion in step 3 is verified then

$$c'x_{k+1} - \gamma_{opt} \le \phi(\gamma_{k+1} - c'x_{k+1})$$
$$\le m\,(\epsilon/m) = \epsilon$$

and the optimal solution is found within the prespecified precision level $\epsilon > 0$ imposed by the designer. Second, the ratio of geometric convergence, such that

$$0 \le c'x_k - c'x_{opt} \le \alpha\beta(\phi)^k$$

for some $\alpha > 0$, is estimated as being

$$\beta(\phi) := \frac{\phi + \theta}{\phi + 1}$$

which is an increasing function of ϕ. The worst estimation is then obtained for $\phi = m$ providing thus $\beta(m)$. It is important to realize that, doing this, the conclusion is that the Method of centers converges geometrically but with a ratio that goes to unity

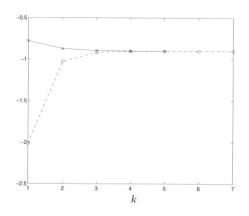

Figure I.3: Convergence behavior

as the dimension of the problem to be solved increases. In other words, it performs, under this worst case analysis, as the Separating hyperplane algorithm. However, it is possible to introduce in the Method of centers a simple modification in order to get much better convergence behavior. Indeed, if in the determination of the analytic center $x_{ac}(\gamma)$ the objective function is changed to

$$-\log \det[\mathcal{A}(x)] - m \, \log(\gamma - c'x)$$

which is nothing more than to redefine the augmented LMI by replacing the scalar $\gamma - c'x$ by the $m \times m$ diagonal matrix $(\gamma - c'x)I$, then the same reasoning used in the proof of Lemma I.3 yields the new estimate for the ratio of geometric convergence

$$\beta(\phi) := \frac{\phi + \theta m}{\phi + m} \leq \frac{1 + \theta}{2} < 1 \qquad (I.8)$$

The worst case for the ratio of convergence is now independent of the problem dimension. It depends only on the parameter $0 < \theta < 1$ to be fixed by the designer.

Example I.3 Consider the LMI of example I.1 and $c' = [0 \ 1]$. The optimal solution of problem (I.1) is found to be $x_{opt} = [0.3132 \ -0.9018]'$ and $\gamma_{opt} = c'x_{opt} = -0.9018$. Figure I.3 shows the objective function per iteration calculated by Algorithm I.2 (dashed line) and by Algorithm I.4 (solid line). As expected, the first evolves through infeasible points and the objective function is increasing. On the contrary, the second is always feasible and the objective function is decreasing. □

Finally it is also to be noticed that the determination of a feasible starting point required for many algorithms can be done with no great difficulty. Given the LMI $\mathcal{A}(x) \geq 0$, the problem is to find (if one exists) a vector x such that $\mathcal{A}(x) > 0$. This is accomplished by solving the auxiliary convex problem

$$\min \{\lambda \ : \ \mathcal{A}(x) + \lambda I \geq 0\}$$

which presents two interesting properties. First, the pair $(x, \lambda) := (0, 1 - \lambda_{min}[A_0])$ satisfies the LMI constraint strictly and second at the optimal solution $\mathcal{A}(x_o) > 0$ if and only if $\lambda_o < 0$. Of course, the optimal solution of the above problem does not need to be exactly calculated. If in some iteration the current value of the auxiliary variable λ becomes negative, the current value of the vector x is strictly feasible and the search process may be stopped.

Bibliography

[1] J. Ackermann. *Robust Control*. Springer Verlag, 1993.

[2] T. Basar and P. Bernhard. H_∞ *Optimal Control and Related Minimax Design Problems*. Birkhäuser, 1991.

[3] B.R. Barmish. Necessary and Sufficient Conditions for Quadratic Stabilizability of an Uncertain System. *J. Opt. Th. Appl.* 46, pp.399-408, 1985.

[4] B.R. Barmish. Stabilization of Uncertain Systems Via Linear Control. *IEEE Trans. Auto. Contr.* AC-28, pp.848-850, 1983.

[5] D.S. Bernstein and W.M. Haddad. LQG Control with an H_∞ Performance Bound: A Riccati Equation Approach. *IEEE Trans. Auto. Contr.* AC-34, pp.293-305, 1989.

[6] J. Bernussou, P.L.D. Peres and J.C. Geromel. A Linear Oriented Procedure for Quadratic Stabilization of Uncertain Systems. *Syst. Contr. Lett.* 13, pp.65-72, 1989.

[7] S. Bittanti, A.J. Laub and J.C. Willems (Eds). *The Riccati Equations*. Springer Verlag, 1991.

[8] P. Bolzern, P. Colaneri and G. De Nicolao. On the Computation of Upper Covariance Bounds for Perturbed Linear Systems. *IEEE Trans. Auto. Contr.* AC-39, pp.623-626, 1994.

[9] S.P. Boyd, V. Balakrishnan and P.T. Kabamba. A Bisection Method for Computing the H_∞ Norm of a Transfer Matrix and Related Problems. *Mathematics of Control, Signals, and Syst.* 2, pp.207-219, 1989.

[10] S.P. Boyd, L. El Ghaoui, E. Feron and V. Balakrishnan. *Linear Matrix Inequalities in System and Control Theory*. Siam Studies in Applied Mathematics, 1994.

[11] S.S.L. Chang and T.K.C. Peng. Adaptive Guaranteed Cost Control of Systems with Uncertain Parameters. *IEEE Trans. Auto. Contr.* AC-17, pp.474-483, 1972.

[12] C.E. de Souza, U. Shaked and N. Fu. Robust H_∞ Filtering with Parametric Uncertainty and Deterministic Input Signal. *Proc. 31st Conf. Dec. Contr.* Tucson, USA, pp.2305-2310,1992.

[13] J.C. Doyle. Robustness of Multiloop Linear Feedback Systems. *Proc. 17th Conf. on Dec. and Contr.* USA, pp.12-18, 1978.

[14] J.C. Doyle. Analysis of Feedback Systems with Structured Uncertainties. *IEE Proc. Pt.D.* 129, pp.242-250, 1982.

[15] J.C. Doyle. Structured Uncertainty in Control System Design. *Proc. 24th Conf. on Dec. and Contr.* Ft. Lauderdale, USA, pp.260-265, 1985.

[16] J.C. Doyle, B.A. Francis, A.M. Tannenbaum. *Feedback Control Theory.* Maxwell MacMillan Int. Ed., 1992.

[17] J.C. Doyle, K. Glover, P.P. Khargonekar and B.A. Francis. State Space Solutions to Standard H_2 and H_∞ Control Problems. *IEEE Trans. Auto. Contr.* AC-34, pp.831-847, 1989.

[18] J.C. Doyle, J.E. Wall and G. Stein. Performance and Robustness Analysis for Structured Uncertainty. *Proc. 21st Conf. on Dec. and Contr.* pp.629-636, 1982.

[19] B.A. Francis. *A Course in H_∞ Control Theory.* Lectures Notes on Control and Information Sciences. Springer Verlag, 1987.

[20] N. Fu, C.E. de Souza and L. Xie. H_∞ Estimation for Uncertain Systems. *Int. J. Rob. and Nonlin. Contr.* 2, pp.82-105, 1992.

[21] T.T. Georgiou and M.C. Smith. Optimal Robustness in the Gap Metric: Controller Design for Distributed Plants. *IEEE Trans. Auto. Contr.* AC-37, pp.1133-1143, 1992.

[22] J.C. Geromel, P.L.D. Peres and J. Bernussou. On a Convex Parameter Space Method for Linear Control Design of Uncertain Systems. *SIAM J. Contr. and Opt.* 29, pp.381-402, 1991.

[23] J.C. Geromel, J. Bernussou and P.L.D. Peres. Decentralized Control Through Parameter Space Optimization. *Automatica* 30, pp.1565-1578, 1994.

[24] J.C. Geromel, P.L.D. Peres and S.R. Souza. $\mathcal{H}_2$ Guaranteed Cost Control for Uncertain Continuous-time Linear Systems. *Syst. Contr. Lett.* 19, pp.23-27, 1992.

[25] J.C. Geromel. On the Determination of a Diagonal Solution of the Lyapunov Equation. *IEEE Trans. Auto. Contr.* AC-30, pp.404-406, 1985.

[26] J.C. Geromel and A.O.E. Santo. On the Robustness of Linear Continuous-time Dynamic Systems. *IEEE Trans. Auto. Contr.* AC-31, pp.1136-1138, 1986.

[27] W.M. Haddad and D.S. Bernstein. Robust Stabilization with Positive Real Uncertainty: Beyond the Small Gain Theorem. *Syst. Contr. Lett.* 17, pp.191-208, 1991.

[28] W.M. Haddad and D.S. Bernstein. Controller Design with Regional Pole Constraints. *IEEE Trans. Auto. Contr.* AC-37, pp.54-69, 1992.

[29] T. Kailath. *Linear Systems.* Prentice Hall, 1980.

[30] E. Kaszkurewicz and A. Bhaya. Robust Stability and Diagonal Liapunov Functions. *SIAM J. Matrix Anal. Appl.* 14, pp.508-520, 1993.

[31] P.P. Khargonekar, I.R. Petersen and K. Zhou. Robust Stabilization of Uncertain Linear Systems: Quadratic Stabilizability and H_∞ Control Theory. *IEEE Trans. Auto. Contr.* AC-35, pp.356-361, 1990.

[32] P.P. Khargonekar and M.A. Rotea. Mixed H_2/H_∞ Control: A Convex Optimization Approach. *IEEE Trans. Auto. Contr.* AC-36, pp.824-837, 1991.

[33] B. Kouvarikatis and A.G.J. MacFarlane. Geometric Approach to Analysis and Synthesis of System Zeros: Part I - Square Systems. *Int. J. Control.* 23, pp.149-166, 1976.

[34] B. Kouvarikatis and A.G.J. MacFarlane. Geometric Approach to Analysis and Synthesis of System Zeros: Part II - Nonsquare Systems. *Int. J. Control.* 23, pp.167-181, 1976.

[35] V. Kucera. The LQG and H_2 Designs: Two Different Problems? *Proc. 2nd Europ. Contr. Conf..* Groningen, The Netherlands, pp.334-337, 1988.

[36] H. Kwakernaak. The Polynomial Approach to the H_∞ Optimal Regulation. H_∞ *Control.* E. Mosca and L. Pandolfi Eds. Lecture Notes in Mathematics, 1496. Springer-Verlag, 1990.

[37] L.S. Lasdon. *Optimization Theory for Large Systems.* Macmillan Publishing Co., 1970.

[38] C.L. Lawson and R.J. Hanson. *Solving Least-Square Problems.* Prentice Hall, 1974.

[39] W.S. Levine and M. Athans. On the Determination of the Optimal Constant Output Feedback Gains for Linear Multivariable Systems. *IEEE Trans. Auto. Contr.* AC-15, pp.44-48, 1970.

[40] D.J.N. Limebeer, B.D.O. Anderson and B. Hendel. A Nash Game Approach to Mixed H_2/H_∞ Control. *IEEE Trans. Auto. Contr.* AC-39, pp.69-82, 1994.

[41] D.G. Luenberger. *Linear and Nonlinear Programming.* Addison-Wesley Publishing Co., 1984.

[42] A.G.J. MacFarlane and N. Karcanias. Poles and Zeros of Linear Multivariable Systems: a Survey of the Algebraic, Geometric and Complex Variable Theory. *Int. J. Control.* 24, pp.33-74, 1976.

[43] J.M. Maciejowski. *Multivariable Feedback Design.* Addison Wesley, 1989.

[44] T. Mita, K. Kuriyamna and K.Z. Liu. Revision of FI Result in H_∞ Control and Parametrization of State Feedback Controllers. *Trans. Soc. Instrum. Contr. Eng.* 28, pp.801-809, 1992.

[45] Y. Nesterov and A. Nemirovsky. *Interior-point Polynomial Methods in Convex Programming.* Siam Studies in Applied Mathematics, 1994.

[46] S. Parrott. On a Quotient Norm and the Sz.-Nagy-Foias Lifting Theorem. *J. Functional Analysis.* 30, pp.311-328, 1978.

[47] P.L.D. Peres and J.C. Geromel. An Alternate Numerical Solution to the Linear Quadratic Problem. *IEEE Trans. Auto. Contr.* AC-39, pp.198-202, 1994.

[48] P.L.D. Peres, J.C. Geromel and S.R. Souza. Optimal $\mathcal{H}_\infty$ - State Feedback Control for Continuous-Time Linear Systems. *J. of Opt. Theory and App.* 82, pp.343-359, 1994.

[49] P.L.D. Peres, J.C. Geromel and S.R. Souza. $\mathcal{H}_\infty$ Guaranteed Cost Control for Uncertain Continuous-time Linear Systems. *Syst. Contr. Lett.* 20, pp.413-418, 1993.

[50] P.L.D. Peres, J.C. Geromel and J. Bernussou. Quadratic Stabilizability of Linear Uncertain Systems in Convex-bounded Domains. *Automatica* 29, pp.491-493, 1993.

[51] I.R. Petersen and D.C. McFarlane. Robust State Estimation for Uncertain Systems. *Proc. 30th Conf. Dec. Contr.* Brighton, England, pp.2630-2631, 1991.

[52] S.C. Power. *Hankel Operator in Hilbert Space.* Pitman, 1982.

[53] R.T. Rockafellar. *Convex Analysis.* Princeton Univ. Press, 1970.

[54] M.A. Rotea. The Generalized H_2 Control Problem. *Automatica* 29, pp373-386, 1993.

[55] W. Rudin. *Functional Analysis.* McGraw Hill, 1973.

[56] M.G. Safonov and D.J.N. Limebeer. Simplifying the H_∞ Theory via Loop Shifting. *Proc. 27th Conf. on Dec. and Contr..* Austin, USA, pp.1399-1404, 1988.

[57] G. Stein and J.C. Doyle. Beyond Singular Values and Loop Shapes. *AIAA J. of Guid. Contr. and Dynam.* 14, pp.5-16, 1991.

[58] A.A. Stoorvogel. The singular H_∞ Control Problem with Dynamic Measurement Feedback. *SIAM J. Contr. and Opt.* 29, pp.160-184, 1991.

[59] A.A. Stoorvogel. *The H_∞ Control Problem.* Prentice Hall, 1992.

[60] M. Vidyasagar. *Control Systems Synthesis.* MIT Press, 1985.

[61] M. Vidyasagar. *Control Systems Analysis.* Prentice Hall, 1993.

[62] D.A. Wilson. Convolution and Hankel Operator Norms for Linear Systems. *IEEE Trans. Auto. Contr.* AC-34, pp.94-98, 1989.

[63] D.C. Youla, H.A. Jabr and J.J. Bongiorno. Modern Wiener-Hopf Design of Optimal Controllers: Part II. *IEEE Trans. Auto. Contr.* AC-21, pp.319-338, 1976.

[64] G. Zames. Feedback and Optimal Sensitivity: Model Reference Transformations, Multiplicative Seminorms and Approximate Inverses. *IEEE Trans. Auto. Contr.* AC-26, pp.301-320, 1981.

[65] K. Zhou and P.P. Khargonekar. Robust Stabilization of Linear Systems with Norm-bounded Time-varying Uncertainty. *Syst. Contr. Lett.* 10, pp.17-20, 1988.

Index